STO

**ACPL ITEM
DISCARDED**

JAN 13 '72

BIOMECHANICS

Its Foundations and Objectives

EDITED BY

Y. C. Fung, N. Perrone and M. Anliker

PRENTICE-HALL, INC. *Englewood Cliffs*, New Jersey

© 1972 by
PRENTICE-HALL, INC.
Englewood Cliffs, N. J.

All rights reserved. No part of this
book may be reproduced in any form
or by any means without permission
in writing form the publisher.

Current printing (last digit):
10 9 8 7 6 5 4 3 2 1

13-077149-X

Library of Congress Catalog Card Number: 78-160253
Printed in the United States of America

PRENTICE-HALL INTERNATIONAL, INC *London*
PRENTICE-HALL OF AUSTRALIA, PTY. LTD., *Sydney*
PRENTICE-HALL OF CANADA LTD., *Toronto*
PRENTICE-HALL OF INDIA PRIVATE LIMITED, *New Delhi*
PRENTICE-HALL OF JAPAN, INC., *Tokyo*

1627463

This volume is the edited proceedings of the Symposium on Biomechanics, its Foundations and Objectives, which was held in La Jolla, California, from July 29 to 31, 1970. This Symposium was sponsored by the University of California, San Diego and the United States Navy, Office of Naval Research.

Preface

All interdisciplinary areas are characterized by a certain vagueness in definition and vastness in literature. Biomechanics as an area common to mechanics, materials, physiology, medicine, surgery, pathology, prosthesis, patient care, dentistry, athletics, and even social and environmental studies, is particularly difficult to define. For this reason, people working in the field have long felt a need to get together to discuss the objectives, review the foundations, and display the current developments. There is a need also to publish in a convenient form an authoritatively filtered bibliography, so that newcomers to the field, as well as old workers, may find their way in the literature. It was for these reasons that the *Symposium on the Foundations and Objectives of Biomechanics* was organized. The present volume is the proceedings of that symposium.

The La Jolla symposium was held on July 29–31, 1970 on the campus of the University of California, San Diego. Approximately 300 scientists, engineers, and M.D.'s attended the meeting. The original idea of organizing the symposium was due to Dr. Nicholas Perrone. The planning was done by Drs. Anliker, Perrone, and myself. Financial support was obtained from the U.S. Navy Office of Naval Research. The labor and love was provided by the staff and graduate students of the UCSD Bioengineering Laboratory, especially Ling Lin, Barbara Johnson, Frank Yin, John Pinto, and Eugene Mead.

No treatment of biomechanics within the bounds of a single book can claim completeness. The present book tends to lay emphasis on physiology and on the blood circulation system. For each topic the accent is on the review of the foundations and hard facts on which boundary–value problems are based. It will become clear, even at a cursory reading, that there exists already a wealth of material in biomechanics. But it is also clear that much remains to be done. It is to the future development that this volume is dedicated.

Y. C. Fung

Contents

PART I
Objectives of Biomechanics

1   Biomechanics and Physiology   **3**
Benjamin W. Zweifach

2   Biomechanics and Surgery   **15**
Richard Peters

3   Biomechanics, Patient Care, and Rehabilitation   **29**
Ernst O. Attinger

PART II
Basic Mechanical Properties of Living Tissues

4   The Rheology of Human Blood   **63**
Giles R. Cokelet

5   The Properties of Blood Vessels   **105**
Derek H. Bergel

6   Properties of Tendon and Skin   **141**
John D.C. Crisp

7   Stress-Strain-History Relations of Soft Tissues in Simple Elongation   **181**
Yuan-Cheng B. Fung

8   Muscle Mechanics   **209**
Bernard C. Abbott

9   Influence of Composition on Thermal Properties of Tissues   **217**
Julia T. Apter

10   Biomechanics of Bone   **237**
Benno Kummer

11   The Interstitial Space   **273**
Curt A. Wiederhelm

PART III

Biomechanics and Physiology, Medicine and Surgery

12   Mechanics of Contraction in the Intact Heart   **289**
James W. Covell

13   Determinants of Cardiac Performance   **303**
Charles Urschel and Edmund H. Sonnenblick

14   Lung Elasticity   **317**
George Lee and Fred Hoppin

15   Toward a Nontraumatic Study of the Circulatory System   **337**
Max Anliker

16   Flow and Pressure in the Arteries   **381**
Thomas Kenner

17   Analysis of Recent Developments in Blood Flow Measurement   **435**
Arnost Fronek

18   Mechanics of the Microcirculation   **457**
Richard Skalak

| 19 | Mechanical Hemolysis in the Flowing Blood | 501 |

Perry L. Blackshear, Jr.

PART IV
Injury and Prosthesis

| 20 | Biomechanical Compatibility of Prosthetic Devices | 531 |

J.P. Paul, J. Hughes, and R. M. Kenedi

| 21 | Fluid Dynamics of Heart Assist Devices | 549 |

Robert T. Jones

| 22 | Biomechanical Problems Related to Vehicle Impact | 567 |

Nicholas Perrone

| 23 | Biomechanics of Head Injury | 585 |

Werner Goldsmith

Index — 635

Contributors

Bernard C. Abbott, Ph. D.
Allan Hancock Foundation and
 Department of Biology
University of Southern California
University Park
Los Angeles, California 90007

Max Anliker, Ph. D.
Department of Aeronautics &
 Astronautics
Stanford University
Stanford, California 94303

Julia T. Apter, M. D., Ph. D.
Laboratory of Biomaterials
 and Biomechanics
Rush Medical College
1753 W. Congress Parkway
Chicago, Illinois, 60612

Ernst O. Attinger, M. D., Ph. D.
Division of Biomedical Engineering
University of Virginia Medical
 Center
Charlottesville, Virginia 22901

Derek H. Bergel, Ph. D., M. B., B. S.
University Laboratory of
 Physiology
St. Catherine's College
Oxford, England

Perry L. Blackshear, Jr., Ph. D.
Department of Mechanical
 Engineering
University of Minnesota
Minneapolis, Minnesota 55455

Giles R. Cokelet, Ph. D.
Department of Chemical
 Engineering
Montana State University
Bozeman, Montana 59715

James W. Covell, M. D.
Department of Medicine
School of Medicine
University of California, San
 Diego
La Jolla, California 92037

CONTRIBUTORS

John D. C. Crisp, M. E.
Department of Mechanical
    Engineering
Monash University
Clayton, Victoria 3168
Australia

Arnost Fronek, M. D., Ph. D.
Department of Engineering
    Sciences
University of California, San Diego
La Jolla, California 92037

Yuan-Cheng B. Fung, Ph. D.
Department of Engineering
    Sciences
University of California, San Diego
La Jolla, California 92037

Werner Goldsmith, Ph. D.
Department of Mechanical
    Engineering
University of California, Berkeley
Berkeley, California 94720

Fred Hoppin, M. D.
Department of Physiology
School of Public Health
Harvard University
Boston, Masachusetts 02110

John Hughes
Bioengineering Unit
University of Strathclyde
Montrose Street
Glasgow Cl. Scotland

Robert T. Jones
NASA AMES Research Center
Mountain View, California 94040

R. M. Kenedi, Ph. D.
Bioengineering Unit
University of Strathclyde
Montrose Street
Glasgow Cl, Scotland

Thomas Kenner, Ph. D.
Division of Biomedical Engineering
University of Virginia Medical
    Center
Charlottesville, Virginia 22901

Benno K. F. Kummer, Dr. med.
Anatomisches Institut der
    Universität zu Köln
5 Köln-Lindenthal
Lindenburg, Germany

George C. Lee, Ph. D.
Department of Civil Engineering
State University of New York at
    Buffalo
Buffalo, New York 14214

Jere Mead, M. D.
Department of Physiology
School of Public Health
Harvard University
Boston, Massachusetts 02110

John P. Paul, Ph. D.
Bioengineering Unit
University of Strathclyde
Montrose Street
Glasgow Cl, Scotland

Nicholas Perrone, Ph. D.
Department of the Navy
Office of Naval Research
Arlington, Virginia 22217

Richard Peters, M. D.
Department of Surgery
School of Medicine
University of California, San Diego
La Jolla, California 92037

Richard Skalak, Ph. D.
Department of Civil Engineering &
    Engineering Mechanics
Columbia University
New York, N. Y. 10027

Edmund H. Sonnenblick, M. D.
Cardiovascular Unit
Peter Bent Brigham Hospital and
 Harvard Medical School
Boston, Massachusetts 02115

Charles Urschel, M. D.
Cardiovascular Unit
Peter Bent Brigham Hospital and
 Harvard Medical School
Boston, Massachusetts 02115

Curt A. Wiederhielm, Ph. D.
Department of Physiology and
 Biophysics
School of Medicine
University of Washington
Seattle, Washington 98105

B. W. Zweifach, Ph. D.
Department of Engineering
 Sciences
University of California, San Diego
La Jolla, California 92037

PART I

Objectives of Biomechanics

1
Biomechanics and Physiology

BENJAMIN W. ZWEIFACH

The science of physiology deals with the study of the functional behavior of living structures, a term which includes all spectra of life from viruses, to plants, to unicellular organisms, to complex mammalian forms, and it is the objective of physiology to establish the physical and chemical relationships that underlie such phenomena. Physiological processes range from mechanisms occurring in the microscopic organellae of the basic unit of life, the cell, to complex interactions at the macroscopic level involving whole organs and literally billions of cells. Despite the diversity of forms and adaptive structures which have evolved to permit living creatures to survive in vastly different environments, the biological materials and principles which underlie the behavior of key physiological systems are surprisingly similar in most species. It is with these basic elements of cell behavior and systems design that physiology is concerned (Dowben, 1969).

Major advances in modern physiology have come from two principal directions—one concerned with molecular interactions within the cell, and another with the application of systems analysis techniques to facilitate working with the extraordinary number of variables influencing the behavior of organic units.

On a working level, the physiologist seeks to analyze the relationship between structure and function. A standard approach is to isolate the system in the hope that a study of its subunits will elucidate the operation of the whole system. From its very beginning, much of physiology has been phenomenological—a careful description of events with emphasis on factors which alter these events in a predictable way.

The field of mechanics, in a not too dissimilar context, seeks to develop mathematical laws which define the common properties of an idealized body in both a static and in a dynamic sense, and on the basis of these laws derives constitutive equations which illustrate their application to different specific situations (Truesdell, 1965). In contrast to physiology, which is fundamentally experimental in its approach, much of mechanics is theoretical in exposition. There is, however, a striking similarity in the basic concerns of the two disciplines (Table 1).

TABLE I

PHYSIOLOGY	MECHANICS
Body as a whole	Bodies
Functional organization	Motions
Energy sources	Forces

Admittedly, the precision that the tools and methods of mechanics demand would seem to preclude their application to the less rigorously defined biological processes. It is, however, a well-known truism that problem solving requires a clear definition of the problem before one can even pose the proper question. The interaction of physiology and mechanics is in its infancy and in all probability the substantial application of mechanics to physiological processes will require sets of axioms which have not yet been developed. The thrust in this direction, that must *per force* come from the theoretician, cannot be achieved by him alone but through a close working relationship with the experimentalist.

In many ways physiology today is in danger of being splintered into highly specialized subunits. In comparison with the field of physiology as a whole, biophysics and neurophysiology have made such substantial strides that they are tending to become encapsulated. As mathematical modeling and the systems analysis approach become more complex, they almost by default have fallen into the domain of biomathematics. The physiologist, who has relied chiefly on the experimental approach, faces an insurmountable task in attempting to reconstruct from molecular phenomena the operational characteristics of large physical entities such as the organs of the body. The number of permutations and combinations are astronomical. In some respects the frontiers of physiology have been advanced as far as they can through the use of the conventional tools of the physiologist. New dimensions, new approaches, a more rigorous analysis of basic phenomena must be brought into the field from the domain of the physical scientist.

It is obvious that the blending of these two sciences carries with it major difficulties. From the point of view of the physiologist, much criticism has appeared: "The mathematics is too complex for any but the expert, and must be accepted at face value. The situation has been idealized to the point where it becomes difficult to conceive of how the model bears any relationship to its living counterpart. The analyses deal with trivial points and the more important features of the problem have been ignored."

Equally sharp criticism has come from the physical scientist. "The physiologist places too much emphasis on descriptive and phenomenological features. Boundary conditions are rarely stated or defined. The number of uncontrolled variables is such that no meaningful analysis can be made. Biological or physiological processes consist of many subunits acting simultaneously and much of the experimental evidence does not distinguish between these."

One is reminded here of the homology that Albert St. Gyorgi is fond of making, using the watch, an instrument designed to portray time accurately in analogue fashion. One can be the world's foremost expert in metallurgy, have intimate knowledge of the chemical and molecular properties of the materials which go into the manufacture of such metals, but still be unable to construct the end product, the watch. It is quite evident that information at both levels are needed, neither of which is *a priori* more important than the other. The organization of cells into discrete aggregates with a common function, and then into successively more complex organs or tissues, creates problems which are of an entirely different order of magnitude and involve forces which are more in the physical domain than in that of elementary particle physics. When 100 cells are aligned to form a membrane barrier or a tubule, the properties of this larger structure are not simply some additive function of a cell prototype but reflect new processes. We find, therefore, the need to approach the long-range solution to such problems from both ends of the spectrum. It is for this intermediate zone that new axioms will have to be developed.

It is obvious that not all physiological processes lend themselves to the tools and methods of mechanics (Table 2). For the most part, four areas have been investigated—the *cardiovascular system* including the heart; the

TABLE II
Physiological Areas

General	Cell membranes Muscle contraction Nerve-bioelectric
Cardiovascular	
Pulmonary	
Muscular system	
Nervous system	
Digestive tract	
Excretory organs	
Sensory	
Endocrines	
Reproductive	

pulmonary system including the airways, the lung proper, and the muscular structures involved in breathing; *muscle mechanics* including a materials analysis of skeletal, cardiac and smooth muscle; and finally, a broad category covering the *supporting tissues* of the body including materials such as collagen, elastin, complex polysaccharides, and structures such as bone, cartilage, tendons, and skin.

If there is any substantive criticism to be levelled at the role which mechanics has taken in physiology, it would be that the approach has been too conservative—being concerned primarily with obvious functions—as in the case of muscle or the circulation. There has been a tendency to rely too much on existing data, with comparatively few attempts made to generate new information which would make the physical analyses much more appropriate. As an example, stress-strain analyses of the properties of the blood vessel wall are useful by themselves, but in terms of disease they are only of limited value. The vessel wall, in a material sense, is made up of several different structures, the most important of which in terms of stress-strtain relationships are collagen and elastin. These fibers are unevenly distributed and frequently have a spiral or helix-like configuration. In calculating the modulus of elasticity of the vessel wall, it is not enough to express the properties of the wall as a percent of the volume or weight occupied by collagen fibers, but more properly to include some reference to the configuration or disposition of such materials. Furthermore, in a living system the properties of the vessel wall may change with time without any change in the relative amount of collagen. There is clearly a need to examine what effects specific physicochemical factors have on the properties of biomaterials. In many ways, it is the subtle changes in the physical properties of materials which characterize, for example, aging as a physiological phenomenon.

Although much of the work of our own group has dealt primarily with the transport and exchange functions of the circulatory apparatus, it can serve as a good example of the kinds of problems which are encountered in the application of mechanics to a major area of physiology. We have, in particular, attempted to generate data for the development of constitutive equations for phenomena such as flow through single capillaries (Zweifach and Intaglietta, 1968) and (Lew and Fung, 1969), the interplay of hydraulic and osmotic forces in fluid movement (the so-called Starling relationship Landis and Pappenheimer, 1963), the compliance of the successively smaller segments of the vascular tree, the physical characteristics of vascular smooth muscle, and so on.

Data for such constitutive equations most readily can be obtained in transparent, almost two-dimensional, tissues such as the mesentery. Before such information can be applied in a general context, it becomes necessary to study the same phenomenon in other tissues. Both quantitative and qualitative differences exist in otherwise comparable systems, depending upon the special needs of each tissue. An especially productive approach for understanding physiological processes is that offered by comparative physiology (Prosser, 1961). The analysis of similar systems in different species provides a unique means of establishing the manner in which a given problem is handled in the face of peculiar circumstances or needs.

Another complicating feature of physiological processes stems from the fact that such systems are usually under the control of overlapping mechanisms because of the need to fulfill both local demands and central requirements for the body as a whole. Many of the mechanical attributes of tissues are intrinsic to the cells proper, for example, skeletal or cardiac muscle. Others depend upon the junctional complexes by which cells are bound together, or upon the material properties of the so-called connective tissue. The conformational characteristics of the macromolecules of both the fibrillar elements and the ground substance are affected by the environment in which they exist and undergo continuous modification and degradation (Schubert, 1964). It becomes necessary, therefore, to recognize that changes in the physico-chemical properties of living materials occur even during the actual process of deformation. Mechanical analyses usually include terms which are referred to as "material constants," quantitative expression of the organizational framework which imparts to the material a unique and invariant structural characteristic. In dealing with biomaterials, the fact that dynamic processes operate to keep the structure alive should in some way be included in the materials constant.

As an example, vascular smooth muscle exhibits a property referred to as "myogenic tone" (Nicoll, 1969). When muscular blood vessels are placed under increased intravascular tension, the vessel at first distends but then begins to narrow spontaneously until it returns essentially to its original state. Conversely, when the pressure in a blood vessel is reduced, the smooth muscle in the wall tends at first to shorten but then relaxes and readjusts to its original disposition (Johnson, 1968). This phenomenon of "autoregulation" appears to be related either to some change in the permeability of the cell membrane or to some intracellular chemical reaction. In this instance, the derivation of a set of material constants should include some energy-dependent processes such as transmembrane ion transport or release.

Most physiological functions are kept in balance by a variety of subsystems which act through separate pathways to achieve the same end result. Perhaps the most elaborate controls exist at the biochemical level, an interesting example of which are the mechanisms concerned with blood coagulation (Guest, 1965). In a physical sense, similar interlocking processes act to control blood flow through the microscopic blood vessels (Renkin, 1968 and Zweifach, 1961). When one inspects blood flow in the small blood vessels through the microscope, the most striking impression is that the velocity at which the blood is moving in the different vessels shifts from moment to moment. The number of blood cells in adjacent capillaries, the so-called capillary hematocrit, likewise appears to be quite variable.

In any given network of capillaries, the blood flow depends upon the differential between the arterial pressure feeding into the network and the venous pressure at the outflow end of the circuit (Palmer, 1965). Inasmuch as the capillaries are relatively nondistensible and are not contractile, the resistance to flow will be determined, for the most part, by the resistance offered by the blood itself; in particular, the red blood cells. As the blood is distributed into successive branches, it is clear that there are considerable differences in the number of red cells entering the various branches (Fig. 1). The uneven distribution of red blood cells can be shown to be related to the velocity difference in two branches (Svanes, 1968). This feature, the number of red cells actually present in a given vessel, will by itself tend to increase or to decrease the resistance to flow, and in turn will affect the actual pressure drop along its course. The resulting readjustment of velocity will determine the red cell hematocrit in the affected vessels, thereby closing the feedback loop.

Another important feature of the normal vascular state is the ability of vascular smooth muscle to be maintained in a tonic or partially constricted state. Attempts to identify the substance in blood which fulfills this function have turned up as many as 10–12 candidates present in tissues (Mellander and Johansson, 1968). A similar situation has arisen out of studies designed

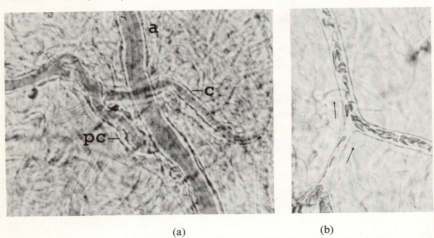

(a) (b)

Fig. 1(a): Flow of blood in arteriole (*a*), precapillary branch (*pc*), and capillary (*c*) in omentum taken at normal exposure (1/250 sec.). Note apparent clear zone at periphery of arteriole. **1(b):** Blood flow in capillary vessels of mesentery taken at 1 : 10,000 second so that individual red cells (rbc) can be seen. Note they are deformed and have a distorted leading edge. Clearance of rbc in capillary is from 1.0–2.0 micra. Velocity is between 150–250 micra per second.

to determine the mechanism responsible for increased blood flow in tissues such as skeletal muscle after exercise or a period of deprived blood flow. Elevated levels of K^+, reduced O_2 tension, increased pCO_2, shifts in pH, release of histamine, polypeptides, and even a simple increase in total concentration of chemical by-products to the point of hyperosmolarity have all been implicated (Skinner and Powell, 1967, and Lundvall et al, 1969). Modelling at this level of organization presents obvious constraints.

Biological structures concerned with a specific function often exhibit such a wide range of diversification that it is difficult to formulate a clear-cut structure/function relationship. This feature would seem to be in conflict with a fundamental concept in physiology, that of homeostasis of the internal environment, by which it is implied that readjustments have to be made continuously to maintain a stable state.

There is no agreement concerning the nature of the controls which maintain tissue homeostasis by regulating local tissue blood flow. Most analyses have assumed that the supply of oxygen is the critical feedback, presumably by widening or narrowing the feeding small arteries or arterioles (Knisely et al, 1969). In order for such controls to be effective, however, many other features must be adjusted over and above a simple increase in volume flow of blood—the surface area for exchange must be increased; different capillaries do not have identical permeability characteristics so that the relative number of each type which is involved will affect the exchange of materials; the pressure within the capillary vessels, as influenced particularly by venous outflow resistance; the effect of increased or decreased blood flow on the tissue pressure; the flow properties of the blood itself.

The manner by which these different processes are brought into play will depend, in final analysis, upon the structural properties and arrangement of the microscopic blood vessels. In some tissues the capillary network has a characteristic pattern or design, and in others the vessels appear to be distributed randomly (Fig. 2). We thus have concepts which envisage on the one hand local control through strategically placed flood gates, or sphincters delivering quanta of blood upon demand, (Folkow, 1960) and on the other a randomized phenomenon which because of the multiplicity of factors involved will tend to operate around a mode compatible with tissue needs (Gaehtgens et al, 1970).

The need for a precise definition of the fine structure in relation to function is well illustrated by examining the wall or barrier between the blood and the tissue compartment. The capillary wall in the living state appears as a thin line about one micron thick except at points where the endothelial nucleus bulges into the lumen. The unusual properties of the blood capillary, by virtue of which it appears to act as a molecular sieve allowing only substances with a molecular weight of 10,000 or less to penetrate into the tissue, have been attributed to the thinness of the barrier. Mathematical

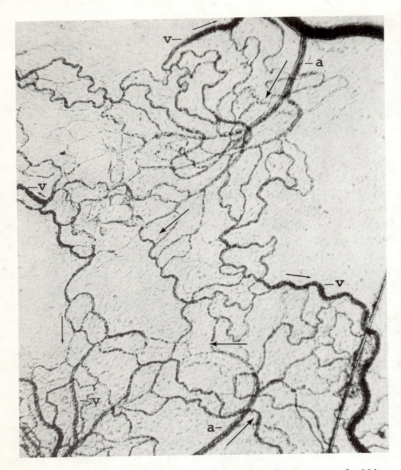

Fig. 2: Low power overview of capillary network in omentum of rabbit indicating pattern of distribution of vessels from two arterioles (*a*) and return by way of three collecting venules. Arterioles and their direct extensions can be traced through almost to venules and present a low resistance series circuit. Majority of capillaries are side branches and thus parallel circuits and present a higher resistance to flow (x 35).

analyses have idealized the capillary network as a cylinder with walls having water-filled paths of a given pore size. Recent ultrastructural studies together with kinetic measurements of exchange have indicated that a number of different physical pathways are available for movement across the capillary barrier (Karnovsky, 1967).

The wall is actually a mosaic of several structures. As seen in cross section, the vessels are lined by from two to three endothelial cells (Fig. 3). The points at which the separate cells abut or overlap present potential paths for fluid

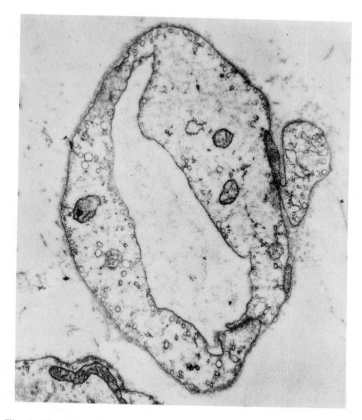

Fig. 3: Electron microscope section of capillary in skeletal muscle of rat (x 15,000). Variations in thickness of wall in region of nucleus. Wall is made up of two endothelial cells. Note tiny invaginations at both surfaces and in interior of cell. These are so-called vesicles. Thin, amorphous basement membrane surrounds vessel and encloses portion of a pericyte.

and solute exchange. The nonselective nature of the exchange, except for molecular dimensions, strongly favors an extracellular pathway of this type. There is some question whether a uniform set of pores are present or whether a small number of larger defects appear from time to time. In addition to the endothelial layer, the small vessels are invested closely on the outside by a thin, amorphous sheath, frequently seen only under the electron microscope. This outer coat is referred to as the basement membrane and serves as a barrier primarily restricting the movement of large molecular complexes and blood cells. In addition, a striking ultrastructural feature of endothelium is the presence of large numbers of small surface invaginations and membrane-bound vesicles (90 A°) in the cell interior which can be labelled with colloidal particles or macromolecules injected into the bloodstream (Karnovsky,

1967). This evidence suggested a transfer mechanism involving the uptake of materials from one side of the endothelial membrane and their transfer to the other side. Still another factor is the presence, at the actual interface between the blood and the lumen surface of the endothelium, of a thin layer of absorbed material, probably a combination of protein and polysaccharides which can serve as a selective filter on the basis of charge, chemically active groups, water content, etc. (Luft, 1965). One must superimpose onto this physical framework the factor of water versus lipid solubility in regard to transport. Apparently gases, which are highly lipid soluble, diffuse through the blood vessel wall much more rapidly than do water soluble substances.

The permeability properties of the barrier are usually defined either in terms of a permeability coefficient *per se*, or in terms of a reflection coefficient relative to the perviousness to some standard macromolecular material (Crone, 1963). Again in formulating mathematical relationships, we are faced with the need to define the physical pathway involved for each type of exchange, the possibility of interactions between separate processes, and the extent to which the relative importance of these processes may change with time.

SUMMARY

We have covered somewhat sketchily a number of broad concerns that have become apparent as the interaction between various aspects of mechanics and physiology broadens and begins to deal with phenomena at the very frontiers of our knowledge in both sciences. The opportunities are great; advances will be spectacular; but pitfalls exist when communications between these disciplines break down. The by-words are: patience, caution, and courage. I am reminded of the statement made by an elderly statesman of the troubled state of Israel. He was asked what was needed most for survival in difficult times. He stated, "There is no alternative. You have to be a realist." He was then asked to explain what one had to do to be a realist. "It is very simple," he replied, "you only have to believe in miracles."

Acknowledgement. *This work supported by a grant from the United States Public Health Service, $HE = 10881$ and U.S. Army Grant DADA 17-67-C-7001.*

REFERENCES

Crone, C. 1963. The permeability of capillaries in various organs as determined by the use of the "indicator diffusion" method. Acta Physiol. Scand. 58: 292–305.
Dowben, R. M. 1969. General Physiology. Harper and Row Publishers, New York.

Folkow, B. 1960. Range of control of the cardiovascular system by the central nervous system. Physiol. Rev. 40 (suppl. 4): pp. 93–101.
Gaehtgens, P., H. J. Meiselman, and H. Wayland 1970. Erythrocyte velocities in mesenteric microvessels of the cat. Microvasc. Res. 2: 151–62.
Guest, M. M. 1965. Circulatory effects of blood clotting, fibrinolysis and related hemostatic processes, pp. 2209–248. In W. Hamilton and J. Dow, (eds.) Waverly Press, Baltimore, Maryland. Handbook of Physiology Vol. III, Circulation.
Johnson, P. C. 1968. Autoregulatory responses of cat mesenteric arterioles measured in vivo. Circulat. Res. 22: 199–212.
Karnovsky, Morris, J. 1967. The ultrastructural basis of capillary permeability studied with peroxidase as a tracer. J. Cell Biol. 35: 213–36.
Knisely, M. H., D. Reneau, Jr., and D. F. Bruley, 1969. The development and use of equations for predicting the limits and the rates of oxygen supply to the cells of living tissues and organs. Angiology 20 (Suppl. 11): pp. 1–56.
Landis, E. M., and J. R. Pappenheimer 1963. Exchange of substances through the capillary walls, pp. 961–1034. W. Hamilton and J. Dow, (eds.) Handbook of Physiology Vol. II, Circulation. Waverly Press, Baltimore, Maryland.
Lew, H. S., and Y. C. Fung 1969. Flow in an occluded circular cylindrical tube with permeable wall. Zeit. f. Angew. Math. Physik 20: 750–66.
Luft, J. H. 1965. The ultrastructural basis of capillary permeability, pp. 121–60. In B. W. Zweifach, L. Grant, and R. T. McCluskey, (eds.), Academic Press, New York, N.Y. The Inflammatory Process.
Lundvall, J., S. Mellander, and T. White, 1969. Hyperosmolarity and vasodilation in human skeletal muscle. Acta Physiol. Scand. 77: 224–33.
Mellander, S., and B. Johansson, 1968. Control of resistance, exchange and capacitance functions in the peripheral circulation. Pharmacol. Rev. 20: 117–96.
Nicoll, P. A. 1969. Intrinsic regulation in the microcirculation based on direct measurements, pp. 89–101. In Winters & Brest, (eds.) The Microcirculation. C. C. Thomas, Springfield, Illinois.
Palmer, A. A. 1965. Axial drift of cells and partial plasma skimming in blood flowing through glass slits. Amer. J. Physiol. 209: 1115–122.
Prosser, C. L. 1961. Circulation of body fluids, pp. 386–416. In C. L. Prosser and F. A. Brown, Jr., (eds.) Comparative Animal Physiology. W. B. Saunders Company, Philadelphia, Pa.
Renkin, E. M. 1968. Neurogenic factors in microcirculatory low flow state, pp. 139–48. In D. Shepro and G. Fulton, (eds.) Microcirculation as Related to Shock. Academic Press Inc., New York.
Schubert, M. 1964. Intercellular macromolecules containing polysaccharides. Biophys. J. 4: 119–38.
Skinner, N. S., Jr., and W. J. Powell, Jr. 1967. Action of oxygen and potassium on vascular resistance of dog skeletal muscle. Amer. J. Physiol. 212: 533–40.
Svanes, K. 1968. Variations in small blood vessel hematocrits produced in hypothermic rats by microocclusion. Microvasc. Res. 1: 210–20.
Truesdell, C. 1965. The Elements of Continuum Mechanics. Springer–Verlag, New York.
Zweifach, B. W. 1961. Functional Behavior of the Microcirculation, pp. 27–49. C. C. Thomas, Springfield, Ill.
Zweifach, B. W., and M. Intaglietta 1968. Mechanics of fluid movement across single capillaries in the rabbit. Microvasc. Res. 1: 83–101.

2

Biomechanics and Surgery

RICHARD M. PETERS

INTRODUCTION

Both engineers and surgeons are applied, not pure, scientists. This implies that they both must combine the art of their profession with the scientific knowledge available. The surgeon's objective in furthering development of the field of biomechanics is to increase the portion of his decisions based on science, so his art may be better fitted to the needs of his patient. Of all the clinical sciences, surgery is the most mechanical; hence, biomechanics is basic to all the tasks of the surgeon. The proper application of biomechanics by surgeons has saved the lives of many patients but unfortunately many have been hurt and some have died when biomechanics was ignored or misapplied. Because biomechanics is so fundamental to surgery, it is only possible at this time to touch on a few of the objectives and to cite some examples of its application to surgical research, diagnoses and therapy.

THE SURGICAL WOUND

The surgeon, to practice his art, must create a wound. Proper healing of this wound is fundamental to patient recovery. During the process of healing stress will be applied to this wound. Knowledge of the mechanical properties of the biological material which is molded to restore continuity to tissue during the healing process is fundamental to surgical science and to most of the therapeutic decisions of the surgeon. There are two important types of information requisite to understanding this fundamental process; the physical properties of the material laid down to repair a wound and the chemical nature and activity of the material.

To date, much interesting work has been done to study the burst strength of wounds and the composition of the chemical components that are recruited to the area of injury as repair proceeds (Van Winkle, 1969). Recent work has shown that the active chemical process of repair takes far longer than earlier simplistic studies showed.

The imperfections of only testing the resistance of wounds to bursting have resulted in wide variations in the measured strength of wounds by different investigators. Most studies show that by 3–4 weeks the "strength" of scar and normal skin and subcutaneous tissue is about equal. Biomechanics suggest that the burst strength is a poor criterion for testing the strength of a wound. A far better method would be measurement of a stress-strain curve and identification of the yield point (Lichenstin, et al 1970). From clinical observation we might expect a much different stress-strain curve in a "mature" wound than in a new wound.

While testing has revealed that full strength is achieved at about 4 weeks, for as long as scars (the permanent evidence of wounding) have been observed, it has been known that they may continue to undergo biomechanical change for at least a year. Recent work by Madden and Peacock (1968) has clearly demonstrated that wounds continue to have prolonged active metabolic processes which correlate with the clinical observations of mechanical change. The turnover of collagen is elevated for many weeks at least, presumably as a result of continued remodeling of the collagen. Presumably it is this remodeling that produces a more pliable scar. However, all scars do not improve their biomechanical properties with time. In some patients scar becomes stiff and restricting, an extreme being scars which form keloids, or excessive scar tissue. In these patients the healing process goes awry and the scar develops very debilitating mechanical properties which may prevent motion. An important question regarding the role of stress in the development of keloids is whether a wound under tension is more likely to develop excessive scar tissue.

Another shortcoming of the assessment of wounds by measuring the burst strength is the destructive nature of the test procedure. We need a nondestructive test. With proper instrumentation perhaps determination of stress-strain curves to identify the yield point could be plotted without disruption of the wound. Modern solid-state transducers might be implanted to determine stress-strain and yield in a functional wound such as a sutured ligament, tendon or fascia of the abdominal wall, or even scar about an intravascular prosthesis.

The need for such a nondestructive method becomes more obvious when looking beyond the properties of scar in connective tissue and skin. In skin or other tissues which can be put to rest, relatively weak sutures can be used to approximate the tissue for the purpose of speeding healing and reducing scar. In some organs such as the heart, function must be restored immediately and cannot await the development of natural healing to resist stress. In such tissues where stress will be immediately applied, breakage of sutures or disruption of tissue due to fatigue may be disastrous. New, more reliable sutures have had to be developed for repair of vital organs such as the heart; in addition, we need studies to determine quantitatively how sutures can be placed to decrease shear force, which tends to tear them loose or fracture them before healing can be complete. These studies must determine the stress resistance of tissue as well as the suture material.

THE BIOMECHANICS OF INJURY

Just as fundamental to surgery as the biomechanics of wound healing is the biomechanics of accidental trauma: first, so that means can be found to decrease crippling and death; second, to permit prediction of the extent of internal organ injury from the type and extent of surface or skeletal injury.

Disruption of the skeleton by trauma is one of the major areas of interest to orthopedic surgeons. There is, unfortunately, very little good biomechanical analysis of the functions of the skeleton and its response to stress. Studies are beginning of the mechanics of the skeleton, but there are disappointingly few orthopedic surgeons with knowledge of engineering principles or engineers who appreciate the problems of analysis in biological systems. As a result, orthopedics is still largely an art practiced by superbly skilled individuals who call on the experience of prior successes and failures. It is only necessary to suggest a few of the problems to illustrate the great potential for progress in this area.

For most types of fractures and joint disruptions a qualitative description of the injuring force leading to the fracture is provided in standard textbooks. These descriptions are incomplete because the mechanics of the region have not been determined and there is little or no quantitative study of bone strength, ligament strength nor of the force exerted by the attached muscles. Such rudimentary information as is available has led to such things as better ski bindings, but has done little to influence our sport promoters and coaches to alter rules to protect the skeletons or ligaments of their valuable star performers. The terrible toll of injuries suffered by star athletes and youths striving for stardom results both from the rudimentary level of knowledge of the mechanism of injury that can be supplied for designing protective equipment and rules and from the failure of the responsible sports officials to listen to what advice is available. It is truly amazing that profit-making industries such as the major league sports enterprises and intercollegiate athletics should not be more interested in some basic research which might lead to methods of protecting their valuable players from crippling injuries.

Parallel to our general ignorance of the biomechanics of skeletal injury is our ignorance of the biomechanics of repair. How do mechanical forces influence the healing of fractures, tendons and ligaments? Again, we have little or no quantitative data and only rudimentary qualitative data. While it is difficult to obtain such information about bone or ligament strength, it presumably should be easy to study materials used to immobilize fractures. However, only sparse studies have been made of the stress-strain characteristics of plaster of Paris casts, the principal material used to immobilize fractures (Chenek, et al 1969). From a mechanical viewpoint it is apparent that this is a relatively poor material, but it is kept as the material of choice because it is cheap and easy to handle. Research to find a material that can withstand the rigors of the adolescent male's activity during fracture healing would be a boon to patient and surgeon. Studies of the most appropriate material for immobilization of fractures must be correlated with determinations of the biomechanical factors which act to increase and decrease fracture healing. The material and method of application can then be designed to minimize the disruptive stress at the fracture site and to enhance those that increase healing.

Engineers and orthopedic surgeons are trying to design treatment for specific fractures which would neutralize the residual stress on the fragments and mechanically assist the resorption of bone at fracture sites which must occur for proper healing to commence. If the healing process could be materially speeded and nonunion prevented, months and years of patient disability could be eliminated and the fantastic cost of such major injuries reduced.

Man among mammals is unique in his assumption of the upright posture. The lower primates tend to crouch and never voluntarily sit in chairs, certainly not in the monstrosities called seats in our high-speed vehicles. The strain applied to the spinal musculoskeletal system by these peculiar habits of man needs careful study.

One of the most disabling acute traumatic injuries we see is fracture of the cervical or the thoracic spine with resultant injury to the spinal cord. The mechanism, acute torsion, on the spinal column is partially understood but practical ways of prevention remain to be widely applied. Though proper general use of seat belts could largely eliminate many of the most serious spinal cord injuries in vehicular accidents, the seat belts themselves may contribute to some lumbar and cervical fractures.

More serious in terms of man-years of disability is the dysfunction of the lumbar spine, so-called low back pain. We know little or nothing about the mechanics of the lower back, nor how to protect it from the stress which leads to this dysfunction. The automobile manufacturer and the designer of seats in jetliners have been concerned about esthetically pleasing seats but have had only minimal interest in studying factors responsible for low back dysfunction.

The back is one of the more complex mechanical parts of the musculoskeleton system (Choffier and Mouks, 1969). However, some of the unsolved mechanical problems of other parts of the skeleton are perhaps surprising. For example, only a start has been made concerning forces applied to various parts of the skeleton.

The need for more effective interdisciplinary research by engineers and orthopedic surgeons prompted the surgical study sections of the National Institute of Health to convene a meeting in September, 1970 to aid both groups in attacking the many unsolved problems. It has been suggested that such studies follow the methods of analysis in mechanical engineering: "(1) The study of the kinematic relationships in the motion of one body part with respect to another and in the intrinsic deformation of each biological material; (2) the quantitation of static force and moment relationships in the musculoskeletal system; (3) the quantitation of dynamic force and moment relationships in the musculoskeletal system; and, (4) the quantitation of the viscoelastic stress-strain relationships in the musculoskeletal system. These four areas correspond roughly to the divisions of study within engineering mechanics and have been traditionally used as dividing lines for research" (E. B. Weis, *personal communication*).

Each subspecialty of surgery must deal with diseases whose diagnosis and treatment depend on accurate application of biomechanical principles. The neurosurgeon's concern with the mechanism of head injury will be presented by one of the participants in this conference. The neurosurgeon is also interested in the flow of cerebrospinal fluid, the dynamics of blood flow in the closed box, the skull, and with increased use of stereotaxic devices, the development of better mechanical localization of anatomical sites in the brain.

Otolaryngologists are concerned with the biomechanics of voice and hearing. The recent advances in understanding the micromechanics of the auditory system have resulted in relief of disability.

The general surgeon and the urologist are both concerned with the mechanism of fluid flow by peristalsis and better knowledge of the biomechanics of blood flow in those complex organs, the liver and kidney. One of the major diseases of our society is alcoholic cirrhosis, which disrupts the pattern of blood flow in the liver. Most of the patients develop varices of the esophagus with severe hemorrhage. Many operative procedures have been devised to deal with this disease, the most popular of which creates a shunt between the portal vein and the vena cava. Knowledge of the biomechanics of the double circulation to the liver in normal individuals is incomplete, and in disease is only rudimentary. To improve effectiveness of operative treatment and to lower the present high operative mortality for this serious diease will require new approaches to the study of this complex circulation.

Rather than trying to accomplish the impossible of describing applications of biomechanics to all areas of surgery, I shall confine the rest of my remarks to some examples of the role of biomechanics in thoracic and cardiovascular surgery, my own field.

CARDIOPULMONARY BIOMECHANICS

Like orthopedics, thoracic and cardiovascular surgery is based on understanding biomechanics. Each step forward in the field has been associated with the acquisition of new information and the application of biomechanic theory to the therapeutic procedure applied by the surgeon. A failure to understand the basic mechanics of respiration, and particularly the mechanics of the mediastinal divider between the two chest cavities, retarded intrathoracic surgery for many years. As with so many of our areas of ignorance, the tragic wars of this century have brought into focus some of our incomplete knowledge of respiratory mechanics. In World War I the influenza epidemic caused many young men to develop streptococcal pneumonia and subsequently empyema, the accumulation of pus in the pleural cavity. A principle of surgery is that pus must be drained, and so many unfortunate young men sick with pneumonia had holes made in their chest walls and open-ended tubes inserted to drain the pus. This resulted in an open pneumothorax and sucking chest wound with air rushing in and out of the opened hemithorax with each breath. The seriously ill soldier was required to produce enough muscular effort to ventilate both his pleural cavity and his lung. As a result untold numbers of men died of exhaustion, the direct result of a treatment which was correct for infection but which disrupted the mechanics of respiration.

Chest surgery owes its background in biomechanics to the careful mechanical analysis of the thorax by Dr. Evarts Graham, the father of pulmonary surgery. Dr. Graham's studies (1925) clearly showed that the mediastinum freely transmitted force from one chest to another and therefore an opening into one chest cavity destroyed the pressure relations in both. The practical corollary of this analysis was that a chest cavity should not be drained without a one-way valve until the infection had caused enough tissue reaction to make the lung firmly adherent to the chest cage.

Injury to the thorax with associated disruption of cardiopulmonary function follows head injury as a cause of accidental death. Injuries to the thorax cause mechanical disruption of the bellows function of the chest and often, in addition, severe injury to the intrathoracic organs: tracheobronchial tree, lungs, heart, great vessels and esophagus. At present we have very little sound knowledge of the mechanism of injury to intrathoracic organs or how to predict their presence. Better studies to correlate the injuring force with the resultant damage are essential.

An example of the bizarre type of injury seen following sudden deceleration, such as occurs in a head-on automobile collision when the forward motion of the body is abruptly stopped, is traumatic rupture of the aorta. This injury can occur in two locations: (1) at the juncture of the aortic arch and the descending aorta and, less commonly, (2) at the point where the aorta passes through the diaphragm. This injury is thought to be caused by sudden deceleration of the chest cage and the descending aorta, which is attached to the chest cage. The transverse aortic arch and heart continue to move, and shearing tear occurs at the juncture between the mobile and fixed portions of the aorta. If this theory is correct a tight chest restraint might aggravate rather than lessen the incidence of this injury. However, if the mechanism is not shearing at the point of transition between mobile and fixed aorta, then constricting protective chest restraints may aid in preventing this injury. An alternative theory is that the mechanical effects of a blow on the chest cause a sudden increase in intraluminal pressure in the aorta and a resultant severe strain on the aortic wall. The juncture of the transverse aortic arch and the descending aorta, as well as being a juncture between fixed and mobile aorta, is a point of acute change in direction and small change in lumen diameter. It could well be a point of stress concentration in the aorta. All individuals with this type of injury have evidence of a severe blow to the chest, usually with rib fractures. The blow on the chest may cause compression of the heart and great vessels resulting in an acute increase in intraaortic pressure with resultant rupture at the point of stress concentration. If this mechanism is correct, a properly designed restraint would prevent such a blow and therefore the aortic rupture.

Though knowledge of the mechanism of the injuring force is important to the surgeon, treatment is more determined by the type of tear and whether complete or only partial rupture with a resultant false aneurysm have occurred. If the latter, what are the factors that lead to the rupture of the aneurysm and resultant exsanguination of the patient? Most such patients have multiple injuries and the repair of such an aneurysm acutely can compromise other treatments and often the life of the patient.

Most aneurysms of the large vessels can be resected and vessel integrity restored. However, such operative procedures entail risk to life, are expensive and cause 6–10 weeks of morbidity. In patients with associated coronary or cerebrovascular disease the risks to life are much greater. The surgeon who diagnoses an aneurysm, whether it be a Berry aneurysm in the cerebral circulation, a traumatic aneurysm, or the common degenerative infrarenal aneurysm of the abdominal aorta, is always faced with the dilemma of balancing the risk of rupture against the risk of operation. The surgeon and biomechanical consultants need to know which aneurysm is likely to rupture. To date, the only criteria for predicting the likelihood of rupture of aneurysms are based on clinical symptoms of pain or the size of the aneurysm. Clinical experience has suggested that as the diameter of an aneurysm increases beyond 4 cm the wall tension induced by the stress of the intraaortic pressure is likely to exceed the tolerance of the tissue of the aneurysm with resultant rupture. The mortality from ruptured aneurysms is more than 50% while that for elective resection is in the range of 15%. If we had better means of predicting the likelihood of rupture, unnecessary operations could be avoided and needed ones could be undertaken electively rather than as emergencies.

Prevention would be more rewarding than resection of aneurysms. This goal depends on a better understanding of the mechanical factors that result in development of aneurysm. While the biomechanics of development of posttraumatic false aneurysms are not clearly defined, there is almost complete lack of biomechanical data to explain the development of aneurysms in degenerative vascular disease. Why do they have a predilection for the aorta between the renal vessels and the bifurcation?

Direct application of biomechanics to surgical therapy is more evident in the problems of diagnosing valvular dysfunction in the intact patient and in designing valvular prostheses. Fundamental to such concepts are analyses of function of the two most frequently diseased valves, the aortic and mitral valves. The aortic valve was the subject of mechanical analysis by the great engineer, Leonardo DaVinci. While more abstract mathematical descriptions of the function of this valve are now being made, one is amazed by the accuracy of this analysis made centuries ago. The unique mechanical attributes of the normal aortic valve with its full orifice flow was a factor in the failure of the first attempts at its replacement, and represents one of the failures to pursue engineering analysis in medical therapy. Since no one has been able to conceive of a valve with an orifice that opens more fully than the normal aortic valve and will also act to divert blood into the coronary arteries during diastole, surgeons hoped to copy its design by simply replacing defective leaflets. The first attempts at replacement were to sew in a cloth prosthesis shaped like the normal valve cusp. The cusps were made, unfortunately, of Teflon cloth, a material which is known by fabric engineers to have very little flex strength. In fact, no fiber is available with the flex strength required of such a prosthesis; all present fibers will ultimately fatigue and fragment. Presumably tissue valves survive only because they are constantly being regenerated.

Subsequent to these original tragic failures, successful replacement became possible when the ball-in-cage type of prostheses were developed. These imperfect valves create some obstruction and all are subject to thrombosis and mechanical failure. To be truly effective they must yield to better mechanical design and improved materials. Among the problems requiring more investigation are studies of the need for turbulence to sweep the prosthesis clean versus the effect of turbulence in traumatizing blood. Can some new surface be found which does not damage blood or must this surface be a natural endothelium, as advocated by Braunwald and Detmer (1968) and others? What is the best design yielding the least constriction of inflow and outflow orifice? What design will minimize prosthesis wear? Is a design feasible that can be quickly and firmly attached to the available tissue without the prolonged process of inserting and tying sutures? How can we test such a valve to be assured that it will withstand the imposed stress? After more than ten years of replacing valves with artificial prostheses, many surgeons are abandoning them for biological grafts from deceased humans or other mammalian species (Gonzalez-Lavin and Ross, 1970). These valves tend to leak and calcify and require much longer time at operation to replace.

One is forced to wonder whether entirely new ideas might be generated by bringing together individuals very knowledgeable about fluid mechanics, materials and surgery. To date, useful but far from ideal valves have been developed through long hours of trial and error by devoted workers. However, if individuals with extensive basic theoretical background in the pertinent disciplines applied their combined knowledge to this problem a valve that approached the effectiveness of the normal valve might result.

At this point there is hope of replacing hearts rather than valves. Consider some of the fundamental problems in biomechanics that must be solved before mechanical substitutes for the heart become effective.

Is pulsatile flow essential to the integrity of the vascular system? What are the factors that must be used to control output of such a heart? How will energy be transmitted to it? What surface must line it? What type of valves should it have? How can it be made durable, small and economical enough to be a practical substitute?

All the technical aspects of cardiac surgery are based on applications of principles of biomechanics. Unfortunately one of the major clinical problems for the cardiac surgeon is the imperfection of these technical procedures and the devices available. It is imperative that devices improve and procedures become more effective so that earlier, complete restoration of function can be advised. At present the cardiologist and cardiac surgeon most constantly attempt to balance the patient's disability against the imperfect operative repairs available. For many patients considerable disability must be suffered before operative therapy is justified.

While the spectacular technical achievements in thoracic surgery get the headlines, they probably are not the major contribution of thoracic surgery to the health of the nation. A successful technical operation from which a patient does not survive is not a triumph. To increase survival of patients undergoing cardiac surgery has required the development of better understanding of the pathophysiology of the postinjury patient. This knowledge has been applied to the care of all types of patients, and the development of theoretical approaches to treatment and training of skilled personnel in their application has probably prevented more disability and saved more lives in patients who have not undergone cardiac surgery than in those that have.

Any analysis of the physiologic systems vital to survival immediately focuses on the heart and lungs. If they do not function adequately without interruption, brain death promptly results and life, even if preserved, becomes meaningless. Therefore first efforts must be focused on maximizing their function. Cardiac dysfunction due to hypovolemia or poor function of the pump was, in the past, one of the most frequent causes of postoperative death. With better knowledge of the mechanical function of the pump and improved pharmacologic agents that can be applied to support it, cardiac failure is less often the critical factor causing the patient's demise. Improvement in this aspect of therapy has focused attention on pulmonary insufficiency, the major cause of death at this time in all types of postoperative and postinjury patients. Often the failure is the result of preexisting mechanical dysfunction of the lungs, but all too often lungs which were entirely normal prior to operation or injury fail in the postoperative period.

The serious toll of postoperative pulmonary insufficiency came to the fore in part as a result of the improvement in cardiac function achieved by a combination of the effectiveness of even the present crude methods of continuously monitoring cardiovascular status in acutely ill patients and better therapy. Another factor may be simply that the lung is the organ most vulnerable to injury when cardiac dysfunction is present. Earlier recognition of changes in cardiac output or ventricular function might lower the high incidence of pulmonary dysfunction induced by cardiac insufficiency. The crudeness of the present transducers and monitoring techniques in appraising cardiac function prevents a fine enough control of therapeutic measures to support the circulation and thus prevent injury to the lungs. One of the major efforts needed in biomechanics is to develop better nontraumatic approaches to monitoring function of the cardiovascular system.

Equally important is better understanding of the mechanics of the lungs and chest cage and better means of assessing early signs of dysfunction. The problems of assessing pulmonary function in the acutely injured patient are different from those encountered in assessing function in ambulatory patients. Also the mechanical dysfunctions encountered are unique. In our laboratories we have been particularly interested in developing methods of diagnosing and treating acute pulmonary insufficiency. We have undertaken the simultaneous development of an automated monitoring system, study of new techniques of analysis of mechanical dysfunction and the development of mathematical models of lung function.

Postoperative pulmonary failure is characterized by a progressive fall in lung compliance, rise in resistance in the lungs and airway, increase in work of ventilation and a fall in arterial oxygen partial pressure (Peters and Stacy, 1964). There is also continued fall in functional residual volume of the lungs, presumably due to progressive decrease in the number of ventilated alveoli.

Since we are dealing with patients, it is necessary to assess which measurable parameters of pulmonary mechanics, lung volumes, distribution of bloodflow and ventilation or respiratory work will give the best trade-off between the information gained and the cost in dollars, discomfort and risks to the patient. At first glance, measurement of dynamic compliance of the lungs would appear to be the most likely predictor. Our studies to date show that it is not altered early in the course of the development of the syndrome despite the fact that acute postoperative pulmonary insufficiency has, as one of its major components, alveolar collapse. An explanation for the lack of early change in dynamic compliance is a change of position of respiration to a lower end tidal volume and the fact that collapse of as many as every third alveolus has a relatively small effect on the pressure required to expand the lung. In addition, the variance that is found in available methods of measurement of lung compliance make the early changes impossible to interpret. Later in the course of the syndrome compliance falls markedly but at that point the diagnosis is obvious.

On the other hand, if total work done on the lungs or work per liter is measured, an earlier indication of respiratory failure may be apparent. When work begins to elevate, respiratory support should be instituted. This knowledge is only useful if an effective way of obtaining information about changes in respiratory work is generally available. To give prompt information to the clinician about respiratory work requires some means of automatic calculation. At this time some progress has been made in this direction. We have developed methods which use either a small special-purpose analog computer or a small digital computer for this computation (Peters, 1969). One of the small analog computers made in our laboratories has been successfully used as a research tool for a year in Viet Nam (Proctor, in press, and Peters, et al, 1969). The major limitation to general applicability of such techniques lies in primary transducers. We need a good, small transducer to use in the esophagus and at the end of an endotracheal tube to measure intrathoracic and intratracheal pressure. We also need an accurate, durable, reliable transducer for measuring air flow in and out of the airway.

At this stage, active therapy is dependent on knowledge of mechanical changes characteristic of the posttraumatic lung syndrome. At end expiration the respiratory muscles are at rest, and the volume of gas in the lungs is determined by the balance between the outward elastic recoil of the chest and the inward recoil of the lungs. If the compliance of the lungs falls, their recoil force increases and the functional residual volume, the amount of air left in the lungs at the end of expiration, falls. Low volume of air in the lungs permits more alveoli to reach a critical size so that they are more unstable and likely to close competely. The most important mechanical therapeutic tool is to increase functional residual volume by preventing the system from coming to complete rest. This can be done in a number of ways: increasing respiratory resistance, decreasing duration of expiration, increasing tidal volume, or putting a stop in the expiratory side of the system so pressure cannot fall to zero. Each of these alternatives leads to other problems. Increased tidal volumes must be compensated for by increasing dead space to prevent hyperventilation of the alveoli. Increased resistance must be associated with adjustments of tidal volume and expiratory time to achieve the proper degree of increased inflation without hyperinflation, which would cut off venous return.

If ideal therapy is to be achieved a mathematical model that could predict changes necessary to prevent further collapse would be useful. We are in the process of testing such a model at this time. Changes in lung inflation pressures must be correlated with blood gases, cardiac function venous pressure, chest wall function, and other factors.

SUMMARY

The best summary for a chapter in this volume on biomechanics and surgery is to state that this symposium could, in fact, assume the title of this chapter. This chapter has only listed a few examples and in doing so each subject obviously overlaps a more specific presentation to follow. The biomechanics of wounds is dependent on basic studies of "Properties of Skin and Tendon" and "stress-strain-history relations of Soft Tissues."

All studies relating to mechanism of injury are pertinent to surgery—"Biomechanical Problems relating to Vehicle Impact," "Biomechanics of Head Injury," "Biomechanics of Bone." In the field of thoracic surgery all of the chapters on blood flow and cardiac function are pertinent and that on lung elasticity is basic.

In brief, as a surgeon, I would like to urge those interested in biomechanics to view surgery as the most likely place for their researches to affect the health of patients. For those of us responsible for the training of bioengineers and surgeons, it is imperative that we create situations where the trainees in both fields can work together. If they sit in the same room as they develop their skills in investigation, they will create a new generation without the communication gap between specialities that is so crippling to progress at present. Unfortunately for those of us now in the position of preceptors and classed as the older generation, the creation of this essential environment will probably increase the generation gap. Our new breed of bioengineer and surgeon will be developing skills that will be hard for us to comprehend. Events of recent months perhaps have taught us that rather than fearing new generations we should nurture and encourage them in their quest for new solutions to old problems. My only fear for the objectives of biomechanics and surgery center on the possibility of a new discipline arising which does as most others have done; only talks to itself. This symposium, so magnificently organized by Dr. Fung, has fostered an interdisciplinary approach to biomechanics. It will, I am sure, achieve this end.

REFERENCES

Braunwald; Nina S., and Don E. Detmer 1968. A critical analysis of the status of prosthetic valves and homografts. Progress in Cardiovascular Diseases, 2: pp. 113–32.

Chenek, T. S., J. H. Somerset and R. E. Porter 1969. Stress in orthopedic walking casts. J. Biomechanics, 2: pp. 227–39.

Choffier, D. B. and E. J. Mouks 1969. An empirical investigation of low back strains and vertebrae geometry. J. Biomechanics, 2: p. 89.

Gonzalez-Lavin L. and Donald Ross 1970. Valve replacement—a five-year experience at the National Heart Hospital, London. J. Thoracic and Cardiovascular Surgery, 60: pp. 1–12.

Graham, E. A. 1925. Some Fundamental Considerations in the Treatment of Empyema Thoracis. C. V. Mosby Co., St. Louis, Mo.

Linchenstin, I. L., S. Herzekoff, J. M. Shore, M. W. Jiron, S. Stuart and L. Mizano 1970. The dynamics of wound healing. Surgery, Gynecology and Obstetrics, 130: p. 685.

Madden, J. W. and E. E. Peacock, Jr. 1968. Studies on the biology of collagen during wound healing. 1. Rate of collagen synthesis and deposition in cutaneous wounds of the rat. Surgery, 69: pp. 288–94.

Peters, R. M. 1969. Mechanical Basis of Respiration. Little, Brown and Co., Boston, Mass., pp. 247–79.

Peters, R. M. and R. W. Stacy 1964. Automated clinical measurement of respiratory parameters. Surgery, 56: pp. 44–52.

Peters, R. M., H. A. Wellons Jr. and T. M. Howe 1969. Total compliance and work of breathing after thoracotomy. J. Thoracic and Cardiovascular Surgery, 57: pp. 348–55.

Proctor, Herbert, T. V. N. Ballantine and N. D. Brousard. 1970. Analysis of pulmonary function following non-thoracic trauma with recommendation for therapy. Annals of Surgery, 71: pp. 172–80.

Van Winkle, Walton Jr. 1969. The tensile strength of wounds and factors that influence it. Surgery, Gynecology and Obstetrics, 129: pp. 819–42.

Weis, E. B. Jr., *personal communication*.

3

Biomechanics, Patient Care and Rehabilitation

E. O. ATTINGER

I. INTRODUCTION

Biomechanics has been defined in many different ways (see Dickson and Brown, 1969) but in the context of this paper, we are concerned primarily with formal and quantitative analysis of the relationships between structure and function of living tissues and the application of the results derived therefrom to man in health and disease. These relationships are a function of the physicochemical properties of tissues and their constituents, changing in time and space as a result of the information that is being transmitted within the system itself and between its "milieu interne" and the "milieu externe." Instead of fixed relationships, as in man-made systems, we are dealing with performance spectra of the multiple subsystems of the "human plant," where adjustment of performance to stress and disease is achieved by appropriate changes of the structure–function relationships in the various subsystems through the influence of hierarchically organized controllers (Attinger, 1970b). One of the most important characteristics of living systems in this connection is their redundancy in design.

Because of the hierarchical structure and the complexity of the interactions between subsystems and components, the behavior of the "living plant" cannot be deduced from the knowledge of the physicochemical properties of its components alone. This implies, of course, that biomechanics, important as it is for the understanding of structure–function relationships has only diagnostic but no prognostic power by itself unless coupled with an understanding of how information flows change under conditions of stress and disease.

I will first try to delineate the role biomechanics can play within the context of clinical medicine and rehabilitation and then give some examples of the implementation of such a role in our own department.

II. THE NATURE OF LIVING SYSTEMS

Living systems are open systems, that is, they require a continuous inflow of energy from their environment to perform their function. They are self-regulating, adaptive and capable of autoduplication, properties for which control and information are essential (Attinger, 1970a). Figure 1 diagrammatically illustrates the interdependencies between an organism and its environment.

As an open system, the very existence of the organism depends on a continous flow of energy $(-\Delta G)$ that permits it to grow, to move and to do work. Morowitz (1968) recently emphasized the general thesis that the flow of energy through a system acts to organize that system. In order to detect, process, retain, and utilize information, a cybernetic state function analogous

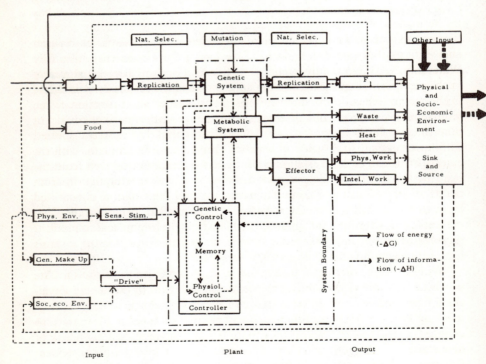

Fig. 1: Interrelations between an organism and its environment, in terms of flow of energy, internal structure and flow of information Attinger, 1970 (P_1 and F_1 represent parental and filial generations respectively.)

to the Gibb's free energy function in the energy domain is required. Such a function has been identified as negentropy, N or H. This analogy and the relation between G and H make it possible to consider the metabolic cost of information and, therefore, the area of inherent overlap between the ergonic and cybernetic components. Whenever information is obtained by making an observation, entropy is decreased in the system which acquired the data and increased in the source of information. It is obvious that this relationship requires an inherent structure, the dearrangement of which is responsible to a large part for systems malfunction.

Figure 1 indicates these multiple and close interrelations between the ergonic and the cybernetic components. The environment plays the roles of a sink and of a source for both the flows of energy and of information. The ergonic component, centered in the functional structure of the metabolic system, has four types of outputs: waste, heat, physical work and intellectual work. It is controlled by "passive" intrinsic mechanisms, such as the law of mass action or Le Chatelier's principle, as well as by energy requiring extrinsic mechanisms arranged in multiple loops and levels. At the intracellular level, these extrinsic mechanisms comprise the genetic mechanism (localized mainly in the nucleic acid domain), the cell membranes (acting as highly selective barriers, as well as ionic pumps), and a number of cytoplasmic boundaries and surfaces that establish a primary level of biological heterogenity. (See Waterman, 1968). At the intercellular level, extrinsic control is vested mainly in chemical regulatory mechanisms, such as hormones, in neurosecretions and in the integrative action of the nervous system. At the multi-organizational level, control depends on behavioral interactions between individuals or populations of the same species or between members of different species. Thus at any level, structure is an essential prerequisite for control, with the controller having at least five types of inputs: information derived from the state of the metabolic system and its effectors, sensory stimulation from the physical and socioeconomic environments and information transmitted through the genetic system. In terms of response times, genetic control represents long-range planning modified by natural selection and mutation; memory acts over an intermediate time span; and the physiological controls deal with relatively short-term phenomena.

Within such a conceptual framework disease can be interpreted as a state in which the control hierarchy has become unable to fully cope with deviations from homeostasis, either because of deficiencies in the control system itself or because of the inability of the "plant" to respond adequately to changes in information flow. If the failure is of limited duration, we speak of acute disease. The control system either reasserts its role with consequent full recovery of the patient, or it may completely fail (death) or it may regain control of the state of affairs only partially (chronic disease). The degree to which control is reasserted thus determines the destiny of the patient.

Some of the most striking examples of the importance of the role of control mechanisms in the evolution of disease are provided by individual case histories of patients with chronic infectious disease, such as tuberculosis before the advent of effective antibacterial therapy (see Mann, 1924). Not only did patients with an apparently identical disease pattern diverge completely in the evolution of their disease, but the treating physician was frequently able to predict these different outcomes from the habits and the attitude of the patient.

III. THE ROLE OF BIOMECHANICS IN MEDICINE

For any application of biomechanics to clinical problems, therefore, alterations of the physical properties of tissues must be considered within the context of the entire organism and its environment, as indicated in Fig. 2. The patient's health (health being defined in its broadest sense) is a function of his physical and mental state as well as of his compatibility with a particular socioeconomic environment. Theoretically, at least, health can thus be expressed by a state vector. However, such a vector is not directly measureable but has to be determined by means of subjective and objective observables which are interpreted as performance indicators of various biological subsystems or combinations thereof. This interpretation is called diagnosis. The effectiveness of a physician depends on his capability to properly weigh the individual components of the confusingly complex symptomatology with

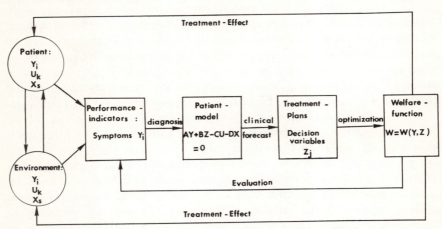

Fig. 2: The relationship between patient, environment and medical care. Diagnostic, therapeutic and prognostic decisions by the physician are based on his choice and interpretation of performance indicators characterizing the patient's health with the goal of improving the well-being of the latter in terms of his own value system.

which he is confronted in each patient. As already stated, the diagnostic and prognostic weight of individual signs depends not only on the underlying pathology in terms of physicochemical principles, but also on the personality of the patient himself. In the most pedestrian case the diagnostic process is based on a taxonomy of disease and in the most inspired case on "intuition." The result of the diagnostic procedure is a patient model, the accuracy of which is related to the validity of the performance indicators which have been chosen as the basis for its construction. Based on this model, a prognosis is established by means of clinical forecasting. This is primarily a matter of clinical experience and could, by the way, be handled very effectively by a computer system with an adequate data bank. Given the constraints imposed by the patient himself and by his environment, an optimal treatment plan is chosen, using appropriate decision variables, and its effectiveness evaluated by means of the observable symptoms. It is clear, therefore, that the validity of diagnosis, prognosis and treatment are primarily a function of the accuracy with which the symptoms chosen represent the state of the patient. Because of the complexity and variability of the symptomatologies encountered, it would be highly desirable, if a general, formalized model were available, which could serve as a guide to the interpretation of symptoms and their weight and predict the relative effectiveness of alternative treatment plans.

The approach outlined in Fig. 2 and the concepts upon which it is based are very similar to that used for the analysis of socioeconomic systems. For example, both are faced with the problems of choosing significant performance indicators and of weighing short-term versus long-term benefits. Policy decisions which may improve short-term forecasts may be detrimental to long-term forecasts. (The history of pharmacology is full of examples of law suits concerning long-term side effects of drugs which had a beneficial effect on some acute condition, e.g. thalidomide.)

It is, therefore, not surprising that medical care can be formalized by means of a model which was originally developed for economic systems by Tinbergen (1952). We distinguish two types of exogenous variables: policy instruments (Z) defining the treatment plan and noncontrollable data (U); two types of endogenous variables: target variables or symptoms, defining the state of health according to the patient model (Y) and irrelevant side effects (X), as well as a welfare function W. Examples of Z are food supply, economic security, drugs, surgery, psychotherapy, and so on; of Y, malnutrition, poverty, fever, physical properties of biological tissues, chemical composition of body fluids, energetics of motion, and similar factors; of U, age, race, sex, hereditary traits, and others. The composition of the group of irrelevant variables depends, of course, on a particular situation, but is usually surprisingly small in any system, (see Millendorfer and Attinger, 1968).

The welfare function W defines the well-being of the patient in terms of his own value system. (There are marked differences in tolerance to pain, in mental stability, in eagerness for work, in personal ambition, and in reactions to adverse environmental conditions between individuals. The importance of local value systems on the means by which population control programs can be implemented is an excellent example of the dependence of a welfare function on a value system.) W is a function of the target variables Y (the symptoms, an $i \times l$ vector) and the instrument variables Z (the components of the treatment plans, a $j \times l$ vector).

(1) $$W = W[Y(Z), Z]$$

Assuming linearity, the quantitative model M relating the different variables can be written as

(2) $$AY + BZ - CU - DX = 0$$

where A, B, C, and D are matrices of coefficients of appropriate orders, U is a $(k \times l)$ vector of noncontrollable data and X an $(s \times l)$ vector of irrelevant side effects. It is clear that this linear formulation represents only a first approximation, and more complex functions may eventually have to be substituted. In the absence of adequate empirical data the additional complexity required by the latter approach cannot be justified at present.

The constraints imposed on the target variables (symptoms or observable performance indicators) and the policy instruments (treatment components) are

(3) $$Y_{min} \leq Y \leq Y_{max}$$
$$Z_{min} \leq Z \leq Z_{max}$$

This simply means that life sets an upper and a lower bound on all the variables upon which it depends. Normally this range is rather narrow and has led to the concept of homeostasis with respect to the Y components (Millendorfer and Attinger, 1968).

The goal of treatment is then to optimize the welfare function W by solving the following set of partial differential equations, which are the necessary conditions for an optimum

(4) $$\frac{dW}{dZ_j} = \Sigma_i \frac{\partial W}{\partial Y_i} \frac{\partial Y_i}{\partial Z_j} + \frac{\partial W}{\partial Z_j} = 0$$

provided that the second order condition for a maximum or minimum is fulfilled. For the case where the number of target variables is larger than the number of instrument variables ($i \geq j$), all targets cannot, in general, be reached simultaneously and their number must be reduced so that $i \leq j$.

For example, in a patient in moderate cardiac vascular failure and acute diabetic coma the primordial target consists in correcting the metabolic disorder rather than the condition of the cardiovascular system. On the other hand, if a diabetic has an accident in which a limb is severed, the first task is to prevent further blood loss.

The efficiency of the treatment plan itself is specified by the term $\partial W/\partial Z$. The impact of the treatment variable Z_j on the performance indicator Y_i is expressed by $\partial Y_i/\partial Z_j$. (For example, corticosteroids affect allergic reactions favorably but have marked side effects on water and electrolyte balance as well as on secondary sex characteristics.) Hence the term $\Sigma_i \partial Y_i/\partial Z_j$ expresses the sum of all marginal effects of a unit change in the given treatment variable Z_j, some of which may be positive (spin-off benefits) and some of which may be detrimental (undesirable side effects).

Because such a scheme may appear very attractive and easy to realize, it is necessary at this time to inject a note of caution. Physiology, which, in essence, represents an attempt to quantify our classical Western medical concepts of well being deals with the analysis of normal systems and only prepares a frame of reference for abnormal systems. Normalty is, of course, a quality standard. Every quality standard, even those pertaining to health or esthetics, is a function of a cost-benefit ratio although the ratio may be defined in some esoteric rather than monetary terms. The setting of quality standards thus always involves a value judgement and a value judgement imputes a cost-benefit ratio. This is the reason why I have stressed the role of the system of human values for the definition of the welfare function and as an inherent limit for the validity of conclusions based on purely physicochemical principles with respect to patient care. Nevertheless, models such as the one discussed in the preceding paragraphs are extremely useful as an aid in establishing a first level of quantitation in patient care commensurate with the present state of knowledge and expandable as the latter increases.

What then is the role of biomechanics with respect to patient care and rehabilitation within this general framework? As I see it this role is four-fold:

1. To quantitatively analyze the relationships between structure and function of living tissue in health and disease.
2. To derive from these analyses appropriate indicators for the mechanical performance of biological systems and thus a more realistic basis for the construction of patient models and for strategies of patient care.
3. To define adequately tissue-transducer interfaces for the purpose of measuring these indicators more reliably by noninvasive or, at least, nontraumatic techniques.
4. To define the physical properties of living systems in such a way that reliable, long-term internal and external prostheses can be designed and produced economically.

IV. AN EXAMPLE OF THE APPLICATION OF BIOMECHANICS TO CLINICAL MEDICINE

In the remainder of this paper, I am going to discuss one of the ongoing research programs in our department in which all these aspects are, at least partially, included. It involves a quantitative analysis of the relationships between structure and function of the cardiovascular, respiratory and renal systems, and the choice of appropriate physiological variables or correlates thereof as sensitive and reliable indicators of systems performance which can meet the criteria for valid diagnostic and prognostic indicators. The concepts and methods used in our approach are first discussed within the general framework of an intensive care unit for critically ill patients and then more fully explored using the oxygen transport system as an example. The latter analysis leads to a group of pathophysiological correlates which seem to meet the requirements for valid performance indicators of gas transport, useful not only as therapeutic and prognostic guides but also for the evaluation of alternative therapeutic plans as discussed in the preceding section in connection with the Tinbergen model.

The clinical use of such indicators requires appropriate sensors and data handling equipment for their continuous monitoring. The development of such sensors raises many biomechanical problems such as questions related to system-transducer coupling, to tissue compatibility of implanted materials and to biomaterials in general. After a short discussion of these problems a development program for specific sensors and data processors is briefly described and finally, the various aspects of the program are integrated in terms of a prototype intensive care unit for critically ill patients.

Since the traditional areas of biomechanics are well covered in other papers of this symposium, I am considering primarily the broader aspects of biological systems as they relate to patient care, wherein mechanics proper represents only one of many important components.

A. THE CONCEPT OF AN INTENSIVE CARE UNIT

This program, which has been established with the goal of developing a prototype intensive care unit for critically ill patients represents a joint effort between the divisions of Biomedical Engineering and of Pediatric Cardiology under Dr. F. Damman (1970). It rests on four basic concepts:

1. In general, the pathophysiology of critically ill patients involves at least three organ systems, namely circulation, respiration and the organs maintaining electrolyte and water balance (central nervous system involvement is usually secondary in the patients under consideration). Hence, effective monitoring requires continous evaluation of these

three systems. Some of the criteria used for such performance analyses are based on biomechanical principles, particularly as they relate to respiratory mechanics, pulmonary function and hemodynamics. The intensive care unit should, however, be designed not only for the continuous monitoring, processing and display of pertinent performance criteria but also for provision of all potentially required therapeutic facilities (including staff). In such a unit patient care should be superior and the operations much more efficient than in the present hospital organizations where respiratory, renal and cardiovascular intensive care as well as shock and trauma units are scattered throughout the institution, thus promoting unnecessary duplication of expensive equipment and inefficient use of scarce medical resources.
2. The critically ill (as well as less sick) patients often show subtle signs of system failure long before the deterioration of their condition can be recognized by conventional means (that is, clinically). The judicious selection of parameters which are valid indicators of performance is constrained, at present, primarily by: a) the level of our understanding of living systems, b) the available technology and c) the safety and the comfort of the patient. Continuous monitoring of a number of variables is carried out in many hospitals, but the choice of the variables to be measured is all too often dictated by the availability and the price of commercial equipment and their compatibility with patient and physician comfort rather than by the pathophysiological significance of the variable they monitor.
3. The third concept recognizes the fact that although an individual patient may temporarily be saved through the combined efforts of all the highly-skilled manpower available within a modern medical institution, he will often be discharged without being properly equipped nor motivated to cope with an environment he has come to regard as hostile and to which he will sooner or later succumb. These latter aspects must be part of a proper rehabilitation.
4. In order to be able to demonstrate the medical, economic and administrative advantages of such a unit, it is absolutely essential that the unit be autonomous and isolated from obstructive interdepartmental competition.

The initial tasks of the program are therefore threefold:

1. To construct performance indicators (the Y's in Fig. 2) which are sensitive and reliable enough to serve as quantitative, diagnostic, therapeutic and prognostic guides (the relation between the Z's and W in Fig. 2) before the deterioration of the patient becomes clinically manifest.

2. To develop transducers and data handling systems, which, while not interfering with the comfort of the patient, are nevertheless reliable, self-testing and acceptable for the environments both of a small community hospital and of a medical center.
3. To prove to the hospital administration, the medical staff and the various government agencies that a centralized autonomous unit is superior to a number of competing departmental intensive care units. Since this particular task is unrelated to the topic of this paper, it will be omitted from further discussion.

B. Performance Indicators

The first task is essentially a systems analysis in pathophysiology, involving a host of biomechanical concepts and methods particularly for the evaluation of the respiratory and circulatory systems. As already mentioned, in medicine, performance criteria for clinical purposes have traditionally been chosen on the basis of convenience (methodology and patient comfort) and of faith. As a result, much of the development in medical instrumentation has been misdirected toward the design of better mousetraps instead of toward new concepts.

The prime variables which are monitored continuously, semicontinuously or as spot samples in many intensive care units (heart rate, respiratory rate, electrocardiogram, arterial blood pressure, cardiac output, and so on), are characterized, in general, by a normal range which is too large to permit any definite conclusion with respect to a specific malfunction. Spot samples of various biochemical parameters (such as blood gases, electrolytes, or components characterizing the acid-base balance) cannot be examined frequently enough to be reliable indicators of impending catastrophe. The following two examples illustrate the unreliability of single measurements of these variables as performance indicators.

Figure 3 emphasizes the wide range of normal cardiac output values for a number of animal species. It is based on more than 500 cardiac output determinations obtained by some fifty investigators between 1950 and 1965 (Attinger et al, 1967). Although body surface has long been the most common standard of reference for purposes of comparison of cardiac output determinations, recent studies indicate that in most species cardiac output (CO) is better related to body mass (BM) than to body surface (the latter is an unreliable estimate anyway). A correlation analysis of these data yields the following results

$$\text{for all data: } CO = 0.1515 \, BM^{0.903}; \quad r = .878, \, n = 531$$
$$\text{for humans only: } CO = 2.55 \, BM^{0.193}; \quad r = .114, \, n = 139$$

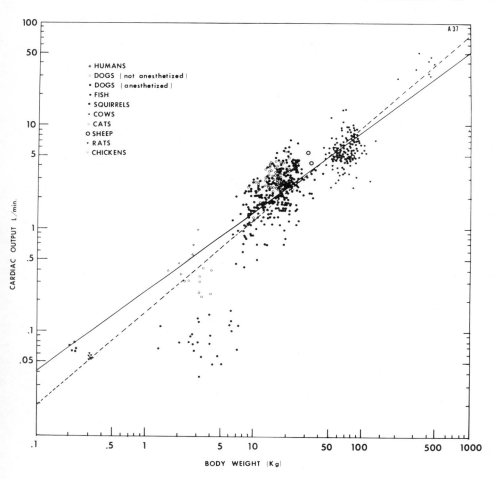

Fig. 3: Relationship between cardiac output and body weight in different animal species Attinger, et al, 1967.

indicating that although such a relationship permits a rough prediction of the cardiac output for a given species on the basis of body mass, the resulting accuracy is clearly inadequate as a prediction for individuals within the species, where the cardiac output may vary by a factor of three (even at rest) from one individual to the next. Similar data have been reported on the normal range of heart rate in man, with a composite regression equation of heart rate (HR) versus age (A) for 432 subjects (316 normal males and 116 normal females) (Jose and Collison, 1970):

$$HR = 118.1 - 0.57 A \qquad r = 0.644$$

Evidence of this type argues strongly against the use of "normal" values as indicators or predictors.

Trend observations of these variables constitute, in general, more reliable criteria but are frequently too slow for the assessment of the effectiveness of therapeutic measures because they are relatively late indicators of individual responses to stress and of incapability to meet increased stress requirements. For these reasons, we believe that properly chosen pathophysiological correlates of variables, which relate structure and function over a wide range of stress states meet more sensitive and more reliable criteria as performance indicators than an array of individual variables. An example of such correlates is discussed in connection with the analysis of the oxygen transport system.

C. Performance Analysis of Biological Systems

As pointed out in section III, biomechanical concepts and methods play an important role in carrying out performance analyses of biological systems aimed at defining performance criteria. Although associated with considerable theoretical and practical problems, a systemic approach appears to offer the most promising strategy for the interpretation and integration of the large amount of knowledge that has been accumulated about specific mechanisms and for the evaluation of their relative clinical significance (Attinger and Millendorfer, 1968). Gradually techniques are being developed which permit new approaches to the analysis and control of complex systems, the properties of which change as functions of time and environment and are thus not precisely specifiable. By means of simulation, the behaviour of such systems can be evaluated when subjected to various types of inputs, taking into account the interactions between various subsystems and components and their effects on the different levels of the system's control hierarchy. The Tinbergen model is an example of such an approach based upon the optimization of a welfare function which can be defined in terms of target variables and policy instruments (Eq. 4).

In the critically ill patient adequate perfusion of the individual organs for meeting metabolic needs (including the removal of waste products and heat) clearly is of first priority. The efficiency (Eq. 4) of this transport function involves the relation between the supply required by the individual tissues under various conditions of metabolic needs and the associated costs of transportation. Along with the latter, the costs for information handling and control must be included. Although traditionally the cardiovascular system is considered the primary plant for these transport functions, the latter depend as critically on the performance of the respiratory and the renal systems; mechanical transport cannot be fully segregated from diffusional and active chemical transport.

A quantitative assessment of the overall transport function requires, therefore, that at a minimum the following data be obtained either directly or indirectly:

1. Minute ventilation and the metabolic costs of the respiratory pump.
2. Rate of oxygen extraction from the lungs as an indicator of the overall ventilation-perfusion ratio.
3. Cardiac output and the metabolic costs of the circulatory pump.
4. The distribution of blood flow.
5. Local oxygen extraction.
6. Blood volume and hematocrit.
7. Renal blood flow and the metabolic costs of the renal pump (Na-reabsorption).
8. Urine output and composition.
9. Acid–base balance.
10. Temperature.

The relative significance of these variables with respect to the overall efficiency of mass transport in the living organism can be evaluated by means of a parametric performance analysis. As an example of such an analysis, I will discuss a model of the oxygen transport system (Pennock and Attinger, 1968) which can be extended to any other biological transport function by an appropriate transformation of variables. This example has been chosen not only because pulmonary and cardiovascular function and their interdependencies represent an important area of biomechanics but also because it illustrates the many relationships between individual biological systems as well as the central role of models in providing patient care (Fig. 2).

D. THE OXYGEN TRANSPORT SYSTEM

The O_2 transport system simultaneously supplies a number of organs and tissues in varying states of activity. Particularly under conditions of stress, the competition for a finite supply of oxygen (the main metabolic energy source) among different organs brings to mind the picture of a competitive economic market. For this reason, it seems attractive for the analysis of optimization processes in biological systems to draw on the store of knowledge accumulated in economics where success and failure may depend on appropriate optimization. A parametric performance analysis of such a system reveals the relative importance of the various parameters for a desired output because, in a model of the system, they can be varied at will. The results of the analysis thus provide an indication of the ranking of variables with respect to their functional significance and an estimate of the level their control mechanisms must occupy within the hierarchy of the overall system

(Attinger and Millendorfer, 1968). Information of this type is essential for the evaluation of the status of a patient and for the choice of the best treatment combination from a variety of alternative possibilities. It defines, for example, the borderline between medical and surgical treatment of vascular heart disease. Because of the large reserve capacity of the system at rest, we have assumed that the features of optimization become apparent primarily during conditions of stress.

Figure 4 shows a diagram of the most recent version of our model that has been the basis of the optimization analysis (Attinger 1970a, Anné, 1970). The three circles represent the three pumps: respiratory muscles, left and right heart. The amount of oxygen being transported can, of couse, be increased by augmenting the output of these pumps. There will be, however, a limit where the oxygen cost of the increased pumping activity becomes equal to the increased amount of oxygen transported so that the oxygen supply available to the periphery begins to decrease (Pennock and Attinger, 1968). In the lungs the oxygen is transferred from the gas to the blood phase, and this transfer can be characterized by the ventilation–perfusion ratio. Three regions with different ratios are indicated in the diagram. In the periphery, the vascular beds supplying the various organs and tissues are arranged in parallel. A diagram of such a bed is indicated in the lower part of the figure. The amount of oxygen available within the tissue at any time is a function of the volume of the organ and its partial O_2 pressure; the consumption is a function of the metabolic rate and of the supply. The blood supply to the organ is divided into two channels. One channel represents the nutritive flow where oxygen fully equilibrates between tissue and capillary blood. The second channel represents a functional shunt where no oxygen exchange occurs. This is, of course, a very simplified model of the real system because it represents only the two limits of a continuous spectrum. In reality, most of the vascular channels occupy an intermediate position so that both oxygen consumption and local blood flow distribution must be represented by continuous distribution functions.

The oxygen available to the periphery can be calculated as a function of the following parameters:

1. The concentration of inspired oxygen. ($F_{I_{O_2}}$, volumes percent.)
2. The ventilation (litres/min) and the energy requirements of the respiratory pump (ml O_2/min).
3. The cardiac output, $\dot{Q}$, (litres/min) and the energy requirements of the heart (ml O_2/min).
4. The local ventilation-perfusion ratios in different areas of the lung ($\dot{V}_{A_j}/\dot{Q}_j$).
5. The oxygen carrying capacity of the blood (Cap_{O_2}, volumes percent).

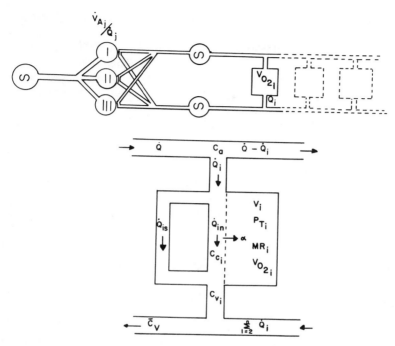

Fig. 4: A distributed parameter model of the oxygen transport system (top) and details of one model block representing the ith peripheral vascular bed (bottom).

$\dot{V}_{A_j}/\dot{Q}_j$ = local ventilation-perfusion ratio
$\dot{Q}$ = Blood flow (l/min)
$\dot{Q}_i$ = Blood flow into the ith vascular bed
$\dot{Q}_{is}$ = "Shunt flow" through the ith bed
$\dot{Q}_{in}$ = "Nutritive flow" through the ith bed
C_a = Arterial oxygen content
C_{ci} = Oxygen content in nutritive capillaries of ith bed
C_{vi} = Oxygen content of venous blood leaving the ith bed
$\bar{C}_V$ = Oxygen content of mixed venous blood
V_i = Volume of tissue supplied by ith bed
P_{T_i} = Tissue oxygen tension of ith bed
MR_i = Metabolic rate of tissue supplied by ith bed
$\dot{V}_{O_{2_i}}$ = Oxygen uptake from ith bed
α = Local extraction rate for O_2

The three circles ⓢ represent the respiratory pump, the left and right heart, respectively; the other three circles represent lung regions with different ventilation perfusion ratios, Attinger, in press.

6. The local blood flow through a peripheral bed ($\dot{Q}_1$, litres/min) and the local extraction rate for O_2 (α, ml/min).

Because of the organizational complexity of the system, we have chosen a state variable approach for its analysis (Anné, 1970). Using the volumes of O_2 (X_1) and CO_2 (X_2) in the tissue as the state vector,* the mass balance equations can be written for a single bed as:

for O_2:

(5) $$\frac{dx_1(t)}{dt} = a_{11}(t)x_1(t) + n_1(t)u_1(t) - \dot{V}_1(t)$$

(6) $$y_1(t) = h_1(t)x_1(t) + s_1(t)u_1(t)$$

for CO_2:

(7) $$\frac{dx_2(t)}{dt} = a_{22}(t)x_2(t) + n_2(t)u_2(t) + \dot{V}_2(t)$$

(8) $$y_2(t) = h_2(t)x_2(t) + s_2(t)u_2(t)$$

resulting in the following state equations for the entire system in matrix form:

(9) $$\dot{\mathbf{X}}(t) = \mathbf{A}(t)\mathbf{X}(t) + \mathbf{N}(t)\mathbf{U}(t) + \mathbf{W}\dot{\mathbf{V}}(t)$$

(10) $$\mathbf{Y}(t) = \mathbf{H}(t)\mathbf{X}(t) + \mathbf{S}(t)\mathbf{U}(t)$$

The first four equations describe the changes in tissue oxygen and CO_2 and in venous outflow in a single bed as functions of metabolic rate ($\dot{V}$), total blood–gas supply (u), shunt flow and nutrient flow, the tissue volume and the dissociation characteristics of the blood. The last two equations formulate these relationships for the entire body in the form of vector matrix differential equations where:

$\mathbf{X}(t)$ = state vector
$\dot{\mathbf{V}}(t)$ = metabolic rate vector
$\mathbf{U}(t)$ = input vector
$\mathbf{Y}(t)$ = output vector
$\mathbf{A}(t)$ = autonomous system matrix ($n \times n$)
$\mathbf{W}$ = constant matrix ($n \times n$)
$\mathbf{N}(t)$ = nutrient matrix ($n \times 2$)

* The notations for the state and target variables used here are different from those used in the Tinbergen model.

$S(t)$ = shunt matrix (2 × 2)
$H(t)$ = output matrix (2 × n)

The elements of these vectors and matrices are, of course, time dependent, as indicated in the equations. The time-domain solution of Eq. 9 is given by

(11) $$X(t) = X(0)\Phi(t) + \int_0^t \Phi(t - \tau)[NU(\tau) + W\dot{V}(\tau)]\,d\tau$$

where $\Phi(t) = e^{At}$ is the state-transition matrix.

The model has been programmed on a hybrid computer, the analog part being used for the representation of the physical system and the digital part for the control of the vector and matrix elements, and hence, of the system parameters.

Figure 5 shows the results of such a simulation. The total oxygen available to the periphery is plotted on the ordinate, cardiac output on the abscissa.

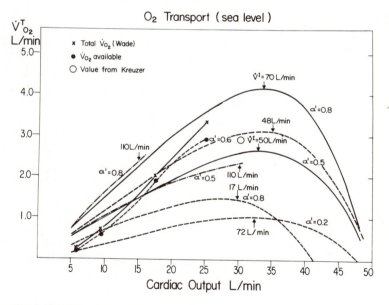

Fig. 5: Performance of the O_2 transport system at sea level for normal man during exercise. The oxygen available to the periphery is plotted on the ordinate, the cardiac output on the abscissa. The family of curves represents all possible combinations for values of ϵ and α' of 0.2, 0.5, and 0.8. (Solid lines, $\epsilon = 0.8$; dash line, $\epsilon = 0.5$; dashed-dotted lines, $\epsilon = 0.2$; the values for α' are indicated on the respective curves; for $\alpha' = 0.2$, the three values of ϵ give identical results.) Values for the ventilation required at the points of peak supply are indicated by the labeled arrows. Note the close correlation between model prediction and experimental data (full circles). Attinger and Millendorfer, 1968.

The family of curves represents a number of combinations for different values of the ventilation-perfusion ratio (ϵ) and of the local oxygen utilization fraction (α'). The latter is a function of local oxygen extraction and shunt flow. The values for the ventilation required at the points of peak supply are indicated by the labeled arrows. Since the energy requirements of the pumps eventually increase more rapidly than the total amount of oxygen taken up from the environment, the oxygen delivered to the periphery reaches a peak beyond which the transport becomes too expensive and thus highly inefficient. This is, of course, one of the mechanisms responsible for dyspnea and cyanosis in patients with chronic respiratory disease, where the metabolic costs of ventilation are so large that the supply of oxygen to the periphery becomes insufficient. For any given ventilation or cardiac output, the amount of oxygen available to the periphery may be increased markedly by appropriate changes in the $\dot{V}_A/\dot{Q}_p$ ratio or the tissue oxygen utilization ratio (α'). Changes in these parameters may increase the peripheral oxygen supply by a factor of more than four without changes in cardiac output or ventilation (compare the points indicated by arrows on the lower-and uppermost curves). The model thus predicts that for any given cardiac output the oxygen supply to the tissue is increased first by increasing α' (represented by a vertical movement through the family of curves) and then by a change in the $\dot{V}_A/\dot{Q}_p$ ratio until the maximal value of $\dot{V}_{TO_2}$ is reached. The most efficient way to increase α' consists in altering the ratio between shunt and nutritive flow. The results of the analysis thus emphasize the importance of peripheral vascular control versus control of cardiac output.

It is not surprising that the most efficient way of altering the local extraction rate of oxygen (α_1) consists in changing the ratio of shunt flow to nutritive flow. There is ample experimental evidence about the nonuniform blood flow distribution in the microcirculation, both from direct observation as well as from measurements of local oxygen tensions. Figure 6 shows a schematic diagram of a capillary network (Renkin, 1964). If all the capillaries have the same cross section, the amount of blood flowing through them is inversely proportional to the flow path length. Hence, the uppermost cells in the figure receive the least blood supply. The blood flow distribution is improved if the resistances in the lowermost flow channels are increased as indicated on the right of the figure. Assuming various distributions of the oxygen consumptions for these flow distriutions, the resulting distribution of local tissue-oxygen tensions can be evaluated and the obtained results compared with experimental data on the local distribution of O_2 tension, such as those reported by Whalen and Pankajain, (1967) or Lübbers, (1969). From such comparisons one is led to the conclusion that the experimental data can only be matched if:

1. The spatial and temporal distribution of O_2 consumption and blood flow are similar and

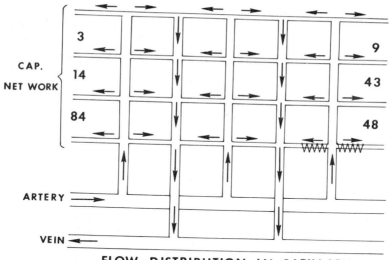

FLOW DISTRIBUTION IN CAPILLARY NET WORK

Fig. 6: Schematic flow distribution in a capillary network. A total inflow of 100 ml/min is assumed. Because of the unequal length of the individual channels, the perfusion is very inhomogeneous (left side). If the resistance of the shorter flow channels is increased (right side) the perfusion becomes more homogeneous. (The numbers in the individual cells of the network indicate the magnitude of local bloodflow.) (After Renkin, 1964.)

2. The total blood flow is adjusted to the oxygen needs of the organ.

Additional evidence for these conclusions is provided by the results obtained by Carter and Rapela (1970), who studied the effects of altering tissue metabolism–flow ratio on vascular reactivity (Fig. 7).

There is now increasing evidence that many of the arterial and venous capillaries are arranged in loops, permitting a countercurrent exchange (Lübbers, 1969). This mechanism has been well recognized as an extremely efficient way for the enhancement of renal function as well as of temperature homeostasis in animals living in cold environments, but seems to be much more universal than originally thought. For example, recent data indicate that a significant fraction of the total O_2 supply to the tissues leaves the vascular channels by diffusion through precapillary vessels (Duling and Berne, 1970). Such an observation would not only be difficult to explain in the absence of a countercurrent mechanism, but also provides the basis for an extremely effective mechanism for local control of intraorgan blood flow distribution. The local controls for intra- and inter-organ blood flow distribution are, of course, in competition with higher level control loops that maintain the integrity of the organism in the presence of excessive local demands.

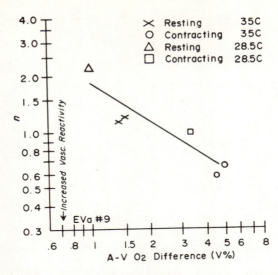

Fig. 7: Effect of altering the control tissue metabolism blood flow ratio on vascular reactivity to change in perfusion pressure. Representative data obtained from an experiment in the autoperfused hindlimb skeletal muscle of the dog. A-V O_2 difference (the metabolism flow ratio) and O_2 consumption were measured just prior to determination of the pressure flow relationships in each of the four experimental conditions. The vascular reactivity to changes in perfusion pressure (n) was derived from determinations of pressure (P) flow ($\dot{Q}$) relationships ($\dot{Q} = kP^n$). Values of consumptions (ml O_2/min·100 g) were 0.18–0.19 for resting muscle at 35C; 0.69–0.72 for contracting muscle at 35C; 0.09 for resting muscle at 28.5C and 0.31 for contracting muscle at 28.5C. The control vascular conductances corrected for effect of temperature on viscosity were about the same for different control experimental conditions (0.10 ml/min·mm Hg 100 g). Carter and Rapela. 1970.

In order to evaluate the balance between local and central control, we compared the effects of local and general hypoxia on the performance of the O_2 transport system in anesthetized dogs (Kenner et al, 1968). General anoxia was produced by exposing the animals to a low oxygen environment (5–10% O_2); local anoxia was achieved by producing a stepwise change in the arterial oxygen concentration of the left hindleg.

Figure 8 illustrates the results of twenty-five such experiments in five dogs. Iliac flow is plotted on the ordinate, arterial pressure on the abscissa. The circles represent the control state and thus express the normal pressure-flow relationship in the hindleg of the anesthetized dog. The crosses represent the flows achieved by a sudden lowering of arterial O_2 tension to values of 20 mm Hg or lower. It is apparent that the flow response is pressure-dependent although the flow change itself was never associated with a pressure change.

During general anoxia, on the other hand, the flow response to a given change in arterial pO_2 is always less than in the case of localized anoxia.

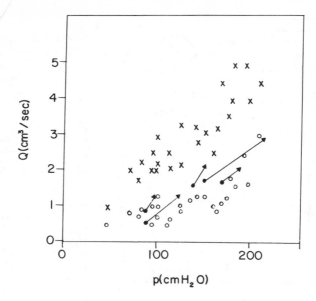

Fig. 8: Relation between the iliac artery flow (ml/sec) and arterial pressure (cm H_2O) from 25 experiments in 5 dogs. The vertical distance between circles (control values) and corresponding crosses represents the increase of flow due to local hypoxia perfusion (blood pO_2 less than 20 mm Hg). Flow is pressure dependent both in the control state and during hypoxia. No measurable change in pressure occurred during perfusion. The arrows indicate the changes in pressure and flow during general hypoxia (breathing 5% O_2). In contrast to the effects of local hypoxia, there is an increase of arterial pressure and a comparatively smaller increase of flow. Kenner et al, 1968.

Furthermore, it is accompanied by an increase in arterial pressure. These data are represented by the arrows. The increase in pressure under this condition is relatively larger than that of local flow. From a series of such experiments we were able to establish the local flow response as a linear function of local arterial pH, as a logarithmic function of local arterial pO_2, as a curvilinear function of arterial pressure and as a function of the catecholamine level, where the effectiveness of the catecholamine is modified by the local pO_2.

Figure 9 pictures a scheme which would explain these results. Oxygen is considered as a vasoconstrictor and a decrease in arterial O_2 tension produces a change in vascular geometry by altering the contractile state (CS) of the vascular wall. The dependence of the contractile state on local pO_2 and catecholamine concentration is given by the equation at the bottom of the figure. K_1 and K_2 represent reaction constants, and A_2 expresses the effects of other factors, such as pH, potassium, blood pressure, and so forth, which are not yet explicitly formulated in this model but which we are investigating at present. The change in the contractile state results in a change in con-

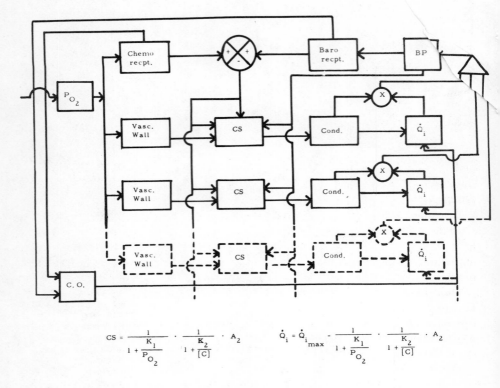

Fig. 9: Local and generalized action of O_2 on blood flow distribution (for details see text).

CS = Contractile state of the vascular wall
$Cond$ = Conductance
$\dot{Q}_i$ = Blood flow in the ith bed
CO = Cardiac output
BP = Arterial blood pressure

ductance and, in the case of purely local changes, in a change of the distribution of cardiac output. The products of the conductances and flows in the various parallel beds are summed, yielding arterial pressure and thus total peripheral resistance. Arterial pressure affects the contractile state by itself through the pressure-volume characteristics of the wall. It also affects the CS indirectly through the baroreceptor reflex and subsequent release of norepinephrine at the level of the resistance vessels. Similarly, the arterial pO_2 counteracts the local response through the local norepinephrine release mediated by the chemoreceptor reflex. The resulting flow response can thus be expressed by the second equation at the bottom of the figure. This equation basically states that at any given pressure the actual flow is equal to the flow during maximal vasodilation minus the modification resulting from local vasomotor effects of O_2 and local catecholamine levels. For the purposes of this analysis, it is irrelevant if the low O_2 tension itself acts as a vasodilator or if it simply represents a sensitive indicator for a vasodilator agent produced by local metabolic processes.

E. APPLICATIONS OF THE RESULTS OF THE O_2 TRANSPORT SYSTEM ANALYSIS TO PATIENT CARE

Theoretical analysis, computer simulation and experimental data all indicate that the most efficient mechanisms for increased O_2 supply to the periphery are, in order of importance

1. Intraorgan blood flow distribution
2. Interorgan blood flow distribution
3. Improvement of ventilation-perfusion ratio
4. Increase in cardiac output and in ventilation

These results have been used to establish tentative levels of a control hierarchy for the O_2 transport system. As already mentioned, such information is essential for proper diagnosis, treatment and prognosis in patient care (Fig. 2). Local control appears to be preferable in those cases where local needs do not conflict with overall needs. Central control is called for when there are conflicting needs or when the local "attitude" is indifferent. This means that the importance of any level in the control hierarchy at a given time depends on the conditions prevailing within the milieu interieur and exterieur at that time.

In terms of the model represented by Eq. 1–4 and Fig. 4 the target variables* are expressed as

(12) $$Y = \Sigma_i \dot{V}_{O_{2_i}} - (\dot{V}_{O_{2_h}} + \dot{V}_{O_{2_r}})$$

with the constraints

$$Y_{min} \geq 0$$

(13) $$Y_{max} \leq \Sigma_i \dot{Q}_i (c_{aO_2} - c_{vO_2{crit}})$$

where

$\dot{V}_{O_{2_h}}$ = O_2 requirements of the cardiac pump
$\dot{V}_{O_{2_r}}$ = O_2 requirements of the respiratory pump
$\dot{Q}_i$ = blood flow to the ith organ
c_{aO_2} = arterial O_2 content
$c_{vO_2{crit}}$ = venous O_2 content corresponding to the critical O_2 tension

* See footnote p. 44.

The policy instruments are defined in terms of

α_i = local extraction rate
$\dot{Q}_i$ = local bloodflow
$\dot{Q}$ = cardiac output
ϵ = O_2 extraction rate in the lungs (ventilation-perfusion ratio)
$\dot{V}$ = ventilation rate
Cap O_2 = O_2 carrying capacity of the blood

This leads to the following tentative formulation of pathophysiological correlates:

1. Metabolism-flow ratio (for crucial organs such as heart, brain, and kidney as well as for the body as a whole).
2. Metabolic cost per unit O_2 transported
3. Contractile state of the vasculature
4. Myocardial function

The first and second are a measure for the efficiency of the O_2 transport, the second and fourth an index of cardiac and ventilatory performance, and the third reflects the adequacy of humoral and neural control of the cardiovascular system.

This example of the analysis of the O_2 transport system has been discussed at some length because it illustrates not only the role classical biomechanics plays in the analysis of biological systems, but also emphasizes a variety of other aspects which must be taken into account in the application of biomechanics to clinical medicine, thus underlining the central role of models in providing patient care (Fig. 2). In addition to providing a better understanding of physiological and pathophysiological processes, it also points out the urgent need for sensors by which the critical parameters can be reliably, accurately, safely and continuously measured. The most crucial lack in instrumentation with regard to the evaluation of the O_2 transport system relates to the estimate of metabolism-flow ratios for local beds as well as for the body as a whole. The role of biomechanics in the development of transducers has been indicated in the introduction and will be explored in more detail in the following sections.

F. THE CRUCIAL NEED FOR MORE ADEQUATE SENSORS

At present no satisfactory methods for direct continuous measurement of blood flow in patients are available (Attinger 1964.) Therefore, a number of indirect methods have been suggested, which are generally based on the fluid-mechanical relationships between stroke volume and pressure pulse.

(For a review of the different methods see Kouchoukos et al, 1970, Wetterer and Kenner, 1968.) Pressure and flow are, of course, related through the vascular impedance which is characterized by elastic, viscous and inertial components. The indirect methods for the determination of cardiac output are usually evaluated by a limited number of calibrations against a direct method (generally a dilution technique) and the resulting "calibration factor" (and hence vascular impedance) are assumed to remain reasonably constant over the period of monitoring. While the latter assumption is probably correct for healthy subjects, it may lead to dramatic errors in the critically ill patient, where the state of the cardiovascular system often changes quite abruptly. But the detection and prediction of these sudden changes represents exactly one of the primary goals of continuous monitoring of critically ill patients and is thus incompatible with the assumption of a constant "calibration factor."

Figure 10 indicates the magnitude of the error which may be associated with such indirect methods, based on single pressure pulses or on ballistocardiography. The data were obtained from simultaneous measurements of blood flow and pressure in the ascending aorta of a dog under conditions of hypovolemia (withdrawal of 30% of the blood volume), after reinfusion of the blood and during slow infusions of epinephrine and acetylcholine. A series of pressure and flow pulses were subjected to Fourier analysis and the pressure pulse resynthetized assuming that either flow or impedance had not changed from the control state. (It will be recalled that the indirect methods assume a constant impedance.) It is apparent from the figure that there are marked differences between these resynthetized pressure pulses and the actually observed waveform, particularly in the case of hypovolemia. The latter is, of course, of particular interest in an intensive care unit.

The state of development of chemical sensors for continuous evaluation of metabolic conditions is equally unsatisfactory. Available sensors for gases suffer from lack of long-term stability, aging effects, inadequate response time, and temperature sensitivity. Their large size renders them sensitive to changes in flow and pressure. The development of microsensors is highly desirable because of performance improvement and of better patient acceptability. Some of the problems associated with the development of such sensors (as well as of prosthetic devices), related specifically to the area of biomechanics and biomaterials, are discussed in some detail in the next section.

G. Biomaterials and Tissue Compatibility

All versions of the electrochemical microsensors (and many prosthetic devices) are in practice exposed to a variety of body fluids; catheter-tip and shunt types must be capable of long-term use in the blood stream. Patient safety requires, therefore, that their surface be as nonthrombogenic as pos-

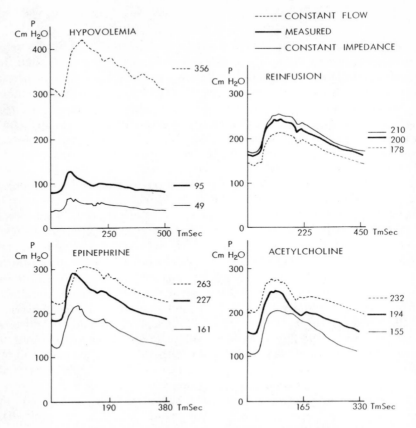

Fig. 10: Aortic pressure pulses under conditions of hypovolemia, after reinfusion of blood and during infusion of epinephrine and acetylcholine. The heavy solid line represents the actually measured pressure, the dotted line and the dashed line are the results of reconstitutions of the pressure pulse from Fourier analyses of flow and vascular impedance. The dotted line was calculated on the basis of the measured flow and the impedance of the control state; the dashed line on the basis of the actual impedance and the flow of the control state, using the relationship:

$$Z_n = \frac{P_n}{\dot{Q}_n} e^{j(\phi_n - \theta_n)}; n = 0, 1, 2, \ldots$$

where Z = vascular impedance
P = Magnitude of the harmonic component of aortic pressure
$\dot{Q}$ = Magnitude of the harmonic component of aortic flow
ϕ = phase angle of the pressure phasor
θ = phase angle of the flow phasor
n = the number of the harmonic under consideration

For $n = 0$, the equation represents the ratio between mean pressure and mean flow, for other values of n, the pressure flow relation for the corresponding harmonic. Note the large differences between calculated and measured values indicating the magnitude of the potential error involved in indirect determinations of cardiac output from stroke volume.

sible. Even without this essential safety consideration, stability of characteristics and long useful life demand that sensor surfaces be biologically inactive and not subject to rapid deposition of such components as fibrin or platelets.

When foreign materials are implanted in the body, a complex set of interactions ensue. They may be organized as follows, in rough order of time scale.*

Effects on Tissue Components
a) Components from the body fluids are promptly deposited on the alien surface; proteins, in particular, are adsorbed on surfaces.
b) The chemical environment near the implant changes, a prominent effect being lowered pH and pO_2; enzyme activities are altered. (In flowing blood these effects are negligible.)
c) The complex mechanism of blood clotting is initiated; commencing with protein (fibrinogen) deposition and platelet adhesion, it proceeds through many stages to build up a massive thrombus containing both a protein network and entrapped cells.
d) Other long-term effects may follow such as inflammatory reactions, tissue necrosis, ultimate replacement of thrombus by scar tissue, or occasionally tumor growth.

Effects on Implanted Materials
a) Adsorption and environmental changes alter the surface characteristics, sometimes profoundly.
b) Most metals are corroded, dissolving or being converted into insoluble oxides or hydrous oxides; a two-phase alloy may be preferentially attacked and mechanically degraded.
c) Polymers are attacked, the type of degradation depending on the chemical linkages which occur in main and side chains. Usual effects are reduction of degree of polymerization, with some cross-linking. These actions, along with loss of any plasticizers, embrittle and weaken the polymer and change its permeability to diffusing substances.
d) Ceramic materials may be etched or dissolved, either generally or selectively. The extent and rate of attack depends strongly on such characteristics as composition, type of chemical bonding involved, occurrence of stressed or metastable regions, and so forth.
e) In a complex structure, diffusion of mobile components occurs, both passively and as modified by alterations of the implant components by tissue action.

Regarding effects on tissue components, the first three are of major importance for the development and use of chemical microsensors. These actions

* I am indebted to my colleague O. L. Updike for this tabulation.

alter sensor characteristics and may even cause complete device failure. Even more seriously, local and embolic vessel occlusion are major hazards from clotting. Heparin anticoagulant therapy prevents these but carries its own risks, usually lesser but of definite concern.

With respect to the effects on the foreign material itself, changes in either the surface or the bulk structures are critical. A number of materials exist, however, which are affected only very slowly by the intracorporeal environment, the outstanding ones being silicone and fluorocarbon polymers. Noble metals, gold and platinum especially, and numerous ceramics are not overtly altered by the body but unfortunately are thrombogenic.

Techniques exist for minimizing the interaction of selected materials with the biological environment. Clotting phenomena have received most attention, although because of their complexity they are still incompletely understood. The electrical charge on surfaces in contact with blood strongly influences clotting rates. Unfortunately, strongly anodic metals are the least thrombogenic but are very rapidly corroded. Imposition of a voltage from an external source is sometimes effective in delaying clotting, but is not an established safe procedure and is impractical with insulators. The thrombogenicity of many materials appears to vary directly as the critical surface tension, an observation consistent with the rate of protein adsorption and platelet adhesion in clot initiation. For polymers, this correlation accurately points to fluorocarbons and silicones as surfaces of low thrombogenicity. (It must be noted that if even low thrombogenicity is associated with very poor thrombus adhesion, it may lead to a barrage of microemboli produced by an ostensibly clot-free surface.) Finally, considerable success has been achieved on some surfaces by techniques of bonding the anticoagulant heparin firmly to the surface, either through layers of adsorbent graphite and cationic benzalkonium salts or through a chemically grafted surface containing quaternary ammonium groups whose cationic sites bind the highly anionic heparin molecules.

H. Development of Specific Sensors and Data Processors

Because of the urgent need for transducers suitable for long-term monitoring of critically ill patients, a research and development program for novel transducers has been intiated with priority given to sensors reflecting changes in the metabolism-flow ratio (Sect. IV, E). Among the prototypes under development is a miniature, disposable tissue or blood oxygen sensor, the design of which is based on semiconductor thin film deposition techniques and a solid electrolyte. Patient safety is assured by optical isolation of the new transducers from any powerline operated equipment and by careful worst-case analysis of any conceivable constellation of failures.

Without proper data handling and display facilities, however, the value of such transducers is severely limited. Furthermore, the requirements for data handling and display are clearly different for the small community hospital and the medical center. We have, therefore, developed a modular system of "HIA" preprocessors (HIA = hardware implemented algorithms) which are foolproof, easy to manipulate, self-testing and self-calibrating. Their outputs can be directly displayed in analog and/or digital form on a modular display system placed at various strategic sites, and fed into larger computers for data storage, development of predictive functions and automated control of therapy. Both the specific modules to be used and the display and transmission of data for a particular patient may be prescribed in detail, as needed. Compact, relatively inexpensive and characterized by extreme reliability through modern solid-state design, these modules can be stored in sufficient numbers for any contingency even in a small hospital. The development of the HIA's as well as of the fiber optic isolator-coupler for transmitting physiological signals with complete freedom from electrical hazards are excellent examples of what can be done with a minimum staff, some good ideas and an enlightened environment which is directly involved with the health care system.

I. A prototype intensive care unit

In the previous sections, the different aspects of a comprehensive approach to the care of the critically ill patient have been discussed. Figure 11 summarizes these concepts for the design of a prototype intensive care unit for a small community hospital (upper-half) or for a medical center (lower-half). The central position is occupied by a model of the patient in terms of the pathophysiology of the critically ill. (The O_2 transport system represents a submodel of this model.) The essential role of such a model has been stressed throughout this paper and is reiterated here for reemphasis:

1. It contributes to the understanding of the relationships between structure and function of living systems in health and disease and is an essential prerequisite for their quantitative analysis.
2. Model simulations contribute heavily to the evaluation of the significance, sensitivity and reliability of individual parameters and pathophysiological correlates as performance indicators of biological systems and thus of better indices for diagnosis, treatment and diagnosis.

In order for these models to reach their full potential it is of crucial importance that they are developed in connection with a clinical setting, are based on hard data and are continuously updated as more and better information becomes available. It is clear that as yet we are only in the initial stages of such a development.

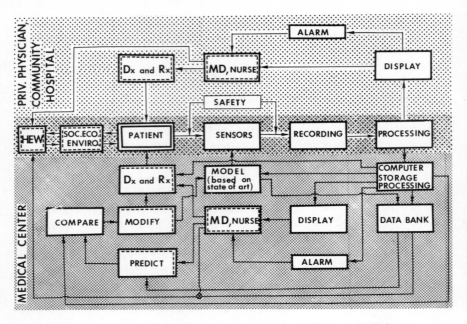

Fig. 11: Prototype design for an intensive care unit for a small clinic or community hospital (upper-half) or for a medical center (lower-half). Note the central role of models of biological systems for the effective function of such a unit and the importance of feedback to government agencies (such as the Department of Health, Education and Welfare) for required modifications of the socioeconomic environment.
Boxes with solid borders indicate fully automated functions. Boxes with both solid and dashed borders imply a significant human interface.
Dx: Diagnosis   Rx: Treatment

V. SUMMARY

The role of biomechanics with respect to patient care and rehabilitation is considered within the framework of systems pathophysiology. Since the traditional areas of biomechanics are well covered in other contributions to this symposium, biomechanics as it relates to clinical medicine is interpreted as part of the broader aspects of biological systems, where mechanics proper represents only one of many important components.

Disease can be characterized by the fact that the control hierarchy of the individual is unable to cope fully with deviations from homeostasis resulting from internal and/or external stresses, either because of deficiencies in the control system itself or because of the inability of the plant to respond adequately to changes in information flow. The patient's state, as seen by an observer, is formalized in a model, in which health status, diagnosis, treatment and prognosis are related through target variables (symptoms) and policy instruments (treatment variables) to the well-being of the patient in terms of his own value system (the welfare function). The model does permit an objective evaluation of the patient's status and the relative effectiveness of alternative therapeutic procedures through optimization of the welfare function.

Within this framework, biomechanics plays a four fold role:

1. To quantitatively analyze the relationships between structure and function of living tissues in health and disease.
2. To derive from these analyses reliable and sensitive indicators for the mechanical performance of biological systems and thus a more realistic basis for the construction of patient models and of patient care.
3. To define adequately tissue-transducer interfaces for the purpose of measuring these indicators more reliably by noninvasive, or at least, nontraumatic techniques.
4. To define the physical properties of living systems in such a way that reliable, long term internal and external prostheses can be designed and produced economically.

These various aspects of biomechanics are discussed specifically using as an example a research program designed to develop a prototype intensive care unit for critically ill patients. The development of better performance indicators and their application to clinical medicine is illustrated on the basis of an analysis of the oxygen transport system. The need for a substantial effort for the design and development of appropriate transducers and data handling facilities for different medical environments is emphasized, with particular consideration of problems involving the areas of biomechanics and biomaterials.

Acknowledgements.

The research referred to in the author's department is supported by NIH grants HE–11747, HE–08938, GM–01919.

I am indebted to my associates, Drs. Anné, Damman, Kenner, Millendorfer, McCartney, Updike, Wright, and to my wife Francoise for their many contributions to the theoretical, experimental and technological development upon which this paper is based.

BIBLIOGRAPHY

Anné, A. 1970. State variable approach to the analysis of biological systems. *In* E. O. Attinger (ed.) Global Systems Dynamics. S. Karger, Basel and New York. pp. 7–12.

Attinger, E. O. 1964. Pulsatile Bloodflow. McGraw Hill Book Company, New York.
Attinger, E. L., H. Sugawara, A. Novarro, T. Mikami and R. Martin. 1967. Flow patterns in the peripheral circulation of the anesthetized dog. Angiologica 4. 1–27.
Attinger, E. O., and H. Millendorfer. 1968. Performance control of biological and societal systems. Persp. Biol. Med. 12: 103–23.
Attinger, E. O. 1970a. Potentials and Pitfalls in the Analysis of Social Systems. In E. O. Attinger, (ed.) Global Systems Dynamics. S. Karger, Basel and New York. pp. 130–44.
Attinger, E. O. 1970b. Hierarchical Organizations and distributive models, In D. Ramsey-Klee, (ed.) Proc. 5th Conf. Information and control processes in living systems. In press.
Bernard, C. 1865. Introduction à l'étude de la mèdecine experimentale. Masson Co., Paris.
Carter, R. D., and C. E. Rapela. 1970. Fed. Proc. 21: 259.
Damman, F. J. Jr. 1970. Intensive coronary care units. Med. Trial Technique Quart.
Damman, F. J. Jr. A look at the problem of the critically ill. Submitted for publication.
Dickson, J. F., and J. H. U. Brown, (eds.) 1969. Future Goals of Engineering in Biology and Medicine, Acad. Press, New York and London.
Duling, B. R., and R. M. Berne. 1971. Longitudinal gradients in periarteriolar oxygen tension, a possible mechanism for the participation of O_2 in local regulation of bloodflow. Circ. Res. I–65–72.
Jose, A. D., and D. Collison. 1970. The normal range and determinants of the intrinsic heart rate in man. Cardiovasc. Res. 4: 160–67.
Kenner, T., M. Ueda, L. Huntsman and E. O. Attinger. 1968. Effects of local and general hypoxia on iliac flow. Angiologica 5: 345–63.
Kouchoukos, N. T., L. C. Sheppard and D. A. McDonald. 1970. Estimation of stroke volume in the dog by a pulse contour method. Circ. Res. 26: 611–24.
Lübbers, D. W. 1969. The meaning of the tissue oxygen distribution curve and its measurement by means of Pt electrodes. Progr. Resp. Res., 3: 112–23. S. Karger, Basel and New York.
Mann, Thomas 1924. Der Zauberberg. S. Fisher Publ., Berlin.
Millendorfer, H., and Attinger, E. O. 1968. Global Systems Dynamics. Med. Care 6: 167–489.
Morowitz, H. J. 1968. Energy flow in Biology. Acad. Press, New York and London.
Pennock, B., and E. O. Attinger. 1968. Optimization of the oxygen transport system. Biophys. J. 8: 879–96.
Renkin, E. M. 1964. Normal Regulation of tissue circulation. In H. L. Price, P. J. Cohen (eds.) Effects of anesthetics on the circultation. C. C. Thomas Publ., Springfield, Ill.
Tinbergen, J. 1952. On the theory of economic policy. North Holland Publishing Co., Amsterdam.
Waterman, T. H. 1968. Systems theory and biology: View of a biologist. In M. D. Mesarovic (ed.) Systems theory and Biology, Springer, New York.
Wetterer, E., and Kenner, T. 1968. Die Dynamik des Arterien Pulses. Springer-Verlag, Berlin.
Whalen, W. J., and Pankajain, N. 1967. Intracellular pO_2 and its regulation in resting skeletal muscle of the Guinea Pig. Circ. Res., 21: 251–61.

PART II

Basic Mechanical Properties of Living Tissues

4

The Rheology of Human Blood

GILES R. COKELET

I. INTRODUCTION

The desire to quantitatively understand both the microscopic and macroscopic behavior of blood when it flows through the vessels of the circulatory system, and other geometries, has supplied the motivation for appropriate investigations. As an aspect of such research, the macroscopic rheological properties of blood have been studied for a long time. The ultimate aim of such studies is to be able to predict from the macroscopic rheological data the microscopic flow behavior of blood in any part of the circulatory system.

Until a few years ago, most efforts were aimed at obtaining the rheological properties of blood and related fluids, and at determining how these properties varied with changes in such independent variables as red cell volume fraction (hematocrit), temperature, plasma protein concentrations, and others. Most of these investigations were made with blood obtained from people and other animals, apparently in normal health.

About four years ago, many workers in this area shifted the emphasis of their work towards attempts to relate the gross rheological behavior to the behavior of the constituents of blood. Thus, we now frequently see mechanistic explanations of the gross properties of blood in terms of red cell aggregation, plasma viscosity, red cell membrane mechanical properties, and the rheological properties of the red cell contents. We also see more thorough attempts to study pathological bloods and to explain non-normal behavior in mechanistic terms.

Concurrently, but perhaps on a more modest scale, there is an interest in determining the limits within which the continuum model and macroscopic properties of blood can be successfully applied. Of course, there is a corresponding interest in attempting to understand blood flow in those cases where the continuum model does not describe the physical situation to the desired degree, and in relating information from studies of the macroscopic rheological properties of blood to those cases.

It seems to me that these recent trends will continue and even accelerate. Application of rheological information will grow. I feel that these changes in research direction are natural and desirable. More biochemists, physiologists, pathologists and other non-rheologists will enter this research area. For such workers there may develop a strong tendency to treat the rheological instrument as a routine device for obtaining numbers, while the excitement of discovery will shift to other aspects of the research. Unfortunately, it is difficult to obtain good experimental data which truly reflect the macroscopic rheological properties of blood, and interpretation of good data, especially in the low shear rate range, is not a simple matter. The picture is further complicated by the commercial availability of very sensitive instruments which require a considerable amount of maintenance and skill for correct usage. Current literature already provides examples where the difficulties of obtaining data and of interpreting and translating raw data (usually unreported) into derived data (reported) go unmentioned. Such a situation is unsatisfactory, not only because it can lead to confusion for the nonspecialist, but also because we are seeing a refinement of the interpretation of rheological information to a scale where the interpretation in physical or mechanistic terms can be highly influenced by relatively small changes in the derived rheological data.

There already exist several recent reviews (Whitmore, 1968; Wayland, 1967; Merrill, 1969; Chien, 1969), which together supply an essentially complete compilation of the blood rheology literature. There is then no current need for an exhaustive literature survey. In view of this, and my concern voiced in the preceding paragraph, this presentation contains considerations of some of the problems and questions which arise in obtaining and interpreting reliable experimental data. Also included are some examples of the general nature of normal human blood's rheological behavior and a brief discussion of the suitability of the continuum model of blood. The behavior of blood when flowing through the smallest blood vessels, where the capillary size is about that of the red cell, is discussed by Skalak (1970).

II. BRIEF DESCRIPTION OF BLOOD

Human blood is a suspension of particles in a complex aqueous continuous phase. This continuous phase is called plasma, and contains inorganic and organic salts, as well as small organic molecules. In addition, about 7%, by weight, of the blood is macromolecules called proteins, which have molecular weights ranging from 44,000 to over 1,000,000. About half of the protein mass is albumin, with a molecular weight of 69,000.

The particles consist of a variety of cells, but the red cells compose about 97% of the total cell volume in the blood. Consequently, the removal of white cells and platelets from blood does not measurably modify the experimentally determined flow properties of these suspensions. Because of this, and because platelets especially can cause experimental difficulties in viscometers, most blood rheological studies are performed with suspensions which have had most of the white cells and platelets removed. The concentration of most red cells in a suspension is generally reported as the suspension volume fraction occupied by the red cells, which is called the hematocrit. Normally, the hematocrit in the human body is about 42–45% at sea level.

The red cells, also called erythrocytes, consist of a thin, flexible, but essentially unstretchable envelope or membrane, and an interior filled with a complex aqueous solution, which is nearly a saturated hemoglobin solution. The membrane is a specialized structure which serves as a barrier to the transfer of macromolecules, is the site of metabolism-linked transfer mechanisms for particular chemical species, and it permits some small molecules, such as water, to transfer very rapidly. Thus, the erythrocyte is a complicated type of osmometer. In addition, the red cells (as well as the protein molecules) all carry a net negative charge.

The red cell is not spherical in shape. In blood which is not flowing the red cell has a biconcave discoid shape, with a major diameter of 8.1 microns and a maximum thickness of about 2 microns. In normal blood the red cells aggregate, face-to-face. Usually, these aggregates contain 6–10 red cells in a stack, such a primary aggregate being called a rouleau. Secondary aggregation of the rouleaux also occurs. When blood is sheared these secondary aggregates and rouleaux break up, and at sufficiently high shear rates, the cells exist as individuals. If the shear rate again becomes about zero, these aggregate structures reform very rapidly.

Considering these factors, the rheological properties of blood might be expected to be rather complex. This is the case, and the complexity grows when one examines pathological bloods. Fortunately, there is no experimental evidence showing that the rheological properties of blood are history or time dependent, and no normal stress effects have been demonstrated.

It must be pointed out that when blood is withdrawn from its natural environment, certain irreversible chemical and mechanical reactions occur, the cumulative effect being called "clotting." To prevent this from occurring various anticoagulants are added to the blood as it is drawn from the animal. The effect of an anticoagulant on the rheological properties of blood are difficult to ascertain because of the difficulty in making reliable rheological measurements in the short time available between the blood drawing and the onset of blood clotting. The best study of this problem is probably that of

Frasher, et al (1968), who found that the natural anticoagulant heparin had no effect on the flow properties of dog blood. There is some evidence (Meiselman, 1965) that calcium-chelating anticoagulants slightly affect high shear rate behavior, but that low shear rate properties are unaffected.

III. SOME EXPERIMENTAL PROBLEMS IN DETERMINING FLOW PROPERTIES

Almost all attempts to obtain the flow properties of blood and related fluids have used one or more of three types of viscometers: (1) the capillary-tube type, (2) the concentric-cylinder type, and (3) the cone-and-plate configuration. Each of these types of devices has advantages and disadvantages, some of which are related to peculiarities of the instrument type, and some of which are mainly due to characteristics of blood. It is my intent to briefly outline in this section some of the problems which arise when trying to get blood rheological data, and to discuss some questions which I feel have been inadequately treated. Some additional problems are considered in Sections V and VI.

A. REDUCTION OF EXPERIMENTAL DATA—IDEALIZED SITUATIONS

For completeness, I include here a brief summary of the general idealized flow equations used to reduce experimental data to the rheological functions for the three commonly used types of viscometers. Since, as far as I know from the literature and my own experience, blood and related fluids have history-independent flow properties, and do not show the effects due to deviatoric normal stresses, this review is devoted to shear stress and rate of deformation relationships.

1. Tube Viscometers. It is possible to calculate, for a simple steady laminar flow in a tube, the corresponding shear stress and shear rate at the tube wall:

(1) $$\tau_w = \frac{D(\Delta P)}{4L}$$

(2) $$\dot{\gamma}_w = \frac{\tau_w}{4} \left(\frac{d \frac{8\bar{V}}{D}}{d\tau_w} \right) + \frac{3}{4} \left(\frac{8\bar{V}}{D} \right)$$

$\dot{\gamma}_w$ = shear rate at the wall
D = tube diameter

ΔP = axial pressure difference for two points an axial distance L apart
L = axial distance
$\bar{V}$ = bulk average velocity

The equation for the wall shear rate, apparently first derived by K. Weissenberg (see Markovitz, 1968), is limited to fluids for which the local shear stress is only a single-valued function of the local rate of deformation for all points in the tube. This may not be true for blood in a capillary whose diameter is less than about 500 microns (0.5 mm). The relationship for the wall shear stress is generally valid for this flow.

2. *Concentric-cylinder Viscometers.* A characteristic feature of the simple, steady, laminar flow found in the space between concentric cylinder surfaces, one of which rotates relative to the other, is that the total torque, T, on a cylinder surface of radius r due to the shear stress on that surface, τ, is independent of r. Therefore, in general, for this flow

$$\text{(3)} \qquad \tau = \frac{T}{2\pi r^2 L}$$

where L is the length of the bob. For the shear rate, the most general relationship available is restricted to fluids for which the shear stress is only a function of the rate of deformation throughout the fluid space. This relationship for a stationary bob and rotating cup, according to Krieger and Elrod (1953), is an infinite series:

$$\text{(4)} \qquad \dot{\gamma} = \frac{\omega}{\ln s}\left[1 + \ln s \frac{d \ln \omega}{d \ln \tau_1} + \frac{(\ln s)^2}{3\omega}\frac{d^2\omega}{d(\ln \tau_1)^2} + \cdots\right]$$

where ω = cup rotational speed
$\dot{\gamma}$ = shear rate at the bob surface
s = r_2/r_1, the ratio of cup radius to bob radius
τ_1 = shear stress on the bob surface.

A similar relationship exists for the case where the bob rotates and the cup is stationary. Krieger (1968) has presented a variation of this method which incorporates the power-law approximation with the exponent determined at each point.

It is essential that simple flow actually exist in the viscometer gap. Thus, it is necessary to insure that turbulent flow is not present, and that Taylor vortices are not formed (in the case where the bob is rotated).

3. *Cone-and-plate Viscometer.* In order that idealized simple flow exist in the fluid space between the cone and plate, it is necessary that inertial effects be negligible and that α, the angle between the cone and plate, be very

small (less than 4°). For the case where the cone rotates at a speed of ω, the shear stress, τ, which is essentially constant everywhere in the fluid, has a value given by

$$\tau = \frac{3T}{2\pi R^3} \qquad (5)$$

and the shear rate, $\dot{\gamma}$, is equal to ω/α, with $T =$ torque required to rotate the cone and $R =$ cone slant height.

In addition to the requirements that inertial effects be negligible and the angle α be very small, it is essential that the cone axis be closely aligned so that it is perpendicular to the plate.

B. End, edge and entrance effects

In deriving the means of converting raw experimental data into the rheological properties of a fluid, we always have a model of the flow field known as simple flow, that is, the local velocity is always directed in just one of the appropriate coordinate system directions and its magnitude is a function of only one of the other coordinates. In our instruments we cannot limit the flow solely to such simple flow, and the non-ideal flow effects are usually referred to as end, edge and entrace effects. Such effects must be eliminated, or shown to be negligible, before we can reduce our raw data.

1. The Capillary-Tube Viscometer. In this instrument the nonuniform flow regions are near the ends of the capillary where the fluid flows from a reservoir into the tube, or out of the tube into a second reservoir. Since the shear stress usually calculated at the tube wall generally makes use of the pressure difference between locations in the two reservoirs, these effects must be considered.

These effects have been studied extensively for Newtonian fluids. The results of these studies can be used for rough guides for the effects in non-Newtonian fluids. However, there does not seem to be general agreement on the magnitudes of some of these effects. Consequently, it seems safest, at this time, to obtain flow data with tubes of the same diameter but different lengths, keeping the upstream and downstream geometries constant. A simple subtraction will then give the wall shear stress free of end effects

$$\tau_w = \frac{D(\Delta P_L - \Delta P_s)}{4(L_L - L_s)} \qquad (6)$$

provided that the shorter tube has a section of appreciable length where the flow is simple. In this equation, ΔP_P and ΔP_s are the pressure differences found for the long and shorter tubes of lengths L_L and L_s, respectively, with diameter and bulk flow rates being the same for both tubes. To insure that the shorter tube is long enough, a third tube of a third length should be used to be sure that τ_w calculated from Eq. 6 is independent of the shortest tube length. Of course, the data from such an experimental procedure can be used to determine if the end effects are significant in any of the tubes.

When using the above method, it must be recognized that the tube lengths should be chosen so that the indicated differences of pressure drops and lengths have the desired precision.

2. The Concentric-Cylinder Viscometer The errors in the torque readings from a concentric-cylinder viscometer can arise in three regions: (1) the fluids in the liquid space above the inner cylindrical surface (the bob); (2) the fluid between the bottom of the bob and the bottom of the container forming the outer cylindrical surface (the cup); (3) the fluid near the edges of the gap formed by the cylindrical surfaces. Various methods have been devised to overcome or correct the effects of the nonsimple flow in these regions. A common computational method assigns to the bob a fictitious length (greater than its real length), which is experimentally determined with Newtonian fluids of known viscosity. We have not found this to be satisfactory, since the so determined bob effective length varies with shear rate and viscosity even for Newtonian fluids. For non-Newtonian fluids, the effective bob length would presumably depend on the flow characteristics of the fluid. Other correctional techniques of this type have been used, but generally cannot be directly validated.

Again, a simple experimental method exists which eliminates the end and edge effects. This technique uses two bobs of identical shape and diameter but different lengths. At a given viscometer rotational speed with the same fluid space geometry and flow above and below the bob, torque determinations are made with each bob. The shear stress at a radial position r, free of end and edge effects is then calculated as

$$(7) \qquad \tau = \frac{(T_L - T_s)}{2\pi r^2 (L_L - L_s)}$$

where T_L and T_s are the torques determined with bobs of length L_L and L_s respectively. A third bob length is needed to guarantee that the short bob is sufficiently long. Again, bob lengths must be chosen so that the differences formed in the equation have the desired significance.

3. Cone-and-Plate Viscometer This instrument has a nonideal flow pattern near the cone edge. Also, if the fluid fills more than just the space between the cone and plate, an additional torque from fluid shearing in this additional space is developed. The best experimental procedure to insure that these extraneous effects are negligible is to take measurements with two cones of equal cone slant height but different cone angle. If both sets of data give the same rheological relationship, there is strong evidence that the method of analysis is correct. Of course, in varying the cone angle we must remember that the theory is valid only if the angle between cone and plate is kept very small.

C. Liquid-Gas Surface Effects

The proteins in blood plasma, and possibly other components as well, act as surfactants. Consequently, when a blood-gas, plasma-gas, or serum-gas interface is present, these surfactants form an interfacial film. This film has mechanical properties which can cause the viscometric determination of the flow properties of (especially) plasma and serum to be seriously in error, particularly when there are low rates of deformation in the instrument test section. Thus, the erroneous literature reports that blood plasma and serum are non-Newtonian at shear rates below about 100 sec^{-1} are most likely due to a failure to recognize the influence of this film.

In rotational instruments this source of error can be removed by mechanically preventing the transmission of a torque through the film to the torque-sensing surface of the viscometer. One method is to impose a cylindrical ring through the liquid-gas interface so that it is mounted independent of the torque measuring surface, but capable of rotating with that surface as it moves in response to different torques. This is not so easy to arrange mechanically. However, if the total angular deflection of the torque measuring surface from the null position is very small, the guard ring can be kept stationary. Thus, Brooks, et al (1970), using a modified Weissenberg rheogoniometer with a total angular deflection of less than 0.5°, found ACD plasma to be Newtonian when a guard ring was used; the same fluid seemed highly non-Newtonian when measurements were made without the guard ring. Contrary to published reports, the guard ring does not prevent the formation of the interfacial film, it only eliminates its effect.

In the concentric-cylinder instrument, use of the two-bob technique described above should cause this film effect to cancel out of the calculations. However, the two-cone method suggested for the cone-and-plate instrument will not eliminate the effect.

In capillary-type viscometers it is customary to assume that there is no pressure drop across any fluid-gas interface in the instrument. However, this apparently is not true for blood-related fluids; the pressure drop across a falling interface is apparently negative (liquid pressure is less than adjacent gas pressure) while an advancing interface has a significant positive pressure drop. One way to overcome this problem is to measure the pertinent pressures in the viscometer without having a liquid-gas interface in the measuring system (for example using isotonic saline filled pressure transducers and connecting lines as done by Benis, 1964). It also seems that the two-tube technique suggested earlier for the capillary-tube viscometer would eliminate these surface-film effects if the experiments were done carefully.

D. RED CELL SEDIMENTATION

The erythrocytes, having a density of about 1.09 gm/ml, settle under the influence of gravity when in blood plasma (density about 1.05 gm/ml). This red cell sedimentation has been studied a long time, and forms a common clinical test since the rate of settling is a strong function of red cell aggregation and such aggregation is a reflection of variables, such as the fibrinogen concentration, which often change with changes in health. The question arises whether or not red cell sedimentation can affect viscometric measurements. The problem would be expected to be most serious at the lowest shear rates where red cell aggregates can form.

The cone-and-plate viscometer, with its essentially horizontal surfaces, seems especially susceptible to any problems which might arise from red cell sedimentation. At low nominal shear rates, below 1–2 sec^{-1}, a layer of essentially cell-free plasma could form at the upper surface, resulting in erroneous torque measurements (which would be lower than they should be). A very thin plasma zone would result in a relatively large error. Apparently, this problem has not been investigated by users of plate-and-cone viscometers.

In the concentric-cylinder viscometer, the cylindrical surfaces are vertical so that cell sedimentation away from these surfaces is not a problem. However, red cells will still settle, leaving the test section at the bottom of the fluid volume between the surfaces and also entering the test section from above. Since the blood above the test section is subjected to a shear rate much less than that of the fluid in the viscometer gap, we would expect a slow accumulation of red cells in the test section due to this sedimentation. With normal bloods, any effect due to this does not seem to be detectable when the nominal shear rate is above several inverse seconds since the torque measured at a given rotational speed is constant over periods of time as long as 20 minutes. At nominal shear rates below 1–2 sec^{-1}, the question is difficult to answer because another mechanism causes the torque response, at a given rotational speed, to be a function of time (see Section V). However, even at these low rotational speeds, sedimentation effects do not seem to show in the torque measurements for 15–20 minutes.

The capillary viscometer, usually operated with the tube in a horizontal position, would also be expected to be susceptible to sedimentation effects, especially at low flow rates where the rates of deformation are correspondingly low.

By microscopic observation, both visual and cinematographic, Meiselman (1965) found that for normal bloods flowing in horizontal tubes with $\bar{U} > 0.1$

−0.5 sec^{-1}, the blood appeared to be an homogeneous, well-mixed suspension without any evidence of red cell sedimentation, whereas below that critical $\bar{U}$ range, sedimentation away from the top wall of the tube was detectable. The plasma space at the top wall increased as the flow rate decreased. Aggregates of red cells were also seen at these low flow rates where sedimentation was detected. Here, $\bar{U} = \bar{V}/D$.

While the criteria for the onset of sedimentation effects may be somewhat dependent on tube length, and certainly should depend on any compositional variable which affects red cell aggregation and thus sedimentation, it seems safe to say that horizontal tube viscometer data obtained with $\bar{U} <$ 1 sec^{-1} should be viewed with suspicion. If the capillary is vertical, red cell sedimentation may still influence the measurements; also, for reasons given in Sections V and VII, data obtained when red cell aggregation is present to any great extent should be questioned. For pathological bloods, red cell aggregation may be so altered from normal values that the $\bar{U}$ criteria will have to be altered.

E. Smooth-Wall ("Vand") Effect

When a smooth wall forms an interface with a suspension of particles, the presence of the solid wall physically prevents the particles from occupying the fluid space next to the wall in the same manner as they do in the bulk of the suspension. For example, in blood, the centers of red cells can never be found in the fluid space immediately adjacent to the wall.

Vand (1948) studied this wall effect with suspensions of glass spheres in one concentric-cylinder viscometer and two different capillary viscometers. He determined that this wall effect was measurable in both types of instruments. For a hypothetical model, which is not physically realistic but serves to predict this wall effect, Vand considered the fluid space to be filled with two fluids: (1) a thin layer of uniform thickness next to each wall which contained only the continuous phase of the supension, and (2) the suspension which occupied the remainder of the fluid space. For such a model and from his data, Vand concluded that the wall-layer thickness was about 1.1 times the particle radius for all viscometers and suspension compositions studied (however, he used only one size sphere).

It would seem likely that a similar wall effect would exist with blood. Cokelet, et al. (1963 *a*, *b*) reported finding the effect in a concentric-cylinder viscometer when data was obtained from experiments with grooved (but hydrodynamically smooth) surfaces and with smooth surfaces. Using Vand's model, the plasma wall-layer for the smooth surface had a calculated thickness of 1–3 micron. Their data indicated that the wall effect is not of great importance for bloods with hematocrits below about 40%. It is of greatest importance at very low shear rates.

Data for blood flow in large tubes, such as that of Merrill, et al (1965a), indicate no experimental evidence for a wall effect. While it must exist it either is small compared to the experimental imprecision of the data, or else it is almost entirely compensated for by additional effects. The analysis of very small tube blood flow data is complicated by many factors (see Section VII).

I am not aware of any reported work seeking to evaluate this effect in a cone-and-plate viscometer.

In recent years, many reports have been made of the flow properties of red cell suspensions with hematocrits near the physically possible maximum level. In general, these data were obtained with smooth-surfaced viscometers. Such data are very likely to be erroneous due to the slipping of the viscometer surface over the suspension. For example, Vand found that suspensions of spheres with "hematocrits" above 50% could not be studied in his Couette viscometer because the wall-effect correction for slippage appeared to be larger than the measured values. Since the maximum idealized packing of spheres would yield a "hematocrit" of about 73% (experimental maximum packing is about 62%), this slippage problem becomes serious at sphere concentrations well below the maximum value. I do not think any criteria can be drawn for blood or red cell suspensions by arguments based on the sphere suspension behavior, but it would seem that the problem needs serious consideration when measurements on red cell suspensions of near-maximum hematocrits are under examination. Again, as far as I know, this particular problem has not been investigated.

It should be noted that the grooved surfaces used with the GDM viscometer at MIT were developed solely to counter this smooth wall effect and, contrary to published reports, was not intended to eliminate the effect discussed in Section V.

IV. NORMAL HUMAN BLOOD: TYPICAL BEHAVIOR

It seems generally accepted that human blood is a non-Newtonian fluid. Attempts to detect normal stress effects have failed; rheological properties of blood also seem to be history and time independent. The nature of normal blood's behavior will be discussed, along with examples of methods used to uncover the physical flow behavior of the constituents of blood.

Figure 1 shows the shear stress–rate of deformation relationships for bloods of various hematocrits at low shear rates. This is a so-called Casson plot: The square roots of the shear stresses and the shear rates are plotted. Casson (1959) derived an analytical relationship for suspensions of particles in which the particles aggregate at low shear rates to form rod-like aggregates.

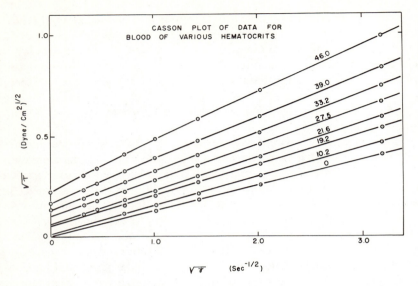

Fig. 1: Casson plot of very low shear rate data for a human blood at a temperature of 25°C. Range of linearity decreases as the hematocrit increases. Plasma is Newtonian. Data from Cokelet (1963a).

This relationship is linear on a plot such as this and has an intercept on the ordinate. Rheologists object to this relationship because it does not show the property of material objectivity, and the shear stress is not just an odd function of the shear rate. Nevertheless, as shown here, it empirically describes the very low shear rate data, although the range of satisfactory fit decreases as the hematocrit increases.

Blood has a yield stress, but a number of investigators dispute this, or consider its measurement to be questionable. Indeed, some workers will question these data for physiological hematocrits and shear rates below one inverse second. These matters will be discussed in later sections.

Accepting for the moment a yield stress for blood, experiments show that it is not a function of temperature. At finite rates of deformation, the apparent viscosity, defined as the shear stress divided by the shear rate, decreases with increase in temperature. With increasing shear rates, the temperature dependency of the apparent viscosity approaches that of water; the slope on an Arrhenius plot of the logarithm of the apparent viscosity of a blood at a given shear rate versus the reciprocal of the absolute temperature approaches the slope shown by water.

Note that plasma is Newtonian, since the shear stress is directly proportional to the shear rate.

Figure 2 shows how the yield stress varies with hematocrit, data for five different human blood sources being shown. Over an appreciable hematocrit

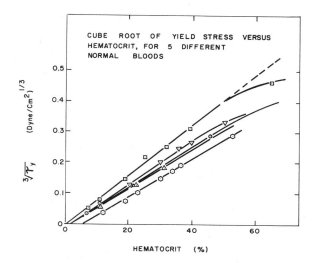

Fig. 2: Variation of the yield stress of blood with hematocrit. Tests on samples with hematocrits less than the applicable abscissa intercept showed no yield stress. Figure from Merrill, et al (1963).

range, the cube root of the yield stress is a linear function of the hematocrit, the slope being almost independent of blood source. The intercept on the abscissa is a critical hematocrit, below which no yield stress is found, presumably because there are an insufficient number of red cells per unit volume to permit bridging of the sample volume by a continuous structure of aggregated red cells, such a structure being required if a yield stress is to exist.

The various bloods in this figure show different values for this critical hematocrit. This state of affairs arises because the plasmas of these bloods differ in composition (especially in fibrinogen concentration), and this variation is reflected by varying efficiencies in constructing a continuous three-dimensional structure of aggregated cells at a fixed hematocrit (the average rouleaux length varies with fibrinogen concentration). A similar relationship between yield stress and particle concentration is shown by other suspensions.

In normal blood, the yield stress is a strong function of the concentration of the protein, fibrinogen. Other macromolecules such as gamma globulins and high molecular weight dextrans, which can cause red cell aggregation, also influence the yield stress. For details, the reader is referred to Merrill, et al (1965b, 1969b) and Chien, et al (1966).

Blood's rheological behavior at higher shear rates is shown in Fig. 3. The dashed curve is the approximate boundary between the regions of Newtonian and non-Newtonian flow behavior. Thus, the $H = 8.25\%$ suspension appears Newtonian (constant viscosity) over the entire shear rate range. The $H = 35.9\%$ suspension shows a Casson equation region at low shear rates, Newtonian behavior at the highest rates of deformation, and a transi-

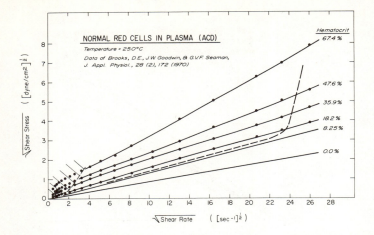

Fig. 3: Casson plot of higher shear rate rheological data for a blood with ACD anticoagulant. The actual data points for plasma and the 8.25% hematocrit red cell suspension are not shown for clarity; these fluids appear to be Newtonian. Viscometer surfaces were smooth.

tion region at intermediate shear rates. Merrill and Pelletier (1967) presented data which supported the idea that the slopes of the Casson equation region and the Newtonian region were about equal. However, this relationship between slopes seems limited to hematocrits below about 40% at 37°C. At higher hematocrits, the Newtonian region commences at higher shear rates and the Casson-equation region decreases in size.

The data of most workers show this qualitative behaior. However, Charm and Kurland (1965) reported that the Casson equation, with just one set of constants, described the rheological behavior of blood over the shear rate range from 0–100,000 sec^{-1}. Merrill and Pelletier (1967), of course, report a behavior which conflicts with the Charm-Kurland finding. Then in 1969, Charm, et al presented data indicating that human blood behavior did not undergo a transition from Casson equation behavior to Newtonian behavior in the shear rate range of 20–100 sec^{-1}. Actually, of the nine blood samples considered, it was judged that three showed the existence of a transition, and six showed no such sign. The technique used for determining the transition point involved the fitting (by a least squares technique) of a straight line to experimental shear stress-shear rate data plotted on a Casson graph, then comparing the slopes and intercepts of the calculated lines found for the data in different shear rate ranges. Part of the discrepancy in the findings of Merrill and Pelletier and of Charm, et al could be due to the fact that often one can fit a single straight line to the Casson plot data in and near the transition region. Perhaps Charm, et al did not cover a large enough shear rate range to detect the transition in their data. This is certainly the case with the data of Merrill, et al (1965*a*) cited by Charm, et al (1969).

The data shown in Fig. 3 were selected from Brooks, et al (1970) because they were obtained with a Weissenberg Rheogoniometer (as a concentric-cylinder viscometer), the same device used by Charm, et al (1969). The data in the shaded region in Fig. 3, at high hematocrits and very low shear rates, may not be as accurate as the rest of the data because (1) the shear rate was calculated using an equation which is strictly valid only for Newtonian fluids, (2) the viscometer surfaces were smooth, and (3) the authors do not indicate how they handled the interpretation problem which arises at very low shear rates due to the torque varying with time (see Section V).

There does not seem to be a single, simple, analytical expression available for describing the rheological properties of a blood over the shear rate range shown in Fig. 3. Piecemeal fitting of simple expressions seems to be the best means of obtaining analytical descriptions of blood's flow behavior.

Plasma is a Newtonian fluid as indicated in Fig. 3. The literature contains numerous reports that plasma is non-Newtonian, but I believe it has now been firmly established that data showing non-Newtonian flow behavior contain an experimental error caused by the presence of a surfactant (protein) film at an air-plasma interface in the viscometer. Figure 4 shows some data for plasma viscosity as a function of temperature. The plasma viscosity varies

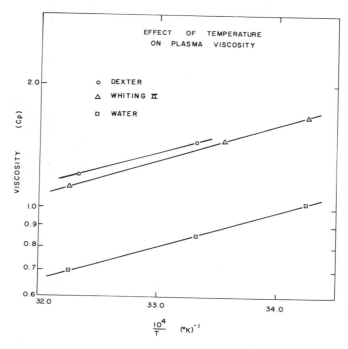

Fig. 4: An Arrhenius plot showing how the plasma viscosity varies with temperature. The two plasmas had different protein compositions, although both were obtained from people in normal health.

with temperature as the plasma solvent (water) viscosity varies with temperature.

The contents of the red cell, mainly a hemoglobin solution in saline, is also Newtonian. Figure 5 shows how this solution's viscosity varies with hemoglobin concentration when diluted with isotonic saline or concentrated by use of an osmotic membrane (dialysis). It is interesting to note that Chien, et al (1970a) found that oxygen-saturated hemoglobin solutions prepared from sickle-cell blood showed the same viscosity-concentration curve as solutions prepared from normal red cells. Bertles, et al (1970) report that deoxygenated sickle-cell hemoglobin solutions show an increased viscosity.

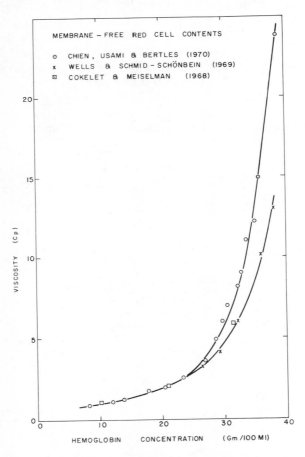

Fig. 5: The viscosity of the human red cell contents at 37°C. Solutions are Newtonian at the concentrations shown. Low concentrations prepared by dilution of hemolysate with isotonic salines; high concentrates prepared by ultrafiltration.

There apparently are no data available upon which to base a decision as to whether or not the solution inside a red cell shows the same flow characteristics as the bulk solution out of the cell.

As indicated earlier, there is now considerable interest in relating the macroscopic flow properties of blood to the properties of the constituents of blood. One useful set of data is shown in Fig. 6, where the relative viscosity of suspensions of aldehyde-hardened red cells in saline and water are shown as a function of hematocrit. Microscopic observation of the suspensions showed no cell aggregation. Such suspensions are Newtonian over the shear rate range from 0.1 to 700 sec^{-1}. The viscosity is seen to rise rapidly as the hematocrit approaches the maximum possible value of about 62%. For comparison, the universal equation for suspensions of rigid particles, propos-

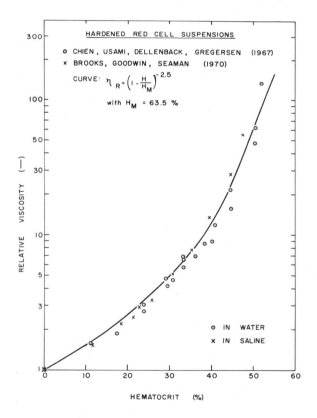

Fig. 6: The relative viscosity of hardened red cell suspensions. The rheological measurements were made in smooth-surfaced concentric-cylinder viscometers and the suspensions appear to be Newtonian. The curve is calculated from an equation proposed by Landel, et al (1965) for suspensions of rigid particles.

ed by Landel, et al (1965) is also shown; the best fit of the data was obtained by assuming a maximum possible hematocrit value of 63.5%.

Normal red cells suspended in isotonic saline also do not aggregate, but these suspensions are not Newtonian, except at very low hematocrits or very high shear rates. Typical data obtained with such suspensions are shown in Fig. 7. At low hematocrits the viscosity varies very little with shear rate while at higher hematocrits the viscosity varies markedly. Also shown here is a curve representing the behavior of suspensions of hardened cells, which was shown in the previous figure. At low shear rates and relatively low hematocrits the behavior of suspensions of normal red cells and of hardened

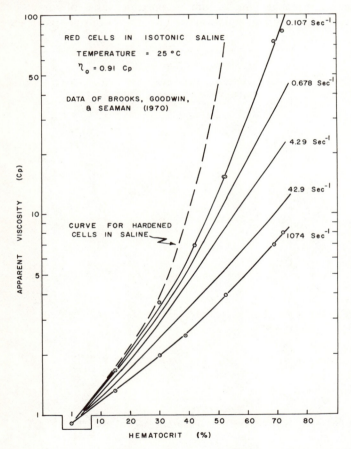

Fig. 7: The apparent viscosity (defined as the shear stress divided by the shear rate) of normal red cells suspended in an isotonic saline. For ease of presentation, data points are shown only at 0.107 and 1074 sec^{-1}. A smooth-surfaced viscometer was used to obtain these data. Dashed curve is solid curve of Fig. 6.

cells closely approach each other. This is usually taken as an indication that in this region of hematocrit and shear rate, deformation of the normal red cells is negligible.

Another comparison often made when the mechanism of blood flow is being considered is shown in Fig. 8. A comparison of flow behavior is made between a suspension of normal red cells in isotonic saline and a suspension

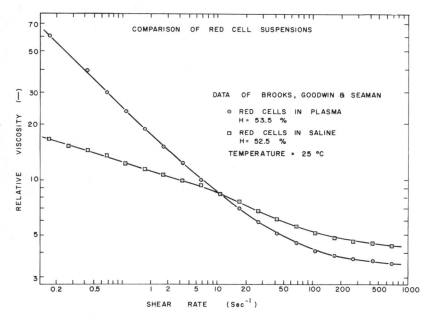

Fig. 8: Comparison of the relative viscosities of normal red cells suspended in plasma and isotonic saline at approximately the same hematocrit. See text for a discussion.

of normal red cells in plasma. At low shear rates there is considerable difference in the flow behavior, and this difference is generally attributed to the fact that the red cells suspended in plasma will aggregate at the lowest shear rates, the aggregate size being a function of the shear rate.

At the higher shear rates the red cells in both suspensions are unaggregated, and the relative viscosities of the two suspensions are much lower than the relative viscosity of about 100 shown by a hardened cell suspension of 53% hematocrit. This dramatic decrease in viscosity is due to the deformability of the cells.

Since the cells in these two suspensions are supposedly the same, we might expect these two suspensions to show the same relative viscosities at the higher shear rates. Indeed, Chien (1907b) has shown a plot similar to this, except that his red cell-saline suspension contained 11% albumin, and his data indicate that the two curves coincide for shear rates above about 10 sec^{-1}. Is there a conflict among these sets of data?

No, there is not. The saline used by Brooks, et al (1970) has a viscosity of 0.91 cp, while the plasma has a viscosity of 1.55 cp. In Chien's experiments, the albumin-saline solution had the same viscosity as his plasma. Plotting the relative viscosity of a suspension of nonaggregating particles frees the graphical relationship of the influence of the continous phase viscosity if the particles are rigid, but for nonrigid particles the relationship is more complicated. If two suspensions of the same deformable particles have continuous phases with different viscosities, then, at a given shear rate, the stress exerted on a particle is greater in the suspension with the more viscous suspending media, and that particle will be more deformed. Based on empirical rheological observations, it would be expected that this would cause the relative viscosity of the suspension with the more viscous continuous phase to be lower. And, this is what we see in this figure.

At the lower shear rates the suspension of red cells in saline shows a lower relative viscosity than the hardened cell suspension. This is interpreted as an indication that at high hematocrits red cell crowding causes cell deformation with the attendant lowering of relative viscosity from that shown by a hardened (undeformable) red cell suspension of the same hematocrit.

Schmid-Schoenbein, et al (1968, 1969a) used a transparent cone-and-plate viscometer to microscopically observe blood flow in the region near the apex of the cone as a function of viscometer rotational speed. They observed that rouleaux existed at very low shear rates and that the aggregate length decreased with increasing shear rate. The shear rates reported in these studies are probably only semiquantitatively correct because (1) the physical geometry of the fluid space differs from the ideal model of the viscometer, and (2) the fluid channel is so small that the continuum model probably cannot be used to calculate local shear rates. Therefore, the numbers reported for the maximum shear rate under which cell aggregates can exist may not be too accurate. Schmid-Schoenbein and Wells (1969b) also report that red cells (suspended in a 35% dextran solution, viscosity of 62 cp) undergo a transition under high stresses to a flow behavior similar to that shown by liquid drops in an emulsion.

Brooks and Seaman (1969) made normal red cell suspensions with 10% high molecular weight dextran-saline soultions as the suspending media. By varying the molecular weight of the dextran, they could change the viscosity of the continuous phase from 4.2 cp–10.0 cp. At this dextran concentration the red cell zeta potential is sufficiently high to prevent red cell aggregation. On shearing these suspensions at 678 sec^{-1}, they found that the relative viscosity-hematocrit relationship was the same for all suspensions. From this they concluded (1) that the effective red cell shape was the same in all suspensions (at the same shear rate), in spite of the variation in dextran solution viscosity, and (2) that "the cells are maximally deformed at this shear rate."

At a shear rate of 0.678 sec^{-1}, the relative viscosity at a given hematocrit was found to decrease with increasing zeta potential (caused by increasing the dextran molecular weight). Since cell deformation has very little effect on the suspension flow properties at this low shear rate, Brooks and Seaman concluded that this relative viscosity data indicated a decrease in cell-cell interactions (such as closeness of approach on "collision") with increasing zeta potential.

This short discussion of the rheological behavior of normal human blood is not intended to be encyclopedic, but rather illustrative of the general nature of normal blood behavior and of the comparative means commonly used to gain an understanding of the physical behavior of blood's constituents in flow and their importance in determining blood's macroscopic behavior.

V. RED CELL MIGRATIONAL EFFECTS AT LOW SHEAR RATES

Viscometric measurements in rotational instruments at shear rates below about one inverse second are complicated by the fact that the measured torque is a function of time. This behavior forms the subject of this section.

Consider a concentric-cylinder viscometer filled with blood, but with both surfaces stationary. At time zero and thereafter, let the rotated surface rotate at some constant speed, and continuously read the torque reading as a function of time. Two types of response curves are obtained, depending primarily on the rotational speed, both when the inner surface is driven [Cokelet, et al (1963b), Chien, et al (1966), Gregersen, et al (1967a)] and when the outer surface is driven (Seaman, 1970). These two types of curves are illustrated in Fig. 9 for a GDM viscometer filled with normal human blood. If the steady state, uniform-fluid shear rate is above about 1–4 sec^{-1}, then the torque time curve will be similar to those shown in the upper part of Fig. 9. Immediately after zero time the torque reading rises to some value and remains there until the viscometer rotation is stopped, then the torque reading rapidly drops to zero. In some instruments, mechanical and electrical damping may be such that upon first rising to the steady state torque value, there is some overshoot in the torque reading. However, for rotational speeds corresponding to nominal shear rates below about 1–4 sec^{-1}, the torque-time curve typically looks like the curve in the bottom part of Fig. 9. Initially the torque reading slowly rises to a peak value, and then more slowly decays to a lower steady value. The first type of curve is rather commonplace, characteristic of pure liquids and solutions. The second type is unusual,

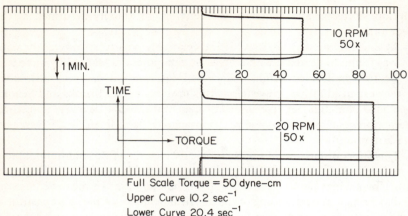

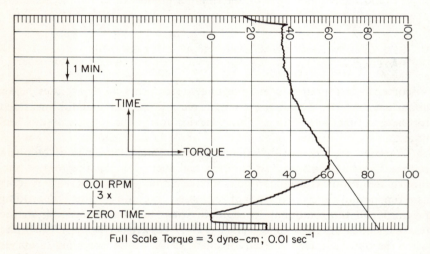

Fig. 9: Torque-time curves obtained from a GDM concentric-cylinder viscometer at high shear rates (upper) and shear rates below about 1 sec^{-1} (lower), with normal blood. At zero time (or zero torque), the viscometer rotational speed is zero. See text. Figure from Cokelet, et al (1963b).

and has been studied in some detail (see references above and Cokelet, 1963a).

Some of the general characteristics of this second, very low shear rate type of blood behavior (for normal blood) are the following:

(a) for a given blood sample, the time required to reach the peak torque value increases rapidly as the nominal shear rate decreases,
(b) the rate of torque decay, immediately after the torque peak, is exponential, and seems to be independent of the nominal shear rate below about 0.1 sec^{-1}. It varies slightly with the rough or smooth

nature of the viscometer surface, with the fibrinogen concentration (increases with increase in fibrinogen concentration), and with the hematocrit (zero below about 10%, maximum around 35%, and decreases as H increases becoming about zero when H is about 55%),
(c) this type of curve is found only when the red cell suspension has a continuous phase which contains those macromolecules which are essential for red cell aggregation at very low shear rates, that is, red cell aggregation at very low shear rates must be present if this second type of behavior is to be found. Thus, this behavior is observed when the continous phase contains fibrinogen, high molecular weight dextrans, and like molecules, but is not found when the continuous phase is serum or saline solutions. In the latter cases, there is no torque decay after the torque rise from zero. Only the first type of torque-time curve is found, although the time required to reach the steady torque value varies with viscometer rotational speed.

Such low shear rate behavior raises the question: What is causing this blood behavior and how do we interpret this torque-time curve in order to get the relevant flow properties?

From several types of experimental observations it is clear that at nominal shear rates below about 1–4 sec^{-1}, the red cells migrate radially away from both cylindrical walls. The evidence for red cell migration away from the outer surface is detailed in Cokelet, et al (1963b), but unfortunately the blood photographs (Figs. 3 and 4) in that publication have lost so much detail in processing that the purpose of showing the photographs is completely lost. Briefly, the results of that study were the following:

(a) The development of a "plasma" layer at the outer surface occurs only at shear rates below 1–4 sec^{-1}, where simultaneously a torque-time curve of the second type is recorded. This layer forms only for that portion of the outer surface which is part of the viscometer gap. It does not form at the portions of the outer surface (above and below the bob) where the shear rate is essentially zero.

(b) If well-stirred blood is placed in the viscometer while the viscometer rotation is kept at zero, the "plasma" layer is not detectable within ten minutes.

(c) Whether the viscometer surfaces are smooth or grooved has no effect on the occurrence of red cell migration. It does influence the torque readings slightly. It should be said that, contrary to reports in the literature, the viscometer surfaces are roughened (grooved) not to eliminate or modify this migrational effect but to eliminate the Vand wall effect (smooth wall physical exclusion of particles from the region next to the wall).

The experimental observations which support the contention that the red cells also migrate away from the inner cylindrical surface were first pointed out by Meiselman (1965). He observed that under viscometer rotational speeds where torque-time curves of the second type are obtained, careful removal of the bob (rotation stopped) from the viscometer after obtaining a curve showed it to be covered with red cell-free, transparent plasma, whereas after high rotational speeds the bob was coated with red, opaque blood.

It thus appears that at low shear rates, when red cell aggregation is appreciable, the red cells move away from both of the concentric-cylinder viscometer surfaces. Deformation of the blood is important, at least as far as the rate of red cell migration is concerned, probably because it increases the frequency of red cell collisions (thereby increasing the rate of aggregation) and because it facilitates the packing of red cells into a smaller volume. The deformation of the blood may also be necessary for the radial migration, since a recent study of the radial migration of a single rigid sphere suspended in a Couette flow (Halow and Wills, 1970) shows that such a sphere will migrate to a position approximately midway between the cylindrical surfaces. However, extrapolation from the behavior of a single rigid sphere to that of a mass of red cells does not seem straightforward.

Having an understanding of what physically occurs in the blood in the viscometer gap when this second type of torque-time curve is found, enables us to proceed to the question of whether or not we can get any meaningful flow properties from such a curve. We must interpret the torque-time curve in terms of what is physically happening in the fluid in the viscometer gap.

The initial part of the torque curve reflects effects due to acceleration of the blood from rest, the change in fluid structure as the blood is sheared, and the dynamic behavior of the instrument. If these were the only effects the curve would look like the dashed curve in Fig. 10; this is the same as the first type of torque-time curve discussed above. However, we have radial migration of red cells and so the real torque-time curve is as shown by the solid curve. After some time, beyond the curve peak, the instrument dynamics and fluid acceleration effects become of minor importance and we see mainly the effects of the red cell migration and of fluid structure changes. *If* we could extrapolate the last of the torque-decaying part of the time curve back to zero time, we would have the torque corresponding to a blood sample where red cell migration had not occurred and which was subject to a certain set of boundary conditions, such as viscometer surface rotational speed. Such an extrapolation is indicated by the dotted curve in Fig. 10. The asymptotic torque of the dashed extrapolation and the zero time-torque value of the dotted extrapolation would be the same since they are for the same uniform fluid subjected to the same boundary conditions. Any extrapolation assumes that the effect on the torque of physical changes in the blood would continue to be the same over the extrapolated period of time. Such an assumption

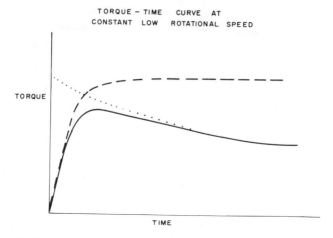

Fig. 10. Schematic drawing of a very low shear rate torque-time curve from a GDM viscometer with normal human blood (the lower part of Fig. 9 shows actual curve). Solid curve is recorded response. Dashed curve represents behavior expected if no red cell radial migration occurs (for red cells suspended in isotonic saline) and represents the transient response from start up of the instrument and the blood. Dotted curve is the hypothetical response if only red cell migration occurred. See text.

needs to be checked by some independent means before the results of any extrapolation are accepted.

The extrapolation indicated in Fig. 10 by the dashed curve is difficult to make because the initial part of the real torque-time curve (which is exponential in nature) is difficult to define precisely enough to permit reliable extrapolation to an asymptote. On the other hand, the extrapolation indicated by the dotted curve in Fig. 10 is easy to define at the point where it coincides with the real curve. A reliable extrapolation to zero time can be made provided, of course that the required extrapolation is not over a large time interval such as fifteen minutes. If the interval from zero time to the peak time is not too large (less than 2–3 minutes), this zero-time, extrapolated torque can be obtained with satisfactory accuracy on the torque recorder chart by linear extrapolation of the torque decay part of the torque recording which is after the peak. This extrapolated zero-time torque is then paired with the viscometer rotational speed as a point in the set of data which will be treated by the Krieger-Elrod method (see Section III A) to obtain the flow characteristics of blood. Such a point describes blood flow for blood with a uniform hematocrit equal to that measured experimentally and which is subjected to the boundary conditions set by the viscometer surfaces.

This technique for obtaining blood flow properties was developed and used in the laboratory of Professor E. W. Merrill. To be accepted, it is necessary that the validity of the extrapolation be proven. I believe this check has been provided by the independent determinations of the yield stress of blood. Experimentally, it is found that this yield stress agrees, within experimental limits, with the intercept of the shear stress-rate of deformation data plot obtained at finite rates of deformation (and using the above extrapolation procedure where needed below 1–4 sec^{-1}). At this time, many questions are raised about a yield stress of blood, and these are treated in the next section.

Workers in the Laboratory of Hemorheology at the Columbia University College of Physicians and Surgeons (which was founded and directed by the late Professor M. I. Gregersen, and is now directed by Professor Shu Chien, not being sure of the validity of the extrapolation procedure outlined above), have now (Chien, et al 1970c) arbitrarily selected the peak torque value as the torque value to pair with the viscometer rotational speed. Such pairs of peak torque and rotational speed are considered a reflection of the flow properties of blood when the nominal shear rate is so low that these torque-time effects are seen. Such a treatment of the viscometer data is to be preferred to reducing the peak torque to a shear stress and the rotational speed to a companion shear rate, since such a reduction is invalid primarily because the blood in the viscometer gap is not uniform in hematocrit. However, such peak torque-rotational speed data will reflect the instrument dynamics and geometry (as well as a transient blood state), and since different instruments have different dynamics and geometries, direct quantitative interinstrument comparison cannot be made.

The radial migration of red cells at low rates of deformation of blood seems to be mainly a result of the ability of the red cells to be attracted to each other and to aggregate. Such behavior would be expected to occur in other types of viscometers, although perhaps overshadowed by sedimentation effects. Rahn, et al (1967) report similar torque-time behavior in a cone-and-plate viscometer, and Schmid-Schoenbein, et al (1968) microscopically observed red cell behavior in a cone-and-plate viscometer consistent with the behavior described above.

This discussion of very low shear rate blood viscometry has been made in the framework of the torque-time response of the GDM concentric-cylinder viscometer. The reason for this is that this phenomenon has only been described in the literature by workers who use this type of instrument. Other types of concentric-cylinder viscometers may show torque-time curves which are quite different from that shown by the solid curve in Fig. 10. An instrument with very little internal damping will produce a curve which shows a sharp torque overshoot followed by a damped oscillatory approach to post-peak part of the solid curve of Fig. 10. On the other hand, an overdamped instrument would show no peak, but rather a very slow asymptotic rise to the torque value shown at the right end of the solid curve in Fig. 10. This latter response would seem to show none of the effects of red cell radial migration, even though such migration occurred. This wide range in viscometer response makes it imperative that the research worker understand the dynamics of his viscometer. While the range of viscometer response dynamics makes it a bit more difficult to extract from the raw data the rheological properties of the fluid, it should also help us in the future when workers derive blood's very low shear rate properties from raw data obtained with instruments other than the GDM viscometer. When the deduced blood properties are the same, regardless of the source of the raw data, then we can have confidence in the derived properties.

To those readers who are concerned mainly with the analysis of blood flow in large tubes or vessels, the discussion presented in the last few paragraphs may seem overemphasized, since the shear rates involved are very small. Such readers may have their feelings on this matter strengthened after considering the section devoted to the questions regarding blood's yield stress, since in any case the yield stress is small. However, the red cell migration at low shear stress values may also occur in blood vessels and play an important role since such a mechanism would greatly influence the pressure drop along a vessel; In large vessels with low flow rates, a pathological condition resulting in increased red cell aggregation could lead to a lower pressure drop than found with normal blood for the same average flow rate. But, more importantly, alterations in the composition of blood or in the properties of one or more blood consituents often cause relatively large changes in blood's flow properties at very low rates of deformation. Hence, an understanding of such low shear rate behavior is needed if we want to turn the procedure around in order to deduce from changes in flow properties the changes which have occurred in the blood constituents (for example, in pathologic bloods).

VI. YIELD STRESS MEASUREMENTS

Before examining the evidence for a yield stress in blood, it seems profitable to first discuss what is meant by the expression "yield stress." An idealized concept of yield stress might result in a definition such as "a yield stress is a critical shear stress and if the material is subjected to a shear stress smaller than this critical value, the material will elastically deform, but will not viscously flow, regardless of how long the stress is applied." Such a definition is idealized because, given enough observational time, any material placed under a stress will flow so as to relieve the stress. The big question really is whether or not we are willing to perform the experiment long enough so that our macroscopic tools can be used to detect and measure the flow. Thus, if we were to subject a material to a small shear stress and we found no detectable movement of the material in some arbitary time interval, perhaps fifteen minutes, we would say that the rate of deformation was zero and the stress was equal to or less than the yield stress. However, if we performed the same experiment for one year, we might get a detectable, viscous movement, from which we could calculate an extremely small rate of deformation. Alternatively, instead of performing the experiment for a year, we might develop displacement-measuring tools to detect and measure extremely small displacements. Then, our fifteen minute experiment would show a very small flow and again we would calculate an extremely small rate of deformation. Thus, our shear–stress rate of deformation diagram might look like that shown in the bottom of Fig. 11.

Needless to say, we do not often see shear stress–shear-rate diagrams where shear rates as small as 10^{-7} sec^{-1} are shown; consequently, most diagrams for substances said to have a yield stress look like the upper part

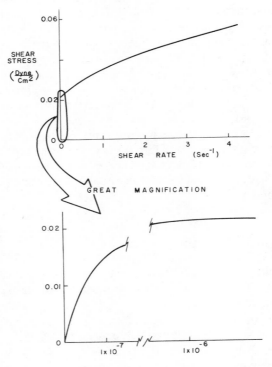

Fig. 11: A typical low shear rate plot of blood's rheological properties (upper), showing as an intercept on the ordinate a yield stress. The lower graph is a greatly magnified (primarily an enormous stretch of the shear rate axis) graph of the region immediately next to the ordinate of the upper diagram. Note that the shear rates are extremely small in this latter diagram and the fluid behavior depicted would appear to coincide with the ordinate axis in the upper diagram.

of Fig. 11. The yield stress which appears as an intercept on the stress ordinate in the top of Fig. 11 is seen, in the bottom of that figure, to actually be the stress where the slope of the stress-rate of deformation curve undergoes a drastic increase as the stress is lowered and the material begins to viscously deform at an extremely slow rate. With this concept of a yield stress, let us consider the question of whether or not blood has a yield stress.

What might be called a classical method of determining a yield stress could be performed in a concentric-cylinder viscometer as follows: the viscometer is operated at a constant rotational speed and then suddenly the rotation is stopped. The torque reading of the instrument as a function of time would look like that shown in the top of Fig. 12: The torque reading at first drops rapidly but then levels out at a nonzero steady value. This steady torque value is converted to a shear stress at the inner cylindrical surface and it corresponds to the yield stress. (The fluid shows no macroscopic evidence of flow when the inner surface stress equals the yield stress.) The fluid

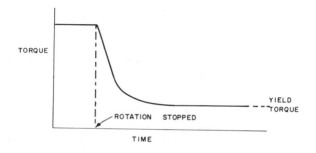

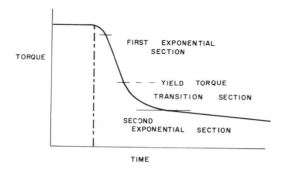

Fig. 12: The torque response of a concentric-cylinder viscometer when the steady state rotation is suddenly stopped. The upper diagram shows the response seen for a fluid possessing a yield stress using a viscometer with a relatively small spring constant and low sensitivity. The lower diagram is for the same fluid in a very sensitive, large spring constant instrument.

is really creeping, but the instrument shows no sign of the motion in several minutes because the viscometer spring constant is relatively small.

If the same experiment is performed in a GDM viscometer (which has a "spring" constant of 7.25×10^5 dyne cm per radian in the following discussion), a torque-time curve such as that indicated in the bottom of Fig. 12 is obtained; the curve consists mainly of two exponential sections joined by a transitional section. By empirical testing of an outdated blood sample and various kaolin suspensions in both a Merrill-Brookfield viscometer (classical type of experiment) and a GDM viscometer (Cokelet, 1963a), it was found that the yield stress, corresponding to the "yield" torque as measured in the classical instrument, was the same as the calculated from the GDM viscometer torque at the end of the first exponential section of the torque-time curve of Fig. 12. For example, for the outdated blood the classical determination gave a yield stress of 0.108 dyne/cm² while the GDM viscometer curve gave a value of 0.110 dyne/cm². This empirical method for getting the yield stress for blood from GDM viscometer data gives values of yield stresses which agree with the extrapolated intercepts of plots of shear stress-shear rate data when the very low shear rate data are determined by the zero time extrapolated stress method described in the previous section.

The second exponential section of the GDM viscometer curve is not a constant torque line, as in the classical instrument response, because the large "spring" constant and sensitivity of this viscometer permit one to detect the creeping motion of the blood when the stress is less than the yield stress.

An example of a GDM viscometer curve is shown in Fig. 13. An analysis of the second exponential section, starting with the equation for the conservation of angular momentum and using the Krieger-Elrod method for evaluation of the rate of deformation of the "solid" blood, indicates that (a) the "solid" blood acts like a Newtonian fluid, (b) has a viscosity of 2.1×10^7 centipoise, and (c) the rate of deformation is about 10^{-8} sec^{-1}. Thus, we can determine that part of the shear stress-shear rate diagram which was sketched in the bottom of Fig. 11 that is, blood behavior below the yield stress. The Newtonian behavior holds over the range where the second section of the torque-time curve is exponential (up to a shear stress of about 0.7×10^{-2} dyne/cm² and a corresponding shear rate of about 3.5×10^{-8} sec^{-1}). These data confirm the existence of a yield stress as defined earlier.

One might object to this analysis, arguing that during the time that the torque takes to decay into the second exponential section of the curve the red cells migrate away from the walls of the viscometer and the viscometer surfaces "slip" over the main body of the blood. If such did occur, the stress calculated earlier would still be valid, but the real shear rate in the blood would be lower, and so the slope of the curve in the bottom of Fig. 11 would be even steeper. Therefore, even this objection does not invalidate this argument for a blood yield stress.

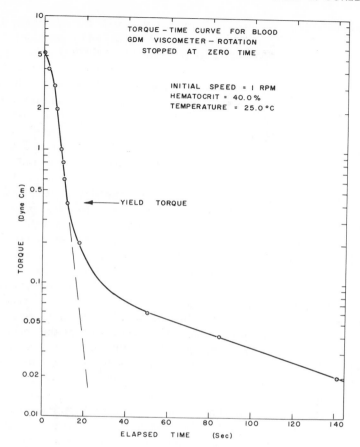

Fig. 13: Torque response curve for a human blood from the GDM viscometer (high sensitivity, very large spring constant) when steady rotation is suddenly stopped. Points are data taken from the original curve tracing by projecting an enlarged version of the original curve on a wall and measuring coordinates with a ruler. Dashed curve shows the exponential decay shown by a Newtonian fluid. Data from Cokelet (1963a).

Other techniques have been used to directly measure blood's yield stress. Merrill, et al (1965a) used a capillary viscometer to see if blood in a capillary could maintain a pressure difference across the tube ends without any detectable fluid flow. Such a pressure difference was detected and the wall-yield stress calculated from it was found to agree with the yield stress obtained by extrapolation of the tube flow data to zero flow rate.

Benis and LaCoste (1968) determined blood's yield stress with the use of a plate balance method. They found that blood showed a yield stress, whereas red cells suspended in saline showed no detectable yield stress. However, their reported values for the yield stresses of bloods were generally lower (by a factor of 2–5) than those reported by Merrill and co-workers. Benis and LaCoste give a concise discussion of possible causes for this discrepancy, although the following additional possible explanation can be added to their list. The calculated shear rate in their experiments at the point that they take to be the yield point is very small (less than 10^{-5} sec^{-1}), but it is difficult to say from the figures in their article precisely what the shear rate is. It seems that their apparatus permits them also to see blood creeping under a stress which is less than the yield stress. The stresses which they report as yield stresses are on the curve in the lower part of Fig. 11 at shear rates below 1×10^{-7} inverse seconds.

Many extrapolations have been made of finite shear stress-shear rate data to a shear rate of zero, such extrapolated shear stresses being positive finite numbers. Perhaps the most convincing extrapolation is that shown by Merrill, et al (1965a) where, on a single plot, data for one blood sample obtained with a concentric-cylinder viscometer and a tube viscometer extrapolate linearly to the same intercept (the yield stress).

In discussions dealing with the question of whether blood does or does not have a yield stress, a 1966 article by Chien, et al is frequently cited as containing data in support of the argument that blood has no yield stress. In particular, this reference contains a plot of the square root of the shear stress versus the square root of the shear rate with data points shown for the shear rate range of 0.01 to about 5 inverse seconds. The curve drawn through the data points extrapolates to zero. However, as discussed in the previous section, for nominal shear rates below about one inverse second, the torque-time curve obtained with a concentric-cylinder viscometer is complex and raises questions of interpretation. In 1966 the cited authors erroneously reduced the peak torques of the torque-time viscometer response curve to shear stresses, and the viscometer rotational speeds to accompanying shear rates. They plotted these reduced data as the flow properties of blood. This representation is incorrect and the data presented in that article for nominal shear rates below about one inverse second should be disregarded. The reader is referred to Chien, et al (1970c) for a more recent method of presenting very low shear rate viscometric blood information by these same workers. However, this newer method of presentation does not separate the effect of the blood flow behavior from the effects of the instrument dynamics, and the blood in the instrument is in a transient, nonhomogeneous condition. Consequently, the very low shear rate information of Chien, et al (1970c) cannot be used to evaluate the yield stress.

It seems to me that the evidence indicates that blood does have a yield stress. The real problem is how to experimentally measure it.

For many applications, the yield stress is so small that it is unnecessary to consider its numerical value. But some applications, such as the study of pathological bloods, can make use of the yield stress determination since it reflects the characteristics of red cell aggregation. At the present time, there is very little good information available on how red cell aggregation influences blood flow in the numerous flow conditions and geometries found in the body. Future studies in this area may uncover additional reasons for developing a simple and reliable means of evaluating blood's yield stress.

VII. VALIDITY OF THE CONTINUUM MODEL OF BLOOD

In trying to apply blood rheological data to a flow problem, we generally consider the blood to be a continuum. Indeed, this assumption of the applicability of the continuum model of blood is implied by the procedures which we use to reduce certain experimental data to the so-called rheological properties of blood. It therefore seems important that we establish the limits of applicability of the blood continuum model.

On first consideration a suitable criterion for establishing these limits of applicability might be the ratio of the smallest dimension of the flow channel to the particle diameter. However, in the case of blood this criterion is insufficient. I have seen some animal bloods, from animals apparently in good health, which formed large aggregates of red cells at low shear rates. In one case, the aggregates had a characteristic dimension of about one millimeter and the clear plasma space between aggregates was about the same size. It seems unlikely that measurements made on such blood in a viscometer with a characteristic dimension of about one millimeter will yield continuum-model, rheological properties at low shear rates. Since the aggregate size is a function of the shear stress and certain properties of the blood's constituents, and since we generally cannot directly measure aggregate size, convenience dictates that the criteria include, in addition to the previously mentioned criterion, a dimensionless group, or groups, which might contain parameters such as the shear stress, the yield stress, the continuous phase viscosity, and so forth. As far as I know, we do not presently have these guidelines.

We might attempt to find the criteria for continuum model failure by making viscometric measurements in a series of concentric-cylinder or cone-and-plate viscometers, in which the gap width or cone-plate angle is gradually reduced. However, with the fluid space thinness required, equipment alignment problems become serious, and it becomes preferable to perform our experiments in capillary tubes or narrow rectangular tubes.

Such experiments have been performed with capillary tubes many times, and with rectangular channels a few times. The classical paper in this area is that of Fahraeus and Lindqvist (1931), where data are presented showing that at high flow rates where the pressure drop is proportional to the flow rate, the calculated viscosity of the blood decreases as the tube diameter decreases from 500 to 40 microns. Two approaches have been used to explain this apparent dependence of calculated viscosity on tube diameter.

The first approach, proposed by Dix and Scott-Blair (1940), assumes that the continuum model of blood has failed where this Fahraeus-Lindqvist effect is found. They proposed a new model, composed of alternating shells of red cells (infinite viscosity) and plasma of equal thickness and arranged in the tube in a concentric fashion. On the basis of this model, they derived an equation showing how the calculated blood viscosity would vary with tube diameter. Using this equation and the available experimental data, Bayliss (1952) found that the computed shell thickness was 3.3-5.5 microns. This "sigma-effect" model has not been used further in recent time.

The second, and most popular, approach assumes that next to the vessel wall there is a cell-free plasma layer, while the rest of the tube is filled with a "core" of blood. One of the simplest models assumes that the core is of uniform hematocrit. Using the rheological properties of blood, as determined in a separate viscometer, to describe the core flow behavior and using the plasma viscosity, one can calculate from pressure drop versus flow rate data the thickness of this plasma wall layer. The literature contains many examples of this approach, for example, Charm, et al (1968) who also considered a core with one special hematocrit-radial position relationship. Almost all such analyses assume that the average hematocrit of the blood in the tube is the same as that of the blood in the reservoir feeding the tube. Such analyses show, for example, that the ratio of the plasma wall layer thickness to the tube diameter is a strong function of hematocrit, being about 0.1-0.2 for a 10% hematocrit and 0.01-0.03 for a 45% hematocrit. As will be shown, the assumption that the average hematocrit in the tube is the same as that of the blood in the feed reservoir is not true for tubes whose diameters are smaller than several hundred microns. The use of this assumption causes the calculated wall plasma layer thickness to be too large.

This model of blood tube flow was possibly fostered by early observations of blood flow in vessels. Such observations give the impression that there is a cell-free plasma layer at the vessel wall. However, high speed microcinematography has shown that cells are in this wall region when the average hematocrit in the tube or vessel is near normal physiological levels (Bloch, 1962). Bugliarello and Hayden (1963) filmed *in vitro* blood flow in glass tubes with nominal diameters of 40, 67 and 83.8 microns. When the average hematocrit was very low, the cells were all located close to the tube centerline, leaving clear plasma in the regions near the walls, and indicating that at least some cells had undergone a radial migration similar to that found by Segre and Silberberg (1962) for very dilute suspensions of spheres. At average physiological hematocrits, such a completely cell-free wall region was not found. Bugliarello and his co-workers characterize the wall plasma layer thickness by determining, from their films, the temporal average perpendicular distance from a spot on the tube wall to the nearest part of the initially encountered red cell. If in a film frame no cell is encountered before the tube center is reached, the tube radius is arbitrarily taken as the layer thickness. Thus, Bugliarello and Hayden report that the average wall layer thickness is about 40 % of the tube radius for a 5% average hematocrit, but is about 10% of the tube radius at 30–40% hematocrit. This has been confirmed by Sevilla-Larrea (1968). This method of determining an average wall layer thickness gives an exaggerated measure of the layer since, in a suspension with uniform distribution of particles, this method of determining a wall layer thickness gives a thickness approaching the tube radius at lower concentrations and a value greater than the particle radius at concentrations of 40% (volume). However, the method does have merit for qualitative comparative purposes.

Using a quick freezing technique, Phibbs (1968) solidified blood flowing through rabbit arteries. The hematocrits were 32–35% and the vessels were about 1 mm in diameter. After freezing, the vessels were freeze-substituted, embedded and thin-sectioned. When the blood flow had been steady, it was found that there was no generalized radial migration of red cells. The cell concentration was approximately constant across the artery except for a slight lowering of the hematocrit immediately next to the wall. The perpendicular distance from the wall to the first cell varied from 0–16 microns, with an average of 2.5–6.5 microns. This means that the cell concentration decrease next to the wall is probably due to the smooth wall effect of physical exclusion of cells. This smooth-wall physical exclusion effect has a profound effect on the flow of a suspension through tubes (Vand, 1948; Karnis, Goldsmith and Mason, 1966). Goldsmith (1968) worked with dilute red cell plasma suspensions and with suspensions of tracer red cells in concentrated ghost cell plasma suspensions. He confirms the red cell radial migration at low average hematocrits. At higher cell hematocrits, he was able to determine the red cell velocity profile. This was blunted somewhat from a parabolic velocity profile, but less blunted than the profiles found for rigid sphere suspensions. Since the rheological properties of the ghost cell suspensions are unknown, it is presently not possible to determine whether or not the velocity profile agrees with the profile one would predict from the rheological properties.

It appears that any attempt to determine when the continuum model fails by studying flow in very small channels must start first with a decision on what is to be taken as a sign of failure of the continuum model. It seems unrealistic to say failure has occurred when the smooth-wall exclusion effect is reflected in the experimental data, since the continuum model may still be satisfactory for the bulk of the suspension. But how do we remove this wall effect from the data? Any model involving a cell-free wall layer or an effective slip velocity at the wall will not enable us to separate a wall effect from a continuum model failure. And, it seems unlikely that we can apply the continuum model approach to the region near the wall, since the fluid region of disturbed cell distribution is only one or two cell diameters thick.

Perhaps the only approach which can solve this problem will involve measuring velocity profiles by Goldsmith's method. With the rheological properties known, one can then test to see if the experimental velocity profile away from the wall region agrees with that calculated with the continuum model. Of course, when the cell size becomes close to the tube diameter, the question of what physical velocity corresponds to the continuum model velocity becomes of primary importance.

Additional problems need to be considered. Phibbs (1968) found that at physiological hematocrits under steady flow, the red cells showed very little preference of a particular orientation at the tube center. As one moved towards the vessel wall, the cells showed a slightly increasing preference for an orientation with a concave cell face parallel to the nearest vessel wall. Goldsmith found that this preferred orientation existed at physiological hematocrits, but very little preferred orientation was shown at very low hematocrits. The preference for orientation parallel to the nearest wall increased with increasing flow rate. The question arises as to whether or not such cell orientation influences blood's rheological properties.

The picture is complicated further by Phibbs' finding that when the tube flow was pulsatile the preference for cell orientation increased markedly. Sevilla-Larrea also found that the plasma wall layer decreased in apparent thickness with a change from steady to pulsatile tube flow. Palmer's 1968 study of steady blood flow in very narrow channels seems to indicate that the hematocrit variation across the slit width is different from that found by Phibbs in circular vessels. Such behaviors need to be included in our considerations.

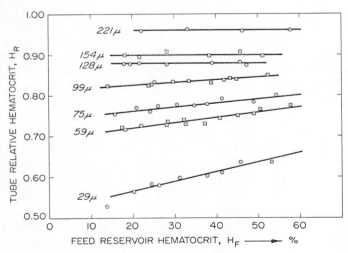

Fig. 14: The Fahraeus effect: when blood flows from a reservoir into and through a small diameter tube, the average hematocrit in the tube is less than that in the reservoir. The tube-relative hematocrit is defined as the average hematocrit of the blood flowing through a tube divided by the hematocrit of the blood in the reservoir feeding the tube. Numbers to the left of the lines are the tube diameters. Figure reprinted from Barbee and Cokelet (1971).

In view of these complications, the simplest approach is probably the practical approach. Therefore, I would like to present for consideration two figures.

Figure 14 shows how the average hematocrit of the blood in a tube (H_T) varies with tube size and the feed reservoir blood hematocrit (H_F) when blood flows steadily from a relatively large reservoir into a small capillary. (H_R is the ratio H_T/H_F.) This plot shows that for a given feed reservoir hematocrit the tube hematocrit decreases as the tube diameter diminishes. This effect was first pointed out by Fahreaus in 1929. It was also found that the mixing-cup hematocrit of the blood leaving a tube 50 microns in diameter, or larger, was the same as the feed reservoir hematocrit; this means that in the tube the hematocrit must vary with radial position.

Figure 15 shows steady flow data for blood in a 29-micron diameter tube. The plot is of the wall shear stress (τ_w) versus $\bar{U}$, the bulk average flow rate

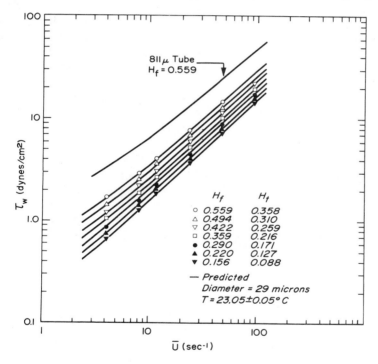

Fig. 15: The flow behavior of blood in a 29-micron diameter tube. The symbols are the actual flow data, recorded as the wall shear stress, τ_w, and the bulk average velocity divided by the tube diameter, $\bar{U}$. The solid curves through the points represent the behavior of the blood predicted from the data obtained in an 811-micron diameter tube when the average tube hematocrit is the same as that experimentally found in the 29-micron tube. In an 811-micron tube, H_F, the feed reservoir hematocrit, and H_T, the average tube hematocrit, are equal. From Barbee and Cokelet (1971).

divided by the tube diameter. The data are plotted this way since it can be shown that for a fluid for which the rate of deformation is just a function of the shear stress, the same function applying at all points in the tube, the data should give one universal curve, even when the data are obtained with tubes of different diameters.

Let us first consider the top curve, labeled 811-micron tube, $H_F = 0.559$, and the circles which represent the experimental data obtained with the 29-micron tube with $H_F = 0.559$. The top curve represents tube flow data obtained with the 811-micron tube with $H_F = 0.559$. These two sets of data should coincide if the shear rate is only a function of the stress. But they do not agree, and this discrepancy is the Fahraeus-Lindqvist effect. One reason they do not agree is that, even though H_F is the same for both tubes, H_T is only 0.358 in the 29-micron tube while it is still 0.559 in the 811-micron tube. If we obtain flow data in an 811-micron tube with $H_F = 0.358$ ($= H_T$), we find that these data are represented by the solid curve which happens to pass through the circles. Thus, it appears that the Fahraeus-Lindqvist effect is due entirely to the Fahraeus effect. The other data points and solid curves represent 29-micron tube data and 811-micron tube data, respectively, when the tube hematocrits are made equal for each pair of data sets. The agreement is surprising, since we know that the assumption used in proving that the wall stress-$\bar{U}$ relationship is a unique function is not true, since the hematocrit must vary with radial position in the tube. Nevertheless, the relationship is found to be unique provided the tube hematocrit is held constant. We have found this to be true for all tube sizes down to 29-micron diameter tubes.

From this discussion, the reader should conclude that the development of criteria for failure of the continuum model of blood is not a simple procedure. Such criteria are presently not satisfactorily determined, and extensive careful experimental and theoretical work in areas where the characteristic flow channel dimension is about equal to that of a few red cell diameters or less is needed.

VIII. CONCLUDING STATEMENT

As indicated in the introduction, my main intent in this presentation was to consider some problems and questions of blood rheology, although I have also sketchily described what seem to be some of the well-established typical rheological properties of normal human bloods and related fluids. I hope that this biased account will alert the reader to the difficulties of blood rheology studies. But, I hope the uncertainties will attract workers to this research field rather than deter them from it. Many large areas are in need of research. As an example, very little is known of the relationships between the macroscopic rheological properties of blood and the microscopic behavior of the constituents of blood during flow through physiological geometries such as bifurcations, capillary networks, and so forth. Such research might be especially useful in understanding pathological situations.

This work was supported by USPHS grant HE-12723.

REFERENCES

Barbee, J. H., and G. R. Cokelet. 1971. The Fahraeus effect. Microvas. Res., *3*, 1–21.

Bayliss, L. 1952. Rheology of blood and lymph, *In* A. Frey-Wissling, (ed.) Deformation and flow in biological systems. pp. 354–418, North Holland Publishing Co., Amsterdam.

Benis, A. M. 1964. The flow of human blood through models of the microcirculation. Sc.D. Thesis, Chem. Eng. Mass. Inst. of Tech. Cambridge, Mass.

Benis, A. M., and J. LaCoste. 1968. Study of erythrocyte aggregation by blood viscometry at low shear rates using a balance method. Circ. Res. XXII: 29–41.

Bertles, J. F., R. Rabinowitz, and J. Döbler. 1970. Hemoglobin interaction: Modification of solid phase composition in the sickling phenomenon. Science, 169: 375–77.

Bloch, E. H. 1962. A quantitative study of the hemodynamics in the living microvascular system. Amer. J. Anatomy, 110: 125–53.

Brooks, D. E., and G. V. F. Seaman. 1969. Role of mutual cellular repulsions in the rheology of concentrated red blood cell suspensions. 2nd. Inter. Conf. on Hemorheology, Heidelberg.

Brooks, D. E., J. W. Goodwin and G. V. F. Seaman. 1970. Interactions among erythrocytes under shear. J. Appl. Physiol., 28: 172–77.

Bugliarello, G., and J. W. Hayden. 1963. Detailed characteristics of the flow of blood *invitro*. Trans. Soc. Rheol., 7: 209–30.

Casson, N. 1959. A flow equation for pigment-oil suspensions of the printing ink type, *In* C. C. Mill (ed.) Rheology of Disperse Systems. Pergamon Press, New York. pp. 84–104.

Charm, S., and G. Kurland 1965. Viscometry of human blood for shear rates of 0–100,000 sec^{-2}. Nature, 206: 617–18.

Charm, S. E., Kurland and S. L. Brown. 1968. The influence of radial distribution and marginal plasma layer on the flow of red cell suspensions. Biorheology, 5: 15–43.

Charm, S. E., G. S. Kurland and M. Schwartz. 1969. Absence of transition in viscosity of human blood between shear rates of 20 and 100 sec^{-1}. J. Appl. Physiol., 26: 389–92.

Chien, S., S. Usami, H. M. Taylor, J. L. Lundberg and M. I. Gregersen. 1966. Effects of hematocrit and plasma proteins on human blood rheology at low shear rates. J. Appl. Physiol., 21: 81–87.

Chien, S., S. Usami, R. J. Dellenback and M. I. Gregersen. 1967. Blood viscosity: Influence of erythrocyte deformation. Science, 157: 827–29.

Chien, S. 1969. Blood rheology and its relation to flow resistance and transcapillary exchange, with special reference to shock. Adv. Microcirc. 2: 89–103.

Chien, S., S. Usami, and J. F. Bertles. 1970a. Abnormal rheology of oxygenated blood in sickle cell anemia. J. Clin. Invest., 49: 623–34.

Chien, S. 1970b. Shear dependence of effective cell volume as a determinant of blood viscosity. Science, 168: 977–78.

Chien, S., S. Usami, R. J. Dellenback and M. I. Gregersen. 1970c. Shear-dependent interaction of plasma proteins with erythrocytes in blood rheology. Amer. J. Physiol. 219: 143–53.

Cokelet, G. R. 1963a. The rheology of human blood. Sc.D. Thesis, Chem. Eng. Mass. Inst. of Tech., Cambridge, Mass.

Cokelet, G. R., E. W. Merrill, E. R. Gilliland, and H. Shin. 1963b. The rheology of human blood measurement near and at zero shear rate. Trans. Soc. Rheol., VII: 303–17.

Cokelet, G. R., and H. J. Meiselman. 1968. Rheological comparison of hemoglobin solutions and erythrocyte suspensions. Science, 162: 275–77.
Dix, F. J., and G. W. Scott-Blair 1940. On the flow of suspensions through narrow tubes. J. Appl. Physics, 11. 574–81.
Fahraeus, R. 1929. The suspension stability of the blood. Physiol. Rev., 9: 241–74.
Fahraeus, R., and T. Lindqvist. 1931. The viscosity of the blood in narrow capillary tubes. Amer. J. Physiol., 96: 562–68.
Frasher, W., H. Wayland, and H. J. Meiselman. 1968. Viscometry of circulating blood in dogs. I: Heparin Injection. II: Platelet Removal. J. Appl. Physiol, 25: 751–60.
Goldsmith, H. L. 1968. The microrheology of red blood cell suspensions. J. Gen. Physiol., 52: 5s–28s.
Gregersen, M. I., S. Usami, S. Chien, and R. J. Dellenback 1967a. Characterisitics of torque-time records on heparinized and defibrinated elephant, human and goat blood at low shear rates (0.01 sec^{-1}): effects of fibrinogen and Dextran (Dx 375). Bibl. anat. 9: 276–81.
Halow, J. S., and G. B. Wills. 1970. Radial migration of spherical particles in Couette systems. A.I.Ch.E. J., 16: 281–86.
Heard, D. H., and G. V. F. Seaman. 1961. The action of lower aldehydes on the human erythrocyte. Biochim. Biophys. Acta, 53, 366–374
Karnis, A., H. L. Goldsmith, and S. G. Mason. 1966. The kinetics of flowing dispersions. I: Concentrated suspensions of rigid particles. J. Coll. Interface Sci., 22: 531–53.
Krieger, I. M., and H. Elrod. 1953. Direct determination of the flow curves of non-Newtonian fluids. II: Shearing rate in the concentric cylinder viscometer. J. Appl. Phys., 24: 134.
Krieger, I. M. 1968. Shear rates in the Couette viscometer. Trans. Soc. Rheol., 12: 5–11.
Landel, R. F., B. G. Moser, and A. J. Bauman. 1965. In E. H. Lee (ed.) Proceedings of the Fourth International Congress on Rheology, Part 2. Interscience, New York. pp. 663–92.
Markovitz, H. 1968. The emergence of rheology. Phys. Today, 21: 23–30.
Meiselman, H. J. 1965. Some physical and rheological properties of human blood. Sc.D. Thesis, Chem. Eng., Mass. Inst. of Tech., Cambridge, Mass.
Merrill, E. W., E. R. Gilliland, G. Cokelet, H. Shin, A. Britten, and R. E. Wells, Jr. 1963. Rheology of human blood, near and at zero flow. Effects of temperature and hematocrit level. Biophysical J., 3: 199–213.
Merrill, E. W., A. M. Benis, E. R. Gilliland, T. K. Sherwood, and E. W. Salzman 1965a. Pressure-flow relations of human blood in hollow fibers at low flow rates. J. Appl. Physiol., 20- 954–67.
Merrill, E. W., W. G. Margetts, G. R. Cokelet, A Britten, E. W. Salzman, R. B. Pennell, and M. Melin. 1955b. Influence of plasma proteins on the rheology of human blood. In A. L. Coply (ed.) Proc. 4th Inter. Cong. on Rheology, Pt, 4, Interscience (Wiley), New York. pp. 601–12.
Merrill, E. W., and G. A. Pelletier. 1967. Viscosity of human blood: Transition from Newtonian to non-Newtonian. J. Appl. Physiol., 23: 178–82.
Merrill, E. W. 1969a. Rheology of blood. Physiol. Rev., 49: 863–88.
Merrill, E. W., C. S. Cheng, and G. A. Pelletier. 1969b. Yield stress of normal human blood as a function of endogenous fibrinogen. J. Appl. Physiol., 26: 1–3.
Palmer, A. A. 1968. Some aspects of plasma skimming. In A. L. Copley (ed.) Hemorheology. Pergamon Press, Oxford. pp. 391–400.
Phibbs, R. H. 1968. Orientation and distribution of erythrocytes in blood flowing through medium-sized arteries, In A. L. Copley (ed.) Hemorheology. Pergamon Press, Oxford. pp. 617–30.
Rahn, A. W. C. Tien and L. C. Cerny. 1967. Flow properties of blood under low shear rate In D. Hershey, (ed.) Chemical Engineering in Medicine and Biology. Plenum Press, New York. pp, 45–84.
Schmid-Schoenbein, H., P. Gaehtgens, and H. Hirsch. 1968. On the shear rate dependence of red cell agregation in vitro. J. Clin. Invest., 47: 1447–454.
Schmid-Schoenbein, H., R. Wells, and R. Schildkraut. 1969a. Microscopy and viscometry of blood flowing under uniform shear rate (rheoscopy).

Schmid-Schoenbein, H., and R. Wells 1969*b*. Fluid drop-like transition of erythrocytes under shear. Science, 165: 288–91.

Scott-Blair, G. W. 1959. An equation for the flow of blood, plasma and serum through glass capillaries. Nature, 183: 613–14.

Seaman, G. V. F. 1970. Personal communication.

Segré, G. and A. Silberberg. 1962. Behavior of macroscopic rigid spheres in Poiseuille flow. J. Fluid Mech., 14: 115–35, 136–57.

Sevilla-Larrea, J. F. 1968. Detailed characteristics of pulsatile blood flow in small glass capillaries. Ph.D. Thesis. Biomed. Eng., Carnegie-Mellon Univ., Pittsburgh, Penna.

Skalak, 1970. Mechanics of the microcirculation. Chapter in this volume.

Vand, V. 1948. Viscosities of solutions and suspensions. J Phys. Coll. Chem., 52: 300–14.

Wayland, H. 1967. Rheology and the microcirculation. Gastroenterology, 52: 342–55.

Wells, R., and H. Schmid-Schoenbein. 1969. Red cell deformation and fluidity of concentrated cell suspensions. J. Appl. Physiol., 27: 213–17.

Whitmore, R. L. 1968. Rheology of the Circulation. Pergamon Press, Oxford.

5

The Properties of Blood Vessels

D. H. BERGEL

"... *the most likely way, therefore, to get any insight into the nature of those parts of the creation which come within our observation, must in all reason be to number, weigh and measure.*" (Hales, 1727).

I. INTRODUCTION

That the blood vessels form a system of pipes conveying blood has been obvious for many centuries. Thus the physical properties of blood vessels have been considered important for a long time. Harvey (1628) recognised that blood flowed through the vessels in one direction and he gave a clear account of the physical function of the venous valves. The quantitative study of the circulation may be held to have commenced with the work of Hales (1733). He discussed arterial and venous elasticity, measured blood pressure and the pressure flow relations of vascular beds, estimated cardiac output and was the originator of the idea that the aorta functions as an elastic reservoir to receive intermittent ventricular output.

We now know a great deal of the passive mechanical properties of blood vessels, and are beginning to describe their active responses in physical terms but, as will become obvious, our descriptions are still couched in the most simple terms. I shall attempt to give a general account of the structure and properties of a variety of blood vessels, employing rather elementary theory, together with a brief discussion of the ways in which these properties are of importance to the function of the circulatory system.

II. DEFORMATION OF ELASTIC MATERIALS

A. Stress and Strain

Many materials, for example, metals, deform when a force is applied and, provided the force has not been too great, return to their original shape when unloaded. Such perfectly elastic materials generally show a linear relation between the applied force and the deformation and are often referred to as Hookean solids. The constants in the force-deformation relationships are the elastic constants or moduli of the material. The moduli relate stress and strain. Stress is the applied force per unit area and strain is the resulting relative deformation. If the strains are very small the deformations may be related to the original unstressed dimensions. Thus in considering linear, tensile strains, as in a wire,

$$\text{stress} \quad s = F/A$$
$$\text{strain} \quad \epsilon = \Delta l/l$$

and the tensile or Young's, modulus of elasticity $E = s/\epsilon$.

Similarly, we may define the bulk modulus K, relating volume changes to hydrostatic compressive forces and the shear modulus G, relating shear stress to shear strain.

Stresses applied to a body may be resolved into components along the three principal axes. For a simple *isotropic* material the relation between stress and strain will be the same in any direction. Such a material may also be *homogeneous* that is, of uniform structure.

A primary strain produced in a material will set up secondary strains, thus a stretched wire will show changes in radius and total volume. If the initial dimensions of the specimen are l, a, b, then a change in l gives a strain

(1) $$\epsilon_1 = \Delta l/l$$

(2) $$\epsilon_2 = \Delta a/a = -\sigma_2 \epsilon_1$$

(3) $$\epsilon_3 = \Delta b/b = -\sigma_3 \epsilon_1$$

The ratio between strains, σ, is termed Poisson's ratio. There are six independant Poisson's ratios corresponding to primary strains in the l, a, and b dimensions. If the material is isotropic these are all equal.

It can be shown that for an isotropic material two elastic constants suffice to decribe its properties and that the three elastic moduli and σ are interrelated. Thus

(4) $$K = \frac{2}{3}\frac{(1+\sigma)}{(1-2\sigma)} \quad G = \frac{E}{3(1-2\sigma)}$$

(5) $$G = \frac{E}{2(1+\sigma)}$$

(6) $$\sigma = \frac{1}{2}\frac{3K-2G}{3K+G}$$

Note that if $\sigma = 0.5$ then $K = \infty$, the material is incompressible. The same can be seen from Eq. 2, 3. A material will undergo no volumetric strain on extension if $\sigma = 0.5$. For most metals $\sigma = 0.3$–0.4, that is their volume increases when stretched.

For anisotropic materials the situation is much less simple, even supposing we retain the assumption of homogeneity and elastic linearity. In the worst case twenty-one independant constants must be evaluated, a task that is virtually impossible with a biological structure. The situation can be improved somewhat by the assumption of orthotropicity or elastic symmetry, meaning that no shearing forces or strains are developed with a linear stress, for then only nine constants remain. In certain cases it is possible to evaluate the three tensile elastic moduli; this will be discussed later.

B. THE VESSEL WALL AS AN ELASTOMER

Classical elastic theory was developed from the study of deformations in metals and other hard materials and is based fundamentally on the idea that the strains observed are extremely small. "In the case of finite or large strains the postulate is no longer applicable and entirely different mathematical techniques... must be introduced" (Treloar, 1958). Soft body tissues deform by relatively enormous amounts, for example, strains of 2–4 as compared to 0.002 or less for metals. This property puts body tissues in the same class as the elastomers; other properties of this group are a relatively low Young's moduli (ca 10^7 dynes/cm^2) and anomalous thermoelastic properties such as shortening when warmed. Other typical elastomers are rubbers and many synthetic polymers. That a piece of artery is more like a rubber band then a metal strip was clearly shown by Roy (1880) who measured length changes on warming. However this realisation has not, as yet, led to any very clear idea of what should be done about it.

It can be shown that many of the elastic properties of elastomers stem from the very long flexible molecules of which they are composed. Thermodynamic forces will lead to a random configuration in which the ends of a molecule will always be less separated than if they were lying in an orderly linear array. This randomness will be increased by warming and the material will shorten; conversely rubber and arteries become warm when stretched. Thus the large scale extensions seen in rubbers are predominantly and initially due to straightening of the molecular chains. Only when the chains are relatively highly organised, either by virtue of intermolecular cross-linkages (vulcanisation) or by the incorporation of rigid intramolecular bonds, or by prestretching, does the strain occur within the molecule. At this point, characterised by the thermoelastic inversion point (no length or tension change on warming) the material acts more like a metal and is relatively inextensible and more highly elastic.

A theoretical treatment of such behaviour, starting from assumptions of molecular length and degree of cross-linking, has been developed (Treloar, 1958) and this predicts the properties of rubbery materials with some accuracy. The materials described are isotropic and homogeneous, the molecules are all the same apart from some assumed distribution of lengths. When applied to data from arteries (Lawton, 1954; King, 1957) the form of the stress-strain curve can be predicted. However, even when smooth muscle is neglected entirely, the properties of vessel walls must be those of a two-phase material containing elastin and collagen. Both of these, especially elastin, may reasonably be taken to be fibrous elastomers. The parameters for an homogeneous elastomer derived from a piece of artery give little insight into what is actually occurring when it is stretched. There is no doubt that if we are ever to explain the mechanical properties of blood vessels in molecular terms, and to describe what is occuring in disease or aging, we badly need a theory based on multiphase anisotropic polymeric materials. However, it is hard to see what can be done on these lines by the physiologist at the present. For those of us who are mainly concerned with the function of the whole vessel it is sufficient now to deal with numbers loosely termed "elastic moduli."

C. VISCO-ELASTICITY

An ideal elastic solid deforms instantly when stressed and returns instantly to its original dimensions when unloaded. In this process elastic energy is stored within the material and may be recovered without loss. It is clear that this is not the true state of affairs, real materials show some degree of retarded response with associated energy loss. It can be imagined that any material shows, to some degree, the properties both of an elastic solid and a viscous fluid. The varieties of behaviour to be expected from viscous solids or elastic liquids is enormous and such materials are of great theoretical and practical interest (Scott-Blair, 1969). There have been attempts to model this *behaviour* by assemblies of Hookean elastic elements and Newtonian viscous elements, but only for very simple materials is it at all possible to extract independent constants relating to elastic and viscous response.

In the case of blood vessels visco-elastic behaviour is well shown. After a sudden stretch tension rises and decays towards some final value, stress relaxation (Stacy, 1957). Following a sudden change in tension a continuing deformation (creep) is observed (Wiederhielm, 1965). The simplest models predict a response which may be resolved into one or more exponentials but this is rarely if ever seen. It has generally been necessary to assume that the material response represents that of an array of relaxing elements with a spectrum of time constants. With the assumption of linearity such spectral distributions may be derived and some information is given on the wall structures predominantly involved in load bearing in a particular instance (Goto and Kimoto, 1966).

An entirely different approach, that is of particular relevance to studies of vascular smooth muscle, has been proposed by Stacy (1957) and applied by Wiederhielm (1965). This is based on relatively simple kinetic equations describing the rate constants in some reversible short-to-long conversion of some unspecified units within the material and will be discussed later when dealing with visco-elastic properties of blood vessels.

D. How should the mechanical properties of blood vessels be described?

In common with other elastomeric materials the elasticity of blood vessels is non-Hookean (Fung, 1967): Indeed it would be extraordinary if this were not so. We know that vessel walls are made up of at least three identifiable components so they are inhomogeneous and likely to be anisotropic. There is some comfort to be taken from the fact that elastomers are relatively incompressible and the high water content of vessel walls (Harkness et al, 1957) suggests that this may be true for blood vessels. Neither the classical ideas about deformation of stiff structures nor elastomeric theory based on simple populations of similar molecules readily lend themselves to the description of tissue elasticity. It will also be necessary to decide some way of describing the time-dependant properties. I can see no immediate prospect of being able to describe these properties in terms of meaningful molecular arrangements; we should aim at a clear description which will allow one to see how vessels will behave as part of the vascular system and also allow dissection of their properties in terms of the behaviour of individual wall components.

A central observation here is that blood vessels *in vivo* are always under considerable longitudinal stress. When removed from the body they shorten by up to 40% (Table 1), and in shortening the diameter increases. This retraction is a most interesting phenomenon, it is seen in arteries (Bergel, 1961a) and veins (Yates, 1969) and decreases in amount with age (Hesse, 1926; Learoyd and Taylor, 1966). Indeed, the arteries of an old subject are in a sense too long for their site. It is uncertain how the forces producing retraction arise, possibly the original polymerisation process is responsible and longitudinal tension is "built-in" at this time.

TABLE I

Vessel	Species	$\gamma \times 100$	Retraction %	Reference
Thoracic aorta	Dog	10.5	32	Bergel, 1961a
Abdominal aorta	Dog	10.5	34	Bergel, 1961a
Femoral A	Dog	11.5	42	Bergel, 1961a
Carotid A	Dog	13.2	35	Bergel, 1961a
Iliac A	Dog	—	40	Bergel, 1961a
Carotid A	Dog	13.6	—	Rees & Jepson, 1970
Carotid A	Cat	14.5	—	Rees & Jepson, 1970
Carotid A	Rabbit	11.2	—	Rees & Jepson, 1970
Carotid sinus	Dog	20.0	—	Rees & Jepson, 1970
Carotid sinus	Cat	16.0	—	Rees & Jepson, 1970
Carotid sinus	Rabbit	12.0	—	Rees & Jepson, 1970
Thoracic aorta	Dog	14	—	Gow & Taylor, 1968
Abdominal aorta	Dog	12	—	Gow & Taylor, 1968
Femoral	Dog	13	—	Gow & Taylor, 1968
Pulmonary artery	Man	2	—	Reid, 1968
Abdominal vena cava	Dog	2.3	30	Yates, 1969
Thoracic aorta	Man	6–9	25–15	Learoyd & Taylor, 1966†
Abdominal aorta	Man	8–13	30–17	Learoyd & Taylor, 1966†
Femoral	Man	12–19	40–25	Learoyd & Taylor, 1966†
Carotid	Man	9–15	25–18	Learoyd & Taylor, 1966†

Values for relative wall thickness (γ) and retraction on excision for a variety of blood vessels.

† Measurements of Learoyd and Taylor are for arteries from young (< 35) and old (> 35) subjects respectively.

In addition it will be seen that vessel dimensions, particularly at rather low pressures, show considerable variation, especially when smooth muscle activity changes. All these facts make it impossible to make an unambiguous definition of strain based on unstressed dimensions. The most satisfactory answer appears to be to study successive deformations with successive increments of load and to define an incremental strain as the ratio of the deformation to the dimensions measured at the start of that deformation. The idea was first used by Krafka (1939) and following its reintroduction (Bergel, 1960) it appears to have been generally welcomed. Biot (1965) deals formally with incremental stress analysis and discusses, *inter alia*, the deformation of anisotropic incompressible materials. Clearly, any incremental modulus of elasticity derived in this way will only be meaningful if the mean stress, or preferably the mean radius or volume pertaining, is stated. Dynamic elasticity may be similarly defined, the complex visco-elastic modulus may then be resolved into real and imaginary parts (Bergel, 1961b).

E. The pressurised isotropic homogeneous tube

The general expression for the static mean circumferential stress in the wall of a tube (Fung, 1968) is

(7) $$S = P'_i R_i - P'_o R_o$$

where P' = pressure and R = radius and the subscripts i and o refer to internal and external, respectively. In addition to the circumferential stress there will be radial (compressive) and longitudinal stresses whose magnitude will depend on the initial conditions, the wall constants and conditions at the tube end.

It follows from Eq. 7 that when $P'_i = P'_o$ the net stress will be compressive. Where relative wall thickness $\gamma = h/R_o$ is about 0.1 (see Table 1), wall stress will become zero when P_i is 1.1 atmospheres, that is when overpressure is 76 mm Hg. Where $\gamma = 0.02$, as it is said to be in the pulmonary artery, this condition is reached at an internal pressure of 16 mm Hg. That the zero-stress pressure is not greatly different from the physiological pressure has been pointed out by Opatowski (1967), but I am not sure that this implies some deep functional principle. It does make the definition of the zero-stress state even more troublesome. It is general practice to ignore P'_o and to redefine P_i as $P'_i - P'_o$, and this convention will be followed here.

Thus we write $S = P_i R_i$ and for thin walled tubes ($\gamma = 0.1$) the stress per unit area can be taken to be $P_i R/h$ where R is mean radius. The implications of this simple expression, frequently dignified by the title "Law of LaPlace," were clearly expressed by Burton (1954). One important result is that wall tension is not simply related to pressure and thus the convenient comparison of vessel wall properties at the same pressure is often misleading.

In a homogeneous isotropic tube the wall forces are not uniform but alter with R, the circumferential tension (S_θ) being greatest where $R = R_i$. The full expression for wall tensions are (Love, 1927)

(8) $$S_\theta = \frac{P_i R_i^2}{R_0^2 - R_i^2} + \frac{P_i R_0^2 R_i^2}{R_0^2 - R_i^2} \cdot \frac{1}{R^2}$$

(9) $$S_r = \frac{P_i R_i^2}{R_0^2 - R_i^2} - \frac{P_i R_0^2 R_i^2}{R_0^2 - R_i^2} \cdot \frac{1}{R^2}$$

(10) $$S_L = \frac{\lambda}{\lambda + G} \cdot \frac{P_i R_i^2}{R_0^2 - R_i^2} + e\frac{3(\lambda + 2G)G}{\lambda + G}$$

where

(11) $$\lambda = \frac{E\sigma}{(1 + \sigma)(1 - 2\sigma)}$$

and

e = longitudinal extension
R = any radius where $R_0 > R > R_i$

Note that if the tube is closed by a flat plate the force on that plate will be $P_i\pi R_i^2$ which will be opposed by the force due to longitudinal wall tension $S_l\pi(R_0^2 - R_i^2)$. Thus we find that

(12) $$e = \frac{1 - 2\sigma}{2E} \cdot \frac{P_i R_i^2}{R_0^2 - R_i^2}$$

Hence for an isotropic tube $e = 0$ if $\sigma = 0.5$. The common observation that an artery with closed ends lengthens considerably when inflated (Fenn, 1957) together with the demonstration that the wall deforms isovolumetrically (Carew et al, 1968) is the simplest evidence that it is not isotropic. In fact, when an artery is inflated to pressures somewhat above physiological, the length lost in retraction is recovered (Tickner & Sacks, 1967) and it has been argued that vessels *in vivo* are held out in this way. However, in life arteries are not closed with flat plates and when pressurised the retractive force will be even greater than that measured after death.

Returning to the circumferential extension, it can be shown (Love, 1927) that the radial displacement, U, of a point on a shell of radius R is of the form

(13) $$U = AR + B/R$$

where

(14) $$A = \frac{P_i R_i^2}{2(\lambda + G)(R_0^2 - R_i^2)} - \frac{\lambda e}{2(\lambda + G)}$$

(15) $$B = \frac{P_i R_0^2 R_i^2}{2G(R_0^2 - R_i^2)}$$

thus when $R = R_0$ and $e = 0$ we have, using Eq. 5 and 11

(16) $$U = \frac{P_i R_i^2 2(1 - \sigma^2) R_0}{(R_0^2 - R_i^2)E}$$

and

(17) $$E_{inc} = \frac{\Delta P_i R_0}{\Delta R_0} \frac{R_i^2 2(1 - \sigma^2)}{R_0^2 - R_i^2}$$

(It will be remembered that R_i and R_0 here refer to the initial radii for the deformation considered. $R_0^2 - R_i^2$ is constant for a given vessel of constant length from the assumption of incompressibility.) This is the expression generally used to compute the incremental modulus (Bergel, 1961a; Gow and Taylor, 1968; Learoyd and Taylor, 1966). Its computation depends on an accurate measurement of wall thickness, generally calculated from R_0 following measurement of the wall volume and assuming incompressibility.

To avoid the introduction of errors from wall thickness determinations, which may be difficult *in vivo*, two varieties of distensibility modulus are frequently used:

(18) $\quad E_P = \Delta P_i R_0 / \Delta R_0$ (Peterson et al, 1960; Patel et al, 1964)

(19) $\quad \kappa = \Delta P V_i / \Delta V_i$ (Bader, 1964)

where

$$V_i = \text{internal volume}$$

If these moduli are to be compared they should be measured under conditions of constant vessel length as *in vivo*, and incremental strains should be used.

The dynamic elastic modulus is similarly defined. Measurements of pressure and radius oscillations are made at a specified frequency and mean pressure and the amplitude ratio and phase difference ($\Delta\phi$) determined. Then, following the notation of Hardung (1952)

(20) $\quad\quad\quad\quad\quad E_{\text{dyn}} = E_{\text{inc}} \cos \Delta\phi$

(21) $\quad\quad\quad\quad\quad \eta\omega = E_{\text{inc}} \sin \Delta\phi$

Since a great deal of published work is based on these isotropic equations we will discuss this before considering anisotropicity.

III. STRUCTURE OF BLOOD VESSELS

The structure of blood vessels follows a simple pattern and is described in the standard textbooks (as in Bloom and Fawcett, 1962). The basic pattern is one of three concentric layers. The innermost tunica intima consists of a thin (0.5–0.1 μm) layer of endothelial cells typically separated from their neighbours by narrow oblique clefts (10–20 μm in "continuous" capillaries) often showing one or more points of narrowing (Karnofsky, 1968). In such capillaries the endothelial cells, together with a basement membrane, form the entire vessel wall. Different capillary types show significant differences in the closeness of fit between adjacent cells which are of great functional significance, but this will not be discussed further here.

Beneath the endothelium lies a thin layer of connective tissue below which is the boundary of the thick tunica media. This is separated from the intima by a prominent layer of elastic tissue, the internal elastic lamina, which is a fenestrated membrane completely encircling the vessel lumen. The media contains collagen and smooth muscle fibres as well as elastin. The outermost layer, the adventitia, merges externally with surrounding tissue and contains a high proportion of collagen. Blood vessel walls, except those of capillaries, contain nerve endings and the larger vessels also possess nutrient vasa vasorum in the adventitia and external media.

The largest arteries are termed elastic because of the relatively great amount of elastic tissue in their walls. The term "elastic" is taken, on the basis of histological appearance, to include the aorta and its major branches. Harkness et al (1957) showed that there was rather a sharp distinction between the intrathoracic aorta and all other arteries, including the pulmonary artery, in terms of chemically determined elastin/collagen ratios. For the thoracic aorta the ratio was two and elsewhere it was reversed, that is 0.5. These results were confirmed, for normal dogs, by Fischer and Llaurado (1966).

When the aorta is fixed while distended and stretched to its natural length, microscopica examination shows that the media is constructed of concentric laminar layers. The divisions are formed by elastic lamellae between which lie collagen fibres and smooth muscle cells. When fixed in this way the specimens show that the waviness of the elastic lamellae seen in conventional preparations, the significance of which was much speculated upon some years earlier (MacWilliam and Mackie, 1908), is absent (Fig. 1) (Bergel, 1960; Wolinsky and Glagov, 1967). Wolinsky and Glagov make the interesting point that the number of lamellar layers increases with wall thickness which is itself a constant proportion (see Table 1) of vessel diameter. Thus each layer, of relatively constant structure and thickness in a large variety of mammals, bears much the same tension. It seems that the lamellar unit is the basic building block for thoracic aortas. Within the lamellar spaces the arrangement is one in which muscle cells are attached between adjacent elastic lamellae and the interstices contain fine collagen fibres which do not appear to be attached to anything (Wolinsky and Glagov, 1964; Pease and Paule, 1960). This arrangement of muscle cells corresponds to Benninghof's (1930) "Spannumuskeln."

In smaller arteries such as the femoral or carotid, the structure is altered somewhat and the relative wall thickness is slightly increased. Elastin is less prominent in the media and with increasing distance from the aorta eventually only the inner and outer elastic laminae can be clearly seen. The amount of muscle increases and this is arranged in quasi-concentric layers with prominent muscle-muscle attachments (Benninghof, 1930; Pease and Molinari, 1960). In addition the arrangement of all wall elements follows a helical pattern which has a finer pitch in the more peripheral vessels.

Thus in the passage from the thoracic aorta to the larger arteries there is a transition from a structure which appears to be predominantly made of elastin to one where the smooth muscle is more prominent and arranged for a better mechanical advantage. As one passes more peripherally still, the muscularity becomes more obvious and the fibrous tissue is reduced to a thin

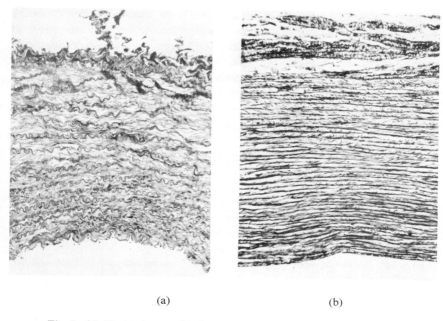

Fig. 1: (*a*) Photomicrograph of cross section of a dog's femoral artery, fixed at excised length and under no distending pressure. Note the extreme waviness of the intimal layer (bottom) and of the dark elastic laminae. (*b*) Photomicrograph of cross section of dog's thoracic aorta, fixed at natural length and at an internal pressure of 100 mm Hg. Note the regular series of straight elastic laminae, between which lie smooth muscle cells which span the gap between adjacent laminae.

internal elastic lamina and fine fibres of elastin and collagen in the media. One would expect the mechanical properties of these vessels to be dominated by the behaviour of the muscle.

In the capillaries only the endothelium remains. Beyond these vessels the blood is collected into veins of gradually increasing size. Generally speaking, the veins accompany arteries and the venous cross-sectional area at any point is larger than that of the arteries, and the velocity of blood flow correspondingly lower.

The basic structure of the veins is similar to that of the arteries. The relative wall thickness is generally lower than that of arteries and the media contains very little elastic tissue. In addition the intimal surface of most larger veins (except the venae cavae, hepatic, renal, ovarian, uterine and portal veins) is thrown into fine folds which form cresentic semilunar valves, generally arranged in pairs and associated with a distinct sinus or local widening of the vessel. The function of the valves, first clearly shown by William Harvey in 1628, is to enforce one-way blood flow. Much more recently it has been shown by Bellhouse and Bellhouse (1968) that the valve sinuses seen at the root of the aorta and pulmonary artery play an important part in ensuring valvular efficiency, and the venous valvular sinuses presumably serve the same purpose.

The adventitia of veins is relatively thick and contains much collagen. In addition threre are prominent longitudinal muscle fibres to be found in the walls of the abdominal vena cava and its main tributaries, and in the mesenteric veins. For a full description of the functional anatomy of veins, Franklin's (1937) monograph is still the best source.

Veins contain relatively high amounts of collagen, the collagen elastin ratio is about 3:1 (Svejcar et al, 1962; Bader, 1964). There are however some differences in composition between the upper and lower segments of human leg veins, the lower segment containing more muscle and less connective tissue. That this is only partly due to hydrostatic pressure differences is shown by the fact that the difference in elastin content is found in the veins of infants who have not yet walked.

The structure of the pulmonary and systemic arteries is very similar; elastic and muscular vessels may be distinguished. In general the relative wall thickness is less than in systemic arteries of similar structure (Table 1). There are no valves in pulmonary veins. It has been shown that relative wall thickness increases rather sharply in the smallest muscular branches of the pulmonary artery (Reid, 1968). That the same is true in the systemic bed is generally held to be the case but there appear to be no measurements on suitably pressurised vessels at their natural length.

Table 1 lists the relative wall thickness (generally h/R_o) for a number of mammalian vessels. For vessels of the same relative thickness the net wall stress per unit area will be the same at the same distending pressure; differences in this ratio between elastic and muscular vessels may be an indication of a change in wall properties. All the figures in Table 1 were obtained with the vessel under physiological conditions except for those quoted by Reid (1968) in which pulmonary arteries were fixed while distended to systemic pressures. This might contribute to the very marked thinness of their walls, but this figure is in line with the relative distensibility of pulmonary and systemic arteries.

IV. ELASTICITY OF BLOOD VESSELS

A. Static Elasticity

I shall generally discuss only those measurements made on intact vessels at constant *in vivo* length, and not the many studies on strips and rings and on vessels free to lengthen.

All studies show that the length-tension relations of vessel walls are non-linear in the sense that elasticity increases with extension (and this is not a result of the use of incremental strains). The relations between pressure and radius, or pressure and volume (see Patel et al, 1960) often show some linearity over a limited range but over a wider range there is elastic nonlinearity. Indeed, if vessel walls had a constant elasticity "blowouts" would occur and

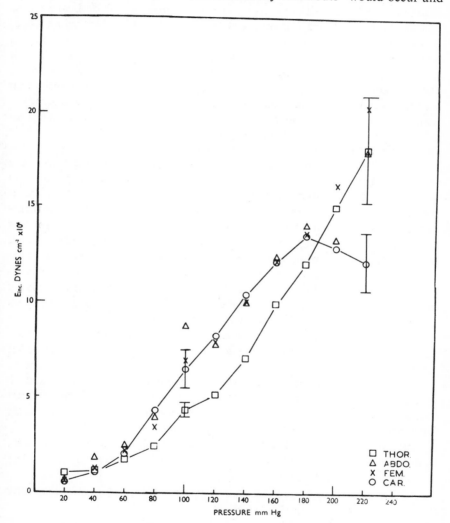

Fig. 2: Static elastic moduli of four different types of dog's artery plotted against pressure. The vertical lines show the standard errors at 100 and 240 mm Hg for the thoracic aorta and the carotid artery. (Reprinted from Bergel, 1961a.)

it is well known that extremely high pressures are needed to rupture a blood vessel: (Hales (1628) burst a dog's carotid artery at a pressure of 5.42 atmospheres.)

Figure 2 shows the average elastic moduli plotted against pressure for dogs' vessels (Bergel, 1961a). These measurements were made *in vitro* (there is evidence that removing an artery does not alter its elastic properties (Lee et al, 1968). Similar values were obtained for human vessels by Learoyd and Taylor (1966) but human femoral arteries are rather stiffer. *In vivo* measurements on arteries cannot be simply compared with these static measurements since they are computed from the changes in pressure and radius with each heart beat and thus are dynamic. However, it has been shown (Bergel, 1961b; Learoyd and Taylor, 1966) that E_{dyn} does not change greatly with frequency above $2Hz$ and it appears that the ratio E_{dyn}/E_{static} is independant of mean stress (see Hardung, 1970). Three groups have reported measurements of arterial elasticity *in vivo* (Gow and Taylor, 1968; Patel et al, 1964; Peterson et al, 1960) using different methods. The results of Gow and Taylor confirm the *in vitro* measurements on dogs' arteries although the details of the frequency dependance of E_{dyn} and $\eta\omega$ could not be verified, presumably because of nonlinearities. In particular, arterial pulse-wave velocities computed from these measurements were in good agreement with actual determinations. The measurements of Patel and his colleagues, who have studied a great number of human and canine vessels, and those of Peterson et al, showed elasticities approximately twice those of the *in vitro* studies. Computed pulse-wave velocities were thus about $\sqrt{2}$ higher than those to be expected (see Eq. 22). This discrepancy has been very puzzling but appears to have been resolved by further work (Patel et al, 1969) in which extreme care was taken in calibration and in the attachment of the radius transducer to the artery. It was found that the isotropic dynamic modulus for the thoracic aorta agreed well with the *in vitro* figures and also with simultaneously measured pulse wave velocity.

Thus it appears that the static elasticities shown in Fig. 2 are representative of canine arteries. It can be seen that E_{inc} varies between about 1×10^6 dynes/cm^2 and 15×10^6. The thoracic aorta is not greatly less elastic than the other vessels, in spite of the fact that the total collagen-elastin rations are so different (Harkness et al, 1957). This serves to emphasise the obvious fact that the total amounts of wall constituents are less relevant than their arrangement on which there is very little information.

Further insight into the behaviour of vessel walls comes from the studies illustrated in Fig. 3a, b. In these experiments the same thoracic aorta was held at a number of different lengths while being inflated. In Fig. 3a, elasticity is plotted against pressure: The short vessel, less stressed longitudinally, (curves 3, 4) was apparently stiffer at all pressures than when it was stretched (1, 5, 6, 7). (Curve 2 was incomplete because of vessel buckling.) This appears

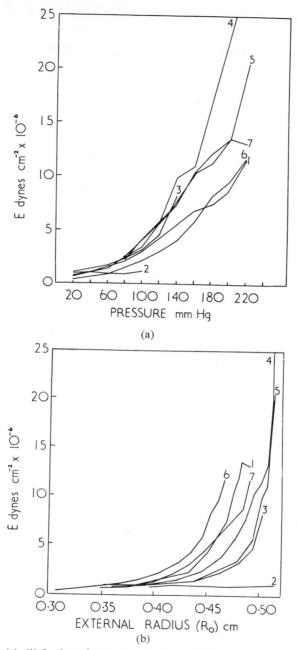

Fig. 3: (*a*), (*b*) In these figures the elastic moduli measured on the same specimen of thoracic aorta are plotted against pressure (*a*), and external radius (*b*). The specimen was held at different lengths during the series of inflations. The curves are identified by numerals; the length (as a percentage of natural length) for each determination was as follows: (1) 100; (2) 66; (3) 75; (4) 90; (5) 100; (6) 110; (7) 100.

very odd at first sight but the discrepancy is resolved when elasticity is plotted against radius (Fig. 3b). It is now clear that stretching the vessel reduced its radius at all pressures and that the true effect of the stretch is revealed as an increase in stiffness at any radius.

It is clear then that the circumferential elasticity of a blood vessel is primarily related to its radius rather than to the pressure. This is in accord with the idea that the deformability of a molecular array depends on the length at which it is held, and this is equally so if we imagine the aortic wall to be made of various extensible materials arranged in parallel. This would not have been apporant if strains had been computed relative to the radius at some reference pressure.

Studies by Hinke and Wilson, (1962) show a similar range of elasticity in arteries as small as 0.5 mm diameter. In Fig. 4 I have compared the elasticities of larger vessels with those computed from Hinke and Wilson's data and from some unpublished studies of one of my dogs' jugular veins. The arteries are rather similar, but the vein shows very low elasticity (0.1 × 10^6 dynes/cm^2) at low pressures and very high elasticity when stretched.

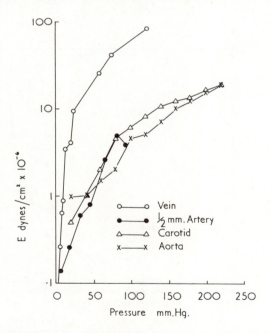

Fig. 4: Static elastic moduli (logarithmic scale) plotted against pressure for four different types of vessel. The values for the aorta, carotid artery and (jugular) vein are from my own measurements on dogs' vessels. The 1/2 mm (radius) artery was from a rat's tail, measured by Hinke and Wilson (1962). (Reproduced by permission of the publishers of "Laboratory Practice").

When pressure-radius curves are plotted for veins (Fig. 5) the slope is very markedly inflected when the muscle is not active.

Yates (1969) has studied venae cavae *in vivo* and also finds that very high moduli are seen at high strains, but these were dynamic moduli derived from wave-speed measurements. The static moduli were considerably lower but reason for this is not clear, the static measurements shown in Fig. 5 are rather

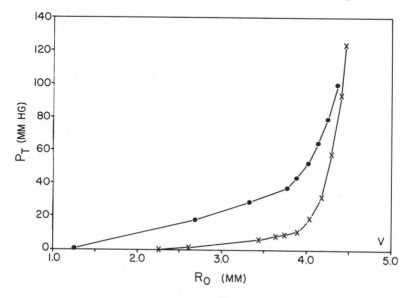

Fig. 5: Pressure-radius relationships of the external jugular vein of a dog *in vivo*. In the first inflation (closed circles) the vessel was in a state of moderate smooth muscle tone induced by dissection. Before the second inflation (crosses) the vessel was irrigated with xylocaine to relax the muscle. Note the increase in radius at low pressures, and that both curves tend towards the same maximum radius. Also note that at 20 mm Hg the slope of the curve for the relaxed vessel is greater than for the constricted state, i.e., the "distensibility" has decreased following relaxation. (Reproduced from Bergel, 1964.)

similar to Yates' dynamic figures. Yates also studied the effect of sympathetic stimulation and adrenaline infusion on the abdominal vena cava. He found decreases in diameter of up to 20% (40% volume decrease) in the face of venous pressure rises; clear evidence of vascular muscle activity. Inflation curves altered in a manner similar to that in Fig. 5. When elasticity was plotted against stress a *decrease* was noted following stimulation. A similar effect of muscular activity in carotid arteries was noted by Dobrin and Rovick (1969) and in veins and arteries by Bergel (1964).

In all these cases when elasticity is plotted against radius (Fig. 6) it is seen that smooth muscle activity increases the elastic modulus at low strains and

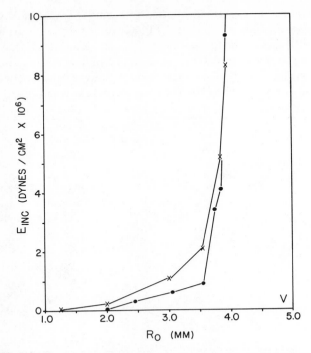

Fig. 6: The same data as were shown in Fig. 5 are here replotted as static elastic modulus against radius. Note that the elasticity of the constricted vein (crosses) is greater than that of the relaxed vein (closed circles) but that the difference is negligible at large radii; thus the effect of smooth muscle activity is a reduction in size and an increase in elasticity at small radii, the greatest increase in modulus being about 1×10^6 dynes/cm². (Reproduced from Bergel, 1964.)

does not appear to alter the maximum possible vessel size. This strongly supports the histological evidence that in veins and muscular arteries the smooth muscle is in parallel with the other wall elements.

There is very little known about pressure-radius relationships of the capillaries and smallest arteries. Lamport and Baez (1962) observed these vessels in the rats mesoappendix and were surprised to find that they acted as if they were essentially indistensible. This property was not abolished by treatment with agents which should have reduced smooth muscle activity. The behaviour of these vessels, (15–30 μm diameter) suggested extreme stiffness of the wall. However, pulmonary capillaries (Glazier et al, 1969) are relatively distensible ($\Delta PR/\Delta R =$ ca 5×10^5 dynes/cm²). The reason for this discrepancy has been suggested by Fung et al (1966) who showed that a distensible tube embedded in a gel-like substance will appear quite indistensible, for the effective capillary wall thickness will be very great. The pulmonary capillaries, on the other hand, are essentially unsupported externally.

The influence of external structures on vessel distensibility must also be considered for the larger vessels. In the case of veins especially, the effect of the surroundings (intramuscular forces, intrathoracic and intra-abdominal pressures) is of extreme importance. It has not been shown that radial expansion of arteries is limited by external constraints but, in virtually all studies of arteries *in vivo*, the actual segment to be studied has to be dissected somewhat. It has even been suggested that this disturbance increases the elasticity of the vessel. Studies on the pressure-diameter relationships of human carotid arteries using ultrasound (Arndt, 1969; Arndt et al, 1968; Kober and Arndt, 1970) show distensibilities very much higher than expected. Thus they quote values for E_p of 0.4×10^6 dynes/cm^2 (cf Greenfield et al, 1964 who give 4.3×10^6 for exposed human carotids and 2×10^6 from my own dynamic measurements in dogs' carotids). There is reason to believe (Patel et al, 1969) that the method used by Greenfield and his colleagues is liable to overestimate elasticity, but Arndt's figures suggest carotid pulse-wave velocities of 4.6 m/s and my own predict 8–10 m/s; while 6–8 m/s has been measured in the dog (McDonald, 1968). Arndt et al, (1968) measured pulse wave velocity in their subject with external transducers and found the velocities they predicted. Furthermore, the ultrasonic method used on human femoral arteries suggested velocities of about 7 m/s (Arndt et al, 1969), the same order as those actually measured. As far as the suggestion that the exposure of a vessel necessarily increases elasticity is concerned, there is no evidence from comparison of elasticities and wave velocities, determined by catheter rather than arterial puncture (McDonald, 1968), that this is so in the dog (Gow and Taylor, 1968).

Arndt's evidence does suggest that the distensibility of human carotid arteries *in vivo* is lower than in the dog. Their predicted and indirectly measured carotid pulse wave velocity is lower than expected and no direct measurements are avilable. Kober and Arndt (1970) produced large changes in carotid diameter by alterations of pressure around the neck without significant change in elasticity, a surprising result which makes it important that more be done to compare predicted and measured pulse wave velocity in vessels more accessible than the human carotid.

However, arteries are externally restrained in the longitudinal direction. That they were tethered was first suggested by reports that the longitudinal motion of arterial walls was very much less than that to be expected from fluid drag (McDonald, 1960). Womersley (1957) described the pressure-flow relations for elastic tubes with external elastic and inertial restraint, and his expressions are widely used. Later studies have shown that the properties of thick-walled viscoelastic tubes, with minimal external restraint, are very much like those of the fully tethered tube described by Womersley, from which it appears that other factors may have the same effect as tethering (Taylor, 1959). Nevertheless, the larger arteries are attached to their surroundings both through the merging of the adventitia with surrounding tissues and by their branches. The influence of tethering in the thoracic aorta has been measured (Patel and Fry, 1966) and shown to involve viscous forces as well as elasticity and inertia. The effect of longitudinal tethering is to make the vessel appear less extensible longitudinally, and this factor contributes 30–40% of the longitudinal modulus *in vivo* (Patel et al, 1969).

Pulmonary arteries have been shown to be very much more distensible than systemic ones (see Patel et al, 1964). It is not certain whether this is explicable solely by the thinner walls of pulmonary arteries. Systemic vessels tend to become more elastic with increasing distance from the heart. This appears to be the case for the pulmonary system of the rabbit (Caro, 1965) and frog (Maloney and Castle, 1969). The pulmonary veins of the rabbit are similar to systemic veins in one respect; the pressure-circumference relationships are also rather sharply divided into a distensible and indistensible region.

B. Dynamic Elasticity

The most direct way to study arterial viscoelasticity is to determine the response to oscillatory stresses. This has been done for human (Learoyd and Taylor, 1966) canine (Bergel, 1961*b*) and bovine (Hardung, 1953, 1970) arteries, though the latter studies were made on vessel rings. There is general agreement that the dynamic modulus (E_{dyn}) is not strongly frequency dependant above 2–4 Hz and that it increases from the static value at quite low frequencies. Hardung (1970) shows that E decreases down to 0.001 Hz. It also appears that the ratio E_{dyn}/E_{static} is higher in the more muscular arteries, though it has not been determined in a very wide range of vessels (Fig. 7). The viscous component, $\eta\omega$, is more complex. In the range 2–20 Hz it is found to increase somewhat, but not linearly with ω so that the derived value for η, the "coefficient of viscosity" falls with ω. In the lower frequency range $\eta\omega$ varies in a complex manner with occasional maxima and minima (Hardung, 1970). An earlier report showing marked resonance effects in a variety of tissues in the range 4–10 Hz (Apter and Marquez, 1968) does not appear to have been confirmed. It seems from Hardung's work that the relative viscosity of the wall ($\eta\omega/E_{dyn}$) decreases with increasing mean stress, for the elastic modulus rises sharply and the viscous modulus is little affected.

It was suggested some years ago that the viscous effects in the arterial wall correlated with the amount of smooth muscle (Bergel, 1961*b*) and the recent evidence is not contrary to this conclusion. It would, however, be helpful to have more figures relating wall viscosity to muscle activity. Wiederhielm (1965) has shown, in elegant studies on the pressure-radius relationships of single arterioles (diameter 100–200 μm), that the time course of distension following a sudden step in pressure is greatly increased when the muscle is active. He reports a forty times increase in the coefficient of wall viscosity.

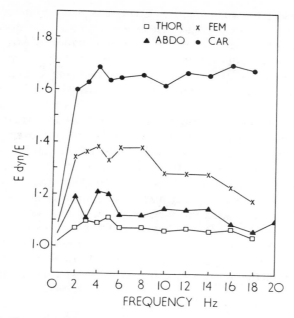

Fig. 7: The ratio of dynamic to static elastic modulus at a mean pressure of 100 mm Hg over the frequency range 0–20 Hz for four different types of dogs' arteries. (Redrawn from Bergel, 1961b.)

The use of sinusoidal stress oscillations lead to predictions on *in vivo* elasticity and on wave velocity and attenuation which have generally been verified (Patel et al, 1969; Gow and Taylor, 1968). Wave attenuation is extremely difficult to measure in the physiological frequency range and the results obtained by a number of workers (McDonald and Gessner, 1967; Maxwell and Anliker, 1968; Booth, 1970) suggest that arterial pressure waves are attenuated by up to 70% per wavelength. The results are quite variable but they suggest that the attenuation is somewhat greater than that calculated from *in vitro* measurements using Taylor's (1959) expressions, which predict 50% transmission per wavelength for the aorta (Bergel, 1961b).

The pattern of arterial viscoelasticity is not that expected from simple spring-dashpot arrays. Similarly, stress relaxation does not follow a simple exponential course, tension decay being a quasi-linear function of log time (Zatzman et al, 1954). The amount of stress relaxation is greater in the muscular umbilical artery than in the carotid, but the time course is the same. Stacy (1957) noted that this pattern of relaxation could be predicted from an array of viscoelastic elements with a wide spectrum of relaxation times, but he felt that a different approach might be more meaningful. Starting from the assumption, generally accepted, that the greater part of the relaxation process was associated with smooth muscle events, Stacy proposed a model in which some element could exist in either a short or a long state. This had been used earlier by Polissar (1952) in an examination of striated muscle mechanics. Such a model is not necessarily incompatible with the sliding-filament model (Huxley, 1969). Stacy assumed that simple reversible kinetic equations could be applied to the short-long conversions, and that the rate constants were functions of the applied tension. This model fitted the stress relaxation data rather well, but the time constants for lengthening were seventy times smaller than for the reverse process, stress recovery. That stress recovery in large arteries is slower than stress relaxation is true in my experience with "relaxed" vessels. It would account for the fact that the first few inflations of an excised vessel result in a gradual increase in vessel size before a stable hysteresis loop is achieved (Bergel, 1961a).

Wiederhielm (1965) used the same model to predict the time course of creep in a contracted arteriole and found this fitted the later part of the curve. His measurements also showed a relatively fast expansion, complete in about twenty seconds, which appeared to be reversible. In Wiederhielm's experiments it was found that the recovery time constants were the shorter for active muscle.

C. Anisotropic arterial elasticity

The simplest demonstration of arterial anisotropicity is the fact that arteries elongate when inflated (Fenn, 1957) although the wall is incompressible. Direct measurements on strips (McDonald, 1960), the determination of radial compressibility (Tucker et al, 1969) and evidence based on wave transmission (Moritz, 1969) support this.

It was shown earlier that the number of independant constants to be evaluated for the fully anisotropic case are impossibly large, but the problem becomes tractable if the material can be shown to be orthotropic and incompressible. An important series of papers (Carew et al, 1968; Patel and Fry, 1969; Patel et al, 1969) have shown how this may be done, and similar ideas have been discussed by Hardung (1964a, b). That the arterial wall is incompressible might be expected from its high water content and was shown by Lawton (1954) and confirmed recently (Carew et al, 1968). Tickner and Sacks (1967) concluded that arterial wall was compressible but this was based on radiographic determinations of wall thickness rather than direct volume measurement. It is possible, though I think it unlikely, that water is extruded from the wall on extension and this would be missed by the method used by Carew et al. This would imply that there was a mobile component of the tissue water which would not bear any part of the load and thus any supposed shifts would not alter the volume of the load-bearing elements.

Patel and Fry (1969) showed that the aorta and carotid arteries developed shearing strains that were very much smaller than the corresponding longitudinal and circumferential strains when inflated to physiological pressures. Thus the vessel is elastically symmetrical or orthotropic and shear strains can be neglected. In this situation the three orthogonal strains, circumferential, longitudinal and radial, are the only important ones and, for an incompressible wall, they are interrelated. In a very elegant series of measurements on *in vivo* aortic elasticity Patel et al (1969) were able to determine the incremental elastic moduli corresponding to these three strains. By making measurements of pressure, radius and the longitudinal forces at different lengths, it was possible to impose incremental longitudinal stresses in the absence of significant stresses in the other directions. From this the longitudinal elastic modulus could be computed, and thence the other two since the material had been shown to be incompressible. They showed that the radial, circumferential and longitudinal moduli, at natural length and pressures of 154 cm H_2O, were 5.5, 7.5 and 10.1×10^6 dynes/cm² respectively. It should be remembered that these moduli will not be constant and need not be in constant ratio at all strains. Indeed, the elongation of a closed vessel shows that E_l is less that E_θ at less than physiological lengths, but the ratio of length changes to circumferential changes is reduced or reversed as the pressure is raised (Fenn, 1957; Tickner and Sacks, 1967). Although the aorta is stiffer longitudinally than circumferentially under normal conditions, there is evidence (Moritz, 1969; Patel, 1970) that the reverse is the case in the dog's carotid. It only remains to determine the dynamic anistropic elasticities and we shall be in a position to predict the gross stresses and strains in arteries *in vivo*. A start in this direction shows that the dynamic radial elasticity of dogs' aortas is rather like that of a parallel spring-dashpot array (Voigt model) in the frequency range 2–40 Hz (Patel et al, 1970) but not at lower frequencies. E_r increases with distance down the aorta and also with increasing mean compressive strain, but the viscous modulus ($\eta\omega_r$) alters in the reverse manner which is not in keeping with measurements of the isotropic viscous modulus (Bergel, 1961b; Learoyd and Taylor, 1966).

D. THE REASON FOR THE SHAPE OF THE DISTENSIBILITY CURVES OF ARTERIES

A paper with this title was published in 1957 by Roach and Burton. They followed Hass (1942) and studied vessel elasticity before and after digestion with formic acid or trypsin to remove elastin and collagen, respectively. They showed that the slopes of the distensibility curves were similar, in their low pressure region, to the trypsin treated specimens, and that the final highly elastic region corresponded to that of the elastin-depleted vessels. This idea has found support from almost every one who has studied vessel elasticity, and something along these lines is immediately suggested by the histology of blood vessels.

Studies of fairly pure elastin in the form of ligamentum nuchae (Krafka,

1939; Reuterwall, 1921) show that it is relatively extensible (200–300% with incremental moduli rising from 1 to 6×10^6 dynes/cm^2). Measurements on single elastic fibres (Carton et al, 1962) showed that they were less extensible, a maximal strain of 1.3 being approached exponentially. Thus elastin itself is not a Hookean material and, when incorporated into a tissue, individual fibres are presumably free to slip and slide across one another allowing high extensions.

Collagen studied as tendon is rather inextensible, with strains up to 0.1–0.3 with elasticity rising up to 10^8–10^9 dynes/cm^1 (Burton, 1954; Reuterwall, 1921). The elasticity of the thoracic aorta at pressures of 50 mm Hg is about 2×10^6 dynes/cm^2 (Fig. 2). If the elastin comprises one-third of the wall material (Harkness et al, 1957), elastin alone could be held responsible for the elasticity of the whole vessel; with increasing strain some other stiffer material must be increasingly involved in tension bearing, yet further strains of 40–60% are possible. Thus the collagen in arterial wall is more extensible than in tendon; we may explain this by analogy with the properties of a net which is more extensible than the threads of which it is made. Even in an arteriole collagen must bear considerable load at low strains (Wiederhielm, 1965). These considerations confirm the ideas put forward by Roach and Burton (1957) but it must be emphasised that there is no sharp inflection on the elasticity curve at which the transition from elastin to collagen may be detected. Observations on elasticity changes in different regions of the stress-strain curves may not be readily explained as alterations in one or another of the fibrous components.

The influence of smooth muscle on the dynamic elasticity has already been mentioned. It is much harder to discover its effects on the static elastic properties of the larger vessels. Much more is known about smooth muscle in general (Bulbring et al, 1970; Lowy and Small, 1970) and vascular smooth muscle in particular (Somlyo and Somlyo, 1968) than ten years ago but there is still considerable ignorance about its mechanical properties. Evidence in the literature suggested (Bergel, 1961a) that static modulus of active muscle was about 1×10^6 dynes/cm^2. More recent figures (Hinke and Wilson, 1962) are of the same order. In a large vessel where elastin is plentiful and the muscle cells are inserted into elastic lamellae, smooth muscle activity will not have a great effect (indeed it has been suggested that the smooth muscle in elastic arteries is responsible for connective tissue formation (Cliff, 1967) so its presence in these vessels does not necessarily presuppose large mechanical effects). When elasticity is plotted against radius for the dog's femoral artery, whose structure is somewhat transitional between elastic and muscular, a small but definite increase in stiffness with muscle activity can be seen (Bergel, 1964). The same is true for a large vein (Fig. 5) but the most noticeable effect is a reduction in size. Gerova and Gero (1969) report that sympathetic activity can alter the diameter of the femoral artery by

−13−+20% and a similar active range has been reported by Jaffe and Rowe (1970).

The influence of muscle activity in smaller vessels is greater, both on account of the "ringmuskeln" arrangement and the relatively greater wall thickness. Van Citters (1966) has shown that small arterioles can constrict to luminal closure. These vessels serve to regulate flow through the bed they supply. The closure does not occur through concentric narrowing of the lumen but by compression and distortion of the intimal tissues lying inside the muscle layers, thus extreme degrees of muscle shortening are not called for.

Another vessel that can actively constrict to closure is the ductus arteriosus which connects the aorta and pulmonary artery in the foetus and which closes shortly after birth. This vessel, with a fairly thick wall ($\gamma = $ ca 0.3) is very muscular. McIntyre (1969) has shown how the active tension in the muscle sums with the passive tension seen in the absence of muscle activity (and which may include forces from muscle cells). At short lengths no active tension may be generated but the slack central muscle cells serve as incompressible material which allows luminal closure while the peripheral muscle is still capable of exerting graded tension.

V. PHYSIOLOGICAL SIGNIFICANCE OF VESSEL PROPERTIES

The most obvious effect of vessel elasticity on the circulation arises from changes in muscle activity in the systemic arteriolar region. The large changes in luminal diameter greatly alter the hydraulic resistance of the vessel. Burton (1954) has proposed that this process is unstable leading to sudden total closure. We do not know enough of the length-tension relations of active smooth muscle to confirm this; noncritical closure can be achieved in the muscular ductus arteriosus (McIntyre, 1969). In addition to the control of hydraulic resistance, smooth muscle activity will affect the contained blood volume of vessels. Because of the relatively great volume of the veins and the marked changes in diameter at low pressures produced by muscle action (see Fig. 5) this affects venous volume significantly. It is clear that veins are innervated but little is known of the reflex control of venous capacitance (Mellander and Johansson, 1968). I do not intend to discuss resistance and capacitance changes in any detail but, rather, to indicate briefly other ways in which vessel elasticity affects the circulation.

A. Pulse-wave velocity

Waves in tubes can be propagated in many modes (Anliker and Maxwell, 1966. The subject is discussed in more detail in this volume by Anliker).

The transmission of radial expansion waves (Young waves) is familiar but axial (Lamb) and torsional waves have also been studied (Moritz, 1969). The impedance of the arterial bed and hence the load on the heart are partly determined by the pulse-wave velocity (PWV). This refers to the velocity of radial waves and is one important parameter that can be measured *in vivo* even without arterial puncture and from which much can be learned.

Radial waves propagate at a velocity given by the Moens-Korteweg equation

(22) $$C^2 = Eh/2R\rho$$

where ρ is the density of blood. The velocity of axial waves is $C_1^2 = E/\rho_w$, where ρ_w is the density of the vessel wall.

This equation requires considerable modification to account for fluid properties, viscoelasticity and other factors (McDonald, 1960; Hardung, 1964c), but it can be seen that PWV is a function of elasticity and relative wall thickness. Since the latter may not always be known we can also express C in terms of the other two elasticity moduli generally used (see Eqs. 18 and 19)

(23) $$C^2 = \kappa/\rho$$

(24) $$C^2 \simeq E_p(1-\gamma)/2\rho \quad \text{(assuming } \sigma = 0.5\text{)}$$

Unfortunately the presence of reflections make it difficult to measure C *in vivo* (McDonald and Taylor, 1959; Gessner and Bergel, 1966; Bergel and Schultz, 1971), but adequate measurements may be made on the velocity at which the foot of the pressure pulse travels (Patel et al, 1969). A more complex method involves the detection of small high frequency waves injected into vessels (Moritz, 1969; Yates, 1969; Anliker et al, 1968). Representative values for PWV in various vessels are shown in Table 2.

In general, the velocity increases with distance from the heart. Taylor (1964) has discussed this phenomenon with respect to the load on the heart. It has been shown (Anliker et al, 1968) that the velocity is increased at increased blood pressures, thus confirming predictions based on elasticity measurements (Bergel, 1961b; Gow and Taylor, 1968). In extreme cases this effect of elastic nonlinearity can lead to marked pressure wave peaking and the development of something akin to a shockwave (Rockwell, 1969) which might occur in disease. It has also been shown that the radial wave is convected with the moving blood (Histand, 1969).

Arterial elasticity may then be inferred from PWV measurements. Thus the increase in PWV with age implies increased elastance. This could be due to higher blood pressure, changes in wall thickness or changes in wall elasticity. It has been known for some years (Bader, 1964) that old arteries are generally larger (and hence likely to be stiffer) and thicker than young ones. However,

TABLE II

Vessel	Species	E_p Dynes cm^{-2} × 10^{-6}	$\Delta\phi$ Radians	C m/s	Reference
Thoracic aorta	Dog	—	—	5.5+	Patel et al, 1969
Thoracic aorta	Dog	1.2	—	7.5	Patel et al, 1964
Ascending aorta	Man	0.8	—	6	Patel et al, 1964
Carotid	Man	6.2	—	17	Patel et al, 1964
Carotid	Man	0.4	—	4.6+	Arndt et al, 1968
Thoracic aorta	Man	0.6–1.0	.12	6–9+	Learoyd & Taylor, 1966++
Abdominal aorta	Man	0.7–1.5	.1	6–8+	Learoyd & Taylor, 1966++
Femoral	Man	2.5–7.0	.15	18–13	Learoyd & Taylor, 1966++
Carotid	Man	2.5–3.0	.1	—	Learoyd & Taylor, 1966++
Pulmonary artery	Dog	—	—	2.75+	Bargainer, 1967
Pulmonary artery	Dog, Man	0.1–0.16	—	2.3–2.9	Patel et al, 1964
Thoracic aorta	Dog	.7	.07	5.5	Bergel, 1961b
Abdominal aorta	Dog	1.6	.07	9.5	Bergel, 1961b
Femoral	Dog	2.0	.11	9.0	Bergel, 1961b
Carotid	Dog	2.0	.11	9.5	Bergel, 1961b
Thoracic aorta	Dog	.7	.08	4.9	Gow & Taylor, 1968
Abdominal aorta	Dog	1.8	.12	8.5	Gow & Taylor, 1968
Femoral	Dog	3.0	.10	10.0	Gow & Taylor, 1968
Thoracic aorta	Dog	—	—	4.5+	McDonald, 1968
Abdominal	Dog	—	—	6–7.5+	McDonald, 1968
Femoral	Dog	—	—	8–10+	McDonald, 1968
Carotid	Dog	—	—	6–8+	McDonald, 1968
Carotid	Dog	—	—	8–11+	Moritz, 1969
Thoracic aorta	Dog	—	—	4–6+	Anliker et al, 1968
Vena cava	Dog	—	—	2–5+	Yates, 1969

Elastic modulus and pulse-wave velocity values collected from the literature. Those values for the PWV marked thus: + were measured, the others have been calculated from dynamic elasticity measurements; ++ measurements for young (< 35) and old (> 35) subjects respectively. $\Delta\phi$ = phase difference between wall stress and strain at 2 Hz. E_p is defined in Eq. 18.

the amount of retraction on excision also alters greatly with age (Hesse, 1926; Learoyd and Taylor, 1966) and this has generally been neglected in studies on excised specimens, resulting in the introduction of unknown age-related length changes. Learoyd and Taylor took account of this and showed that wall thickness, vessel diameter, static and dynamic elasticity altered with age in a complex manner. The net result is reduced distensibility of old arteries. The increase in predicted PWV with distance from the heart was still present in the old, but it was less marked than in the young. When considered in terms of elasticity at any diameter, the old vessels were significantly weaker than the young ones, a finding in accord with the histological evidence which shows that although the total amount of connective tissue is increased individual fibres are broken and appear degenerated.

The PWV is also increased in hypertension, apparently to a greater degree than can be accounted for by elastic nonlinearity (Feigl et al, 1963). In hypertensive dogs there is some decrease in collagen-elastin ratios and a definite increase in water content of the wall (Fischer and Llaurado, 1967; Fiegl et al, 1963). Similarly circumferential aortic strips from hypertensive rabbits are thicker and longer than normal and show significantly higher elasticities, these changes being apparent within 3–15 days after induction of renal hypertension (Aars, 1968*b*). These were static tests on excised aortic strips but a later *in vivo* study (Aars, 1969) gave rather different results. The aortic diameter was increased and the elasticity/diameter curve shifted to the right by about 1 mm in nine (the pressures in this study were greater than in the earlier one). These changes suggest the collagen in the wall has "given" so that the limiting diameter is reached at a higher pressure. This is more like the changes seen in aging and does not agree with the results of Feigl et al (1963); however the rabbit studies were made with implanted ultrasonic diameter gauges and the distensibility of the normal aortas was much lower than that found in other species. These results might reflect the effects of the implantation.

B. Arterial input impedance and the load on the heart

Pressure waves generated by the heart travel through the arteries at a steadily increasing velocity. There is partial reflection from the region of the smaller peripheral vessels, from branch points and other points where mechanical or dimensional changes occur.

The hydraulic input impedance of the arterial system approximates that of a branching elastic line whose stiffness increases with distance (elastic taper) and which terminates at different lengths in a relatively high hydraulic resistance. Since the arterial PWV (Table 2) is of order 5–15 m/s and normal heart rates are around 1–2 per second, the system is much less than a wavelength long and thus reflecting conditions at the termination will influence the impedance. It is an advantage, from considerations of cardiac work, that the impedance at the origin be low and that the peripheral mismatch between small arteries and arterioles be low also (Taylor, 1964). This is achieved in the systemic system by the gradual increase in arterial stiffness and PWV along the line (McDonald, 1968). The load on the heart can then be divided, if linearity is assumed (see Milnor et al, 1966) into a steady resistive load (peripheral resistance) and a pulsatile load determined by the complex frequency-dependant impedance.

The impedance of a single tube with terminal reflections is markedly frequency dependant. In the systemic bed the distance to the innumerable terminations is not constant and the elastic taper reduces the effective reflection coefficient (Taylor, 1966). The effect of this is to make the impedance less variable in the face of heart rate changes (O'Rourke and Taylor, 1967), an effect complicated by the fact that the bed may be considered composed of a long portion to the lower body and a shorter one to the head and neck (O'Rourke, 1967). At low frequencies the impedance rises rather sharply and at slow heart rates (<100 per minute) the pulsatile component of cardiac work increases from about 10% of total to 30%. The relative amount of pulsatile work is also increased by vasodilatation and by an increase in stiffness of the aorta. Thus, although the steady power output of the heart is increased in age, hypertension and exercise, the pulsatile component probably increases even more, so that it becomes a very significant fraction of the total.

The pulmonary system is much more uniform than the systemic in that all parts of the lung have the same function and little useful function would be served (in the normal condition at least) by shunting blood from one part to another, a function served by the systemic arterioles which are rather poorly developed in the pulmonary bed. In order to avoid fluid transudation into the alveoli, a low capillary pressure is desirable. For these reasons the pulmonary artery pressure is much lower than the systemic. One result is that the behaviour of the pulmonary circulation is greatly affected by gravitational forces. Although there is some evidence of elastic taper in the pulmonary arteries (Caro, 1965) this is much less marked than in the systemic system. Indeed the gross distensibility of the pulmonary capillaries is about the same as the main pulmonary artery (Glazier et al, 1969). Although the distance from artery to capillary is shorter in the pulmonary bed, the PWV is also less than in the systemic bed. In fact, in terms of wavelengths the two beds are about the same length (Bergel and Milnor, 1965; Bargainer, 1967).

C. Sense Organs in Blood Vessel Walls

It is well known that pulmonary and systemic arteries, the cardiac chambers and possibly some veins contain nerve endings in their walls which respond to mechanical forces (see Heymans and Neil, 1958). It is clear that the output of these receptors will be greatly influenced by the properties of the substance in which they lie. The most important concentration of receptors lies in the carotid sinus, about which much is known; another group lies in the wall of the aortic arch, these are the arterial baroreceptors.

The main features of baroreceptor function may be summarised briefly. They respond to pressures above a threshold (generally about 60–80 mm Hg), the response is essentially linear but tails off at high pressures. They respond to a steady pressure and more vigorously to an oscillating pressure; following a step change of pressure the output rises sharply and accomodates to a lower level. The output of at least some receptors is increased when the local arterial smooth muscle is stimulated chemically or nervously (Sampson and Mills, 1970).

Almost all of these properties could result, at least in part, from the nonlinear viscoelasticity of the arterial wall, but there is little direct evidence to show to what extent this is so.

The carotid sinus wall is relatively rich in elastin and poor in muscle; the relative wall thickness is similar to the adjacent artery but the medial layer is very much thinner in the sinus (Rees and Jepson, 1970). The receptor terminals are concentrated in the medial-adventitial boundary and in the deeper adventitia. Since the carotid sinus appears (Landgren, 1952) rather distensible it is possible that the thick adventitia is relatively extensible. It is hard to see the functional significance of the atypical structure of this baroreceptor region. Although the radial compressive forces would be high at the junction if the thin media were more extensible than the thick adventitia, this mechanism has been ruled out by experiments in which a sinus encased in rigid material did not respond to luminal pressure changes (Heymans and Neil, 1958). It is also possible, since the adventitia of the sinus varies in thickness around the circumference (Ress and Jepson, 1970), that its distension would not be axisymmetric; this could expose nerve endings at the adventitial-medial border to considerable shearing stress.

It is known that the baroreceptors "reset" in hypertension (Page and McCubbin, 1965) that is the output/pressure curve is shifted along the pressure axis. This change has been studied in the rabbit aortic baroreceptor region by Aars (1968*a*, *b*) who has related the reduced firing of the baroreceptors to reduced aortic distensibility. Another method of altering vessel characteristics is the production of post-stenotic dilatation. (This fascinating phenomenon in which the vessel below a constriction dilates has been studied by Roach (1963). It appears that the change is induced by exposure to vibrations generated by the stenosis.) Aars (1968*c*) produced post-stenotic dilatation in the aortic baroreceptor area of rabbits. This produced dilation to diameters which would normally be accompanied by intense nerve discharge, but the pressure-activity relations were unaltered. The enlarged aorta showed normal elasticity when strain was related to dimensions at a standard load. Thus the dilated aorta showed some increase in incremental elasticity but stretched to greater diameters. If the upper limit to arterial size is set by the collagen then there must have been some alteration of the aortic collagen.

REFERENCES

Aars, H. 1968*a*. Aortic baroreceptor activity in normal and hypertensive rabbits. Acta Physiol. Scand. 72: 298–309.

Aars, H. 1968*b*). Static load-length characteristics of aortic strips from hypertensive rabbits. Acta Physiol. Scand. 73: 101–10.

Aars, H. 1968c. Aortic baroreceptor activity during permanent distension of the receptor area. Acta Physiol. Scand. 74: 183–94.

Aars, H. 1969. Relationship between blood pressure and diameter of ascending aorta in normal and hypertensive rabbits. Acta Physiol. Scand. 75: 397–405.

Anliker, M., and Maxwell, J. A. 1966. The dispersion of waves in blood vessels, *In* Y. C. Fung, (ed.) Biomechanics. ASME, New York. pp. 47–67.

Anliker, M., Histand, M. B., and Ogden, E. 1968. Dispersion and attenuation of small artificial pressure waves in the canine aorta. Circulation Res. 23: 539–51.

Apter, J. T., and Marquez, E. 1968. Correlation of viscoelastic properties of large arteries with microscopic structure. Circulation Res. 22: 393–404.

Arndt, J. O. 1969. Uber die mechanik der intakten A. Carotis communis des Menschen unter verschiedenen Kreislaufbedingungen. Arch. Kreisl. -Forsch. 59: 153–69.

Arndt, J. O., Klauske, J., and Mersch, F. 1968. The diameter of the intact carotid artery in man and its change with pulse pressure. Pflüger's Arch. 301: 230–40.

Arnt, J. O., Kober, G., and Mersch, F. 1969. Die Elastizitätkoeffizienten menschlicher Arterien *in vivo*, abgeleitet auf Grund unblutiger Messung des Arteriendurchmessers mit dem Ultraschall-Echo-Verfahren. Verh. Dtsch. Ges. Kreisl.-Forsch. 35: 321–26.

Bader, H. 1964. The anatomy and physiology of the vascular wall. *In* Handbook of Physiology, Sect. 2, Vol. II, pp. 865–89. American Physiological Society, Washington.

Bargainer, J. D. 1967. Pulse wave velocity in the main pulmonary artery of the dog. Circulation Res. 20: 630–37.

Bellhouse, B. J., and Bellhouse, F. H. 1968. Mechanism of closure of the aortic valve. Nature, 217: 86–87.

Bergel, D. H. 1960. The visco elastic properties of the arterial wall. Ph. D. Thesis, University of London.

Bergel, D. H. 1961a. The static elastic properties of the arterial wall. J. Physiol. 156: 445–57.

Bergel, D. H. 1961b. The dynamic elastic properties of the arterial wall. J. Physiol. 156: 458–69.

Bergel, D. H. 1964. Arterial viscoelasticity, *In* Attinger, E. O. (ed.) Pulsatile blood flow. McGraw-Hill Book Company, New York. pp. 275–89.

Bergel, D. H., and Milnor, W. R. 1965. Pulmonary vascular impedance in the dog. Circulation Res. 16: 401–15.

Bergel, D. H., and Schultz, D. L. 1971. Arterial elasticity and fluid dynamics. Progress in Biophys. and Mol. Biol. 22: 1–36.

Benninghof, A. 1930. Blutgefässe und Herz. *In* Handbuch der Mikroscopischen Anatomie des Menschen, Band VI. Springer, Berlin.

Biot, M. A. 1965. Mechanics of incremental deformations. John Wiley & Sons, Inc., New York.

Bloom W., and Fawcett, D. W. 1962. Textbook of Histology. W. B. Saunders Company, Philadelphia.

Booth, F. M. V. 1970. Arterial wave propagation *in vivo*. Fluid dynamics of blood circulation and respiratory flows. AGARD (Naples) Proc.

Bulbring, E., Brading, A., Jones, A., and Tomita, T. (eds.) 1970. Smooth muscle. Edward Arnold (Publishers) Ltd., London.

Burton, A. C. 1954. Relation of structure to function of the tissues of the walls of blood vessels. Physiol. Rev. 34: 619–42.

Carew, T. E., Vaishnav, R. N., and Patel, D. J. 1968. Compressibility of the arterial wall. Circulation Res. 22: 61–68.

Caro, C. G. 1965. Extensibility of blood vessels in isolated rabbit lungs. J. Physiol. 178: 193–210.

Carton, T. W., Dainauskas, J., and Clark, J. W. 1962. Elastic properties of single elastic fibres. J. Appl. Physiol. 17: 547–51.

Cliff, W. J. 1967. The aortic tunica media in growing rats studied with the electron microscope. Lab. Invest. 17: 599–615.

Dobrin, P. B., and Rovick, A. A. 1969. Influence of vascular smooth muscle on contractile mechanics and elasticity of arteries. Amer. J. Physiol. 217: 1644–651.

Fenn, W. O. 1957. Changes in length of blood vessels on inflation. *In* J. W. Remington, (ed.). Tissue elasticity. American Physiological Society, Washington. pp. 154–60.
Feigl, E. O., Peterson, L. H., and Jones, A. E. 1963. Mechanical and chemical properties of arteries in experimental hypertension. J. Clin. Invest. 42: 1640–347.
Fischer, G. M., and Llaurado, J. G. 1966. Collagen and elastin content in canine arteries selected from functionally different vascular beds. Circulation Res. 19: 394–99.
Fischer, G. M., and Llaurado, J. G. 1967. Connective tissue composition of canine arteries. Effects of renal hypertension. Arch. Path. 84: 95–98.
Franklin, K. J. 1937. A monograph on vieins. C. C. Thomas, Publisher, Springfield, Ill.
Fung, Y. C. 1967. Elasticity of tissues in simple elongation. Amer. J. Physiol. 213: 1532–544.
Fung, Y. C. 1968. Biomechanics. Applied Mech. Rev. 21: 1–20.
Fing, Y. C., Zweifach, B. W., and Intaglietta, M. 1966. Elastic environment of the capillary bed. Circulation Res. 19: 441–61.
Gerova, M., and Gero, J. 1969. Range of the sympathetic control of the dogs femoral artery. Circulation Res. 24: 349–59.
Gessner, U., and Bergel, D. H. 1966. Methods of determining the distensibility of blood vessels. I.E.E.E. Trans. BME. 3: 2–10.
Glazier, J. B., Hughes, J. M. B., Maloney, J. E., and West, J. B. 1969. Measurements of capillary dimensions and blood volume in rapidly frozen lungs. J. Appl. Physiol. 26: 65–76.
Goto, M., and Kimoto, Y. 1966. Hysteresis and stress relaxation of the blood vessels studied by a universal tensile testing instrument. Jap. J. Physiol. 15: 169–84.
Gow, B. S., and Taylor, M. G. 1968. Measurement of viscoelastic properties of arteries in the living dog. Circulation Res. 23: 111–22.
Greenfield, J. C., Tindall, G. T., Dillon, M. L. and Mahaley, M. S. 1964. Mechanics of the human common carotid artery in vivo. Circulation Res. 15: 240–46.
Hales, S. 1727. Vegetable Staticks. Reprinted, 1961. Scientific Book Guild, London.
Hales, S. 1733. Statical essays, containing haemastaticks. Reprinted, 1964. Hafner Publishing Co., Inc., New York.
Hardung, V. 1952. Über eine Methode zur Messung der dynamischen Elastiizatät und Viscosität kautschukähnlicher Körper insbesondere von Blutgefässen und andere elastischen Gewebeteilen. Helv. Physiol. Acta 10: 482–98.
Hardung, V. 1953. Vergleichende Messungen der dynamischen Elastizität und Viskosität von Blutgefässen, Kautschuk und synthetischen Elastomeren. Helv. Physiol. Acta 11: 194–211.
Hardung, V. 1964*a*. Die Bedeutung der Anisotropie und Inhomogenität bei der Bestimmung der Elastizatät der Blutgefässe, I. Angiologica 1: 141–53.
Hardung, V. 1964*b*. Die Bedeutung der Anisotropie und Inhomogenität bei der Bestimmung der Elastizität der Blutgefässe, II. Angiologica 1: 185–96.
Hardung, V. 1964*c*. Propagation of pulse waves in visco elastic tubings. *In* Handbook of Physiology. Section 2, Vol. II, American Physiological Society, Wahington. pp. 107–35.
Hardung, V. 1970. Dynamische Elastizität unde innere Reibung muskularer Blutgefässe bei verschiedner durch Dehnung und tonische Kontraktion hervorgeruferer Wandspannung. Atch Kreisl-Forsch. 61: 83–100.
Harkness, M. L. R., Harkness, R. D., and McDonald, D. A. 1957. The collagen and elastin content of the arterial wall in the dog. Proc. Roy. Soc. B. 146: 541–51.
Harvey, W. 1628. Exercitatio anatomica de motu cordis et sanguinis in animalibus. [Trans. Franklin, K. J., 1957], Wm. Blackwell & Sons, Ltd., Oxford.
Hass, G. M. 1942 Elastic tissue 2. A study of the elasticity and tensile strength of elastic tissue isolated from the human aorta. Arch. Path. 34: 971–81.
Hesse, M. 1926. Uber die pathologische Veranderungen der Arterien der obseren Extremitat. Virchow's Arch. 261: 225–52.
Heymans, C., and Neil, E. 1958. Reflexogenic areas of the cardiovascular system. Churchill, London.
Hinke, J. A. M., and Wilson, M. L. 1962. A study of elastic properties of a 550μ artery *in vitro*. Amer. J. Physiol. 203: 1153–160.
Histand, M. B. 1969. An experimental study of the transmission characteristics of pressure waves in the aorta. *In* SUDAAR report 369, Stanford University, Calif.

Huxley, H. E. 1969. The mechanism of muscular contraction. Science, 164: 1356–366.
Jaffe, M. D., and Rowe, P. W. 1970, Mechanism of arterial dilatation following occlusion of femoral arteries of dogs. Amer. J. Physiol. 218: 1156–160.
Karnovsky, M. J. 1968. The ultrastructural basis of transcapillary exchanges. *In* Biological Interfaces: Flows and Exchanges. Little, Brown and Company, New York. pp. 64–93.
King, A. L. 1957. Some studies in tissue elasticity. *In* J. W. Remington (ed.) Tissue elasticity. American Physiological Society, Washington. 00. 123–29.
Kober, G., and Arndt, J. O. 1970. Die Druck-Durchmesser-Beziehung der A. Carotis communis des wachen Menschen. Pflüger's Arch. 314:27–39.
Krafka, J. 1939. Comparative studies of the histophysics of the aorta. Amer. J. Physiol. 125: 1–14.
Lamport, H., and Baez, S. 1962. Physical properties of small arterial vessels. Physiol. Rev. 42: Suppl. 5: 328–45.
Landgren, S. 1952. The baroreceptor activity in the carotid sinus nerve and the distensibility of the sinus wall. Acta Physiol. Scand. 26: 35–56.
Lawton, R. W. 1954. The thermoelastic behavior of isolated aortic strips of the dog. Circulation Res. 3: 403–408.
Learoyd, B. M., and Taylor, M. G. 1966. Alterations with age in the viscoelastic properties of human aortic walls. Circulation Res. 18: 278–92.
Lee, J. S., Frasher, W. G., and Fung, Y. C. 1968. Comparison of elasticity of an artery *in vivo* and in excision. J. Appl. Physiol. 25: 799–801.
Love, A. E. H. 1927. A treatise on mathematical elasticity. Cambridge University Press, London.
Lowy, J., and Small, J. V. 1970. The organisation of myosin and actin in vertebrate smooth muscle. Nature 227: 46–51.
McDonald, D. A. 1960. Blood flow in arteries. Edward Arnold (Publishers) Ltd., London.
McDonald, D. A. 1968. Regional pulse-wave velocity in the arterial tree. J. Appl. Physiol. 24: 73–78.
McDonald, D. A. and Gessner, U. 1967. Wave attenuation in viscoelastic arteries. *In* A. L. Copley, (ed.) Hemorheology. Pergamon, London. pp. 113–25.
McDonald, D. A., and Taylor, M. G. 1959. The hydrodynamics of the arterial circulation. *In* J. A. V. Butler and B. Katz, (eds.) Butler Progress in Biophysics 9. Pergamon, London. pp. 105–73.
McIntyre, T. W. 1969. An analysis of critical closure in the isolated ductus arteriosus. Biophys. J. 9: 685–99.
MacWilliam J. A., and MacKie, A. H. 1908. Observations on arteries, normal and pathological. Brit. Med. J. 1908-2: 1477–481.
Maloney, J. E., and Castle, B. L. 1969. Pressure-diameter relations of capillaries and small blood vessels in frog lung. Resp. Physiol. 7: 150–62.
Maxwell, J. A., and Anliker, M. 1968. The dissipation and dispersion of small waves in arteries and veins with viscoelastic wall properties. Biophys. J. 8: 920–50.
Mellander, S., and Johansson, B. 1968. Control of resistance, exchange, and capacitance functions in the peripheral circulation. Pharm. Rev. 20: 117–96.
Milnor, W. R., Bergel, D. H., and Bargainer, J. D. 1966. Hydraulic power associated with pulmonary blood flow and its relation to heart rate. Circulation Res. 19: 467–80.
Moritz, W. E. 1969. Transmission characteristics of distension, torsion and axial waves in arteries. *In* SUDAAR report 373, Stanford University, Calif.
Opatowski, I. 1967. Elastic deformations of arteries. J. Appl. Physiol. 23: 772–78.
O'Rourke, M. F. 1967. Steady and pulsatile energy losses in the systemic circulation under normal conditions and in simulated arterial disease. Cardiovascular Res. 1: 312–26.
O'Rourke, M. F., and Taylor, M. G. 1967. Input impedance of the systemic circulation. Circulation Res. 20: 365–80.
Page, I. H., and McCubbin, J. W. 1965. The physiology of arterial hypertension, *In* Handbook of physiology Vol. 2 Section III. American Physiological Society, Washington. pp. 2163–208.
Patel, D. J., and Fry D. L. 1966. Longitudinal tethering of arteries in dogs. Circulation Res. 19: 1011–21.

Patel, D. J., and Fry, D. L. 1969. The elastic symmetry of arterial segments in dogs. Circulation Res. 24: 1–8.
Patel, D. J., Greenfield, J. C., and Fry, D. L. 1964. In vivo pressure-length-radius relationships of certain vessels in man and dog. In E. O. Attinger (ed.) Pulsatile blood flow. McGraw-Hill Book Company, New York. pp. 293–302.
Patel, D. J., Janicki, J. S., and Carew, T. E. 1969. Static anisotropic elastic properties of the aorta in living dogs. Circulation Res. 25: 765–79.
Patel, D. J., Schilder, D. P., and Mallos, A. J. 1960. Mechanical properties and dimensions of the major pulmonary arteries. J. Appl. Physiol. 15: 92–96.
Patel, D. J., Tucker, W. K., and Janicki, J. S. 1970. Dynamic elastic properties of the aorta in the radial direction. J. Appl. Physiol. 28: 578–82.
Pease, D. C., and Molinari, S. 1960. Electron microscopy of muscular arteries; pial vessels of the cat and monkey. J. Ultrastructure Res. 3: 447–68.
Pease, D. C., and Paule, W. J. 1960. Electron microscopy of elastic arteries; the thoracic aorta of the rat. J. Ultrastructure Res. 3: 469–83.
Peterson, L. H., Jensen, R. E., and Parnell, J. 1960. Mechanical properties of arteries *in vivo*. Circulation Res. 8: 622–39.
Polissar, M. J. 1952. Physical chemistry of contractile process in muscle. I: A physiochemical model of contractile mechanisms. Amer. J. Physiol. 168: 766–81.
Rees, P. M., and Jepson, P. 1970. Measurement of arterial geometry and wall composition in the carotid sinus baroreceptor area. Circulation Res. 26: 461–67.
Reid, L. 1968. Structural and functional reappraisal of the pulmonary artery system, *In* Scientific basis of medicine. Annual Reviews, 1968. pp. 289–307.
Reuterwall, O. P. 1921. Über die Elastizität der Gefässwande und die Methoden ihrer näheren Prüfung. Acta Med. Scand. Suppl. 2: 1–175.
Roach, M. R. 1963. An experimental study of the production and time course of poststenotic dilatation in the femoral and carotid arteries of adult dogs. Circulation Res. 13: 537–51.
Roach, M. R., and Burton, A. C. 1957. The reason for the shape of the distensibility curves of arteries. Canad. J. Biochem. Physiol. 35: 681–90.
Rockwell, R. L. 1969. Nonlinear analysis of pressure and shock waves in blood vessels. *In* SUDAAR report 394, Stanford University, Calif.
Roy, C. S. 1880. The elastic properties of the arterial wall. J. Physiol. 3: 125–62.
Sampson, S. R., and Mills, E. 1970. Effects of sympathetic stimulation on discharges of carotid sinus baroreceptors. Amer. J. Physiol. 218: 1650–653.
Scott Blair, G. W. 1969. Elementary Rheology. Academic Press, London.
Somlyo, A. P., and Somlyo, A. V. 1968. Vascular smooth muscle 1. Normal structure, pathology, biochemistry and biophysics. Pharm. Rev. 20: 197–272.
Stacy, R. W. 1957. Reaction rate kinetics and some tissue mechanical properties, *In* R. W. Remington, (ed.) Tissue Elasticity. American Physiological Society. Washington. pp. 131–37.
Svejcar, J., Prerovsky, I., Linhart, J., and Krumi, J. 1962. Content of collagen, elastin and water in walls of the internal saphenous vein in man. Circulation Res. 11: 296–300.
Taylor, M. G. 1959. An experimental determination of the propagation of fluid oscillations in a tube with a visco elastic wall; together with an analysis of the characteristics required in an electrical analogue. Phys. Med. Biol. 4: 63–82.
Taylor, M. G. 1964. Wave travel in arteries and the design of the cardiovascular system. *In* E. O. Attinger, (ed.) Pulsatile blood flow. McGraw-Hill Book Company, New York. pp. 343–68.
Taylor, M. G. 1966. The input impedance of an assembly of randomly branching elastic tubes. Biophys. J. 6: 29–51.
Tickner, E. G., and Sacks, A. H. 1967. A theory for the static elastic behavior of blood vessels. Biorheology 4: 151–68.
Treloar, L. R. G. 1958. The physics of rubber elasticity. Oxford University Press, Inc., New York.
Tucker, W. K., Janicki, J. S., Plowman, F., and Patel, D. J. 1969. A device to test mechanical properties of tissues and transducers. J. Appl. Physiol. 26: 656–58.

Van Citters, R. L. 1966. Occlusion of lumina in small arterioles during vasoconstriction. Circulation Res. 18: 199–204.

Wiederhielm, C. A. 1965. Distensibility characteristics of small blood vessels. Fed. Proc. 24: 1075–84.

Wolinsky, H. and Glagov, S. 1964. Structural basis for the the static mechanical properties of the aortic media. Circulation Res. 14: 400–413.

Wolinsky, H., and Glagov, S. 1967. A lamellar unit of aortic medial structure and function in mammals. Circulation Res. 20: 99-111.

Womersley, J. R. 1957. An elastic tube theory of pulse transmission and oscillatory flow in mammalian arteries. Wright Air Development Center Technical Report WADC-TR-56-614.

Yates, W. G. 1969. Experimental studies of the variations in the mechanical properties of the canine abdominal vena cava. *In* SUDAAR report 393, Stanford University, Calif.

Zatzman, M., Stacy, R. W., Randall, J., and Eberstein, A. 1954. The time course of stress relaxation in isolated arterial segments. Amer. J. Physiol. 177: 299–302.

6

Properties of Tendon and Skin

JOHN D. C. CRISP

I. INTRODUCTION

A. Tendon as Structure

The primitive function of a tendon is to transmit force. The tendon is, at least, well-suited to the transmission of tensile force, although inept in responding to compressive, shear and bending effects. But this function is better illumined by subsidiary statements about the morphology of the tendon-muscle system, about the kinematic capability it makes possible, and about the tendon's kinetic influence on the control of bodily movements. It is usually implicit, for example, in studies of the structural efficacy of the tendon, that its mere existence stems from a natural optimization of sizing, shape and siting of its associated muscle belly and adjacent like structures. There appears to be a long-standing acceptance of the notion that the geometrical details of size and of dimension of a tendon, bear a direct relationship to the effort required of that tendon; and, moreover, that this congruence can be traced to the appropriate muscle, too. Attempts to discover this conformity and to give meaning, in rigorous mechanical terms, to the word effort make it quite clear, if that is necessary, that the tendon cannot be studied as a structural component, in isolation; it cannot be considered apart from its muscular agent.

This symbiosis applies also to the kinematic role of the tendon for nature thereby allows an effective spatial kinematics; the tendon (in association with contiguous retinacula or "pulleys" and appropriately located synovial sheaths) makes possible a precise insertion of force and, therefore, of movement to the mobile part. And the complexity and range of movements of skeletally based mechanisms (using the word in its engineering sense) rely on these kinematic arrangements. Yet the tendon, as such, is ignored in studies of joints, of articulation and of mobility except as an essentially incompressible, perfectly frictionless, absolutely inextensible, infinitely flexible and quite unbreakable tie between an origin and an insertion. In fact, in certain measure the attributes of tendon and its performance may well be so superior as to discourage serious study, as Elliott (1965) has suggested; it is true that tendon, as a structural component, is much ignored. Elliott also notes a general acceptance of a more active role for the tendon in respect of dynamic loads because it is, after all, not inextensible but prone in its physiological state to strains of about 3%; and *in extenso* to strains well in excess of 10% without fracture. It is then ascribed shock absorbing properties or at least a shock attenuation capability. But to attribute to the tendon, discriminately, an action as a dynamic amplifier, or as an energy dissipator seems implausible in view of the relative mechanical impedances of passive muscle and the embedding connective tissues and ligaments. Its dynamic compliance in relation to active muscle performance needs study. There is, in any event, virtually no recent work on dynamic tendonous influences.

B. Skin as structure

Whereas, in mechanical terms, the tendon is a cable structure, skin, on the other hand, is a structural cladding, a flexible and extensible casing which effects its containing function of resisting lateral pressures by sustaining membranous stresses. These are commonly tensile and of a shearing type but are frequently, as at points of articulation, compressive so that "buckling" or wrinkling and folding are common and normal. The whole deformation environment of skin is very much more complex, a circumstance which is reflected in its more complex microarchitecture. Indeed, its structure is visibly arranged in a preferential manner to facilitate certain large movements, stretches and foldings so that anisotropy of its elastomechanical properties is a salient feature. This anisotropy persists totally over the body for functional reasons that remain obscure but that continue to exercise the minds of investigators of Langer's lines, and it cannot be ignored in restorative surgery. But as an integumental covering, or interface between a physiological and physical environment—Montagna (1962) writes of the epidermis as a protective barrier—the mechanical nature and behavior of skin seems quite unrelated to that of the bodily tendons. The coupling of an interest in tendons with one in skin does not derive from macroscopic functional considerations, however; but is encouraged by both the likenesses and the dissimilarities of their material composition and skin's microarchitectural configuration.

C. Tendon and Skin as Collagen

Tendon excites attention because it is virtually synonymous with collagen, a prime constituent protein of all connective tissue, including skin. The tendons of man are 75% collagen by dry weight; in cattle the figure rises to 85% while in the female rat it drops to 51% (Elliott, 1965). Moreover, the collagenous structure of tendon is axially fibrous, being almost fully orientated in alignment with the tensile stresses that the tendon suffers. Skin, in relaxed state, is not so ordered but becomes so under imposed stress (Craik and McNeil, 1966). Its collagen content is of similar magnitude to that of tendon, varying between about 60 and 75% depending on sex, age and site (Ridge and Wright, 1966b). The relationship between tendon and skin is thus a collagenous one and either tissue serves as a vehicle for the study of collagen as a material; the differing stress-sensitivities of the collagenous microstructure afford opportunities of interesting comparative macrostructural studies. Skin is attractive as a test tissue simply because, of all connective tissue, it is readily accessible. The literature is rich with reports of the effects on collagen of growth, of ageing, of chemical degradation, of thermal shrinkage and denaturation, of depolymerization mechanisms and so on. For these purposes tendonous material has been intensively used but the mechanical properties of the tendon appear then only as of secondary interest, as an indicator of the way some physiological event or property changes. Little attention is given here to these matters.

D. Tensility and a Continuum Mechanics

Whether or not interest in the past has lain in the macrostructural performance or in the biochemical significance of each of these elements, tendon and skin, one thing emerges overwhelmingly—the predilection to the use of the tensile test. The "simple elongation test" is clearly well-suited to describing the usual mechanical properties like stiffness, elasticity and strength either in absolute terms or as relative, comparative monitoring parameters, or tracers, of some physiological, clinical or diagnostic condition. The extensibility of skin, for example, is obviously not to be ignored in designing and executing surgical restorations (Gibson, 1964); but, equally, it can serve as an indicator of venous disorders (Molen, 1966), or of the state of the subcutaneous vasculatory (Gibson and Kennedi, 1967). Factors other than tensile states of deformation are, of course, widely used in testing and these are reviewed below. But they are seldom directed to the explicit determination of meaningful mechanical constants. There is, in fact, no satisfactory mechanical constitutive law for any of these materials. To focus attention on ways in which the behavior of skin and tendon might usefully be clarified is to do so for all soft tissues. This broad view is taken here; and to avoid complete superficiality as well as duplication of material presented elsewhere (Fung, Chapter 7), special attention is given only to the elastomechanics of these tissues, leaving aside the details of time dependent and inelastic effects.

The view adopted, then, in this Chapter is that a realistic determination of the constitutive law of tendon, skin and other connective tissue cannot indefinitely be deferred. The formulation of the continuum mechanics necessary to allow this, and its relation to statistical network descriptions of mechanical response is outlined. Experience in dealing with the rubbery polymers affords the sole basis of such a discussion. The complementary nature of the role of the tensile test is elaborated, in particular, in contrast with its present preeminent status. Experimental uncertainties about the performance of tensile tests, too, are emphasised. The importance of studying a minimum range of deformation states is highlighted. The need for such a comprehensive formulation has already been argued (Crisp, 1968) and its continuum foundation already outlined (Saunders, 1964). But a significant inhibition emerges which exemplifies the claim of Zweifach, Chapter 2, that the successful application of mechanics to problems that are essentially or, at least, in the ultimate, physiological, requires the development of new axioms. It requires, certainly, the extension of the analytical and descriptive tools (both theoretical and experimental) of mechanics, and the invention of new mechanical techniques and the solution of novel theoretical problems. Most of this Chapter is devoted to illustrating this, first, in respect of the macrometrology of the tensile testing of tendons; secondly, in relation to the assurance of a tensile state of deformation in both skin and tendon test specimens; thirdly, in respect of tendon-muscle systems; and, finally, by suggesting the need for theoretical problems whose solutions make possible the interpretation of a range of experimental types wide enough to lead to a generally applicable elastomechanical constitutive law.

E. SOME REVIEW SOURCES

As a preliminary to the main body of the argument, a brief account is offered of the tearing, strength, tensile elasticity, creep, hysteresis and permanent set properties of tendon and skin to the extent that they are known. Experimental techniques are also catalogued if only to throw into relief the stringent requirements of methods aimed at establishing absolute finite elastic constants. No attempt is made to present either exhaustively complete data or sources, the intent being rather to sketch the existing state of knowledge and to infer new directions. In any event, fairly recent and complete reviews already exist and these are mentioned now; and those other sources quoted are in most cases meant to be illustrative or to lead to comprehensive or significant bibliographies. The tendon as a structural component, as a stressed material, and as a part of an optimal tendon-muscle system is thoroughly reviewed by Elliott (1965, 1967) who refers extensively to the earlier account of Rigby, et al (1959). The biomechanical properties of skin are described by Gibson and Kennedi (1967) who orientate their discussion toward surgical needs and discuss the futility of designing skin flaps on the basis of inextensible, developable surfaces and also describe the effect of skin tension in producing blanching, dermal rupture and so on. Their paper includes a comprehensive bibliography. The microstructural response of skin to mechanical stress is discussed by Craik and McNeil (1966) and fully exposed visually by Finlay (1969). The well-established continuum theory of finite elasticity is not easily accessible nor simply expounded. The review by Saunders (1964), already mentioned, although not recent, gives an excellent basic account and includes references to the classical sources. Applications of this to biological tissues have been hinted at by Fung (1967), proposed by Crisp (1968), given some specificity by Hildebrandt, et al (1968), attempted by Blatz, et al (1969), by Lee, et al (1967), and more recently by Veronda and Westmann (1970). The subjugation of tensile tissue properties to physiological ends is exemplified by the reports of Adams et al (1969) who are intent to relate skin frictional changes to glandular activity that influences hydration; of Apter et al (1968); of Elden (1963) who recounts the influence on tensility of hydration; of Grahame and Holt (1969) who are concerned with ageing; of Verzár (1963) who exposes the ageing of collagen itself, and of Harkness and Harkness (1968), who by comparative tests assess, for example, the disulphide links in the dermis of the tail skin of rats. Rigby (1964) speculates as to the relation between the ageing of collagen and cyclic extension. An extensive bibliography, of nearly four hundred sources, is given by Ayer (1964) on elastic tissue as a component of connective tissue.

II. MICROBEHAVIOR AND MACROBEHAVIOR

The microstructure of fibrous materials such as compose tendon and skin, as well as the other connective tissues cartilage and ligament, is variously described as that of a woven mesh, an interlacing network, a helicoidal intermeshing and a rope-like twisting. The imposition of external load induces distortion of this microstructural, disordered network which responds by fibrous reorientation and preferential alignment to a greater or lesser degree. The mechanical correspondence between the configurational changes of the microstructure and the macroscopically observed deformation depends on three factors. Clearly the deformation or elastic properties of the fibers themselves must be relevant. Then, the proportioning of the amount of the different fibrous constituents can be expected to be influential. And, finally, the geometrical arrangement of the mesh itself must affect the evolution of a reorientating architecture. Different connective tissues are distinguishable on the basis of these criteria; and it is possible now, with some confidence, to relate the macroscopic mechanical behavior to the network distortion. Collagen is the dominant extracellular elemental fiber and skin and tendon are about equally endowed, when assessed with respect to specific dry weight. Elastic fibers include elastin, having low unit stiffness and being of small amount. Apart from these proteins is the gelatinous ground substance with reticular fibers (Montagna, 1962). In composition and fibrous concentration skin and tendon do not markedly differ. The marked difference in the non-linear tensile characteristic curve of the two materials can thus be ascribed essentially to microscopic geometrical effects and not so much to an effect of differences in material nonlinearity. (The tracing of the macroscopic behavior to a source in a microscopic network is treated at length in another chapter.)

It is of interest, now, to offer only a brief explanation (Sections IIA, B) of this chapter. The main mechanical aspects which purport to describe things like strength, elasticity, rupture, and so on are discussed in Section IIC and Section IID comments on certain functional aspects of both tendon and skin which might be greatly illuminated by appropriate studies in mechanics.

A. Structure and Composition

A definitive report as well as a comprehensive bibliography of the microscopic physical aspects of tendon is given by Elliott (1965) who relied in part on the earlier work of Rigby (1959). More recently, Haut and Little (1969) have written on tendon mechanics in a wider ligamentous context. Accounts of the microstructural arrangements of the protein mesh of skin are found in Tregear (1966); Craik and McNeil (1966); Kennedi, Gibson and Daly (1964); and Finlay (1969).

1. Tendon. The most striking feature of the collagenous mesh is the bundling of fibers into almost parallel arrays to define the principal axis of the tendon itself. The bundles are retained loosely by connective tissue (the paratenon) which include elastic fibers. Where lubrication is necessary, the tendon is constrained within a sheath of small or large extent, catering for one or several tendons. The tendon is contained by a delicate covering layer, the epitenon, which appears silky in texture and which exhibits a transverse pattern which is identified with a physical periodicity of the underlying microstructure. The primary, parallel-running fibrils of collagen form (by plaiting) secondary bundles, the fasciculi, which intermesh to produce helicoidal bundles of fibers. The mammalian fasciculi have crosssectional areas between about 1–3/8 mm^2. It follows that human tendons, in all their variability, are comprised of as few as about ten to as many as 1000 such bundles, the total collagenous tissue making up about three-quarters of the dry weight although this figure varies substantially with age, being higher at maturity. Tendon is deficient in elastic fiber; Elliott (1965) gives, for the tendo calcaneus of man, a figure of 2% of dry weight, or 0.8% of wet weight. In the mature man, the water content of tendonous material ranges from 58–70%.

2. *Skin.* Unlike the microarchitecture of the tendon, that of the relaxed skin is characteristically disordered, the collagenous arrays featuring random orientations and helically coiled fibrous bundles. In addition, the lamellar nature of skin, unlike the cylindrical conformation of the tendon, makes inevitable variations within different dermal layers. The complex spatial interweaving of fibers is identifiably looser or tighter in packing in each of three distinct layered regions. This configuration, in conjunction with any weakly preferential ordering of fibers such as seems to occur, clearly opens the way to anisotropy of macroscopic response arising from kinematic adjustments of the microstructural geometry. But the correlation of the mechanisms is by no means yet worked out. Tendon, too, must have pronounced anisotropy of mechanical properties because of its very pronounced axial fibrosity. But, unlike skin upon which loads might be imposed laterally and in any direction in the integumental plane, lateral or nonaxial tendon loadings will interact with axial or longitudinal force transmission through gross kinematic changes of configuration rather than by local tendon deformation. Elastic fibers are dominated by collagenous material in skin too: Ridge and Wright (1966a) quote contents of 66 and 72% of dry weight, and 77% of weight for dry skin. Hult (1965) has measured elastin in skin.

B. Correlations under quasi-static stress

The macroscopic tensile characteristic curve in connective tissue of load (P) versus extension (η) possesses three features which have been qualitatively explained in terms of the tissue microarchitecture. For stresses that do not approach those of a fracture or rupture condition, the curve is invariably nonlinear, being concave upwards when the load (or stress) is plotted as ordinate. Three characteristic regions of this curve can be identified. At small load there exists a deformation region represented by very low, virtually zero, *stiffness* (the local slope $\partial P/\partial \eta$ of the curve, which is not to be confused with the so-called tangent elastic modulus, $\partial \sigma/\partial \epsilon$ of the stress (σ)-strain (ϵ) curve) and small, virtually zero, curvature. At moderate load there is a contiguous region of moderate stiffness and moderate curvature which terminates in and conjoins the third region having relatively high stiffness and moderate curvature which is asymptotically zero. The appearance is not unlike that of an appropriate hyperbola.

The initial regime in which virtually no effort is required to extend the tissue specimen is thought to correspond merely to the uncoiling of coiled fibers of collagen, the unbending of bent arrays, the kinematic rearrangement of the mesh, and the reorientation of the microstructural elements. The prior disarray of the relaxed, unloaded state is considered to be induced and sustained by the mesh of elastic fibers which must, therefore, in that state be prestressed. So is the "waviness" of the tendon surface explained. The alignment of the collagen fibers is also, then, necessarily accompanied by some opposition emanating from the elastic fibers, which is shortly manifest as the increasing stiffness of the second regime. Finally, the collagenous mesh being fully orientated and the fibrous structure having taken on a denser packing, the response effectively reduces to that of the collagen fibers themselves whose greater stiffness is then mobilized (Ridge and Wright, 1964, 1965). Indeed, the third part of the tensile curve is reckoned to be synonymous with that of collagen itself (Kennedi, et al, 1964; Abrahams, 1967; Mendosa and Milch, 1965), in respect at least to skin. In the process of stretching, the elastic and reticulin fibers are essentially unaffected but suffer displacement and compression (Craik and McNeil, 1966). In skin, the first region is very pronounced (Kennedi, 1967; Kennedi, et al, 1966), giving a characteristic not unlike the curves of Fig. 7. In tendon, the first region, by contrast, is absent and the second region only weakly present, giving rise to what is termed by some as the toe of the curve. It is illustrated in poorly-developed form by the loading curve 1 of Fig. 1 and to better effect by the subsequent loading cycles 2 and 3 of that Figure, which relate to formalin fixed and washed human cadaveric tendon (tibialis anterior). The toe of the curve is enhanced by low rates of straining (Haut and Little, 1969). But Harris et al (1966) report numerous tensile stress-strain curves of which many neither display this toe nor exhibit an upward concavity but, in fact, the opposite curvature and which are, therefore, not reconcilable with the carefully annotated description of Elliott (1967). Indeed, the curves of Fig. 1 confirm this tendency to the sigmoid, as also do the segments *ab* of Fig. 2 in which the tendon strains are somewhat greater. The sigmoid shape evolves more fully in those tests which reach rupture (Elliott, 1967). Benedict et al (1968) distinguish the flexor tendons and their sigmoidal curve from the extensor tendons having a curve slightly concave downwards.

These microstructural evolutions under stress are now described and correlated with mechanical observation with confidence enough to justify the exploration of theoretical, statistical network models aimed at producing a useful macroscopic constitutive law and delineating appropriate material elastic parameters or constants. It is regrettable that this new understanding relates only to the uniaxial tensile state of deformation which, in the long run, is not wide enough a base for an elastomechanical constitutive law. Other deformation states have not been totally neglected, of course, but the

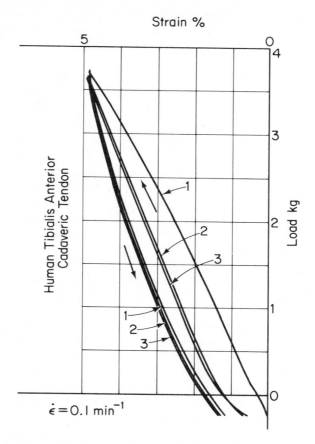

Fig. 1: Uniaxial tensile test of human tibialis anterior tendon, formalin fixed and subsequently washed. Extension was cycled between fixed strain limits at a strain rate $\dot{\epsilon} = 0.1$ min^{-1}, showing hysteresis and subsequent cyclic conditioning. (Acknowledgement is made to Dr. G. G. Carmichael, Department of Anatomy, Monash University, who supplied the test specimens.)

intricacies of the micromesh response with or without ground substance interaction, as in compression (see Hickman, et al 1966) is by no means well explored.

C. Mechanical indicators

The main mechanical features of tendon and skin as materials are well-established and the usual mechanical indicators are in use. Wide ranging investigations report on stiffness, on stress-strain characteristics (although

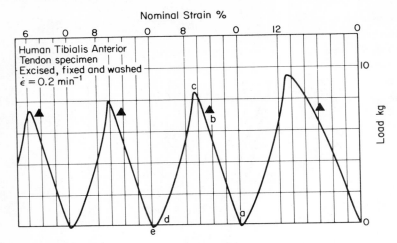

Fig. 2: Uniaxial manual load cycling tensile test of human tibialis anterior tendon, as in Fig. 1, at strain rate $\dot{\epsilon} = 0.2 \, \text{min}^{-1}$, showing intermittent slip at points b, continuous slip at points c prior to load relief. (Acknowledgement is made to Dr. G. G. Carmichael, Department of Anatomy, Monash University, who supplied the test specimens.)

the meaning of both stress and strain is at times made obscure or uncertain by doubtful areal and displacement frames of reference), on Young's modulus (which has no rigorous meaning for these tissues, the terms tangent modulus, or specific stiffness being preferred), on rupture strength, on creep, on stress-relaxation, on hysteresis, on permanent set; and also on some less familiar properties like resiliency, contractility and so on. Quantitative statements are notable, however, both for their diversity in values and for their specialization to a small sample of skin or tendon types and sites; or to one particular mechanical property.

The disparity of numerical values arises in part form the inherent variability of the tissues themselves, the variety of the biophysical and biochemical states of specimens under test, and probably also from the incomparability of the test technique, there being no generally accepted test methods, nor standard test specimens. Little attention seems to have been given *systematically* to comparisons of tests *in vivo* and those on preserved specimens. The usefulness of tests of cadaveric tendons, for example, is still in doubt although Anderson and Williams (1970) report that preservation with formalin and subsequent storage in sterile saline for periods of up to ten weeks does not adversely affect the mechanical properties of fresh human tendon. Blanton and Biggs (1970) report that embalming enhances the ultimate tensile strength. Demichev and Tupitsyu (1966), motivated by the need to preserve tendon for homotransplants, used tensile tests of dog tendons to monitor the specific stiffness of fresh, chilled, frozen and lyophilized specimens. Their conclusions favor chilling. Van Brocklin and Ellis (1965) find no disparity between fresh and frozen extensor digitorum tendons when tested under cyclic tension. Conditions of hydration (Elden, 1963; Harris, et al, 1966) during test can be critical.

But there emerges a picture of finitely deforming, nonlinear, inhomogeneous, anisotropic, creeping, relaxing, hysteretic materials to which elastic concepts can still usefully be applied. Many have reported results from which may be taken values of tangential modulus for both skin and tendon. By way of illustration, Harris, et al (1966) give for the human cadaveric plantaris and extensor digitorum the same average value of 1.75×10^5 lb. in.$^{-2}$ (approx 12,000 kg. cm^{-2}), and for the flexor digitorum superficialis and profundus a figure of 1.10×10^5 lb. in.$^{-2}$ (approx 8000 kg. cm^{-2}). But it might be argued that the embalming resulted in dehydration. Haut and Little (1969) report a tangential elastic modulus of between 1 and 2×10^3 kg. cm^{-2} for canine anterior cruciate ligaments. Values for collagen fibers range between 100–1000 kg. cm^{-2}. (See also Walker, et al (1964); Kirk and Kvorning (1949); Vlasblom (1967); Van Brocklin and Ellis (1965) for further data relevant to uniaxial elasticity. Comments about ultimate tensile strength, linearity, and reversibility are to be found amongst these and the earlier quoted writers, too.)

A strain of 4% is regarded as especially significant in tendons and corresponds to the limit of reversibility (and therefore of elasticity) and to the disappearance of surface waviness (and therefore to the incipient mobilization of the fully aligned densely packed collagen). The general picture for tendons is sketched by Elliott (1965). Enormous variations (500–1000 kg/cm^{-2}) in tensile fracture stress occur from author to author and, indeed, from test to test. Those deriving from formalin fixed embalmed specimens should propably be ignored. No doubt there are inherent variabilities but those induced by test methods are probably dominant as will be emphasized later. Blanton and Biggs (1970) summarize the strengths of nearly fifty different human tendons giving a range of about 15,000–17,000 lb. in.$^{-2}$.

Viscoelastic effects have been widely studied through creep and stress-relaxation tests. A comparison of Figs. 1 and 2, for example, reveals a pronounced strain-rate effect in the cadaveric tendon. Figure 1 also reveals the characteristically large hysteresis associated with the initial load cycle and its subsequent diminution and stabilization; and the repeatability then afforded by this preliminary conditioning. An identical phenomenon is found in the permanent set behavior of rubber (Hart-Smith, 1966), where it is empirically absorbed into elastic theory by referring the finite strain, or extension ratios, to the initial or natural state rather than the current relaxed state. But it is indicated in Fung, Chapter 7 that time-dependence and elasticity, or quasi-static effects, can realistically be separated. It suits later argument to remark that support for this expedient comes from the behavior of the rubbers, for which a constitutive equation has been devised by Valanis and Landel (1967a). This equation indicates three relaxation functions each independent of the deformation state and all having the same functional time-dependence: strain and time are separable. Amongst those who have typically investigated, in a mechanical way, the influence of time on skins and tendon or similar tissues are Apter et al (1968); Frisén et al (1969); Fung (1967); Higuchi and Tillman (1965) and Viidik and Mägi (1967).

The significance of the tensile strength of tendon, in particular, might be questioned inasmuch as failures under tendonous forces invariably produce rupture at the origin or insertion, at least in rabbits and human fresh cadavers (Elliott, 1965). Confirmation is given for the flexor tendon of the thumb by Tylicki and Zbrodowski (1966) who tested fresh cadaverous tendons retained only by two mesotendineum ligatures, under both outward and inward tensile forces. Invariably, ligamental tearing occurred under loads between 8–17 kg according, *inter alia* to sex; and they remark that this rupture can be induced, normally, by violent reflex contraction of the flexor muscle belly. They comment also on the identity between left and right hands in this matter. It is of interest, then, that the normal physiological range of performance lies well within the "perfectly elastic" toe of the tensile curve which is delineated by a reversible strain of approximately 4%; the working strain mounts to about 2.5% (Harris et al, 1966) and the working stress ranges up to about 100 kg. cm^{-2} (Elliott, 1965). A rule of thumb now current is that the working stress is about one quarter of that at ultimate tendon failure. Tendon strength exceeds that of its muscle by a factor of two (Elliott, 1967).

Other traumatic effects have been studied in connection with surgical wounds of skin (Geever, et al, 1966; Glaser, et al, 1965), but little is recorded (see, De Muth, 1968) in connection with accidental trauma. Mention might also be made of a miscellany of interests in skin: its striation during pregnancy (Main, 1969; Gibson and Kennedi, 1967), and its impedance (Tregear, 1965).

D. Functional morphology

1. Tendon-Muscle systems. It is clear that the functional aspects of tendon performance cannot reasonably be divorced from those of the total system in which the tendon operates. The rupturing proclivites of manual ligatures ensure this view, if nothing else. Several investigations have attempted to elucidate the functional justification for tendon sizes and shapes, tendon-muscle architecture and, generally, to demonstrate the working of some natural design optimization which relates growth and hypertrophy to quantitative mechanical performance. These attempts have tended to proliferate the mysteries. There seeems to be no close correlation between the cross-sectional dimensions (or, rather, the girths) of certain tendons and the strength of the corresponding muscle (Elliott and Crawford, 1965). Elliott (1965) suggested an interdependence between the collagenous sectional density and the muscle fasciculi cross section. In mature rabbits, Elliott (1967) subsequently concluded that for both fusiform and penniform muscles under maximum isometric tetanic tension, tendon strength is determined essentially and simply by its thickness and collagen content which seem quite unrelated to the maximum force of which the muscle is capable. Blanton and Biggs (1970) find no sensible difference between the strengths of tendons in upper and lower limbs; and no great differences between flexor and extensor tendons. Benedict et al (1968) find a contrary result. Harris et al (1966) remark on the apparent arbitrariness of system design. In respect of embalmed cadaverous human digital extensor and plantaris tendons as compared with the digital flexors, those apparently taking the greater loads have the smaller elastic moduli. This some consider paradoxical, on the assumption (which remains untested) that different tendons are equally stressed. In addition, the normal working loads of human muscle-tendon systems have not systematically been discovered. The plastic surgeon notes this inhibition in tendon graft restorations of manual function made necessary by accidental severance, usually of the palmar flexor tendons. In excising and substituting a tendon segment between muscle belly and insertion, he must judge intuitively the natural rest (passive) force and extension of the affected muscle-tendon pair. In a continuing series of *in vivo* tensile tests on human flexor tendons this data is now being gathered with results typically like the curves of Fig. 3, which reveal essentially a *muscle* characteristic because the passive muscle has an extensibility between 20–100 times that of tendon. Of particular interest is the comparison between the characteristics for the ring and little fingers tested separately and that of these two tendons tested in parallel, in the manner of Fig. 4. The parallel combination, in coinciding with the characteristic for the ring finger, reveals that the two are not acting as additive springs (even nonlinear ones) for the two ordinates should then be additive. But it does seem to confirm the anatomical fact that all three flexores digitorum profundis V, IV, III, conjoin by fusing above the wrist after their passage through a common palmar synovial sheath and below the tendon-muscle belly junction. On the other hand, the individual three fingers do have differentiable mobility due to profundus tendon stimulation; so that the mechanics of the tendon-muscle system, which must include other constraining and interacting retinacula, is obscure. This simply illustrates a remark attributed to Paul (1967) that in respect of human articulation the joints are kinematically overconstrained. The tendonous or ligamentous conformation with attached musculature giving a joint its mobility contains redundancies of connection which exceed in numbers those demanded of mechanical equilibrium. What constitutes a unique connective configuration

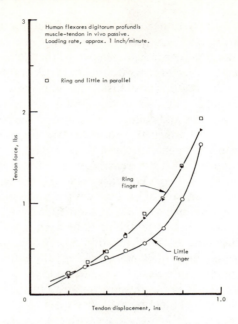

Fig. 3: Force-extension characteristic curve for two human flexores digitorum profundus muscle-tendon systems, tested passive *in vivo* during tendon graft surgery; showing the curves for the little and ring finger tendons separately and the data points (hollow squares) for the two in parallel, as in Fig. 4.

and what ligamentous and tendonous forces actually operate to match the purely mechanical demands of gait (Paul, 1966; Contini, et al, 1964), are being explored (Morrison, 1968, 1970; Wright, 1969) in a way outlined in Paul, Chapter 20. There needs to be, however, substantial correlation between the static and dynamic environment and the individual tendonous loading (Williams and Svensson, 1968).

2. *Skin anisotropy and Langer's lines.* Langer (1861) punctured the skin surface of human cadavers with small circular holes which had the effect of relieving the natural prestress of the integumental membrane. The circular holes relaxed to form what are usually described as elliptical holes but are more often lenticular, having strains in the direction of the minor axis less than about 10%. The major axes of these ellipse-like excisions when tangentially enveloped form, over the whole body, trajectories known as Langer's lines which do not represent principle tensile stress trajectories but rather those of fiber orientation (Cox, 1941). Ridge and Wright (1965, 1966*a*, 1966*b*) confirmed the anisotropy of skin extensibility in relation to the

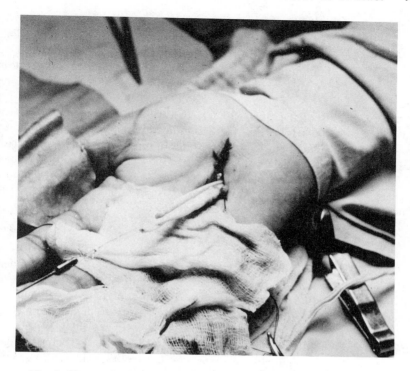

Fig. 4: The tensile testing of the ring and little finger profundus flexor tendons in parallel during tendon graft surgery. (Courtesy Mr. John Snell, Honorary Plastic Surgeon, Alfred Hospital, Melbourne.)

orientation of Langer's lines and also studied (Ridge and Wright, 1966c) the relaxation of a square incision. Consequently, they proposed a plane lattice model of fiber orientation to explain qualitatively the known and measured anisotropy. Gibson, et al (1969) subsequently confirmed the correlation between Langer's lines and dermal fibrosity, the extensibility being greatest normal to the lines; and they remark on the significance of this in surgical restoration of the skin wounds. A summary of the relationship between Langer's lines and macroscopic anisotropy and microstructural fibrous patterns is offered by Finlay (1969). But the functional justification of skin anisotropy is not clear.

III. EXPERIMENTAL PROCEDURES

There is a need to develop experimental mechanics beyond its present capability, and this is apparent from the difficulties and deficiencies in the measurement of the size of tendons. It seems remarkable that such a structure which is dimensionally commensurate with those commonly employed in mechanics suffers from not being easily or, perhaps, properly measured. This same theme is taken further by reference to the graver problems of ensuring the state of deformation intended in a test piece and by highlighting a certain lack of attention to this matter to which experimental mechanics has long been very sensitive.

A brief catalog of the classes of testing variously employed (mostly for skin) follows, mainly with a view towards indicating that in the majority they cannot decently be adapted to yield absolute information about the elastomechanical properties of the tissues. Some proposals are made for new lines of attack, mainly in relation to tendon testing; and although in mechanics the matters treated here are trivial and standard, in application to biological tissues the techniques are by no means perfected.

A. Macrometrology of Tendons

Tendons are far from being geometrically perfect prismatic cylinders and commonly vary greatly, both in cross-sectional shape and natural longitudinal curvature. The precise determination of tendon geometry is rendered difficult by ordinary mechanical means because of the compliance of the surface (making precise contiguity virtually impossible), of the low transverse specific stiffness of the fresh tendon (readily allowing shape changes under gravity), and because of the incompressibility of the material (requiring volume constancy and strong triaxial coupling of distortion errors). The peroneus brevis tendon of the human, for example, displays an approximately elliptical cross section at one station with a ratio of major to minor axis of about 2.2. Five centimeters distant it is essentially the same. But at a station yet a further 5 cm it has a distinctly triangular (isosceles) cross section of an area (5 mm^2), calculated to be closely the same as that at the other two stations. Again, the human tibialis anterior tendon over a 7 cm span retains a roughly elliptical cross section albeit with a change in the ratio of major to minor axis from about 3.2 to 1.9. The sectional area remains constant at approximately 18 mm^2 which is three times that of the earlier tendon cited. Areal constancy with axial distance suggests, of course, that the tendon shape is optimal in the sense of constancy of stress with longitudinal position—the tendon, like its engineering counterpart the tie cable or rod is an equistressed unit. These sorts of variations of tendon shape and size with position and from tendon to tendon have never been systematically reported. And this may, in part, be attributed to the lack of a suitable metrological method. For methods currently in use do not always inspire confidence and certainly do not lead to identity of results (Ellis, 1969).

The difficulties have been evaded usefully at times by dealing with what might be called the physiological cross section which measures that part of the cross section thought to be mechanically significant, and avoids the question of shape altogether by the measurement of the dry weight per unit length (Elliott and Crawford, 1965). Such a measure can only be properly applied to the estimation of stress, of course, upon the demonstration that moist ground substance is incapable of influencing the stress-bearing capacity of the dried components. Another measure of the sectional area is the specific displaced volume per unit length of dry tendon; and either of these measures can be represented alternatively in terms of moist tendon. Special areal micrometers (Walker et al, 1964) have been devised. Probe micrometers (Haut and Little, 1969) have been used in conjunction with a cross-sectional shape assumed to be elliptical or circular (Rigby et al, 1959). Photomicrographs of histological specimens or magnified projections of thin transverse slices, after Blanton and Biggs (1970), offer an easy application of planimetric methods. The amplitude contour of a projected shadow of the illuminated specimen as it rotates allows a graphical reconstruction of the cross section. But the comparison by Ellis (1969) of most of these methods shows differences up to 100%. And whereas the areal micrometer and moist weight per unit length give essentially the same measure, these measures are not most easily repeated. For this and earlier reasons, some attention, it seems, should be directed toward encouraging more reliable methods of areal measurement. Indeed, constitutive laws deal with stress which require, in a continuum sense, a geometric specification so that the mechanical shape cannot be ignored. Invasive or tactile methods are not simple and seldom certain with incompressible materials and a surface metrology suggests itself. Modern optical methods of interferometry and holography are already being used in this connection for nonbiological materials.

B. Nontensile test configurations

Practically all work aimed at any quantitative description of skin elasticity relies on the tensile test, and this emphasis has been decried in Section I. At the same time, tensile properties have been very extensively employed as qualitative monitors. There is a wide range of devices, mainly clinical and diagnostic tools, which are used for sensing the physiological state of the animal system in terms other than tensile deformation states. Rothman (1954) describes some of the earlier means of assessing skin health and mentions the Schade elastometer which registers the penetration of an indenting force, and its adaptation by Kirk and Kvorning (1949). Like all such instruments the subcutaneous tissues and fat contribute significantly to the response, and it is only for other than absolute quantitative results that this is of no consequence.

The nontensile tests can be conveniently classified as of either the pressor or the distensile type.

1. Pressor tests. These, in turn, are distinguishable according to the load as it be applied, the deformation response being characteristically different. The *compressive tests* feature widely distributed compressive strains as in tests of skinfold compressibility (Brozek and Kinzey, 1960; Clegg and Kent, 1967; Parot and Bourliére, 1967) induced by calipers. Pressure pads used by Hickman et al, (1966) allow a qualitative interpretation of skin as a viscoelastic creeping material but no real control can be exercised over the boundary conditions of the deformation. Flattening under pressure, or applanation, is common in tests of the eye, and Mucri and Brubaker (1969) provide an extensive review of such noninvasive techniques of tonometry. The *indentation tests* are characterized by rather concentrated local deformation and penetration of the tissue as in the compressometer of Sokoloff (1966) which is claimed to provide a "Young's modulus" for the instantaneous elastic response. An arbitrary but standard anvil is used to measure a load (or pressure)-deflection characteristic which reflects, of course, not only the properties of the material immediately strained, but also those of adjacent and underlying structures.

In the *diaphragm tests* a membraneous layer of tissue is inflated by lateral pressure of fluid or gas. Dick (1951) reportedly employed this method and Tregear (1966) too gives an account including an analysis based, unrealistically, on the mechanics of a spherical bubble of a material having a constant surface stress-resultant. It is doubtful whether such a theory can be properly applied for reasons outlined below in Section IV. Grahame and Holt (1969) have applied this method, using a suction cup but still interpret the distension curve unrealistically. The formation of blisters on the skin by friction, for example, affords a measure of the elastic property of the epidermal layer (Kiistala, 1968; Lowe and van der Leun, 1968; Sulzberger, et al, 1966), by producing a blister roof that is viable full-thickness epidermis. *Puncture tests*, in which the relaxation of a slit or a circular hole displays qualitatively the principal stress trajectories of the natural state, have already been mentioned in connection with Langer's lines in Section II D. Mackie (1967) employed a punching technique in gaining measures of dry epidermal density which he defined as a mass per unit surface area. Using a biopsy punch (of area 10 mm^2) he cut specimens by rotation of the punch.

2. Distensile tests. In this type of test whole vessels are distended under pressure, the results being presented as a characteristic pressure-size curve. For a long time curves such as this have been used interchangeably as "stress-strain curves" which they are not, the clarification afforded by the warnings of Burton (1961) notwithstanding. There is, of course, no simple direct relationship, and tests of this sort have relevance only for the particular vessel being studied. They cannot yield data about material elasticity without foreknowledge of the general functional form of the constitutive law and its finite elasticity parameters. A reliable knowledge of this form, however, makes distensile tests extremely valuable for elasticity determinations as has been carefully demonstrated for rubbery materials. And it is for this reason that it is mentioned at all, for clearly, such tests are not applicable to tendon structures although skin tissues could conceivably be wrought into a vessel-like conformation.

3. Other tests. Indentation and suction tests really monitor the properties of the subcutaneous layers, and Vlasblom (1967) sought to overcome this disability by conducting static torsion tests of the skin in the expectation that dermal layering would effectively decouple the skin from its base. Duggan (1967), too, chose torsional deformation states as a clinical sensor of the skin elasticity condition with the novelty of oscillatory loads. Also there is the "bow-string" or "slingshot" instrument devised by Lissner (1965) to measure forces *in vivo* in those human tendons of the foot and ankle that are superficially accessible. (A similar device was used in a different context by Lee et al, 1967.)

C. Boundary conditions in tensile testing

In experimental mechanics exquisite care is given to arranging the details of load application to a specimen to ensure that a specified state of deformation exists. In respect of biological specimens, tendon for example, this attitude has not yet prevailed nor yet produced a standard technology. In particular, the gripping or clamping of the tensile test specimen continues to be a problem, to the extent that stress nonuniformities and concentrations produce premature failure in or about the grips. Slipping within the grips cannot be prevented, and gage length cannot be precisely set in those circumstances where extension transducers cannot be applied to the tissue itself and elongation is measured directly as crosshead, or grip, travel. Any locally intense deformation at the point of application of the tension immediately invalidates any attempt accurately to measure stress, strain or displacement based on grip movement. Elliott (1965) recounts some of the difficulties in gripping tendons and lists methods which include splitting and winding the tendon ends into a built-up lump to prevent excessive squashing in the clamps. Alternatively, a specimen having a reduced cross section can be planed by tapered excision; or if the specimen must break at other than a grip site, a stress-concentrating hole can be provided at its midspan. More reasonably, flat strips of very large aspect ratio might be cut. Rigby et al, (1959) describe a method of jamming the two ends of a *looped* (rat tail) tendon into a hole in the loading block by a bakelite plug. Harkness and Harkness (1968) avoid gripping by employing rings cut from the skin of young rat tails. Bruchle and Moll (1968) whose gripping clamp was faced with coarsely pitched corrugations, and Ridge and Wright (1966d), designed special extensometers. Abrahams (1967) mentions the failure in the grips of formalin fixed horse extensor tendons. Harris, et al, (1966) investigated the gripping problem at length and tried and abandoned a method of knotting the specimen and casting its ends into plastic blocks, a split-tapered wedge in a thick cylinder, and flat blocks. Prevention of slip dominated the technique and allowed undesirable stress disconformities. A drill chuck offered promising control of the end conditions. Simple tapered plastic jaws, without serration which were "designed to hold without slipping" were used by Blanton and Biggs (1970) with success. Pneumatic clamps also offer fairly versatile control but require moulded packing (with rubber sheet, for example) and careful selection of lateral pressure level. In Fig. 2, for example, the arrows b indicate the initiation of intermittent slip, and the crests c represent arrested continuous slip of the tendon specimen held pneumatically in rubber-packed clamps with a total lateral force of about 7 kg. In Fig. 2 the slip is avoided by increasing this, but the incompressible tendon material is now extruded from the grips increasing the free length of the specimen and calling into question the meaning of the strain figures. The deformations being always finite, this effect is inevitably and continuously operable during a tensile test with a severity that varies in an unknown manner with the extent of tensile elongation. Ideally, then, the loading grip must adapt itself in a consistent and known manner to the finite changes in size and shape of the specimen it retains, without influencing the uniformity of tensile strain within the gague length of the specimen. The braided cylinder (Fig. 5) seems to offer this self-adaptive capability and is now under test and development. It has the property that it can occupy a whole range of configurations characterized by different radii according to how the wound helix angle is varied. In particular, a solid cylindrical prism inserted partially in one end is powerfully held against axial tension when the braid mesh is closed upon it; and, once mobilized, the frictional gripping increases with increased tension, the hollow cylinder adapting itself to the shape of the inserted prism. The action is essentially kinematic such that contrawound helices pivot at their point of contact, as the measurements of Fig. 6 confirm. For, kinematically, the helix pitch and the cylinder radius r are constrained by the specification of the fiber length L, this result being independent of the number of helix starts. The gripping action is enhanced by a favorable mechanical advantage which applies at the large values of the pitch m where $dm/dr \rightarrow 0$. If the fibers of the cylinder are inextensible and of no thickness, if fiber-fiber friction is absent, and if the clamped specimen is rigid, the force characteristic is like that displayed in Fig. 7 and annotated there as ideal. In practice, the fiber extensibility, the number n of helix starts, and the transverse shape and size of the wound fibers will, intuitively, influence the clamping and axial force characteristic in the manner suggested by the curves for

Fig. 5: A braided hollow cylinder conformation shown, at left, in fully extended state and, at right, contracted and expanded, the transition being purely kinematic. A prototype of a self-adaptive tensile test clamp.

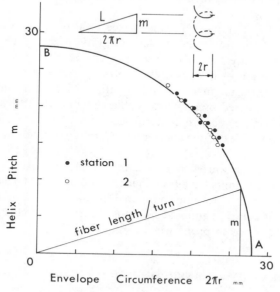

Fig. 6: The kinematics of the helicoidal braided cylinder showing the theoretical Pythagorean relationship between the pitch m and the cylinder radius r together with measured configurations at two stations of cylinder, confirming the kinematic pivoting of the fibers.

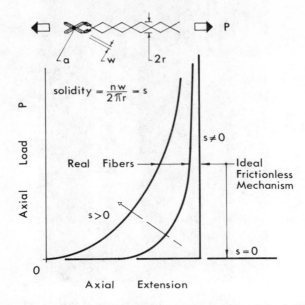

Fig. 7: The hypothetical load-extension characteristic of the helicoidal, braided cylindrical tension clamp, showing the ideal characteristics when the fibers have no thickness ($s = 0$) and the effect of fiber width.

which the solidity is non-zero. This device is also being considered as a force-transducing suture for use in tendon graft surgery. But the point of immediate significance is that the successful development of such a clamp involves the solution of new theoretical problems in mechanics of a somewhat complex character involving finite deformation of an incompressible material as it is gripped by an elastic cable network. And this is demanded because experimental mechanics itself needs further development to meet the special complexities of the soft tissues. Careful attention to these aspects is overdue.

Similar difficulties are encountered in testing skin (Ridge and Wright, 1966d) although attempts have been made to standardize (Glaser et al, 1965) test specimen, strain rate, and measured parameters. Standardization of itself cannot, of course, eliminate spurious deformation fields to which soft tissued specimens are so prone when compressed.

IV. FINITE ELASTOMECHANICS AND SOFT TISSUE

Attempts to devise symbolic statements about the nonlinear, elastic response of skin-like tissues may be variously based. Empirical curve-fitting, firstly, is severely limited because physical constants found in this way apply only to the deformation state (invariabley, tensile) that provides the constants. Ridge and Wright (1966a), for example, fit the tensile load-extension curve stepwise, one part being matched by an exponential relation, the other by a power law. Glasser et al, (1965), too, propose a power law fit of the tensile stress-strain characteristic. More generally and usefully, Fung (1967) derives an exponential law modified by a nonlinear function originating in finite elasticity mechanics (Fung, Chapter 7). It successfully describes a wide range of soft tissues, including skin, and this *universality* of tissues is important in justifying more powerful, general continuum representations.

Simulation by element-modelling of measured response characteristics is common especially where inelastic, time-dependent effects are of concern. Viscoelastic spring-dashpot models abound, (see Viidik, 1968). Some models like that of Frisen, et al, (1969) include dry-frictional elements to model tissues, like tendon and ligament, that are predominantly parallel-fibered collagen. Apter (Chapter 9) outlines uses of such models. And it is important to realize that because no uniqueness condition can be written for the degree of freedom of the elemental array in phenomenological modelling, some arbitrariness can always be ascribed to such representations which do not, therefore, offer a rigorous approach to the constitutive law. Fung (Chapter 7) argues in like manner at greater length.

Network theories can be devised that are based on microstructural idealizations of fibrous architecture. These have proved very successful for the rubbery polymers, and the soft tissues, including skin, should be amenable to such description. Two approaches are possible, neither of which has been at all developed. The first is *deterministic* and exemplified by the uniform plane mesh notion of Ridge and Wright (1966c) invoked to explain skin anisotropy. The braided cylinder mechanism (Section IIIC) clearly enough could form the basis for a generalization of their mesh into a three-dimensional layered tissue, like skin. But it could also serve as an elastokinematic model of fibrous units of a *statistical*, mechanical network theory after the manner of the statistical, thermodynamic theories so well developed for rubber. The new understanding of the quite direct relationship, in skin for example, between collagenous structural response and imposed load might at first sight seem to make imperative (Kennedi, et al, 1964) the use only of network type theories, for skin is clearly not phenomenologically like the common elastomers—a homogeneous elastomeric theory seems to be ruled out. But this is not to say that continuum descriptions are not apt; and there is every reason to explore continuum descriptions whether based in classical elastomeric network theory or in classical finite elasticity continuum mechanics. The basic and prerequisite experimentation that might allow this is the subject matter of this section. Insofar as neither skin nor tendon has yet been subject to the type of scrutiny demanded of a finite elastic representation, the discussion necessarily refers to other tissues and is given some reality by reference to the rubbery high polymers, or elastomers, in the expectation that the more complicated biological polymers will show parallelisms of behavior. The basis of both the statistical and phenomenological continuum theories is excellently exposed by Saunders (1964) and is not repeated here. Veronda and Westmann (1970) give a very brief account of continuum aspects. Attention is rather given to showing the roles of each theory and of arguing the need for a proper experimental regimen.

A. BIOLOGICAL TISSUE AND THE POSTULATES OF ELASTICITY

By invoking the principles of solid mechanics to assist in the description of the mechanical properties and behavior of biological materials, questions of isotropy, homogeneity, compressibility, elasticity itself and, finally, universality, cannot be left unresolved when the tissue material is viewed macroscopically. The components of the soft connective tissue are invariably the same in respect to mechanical make-up, consisting of differently combined aggregates of collagen, elastic fiber and gelatinous ground substance in a complex structural network. The gel state consists of a sparse minute collagenous framework retaining watery fluids and as such exhibits properties of both solids and fluids, transmitting shear stress in the static state. Soft tissues such as skin and tendon are clearly a composite material, preferentially structured and, microscopically, far from isotropic and homogeneous, although this need not militate against effective isotropy and homogeneity at the macroscopic level. But for skin, anisotropic elastic theory must be applied sooner or later; the base for it exists (Green and Atkins, 1960), but its application is formidable and is not considered here. The question of compressibility does seem to have been resolved, and it is now generally accepted that the soft tissues can be regarded as of incompressible material. Some states of compression, however, result in deformation with volume decrease (Gibson and Kennedi, 1967) due to exudation or extrusion and flow of fluid (Hickman, et al, 1966). The whole question of exploring, as suggested by Saunders (1964), the relevance to biological tissue of new theories that describe the elastic liquids, has been ignored. They incorporate molecular descriptions of permanent set phenomena in rubber, for example, which have otherwise (Hart-Smith and Crisp, 1967) been catered for phenomenologically. Questions of creep, stress relaxation, strain-rate sensitivity have already been excluded here and the relevance of the concept of elasticity accepted. The desirability of generality or universality in any elastic description is, finally, reiterated. Just as all engineering metals are usefully described in linear, elastic terms by two elastic constants and just as all rubbers and high polymeric elastomers seem capable of being rendered by three finite elastic constants, so might a *universal representation* of the elasticity of soft tissues be capable of realization. Experimental data already at hand encourages this viewpoint. Notwithstanding the variety of composition of skin, tendon, ligament, cartilage, the commonality of their constituents should allow an elastic description through a single set of finite elastic constants, each set being numerically specific according to the particular tissue is more or less endowed with one or other of collagen, elastin or gel. Here is one reason to insist upon a comprehensive finite mechanics, one realistically catering to all strain states, one relating back to fundamental quantities—the strain energy of deformation and the finite strain invariants.

B. Constitutive Laws for Elastomeric Polymers

In describing nonbiological materials, three approaches have been developed. The polymer physicists, beginning in the 1930s, developed statistical molecular theories based on the probability of networks of long-chain molecules possessing certain lengths and configurations. These theories are variously called kinetic, statistical, or thermodynamic, network theories and have been shown to work well for limited finite deformations. From the 1940s, the solid mechanists (physicists, applied mathematicians, engineers) have successfully devised macroscopic continuum theories which, in varying degrees, fit the facts of rubbery behavior. They are based on postulates which constrain the form of the elastic strain energy function and lead to necessary and sufficient conditions for the number of finite elastic constants, as well as the functional form of the finite elasticity parameters. The most widely used formulation is that of Mooney (1940) although it is demonstrably limited to smaller values of the principal strain invariants, both theoretically and experimentally. The modern empiricists place emphasis on realism, that is to say, on representations which accord with physical reality, in such a way as to favor generality over consistency with the polymeric and continuum theories although the features of the latter are, wherever possible, incorporated. And indeed, these formulations are set in the framework of the continuum theories which hinge upon the construction of an appropriate strain energy function. This basis is reviewed briefly as it pertains to membranes, and a few special forms of the elasticity parameters are presented.

Because the deformations of interest are "large" (finite), the strain measures of classical infinitesimal elasticity theory must be discarded. Tendon extensions normally range up to 3% whereas, under normal conditions, skin suffers strains between 30–50%. The principal extension ratios (stretch ratios) λ_i are, therefore, introduced (Fig. 8) to define the distortion. The elastic state is expressed in terms of the recoverable energy associated

PROPERTIES OF TENDON AND SKIN

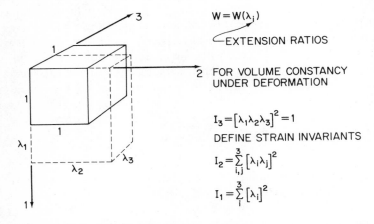

Fig. 8: The definition of the principal extension ratios λ_i for finite homogeneous deformation of an elastic cube.

with this deformation. This strain energy W, is then supposed to be a function only of the state of deformation, that is, of the ratios λ_i; W then characterizes an elastic material.

For infinitesimal elasticity, then, a strain energy function is taken to exist and to depend on strain in some way. Thus, for isotropic materials one or other of the alternative forms

(1) $$\varphi = \varphi(E_i); \quad W = W(I_i)$$

of strain energy is postulated, either of which is only useful if the stresses can be derived from it in the usual way for a potential function. This is so if E_i, I_i are certain strain invariants

(2) $$\begin{cases} E_1 = \epsilon_{11} + \epsilon_{22} + \epsilon_{33} \\ E_2 = \epsilon_{11}\epsilon_{22} + \epsilon_{11}\epsilon_{33} + \epsilon_{22}\epsilon_{33} - \tfrac{1}{4}(\epsilon_{12}^2 + \epsilon_{13}^2 + \epsilon_{23}^2) \\ E_3 = \epsilon_{11}\epsilon_{22}\epsilon_{33} + \cdots \end{cases}$$

where ϵ_{ij} are six *finite-strain* components; or

(3) $$\begin{cases} I_1 = 3 + 2E_1 = \lambda_1^2 + \lambda_2^2 + \lambda_3^2 \\ I_2 = (\lambda_1\lambda_2)^2 + (\lambda_1\lambda_3)^2 + (\lambda_2\lambda_3)^2 \\ I_3 = (\lambda_1\lambda_2\lambda_3)^2 \end{cases}$$

where λ_i are the three *principal extension ratios* (Fig. 8) relative to the undeformed state. For an incompressible material, one having volume constancy under deformation

(4) $$\lambda_1 \lambda_2 \lambda_3 = 1$$

and the set (3) reduces to

(5) $$\begin{cases} I_1 = \lambda_1^2 + \lambda_2^2 + 1/\lambda_1^2 \lambda_2^2 = I_1(\lambda_1, \lambda_2) \\ I_2 = 1/\lambda_1^2 + 1/\lambda_2^2 + \lambda_1^2 \lambda_2^2 = I_2(\lambda_1, \lambda_2) \\ I_3 = 1 \end{cases}$$

Note that the finite strains and extension ratios are connected by the relationships

(6) $$\lambda_i = 1 + 2\epsilon_{ii} = (1 + e_{ii})^2$$

where e_{ii} are principal linearized strains. It is convenient to use the set (2) for infinitesimal deformation because then $E_3 \to 1$, whereas for finite deformation of an *incompressible* material, because $I_3 \to 1$, the set (5) is preferred.

Stresses are then derived from the strain energy function as

(7) $$\sigma_{ii} = \frac{\partial \varphi}{\partial \epsilon_{ii}} \;;\quad \sigma_{ij} = \frac{\partial \varphi}{\partial \epsilon_{ij}}$$

and these are *generalized stresses* which relate to actual stresses t_{ij} as

(8) $$\sigma_{ij} = \frac{s_i^* t_{ij}}{s_i \lambda_j}$$

The ratio s_i^*/s_i is the ratio of the deformed local area normal to the direction i to the corresponding undeformed local area. For example (Fig. 8)

$$\sigma_{11} = \frac{s_1^*}{s_1} \cdot \frac{t_{11}}{\lambda_1} = \frac{A \lambda_2 \lambda_3}{A} \cdot \frac{t_{11}}{\lambda_1} = \frac{t_{11}}{\lambda_1^2}$$

$$\sigma_{12} = \frac{s_1^*}{s_1} \cdot \frac{t_{12}}{\lambda_2} = \frac{t_{12}}{\lambda_1 \lambda_2}$$

for an incomprexssible material, that is, subject to the constraint $\lambda_1 \lambda_2 \lambda_3 = 1$, where A is the deformed area of the plane normal to the direction 1 of the stress σ_{11}.

The "elastic constants" (*finite elasticity parameters*) are contained, as usual, as coefficients of the strain energy function φ or W; and appear explicitly in expressions like (7) for the stresses. Thus, in terms of W, the principal stresses prove to be of the type

(9) $$\sigma_{ii} = 2\frac{\partial W}{\partial I_1} + 2\frac{\partial W}{\partial I_2}(\lambda_j^2 + \lambda_k^2) + 2\frac{\partial W}{\partial I_3}(\lambda_j^2 \lambda_k^2)$$

involving the elasticity parameters $\partial W/\partial I_1$, $\partial W/\partial I_2$, and $\partial W/\partial I_3$ which, in general, themselves depend on I_1, I_2, I_3 the strain invariants and, therefore, on the state of deformation. For membranes, the stress-resultants, incorporating the membrane thickness h, are preferred over the stresses; and taking, as is conventional, the third principal direction ($i = 3$) to be normal to the membrane surface, the surface stress-resultants are

(10) $$\begin{cases} T_1 = 2h\lambda_3(\lambda_1^2 - \lambda_3^2)\left(\frac{\partial W}{\partial I_1} + \lambda_2^2 \frac{\partial W}{\partial I_2}\right) \\ T_2 = 2h\lambda_3(\lambda_2^2 - \lambda_3^2)\left(\frac{\partial W}{\partial I_1} + \lambda_1^2 \frac{\partial W}{\partial I_2}\right) \end{cases}$$

with $\lambda_3 = 1/\lambda_1 \lambda_2$ and $\partial W/\partial I_3 = 1$ for an incompressible material with prescribed lateral loading.

The essential problem of the representation of elasticity then reduces to discovering the *finite elasticity parameters* $\partial W/\partial I_1$ and $\partial W/\partial I_2$ and, in particular, their dependence on I_1 and I_2 or, alternatively, on λ_1 and λ_2 in accordance with the fundamental constitutive relationships (10). This determination is, ultimately, an experimental one aimed at measuring any *finite elasticity constants* which are properly contained in the functions $\partial W/\partial I_1$, $\partial W/\partial I_2$. Ideally, for greatest simplicity the elasticity parameters are themselves constants, but realistic representations seldom allow this. The dependence of the parameters $\partial W/\partial I_i$ on the state of deformation can be expressed either as functions of the strain invariants I_i or of the principal extension ratios λ_i because of the connections (5). Both methods are used, with different advantages accruing.

C. The Strain Energy Function and Its Derivatives

The equations (10) afford a basis for planning experiments directed to revealing $\partial W/\partial I_1$, $\partial W/\partial I_2$ as functions of the state of deformation. This clearly demands the measurement of T_1, T_2 and λ_1, λ_2 for a sufficiently large array of λ_1, λ_2 pairs. Facility must be provided to control λ_1 or λ_2 separately or, alternatively, to hold either of I_1, I_2 constant while the other is varied; or to arrange some appropriate cover of the I_1-I_2 parameter plane to encompass as wide as possible a variety of deformation states. [For the relations (5) proscribe arbitrary pairing of λ_1, λ_2 for a prescribed pair

I_1, I_2, see Hart-Smith (1966).] The whole experimental determination is eased by a foreknowledge or even a guide as to the functional forms of $\partial W/\partial I_1$ and $\partial W/\partial I_2$. For then the problem reduces to the experimental evaluation of the constants of these functional forms. For this reason, mention is now made of some of the theoretical formulations of the finite elasticity parameters $\partial W/\partial I_1$ and $\partial W/\partial I_2$ or, as is often the case, of the strain energy function W itself. Much of the earlier work developed from kinetic theories which consider the probability of the existence of a given molecular length. In a consolidation of these ideas, Treloar (1943) assumes the presence of N molecules, all of the same length, and a distribution of molecular lengths in the undeformed state given by Kuhn's statistical formula. He assumes an isochoric deformation, that is, one which incurs no volumetric change, such that the extension ratios of individual molecules are the same as the bulk or macroscopic extension ratio. As a result he derives one constant strain energy form

(11) $$W = \tfrac{1}{2} Nkt(I_1 - 3)$$

k being the Boltzmann constant, T the absolute temperature and I_1 the strain invariant of the set (5). The analysis becomes invalid when a significant fraction of molecules have lengths of the same order as the fully extended chain length which means that limits exist on the extension ratios $\lambda_1 \lesssim 5$. The range of validity of elasticity parameters based on (11) is, in fact, rather less than this for rubbery materials. Note that, according to this polymeric theory, $\partial W/\partial I_1 =$ constant and $\partial W/\partial I_2 = 0$.

Contrast this with the Mooney (1940) continuum formulation which rests on four postulates:

(a) there exists isotropy in the undeformed state and in a plane normal to stretch;
(b) volumetric changes and hysteresis are negligible;
(c) in simple shear, the strain is linear with shear stress;
(d) a strain energy function W exists.

The condition (c) is supported by experimental evidence for many but not all rubbers; and the condition (d) implies either isothermal or adiabatic circumstances—these are thought to be the same for rubbers. The condition (a) demands symmetry of W with respect to the principal extension ratios λ_i as reflected in the form derived by Treloar from these postulates:

(12) $$W = \tfrac{1}{4} G \sum_{i=1}^{3} \left(\lambda_i - \tfrac{1}{\lambda_i} \right)^2 + \tfrac{1}{4} H \sum_{i=1}^{3} \left(\lambda_i^2 - \tfrac{1}{\lambda_i^2} \right)$$

in which G is the modulus of rigidity and H is a "coefficient of asymmetry."

The form (12) can also be rendered in terms of the strain invariants as

(13) $$W = C_1(I_1 - 3) + C_2(I_2 - 3)$$

where $C_1 = 1/4(G + H)$, $C_2 = 1/4(G - H)$. Consistent with the four postulates it can be shown that (12) is the most general from that W can take and that the two finite elastic constants G and H (or C_1 and C_2) are both necessary and sufficient for the prescription of W. Treloar's form (11) is seen to be a special case of (13), and Mooney notes the *complete consistency of the statistical and continuum formulations* and remarks that these molecular and thermodynamic theories merely demonstrate or corroborate the postulate of shear linearity as well as indicating its temperature sensitivity. The Mooney form has been the most persistent theory because it has worked well for smallish deformations ($\lambda < 2.5$ in a uniaxial condition; $I_1, I_2 < 15$ approximately for $C_2/C_1 = 0.1$ in more general states of deformation) and, more especially, because it is simple for both $\partial W/\partial I_1$ and $\partial W/\partial I_2$ are constants. The appeal of this simplicity has, indeed, often led to its improper use, for anything giving more realism at higher values of I_1, I_2, or larger values of λ_i inevitably means extreme complication. Table I makes this clear. The later more complex and more realistic (and, therefore, more useful) forms are empirically based. But even so, they reduce to the earlier, simpler, con-

TABLE I

Author	$\partial W/\partial I_1$	$\partial W/\partial I_2$	Valid for I_1	Valid for I_2
Treloar (1943)	constant	zero	< 12	> 50
Mooney (1941)	constant	constant	< 15, $\Gamma = 0.1$	< 20, $\Gamma = 0.1$
Rivlin (1948)	$f_1(I_1, I_2)$	$f_2(I_1, I_2)$	All	All
Isihara, Hashitsume & Tatibana (1951)	$C_1 + 2C_3 I_1$	$C_2 - 2C_3$	5–20	< 10
Rivlin & Saunders (1951)	constant	$f_3(I_2)$	< 12	All
Gent & Thomas (1958)	constant	constant/I_2	< 12	All
Klingbeil & Shield (1964)	constant	CI_2^m	< 12	< 100, $m = -2/3$
Hart-Smith (1966)	$G \exp k_1(I_1 - 3)^2$	Gk_2/I_2	All	All
Veronda & Westmann (1970)	$K_1 \exp A(I_1 - 3)$	$-K_2$	uniaxial tension only	

Some forms of the finite elasticity parameters $\partial W/\partial I_1$ and $\partial W/\partial I_2$ for elastomers and for biological tissue (last entry). I_1, I_2 are strain invariants and $C_1, C_2, C_3, C, G, k_1, k_2, K_1, K_2, A$ are all positive constants.

ceptually-based formulations. Thus, the Hart-Smith form yields the Treloar form when $k_1 = k_2 = 0$. It has become common now to speak of a *neo-Hookean material* which is characterized by the vanishing of $\partial W/\partial I_2$ and the constancy of $\partial W/\partial I_1$ as in the Treloar and simpler Rivlin representations. When, more gnerally, $\partial W/\partial I_2$ is a finite constant, then we speak of a *Mooney-Rivlin material*. These representations are of limited value for rubbery materials which are better described as *Hart-Smith materials* or materials having exponential hyperbolic elasticity (or, less precisely, having an exponential-logarithmic strain energy function).

The elasticity parameters $\partial W/\partial I_i$ are equally well expressed explicitly in terms of the extensions λ_i, the coefficients of this function being then the determinative elasticity constants. Valanis and Landel (1967b) provide such a form which has the advantage of being readily correlated with experimental procedures in which the extension ratios λ_i, not the conceptual invariants I_i, are subject to measurement. And Blatz et al, (1969) adopt this course.

D. Various states of deformation and prerequisite experiments

The equations (10) take several forms for each of several special states of deformation. And the form of the "stress-strain law" differs in each case, particularly, as to the appearance of the quantity

$$\frac{\partial W}{\partial I_1} + f(\lambda_i)\frac{\partial W}{\partial I_2}$$

in the last bracketed term. Thus, for

uniform two-dimensional extension $(\lambda_2 = \lambda_3;\ \lambda_1 < 1)\qquad f(\lambda_i) = \lambda_i^2$

pure shear $(\lambda_2 = 1)\qquad f(\lambda_1) = 1$

simple elongation $(\lambda_2 = \lambda_3)\qquad f(\lambda_1) = 1/\lambda_1$

for an incompressible material $(\lambda_1\lambda_2\lambda_3 = 1)$. It is obvious that any experimental determination of the elasticity parameters from tests based on pure shear alone can yield only the sum $\partial W/\partial I_1 + \partial W/\partial I_2$ because then $f(\lambda_1) = 1$, and cannot separate the two. Such a determination cannot therefore be properly applied to other states of deformation for which the two parameters $\partial W/\partial I_1$, $\partial W/\partial I_2$ have a distinguishable influence. It becomes quite essential, therefore, to design experiments that cater to all states of strain and so to range over the full gamut of I_1 and I_2. This is accomplished (Hart-Smith, 1966) by conducting tests under all three states: simple elongation, pure shear and two-dimensional extension (or uniaxial compression). The exponential tensile law of Fung (1967) which implies a strain energy function that is exponential in elongation ratio, thus cannot be expected to fit data from tests of other types and, in this sense, cannot be generalized. The determinations of Blatz, et al, (1969), on the other hand, rest in both uniaxial tensile and biaxial tests and lead to a simple statement that warrants further testing. That of Verona and Westmann (1970), however, relies only on tensile tests.

But Blatz, et al, (1969) consider an alternative form of biaxial tesing based on a fan segment of membranous tissue and described in terms of radial and circumferential stress resultants.

E. New problems in mechanics

Such boundary value problems must be sought out and solved in order to allow a convincing and rigorous interpretation of experimentation. It is easy to insist on a full range of experimental states; the uniaxial tensile test, the "simple elongation test," is not easily applied to soft tissue materials whilst in biaxial tests the enormous difficulties of confidently controlling the boundary loading and deformation conditions are greatly magnified. Compression testing is easier to undertake but uncertain of interpretation. Distension tests offer a powerful and tractable technique but require elaborate and complex theoretical interpretation. The inflation of a finitely deforming membranous sphere for example, is governed by a pressure-size relationship

$$(14) \qquad p = \frac{4h}{A\lambda} \cdot \frac{\partial W}{\partial I_1} \left[1 - \frac{1}{\lambda^6} \right] (1 + \lambda^2 \Gamma)$$

where h is the undeformed thickness, A is a reference area, λ is the radial extension ratio and $\Gamma = [\partial W/\partial I_2]/[\partial W/\partial I_1]$ is the ratio of elasticity parameters (Adkins and Rivlin, 1952). For a simpler membrane of linear elasticity having a surface stress resultant that is independent of deformation, the corresponding relationship is

$$(15) \qquad p = \frac{4h}{A\lambda} \cdot C$$

where C is a material constant (the "surface tension"). To expect the two forms equally to describe distensile behavior is not reasonable. For a more practical and final example, refer to the distension of a flat circular membrane by lateral pressures in the manner of the diaphragm tests mentioned in Section III, and as expounded by Hart-Smith and Crisp (1967) for rubber polymers. The pressure-distension characteristic takes on the calculated appearance of Fig. 9 in which P' represents the distending pressure and

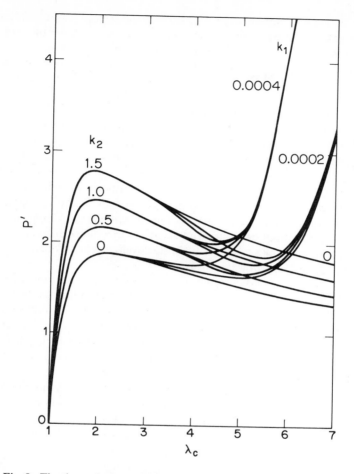

Fig. 9: The theoretical pressure-distension characteristic for the inflation of thin rubber membranes, initially flat and circular showing the variation of the inflation pressure P' with crown extension ratio λ_c for a variety of materials having finite elasticity constants k_1 and k_2, from Hart-Smith and Crisp (1967). (With permission of Intl. J. Eng. Sci., Pergamon Press, Ltd.)

λ_c the extension ratio at the crown of the bubble. Each pair, k_1, k_2, of elastic constants defines a particular material, and the salient feature predicted by the theory is that k_1 and k_2 are separately influential, each in its own regime of distension. Moreover, the maxima and minima occur sensibly at consistent values of λ_c, or bubble size, so that pressure measurements alone at these critical configurations are sufficient to determine each of k_1 and k_2. Such a distended configuration places the membrane in a very complicated, nonuniform biaxial state of deformation; and its description is revealed only by lengthy, complex numerical solutions of nonlinear, coupled differential equations. But the experimental method is simple and reliable and repeatable. In particular, the behavior is insensitive to the boundary conditions, which involve sharp bending of the membrane at its periphery, and which are mounted without undue difficulty. Experimental curves for skin distensibility were elegantly provided in this manner by Grahame and Holt (1969) who applied suction to inflate the skin. But their results are displayed on the basis of two unreasonable hypotheses. First, the membrane is taken to act with a constant surface stress-resultant; secondly, the innate nonlinearity is represented as the gradient of a straight line "of best fit." It is doubtful whether a definitive statement of absolute elasticity is possible from such a framework. It seems certain that experimentation typified by the example of Fig. 9, which may have no direct relevance to skin, cannot be avoided and that it is inevitable that experimental and theoretical mechanics itself needs subtle development before it can match the subtlety of biological tissues.

But there are still old avenues to be explored, which emanate from unresolved anomalies between molecular and continuum formulations of the strain energy function. Inconsistencies seemingly exist between the statistical thermodynamic theories and the continuum theories, on the one hand, and the empirical determinations of W on the other, the latter being the more reliable. Mooney's contention that the one-constant statistically based forms of the strain energy function W simply confirm the shear linearity implicit in his own two-constant expression for W have been mentioned. These two classes of theory are thus reconcilable: the two-constant form offering a more general result without being actually a generalization of the one-constant form. Yet the well-known deficiencies of Mooney's representation force it into disfavor with the availability of the empirical formulations which, alone, satisfactorily describe a wide range of actual behavior of elastomers. The inadequacies of the Mooney description have, of course, spawned several attempts at refinement which take a form of the strain energy function

(16) $$W = K_1(I_1 - 3) + K_2 f(\lambda_1, \lambda_2)$$

where the function f is, in one case (Gent and Thomas, 1958), a rather complicated incomplete elliptical integral. The complexity of this expression, which derives from a modified network theory, seems to have militated against its experimental corroboration and, certainly, against its acceptance as a useful tool. Even here we note that the dependence of W on I_1, in the first term of (16) is identical to the Mooney formulation and cannot, therefore, be reconciled with the more realistic empirical Hart-Smith derivation. On the other hand, an empirical version of $K_2 f$ in Eq. (16) has been proposed that is exactly that of Hart-Smith for $\partial W/\partial I_2$. In this sense then, the empirical representation accords with the analytical theories but is not formally derivable from them. It seems necessary to rely on empiricism and no doubt this is the present safest course in approaching the soft tissues. No attempts have yet been made, in respect of the tissues, at bettering the theoretical approach of King and Lawton (1944) to polymers which parallel with similar developments regarding rubbery materials. The alternative to empiricism is the questioning and reappraisal of the postulates of theories such as Mooney's. And it appears that the most fruitful effort of this kind could be devoted to assessing the consequence of a nonlinear shear relation. For on the one hand, Fung, et al, (1966) have demonstrated experimentally the nonlinear nature of the shear modulus in torsion tests of a mesentery of the rabbit. Thus they give, for the shear modulus

$$(17) \qquad G = G_0 + f_1(I_1) + C_2\sqrt{I_2}$$

or, by *assuming* f_1 to be linear in I_1

$$(18) \qquad G \approx \mu + C_2 |\tau|$$

where μ depends on the initial biaxial tension and $|\tau|$ is the absolute shear stress. On the other hand, Mooney's original work did contain an expression for the strain energy function on the basis that G is deformation dependent, a fact which seems to have been overlooked. Thus, W can be manipulated to give

$$(19) \qquad W = \sum_{r=1}^{\infty} a_r(I_1 - 3)^r + \sum_{r=1}^{\infty} b_r(I_2 - 3)^r$$

of which the simpler Mooney form, linear in $I_1 - 3$ and $I_2 - 3$, is seen to be a degenerate case. It is remarkable that an appropriate truncation of (19) has not been explored in view of the well-known fact that in many harder rubbers the shear linearity condition cannot be tolerated. And so, Eq. (19) provides an obvious base from which to mount one attack on biological soft tissues.

In summary, it is to be noted that any attempt to determine the elastomechanical secrets of the soft tissues requires (a) experiments over a wide range of deformation states, (b) a basis in *finite* mechanics and (c) probably a healthy empiricism.

REFERENCES

Abrahams, M. 1967. The mechanical behavior of tendon collagen fibres under tension *in vitro*, Dig. 7th Intl. Conf. on Med. Biol. Eng. Int. Fed. Med Biol. Eng. Stockholm. p. 509.

Adams, T., and Hunter, W. S. 1969. Modification of skin mechanical properties by eccrine sweat gland activity. J. Appl. Physiol. 26: 417–19.

Adkins, J. E., and Rivlin, R. S. 1952. Large elastic deformations of isotropic materials. IX The deformation of thin shells. Phil. Trans., Roy. Soc. London, A244: 505–31.

Anderson, B. E., and Williams, J. F. 1970. Structural behavior of fresh and preserved tendon. Private communication. (Presented at 10th Conf. on Phys. in Med. and Biol., Melbourne, 24–28 August, 1970.)

Apter, J. T., and Marquez, E. 1968. A relation between hysteresis and other visco-elastic properties of some biomaterials. Biorheology 5: 285–301.

Ayer, J. P. 1964. Elastic tissue. Int. Rev. Connect. Tissue Res. 2: 33–100. (383 refs.)

Benedict, J. V., Walker, L. B., and Harris, E. H. 1968. Stress-strain characteristics and tensile strength of unembalmed human tendon. J. Biomech. 1: 53–63.

Birkhoff, R. D., Abu-Zeid, M. E., and Carter, A. E. 1967. Physics of tissue damage. U.S.A.E.C. Oak Ridge Natl. Lab. Rep. No. 4168. pp. 172–82.

Blanton, P. L., and Biggs, N. L. 1970. Ultimate tensile strength of fetal and adult human tendons. J. Biomech. 3(2): 181–89.

Blatz, P. J., Chu, B. M., and Wayland, H. 1969. On the mechanical behavior of elastic animal tissue. Trans. Soc. Rheol. Vol. 13, Part I, 83–102.

Brozek, J., and Kinzey, W. 1960. Age changes in skinfold compressibility. J. Gerontol. 15: 45–51.

Bruchle, H., and Moll, W. 1968. [A method for the investigation of the mechanical properties of tendons] Eine Messunordnung zur Prüfung des mechanischen Verhaltens von Sehnen. Z. Ges. Exp. Med. 147: 23–28.

Burton, A. C. 1961. Physical principles of circulatory phenomena: the physiological equilibrium of the heart and blood vessels, *In* W. F. Hamilton, (ed,) Handbook Physiol. I(2). Amer. Physiol. Soc., Washington. pp. 85–106.

Clegg, E. J., and Kent, C. 1967. Skinfold compressibility in young adults. Hum. Biol. 39: 418–29.

Contini, R., Gage, H., and Drillis, R. 1964. Human gait characteristics, *In* R. M. Kenedi, (ed.) Biomechanics and Related Bioengineering Topics. Pergamon, Oxford. pp. 413–31.

Cox, H. T. 1941. The cleavage lines of the skin. Br. J. Surg. 29: 234–40.

Craik, J. E., and McNeil, I. R. R. 1966. Micro-architecture of skin and its behavior under stress. Nature 209: 931–32.

Crisp, J. D. C. 1968. Biomechanics—notes of a graduate seminar series. Ind. Inst. Sc., Dept. Aero Eng. Rep. No. AE 223S.

Demichev, N. P., and Tupitsyn, I. O. 1966. [On changes in the elastic properties of tendons during preservation]. In Russin. Ortop. Traum. Protez. 27: 43–48.

De Muth, W. E. 1968. High velocity bullet wounds of the thorax. Amer. J. Surg. 115: 616–25.

Dick, J. C. 1951. Stretch resistance of skin. J. Physiol. 112: 102–13.

Duggan, T. C. 1967. Dynamic mechanical testing of living tissue, Dig. 7th Intl. Conf. on Med. Biol. Eng. Int. Fed. Med. Biol. Eng. Stockholm. p. 368.

Elden, H. R. 1964. Hydration of connective tissue and tendon elasticity. Biochim. Biophys. Acta 79: 592–99.

Elliott, D. H. 1965. Structure and function of mammalian tendon. Biol. Rev. 40: 392–421.

Elliott, D. H. 1967. The biomechanical properties of tendon in relation to muscular strength. Ann. Phys. Med. IX, 1: 1–7.

Elliott, D. H., and Crawford, G. N. C. 1965. The thickness and collagen content of tendon relative to the strength and cross-sectional area of muscle. Proc. Roy. Soc. B162: 137–46.

Ellis, D. G. 1969. Cross-sectional area measurements for tendon specimens: a comparison of several methods. J. Biomech. 2: 175–86.

Fazekas, I. G., Kosa, F., and Basch, A. 1967. [The influence of body constitutional factors (height) on the resistance of human skin to laceration.] In German. Zacchia 3: 512–21.

Fazekas, I. G., Kosa, F., and Basch, A. 1968. [On the tensile strength of skin in various body areas.] In German. Deutsch. Z. Ges. Gerichtl. Med. 64: 62–92.

Fazekas, I. G., Kosa, F., and Basch, A. 1969. [On the effect of constitutional factors (body height) on the tensile strength of the human skin.] In German. Gegenbaur Borph. Jahrb. 113: 295–302.

Finlay, B. 1969. Scanning electron microscopy of the human dermis under uniaxial strain. Bio-Med. Engrg. 4, 7: 322–27.

Frisén, M., Mägi, M., Sonnerup, L., and Viidik, A. 1969. Rheological analysis of soft collagenous tissue. J. Biomech. 2: 13–28.

Fung, Y. C. B. 1967. Elasticity of soft tissues in simple elongation. Amer. J. Physiol. 213, 6: 1532–544.

Fung, Y. C. B., Zweifach, B. W., and Intaglietta, M. 1966. Elastic environment of the capillary bed. Circulation Res. 19: 441–61.

Geever, E. F., Levenson, S. M., and Manner, G. 1966. The role of noncollagenous substances in the breaking strength of experimental wounds. Surgery 60(2): 343–51.

Gent, A. N., and Thomas, A. G. 1958. Forms for the stored (strain) energy function for vulcanized rubber. J. Polymer. Sci. 28: 625–28.

Gibson, T. 1964. Biomechanics in plastic surgery. In R. M. Kenedi, (ed.) Biomechanics and Related Bio-Engineering Topics. Pergamon, Oxford. pp. 129–34.

Gibson, T., and Kenedi, R. M. 1967. Biomechanical properties of skin. Surg. Clin. N. Amer. 47(2): 279–94.

Gibson, T., Stark, H., and Evans, J. H. 1969. Directional variation in extensibility of human skin *in vivo*. J. Biomech. 2: 201–204.

Glaser, A. A., Marangoni, R. D., Must J. S., Beckwith, T. G., Brody, G. S., Walker, G. R., and White, W. L. 1965. Refinements in the methods for the measurement of the mechanical properties of unwounded and wounded skin. Med. Electron. Biol. Eng. 3: 411–19.

Grahame, R., and Holt, P. J. L. 1969. The influence of ageing on the *in vivo* elasticity of human skin. Gerontologia 15: 121–39.

Green, A. E., and Adkins, J. E. 1960. Large Elastic Deformation and Nonlinear Continuum Mechanics. Clarendon Press, Oxford. pp. 10, 23–25, 51–61, 328.

Harkness, M. L. R., and Harkness, R. D. 1968. Mechanism of action of thiols on the mechanical properties of connective tissues. Biochim. Biophys. Acta 154: 553–59.

Harris, E. H., Walker, L. B., and Bass, B. R. 1966. Stress-strain studies in cadaveric human tendon and an anaomaly in the Young's modulus thereof. J. Med. Biol. Engng. 4: 253–59.

Hart-Smith, L. J. 1966. Elasticity parameters for finite deformations of rubber-like materials. J. Appl. Math Phys. (ZAMP), 17(15): 608–26.

Hart-Smith, L. J., and Crisp, J. D. C. 1967. Large elastic deformations of thin rubber membranes. Int. J. Eng. Sci. 5: 1–24.

Haut, R. C., and Little, R. W. 1969. Rheological properties of canine anterior cruciate liagments. J. Biomech. 2: 289–98.

Hickman, K. E., Lindan, O., Reswick, J. B., and Scanlan, R. H. 1966. Deformation and flow in compressed skin tissues, In Biomedical Fluid Mechanics Symposium. ASME, New York pp. 127–47.

Higuchi, T., and Tillman, W. J. 1965. Stress-relaxation of stretched callus strips. Arch. Environ. Health 11: 508–21.

Hildebrandt, J., Fukaya, H., and Martin, C. J. 1968. Completing the length–tension curve of tissue. J. Biomech. 2: 463–67.

Hult, A. M. 1965. Measurement of elastin in human skin and its quantity in relation to age. J. Invest. Derm. 44: 408–12.

Kenedi, R. M. (1967). Biomechanics in the modern world. Chartered Mech. Eng. 14(1): 2–9.

Kenedi, R. M., Gibson, T., and Daly, C. H. 1964. Bioengineering studies of the human skin, In R. M. Kenedi, (ed.) Biomechanics and Related Bioengineering Topics. Pergamon, Oxford. pp. 147–58.

Kenedi, R. M., Gibson, T., Daly, C. H., and Abrahams, M. 1966. Biomechanical characteristics of human skin and costal cartilage. Fed. Proc. 25(3): 1084–1087.

Kiistala, U. 1968. A suction blister device for separation of viable epidermis and dermis. J. Invest. Derm. 50: 129–37.

King, A. L., and Lawton, R. W. (1944). Elasticity of body tissues, In O. Glasser, (ed.) Medical Physics, 2. Year Book Publ., Chicago. pp. 303–16.

Kirk, E., and Kvorning, S. A. 1949. Quantitative measurements of the elastic properties of the skin and subcutaneous tissue in young adults and old individuals. J. Gerontol. 4: 273–84.

Langer, A. K. 1861. Zur Anatomie und Physiologie der Haut. Sitzurgeber Wissenach. D. K. Akad. D. 19: 179.

Lee, J. S., Frasher, W. G., and Fung, Y. C. B. 1967. Two-dimensional finite deformation experiments on dog's arteries and veins. AMES Dept., University of California, San Diego, AFOSR Rep. 67-1980.

Lissner, H. R. 1965. Introduction to biomechanics. Arch. Phys. Med. & Rehab. 46: 2–9.

Lowe, L. B., and van der Leun, J. C. 1968. Suction blisters and dermal-epidermal adherence. J. Invest. Derm. 50: 308–14.

Mackie, B. S. 1967. Dry epidermal density. Brit. J. Derm. 79: 411–15.

Main, R. A. 1969. The skin in pregnancy. Brit. J. Clin. Pract. 23: 47–50.

Mendosa, S. A., and Milch, R. A. 1965. Tensile strength of skin collagen. Surg. Forum. 15: 433–34.

Molen, H. R. van der 1966. Elastizitätsveränderungen der Haut und ihre Bedentung für die Entsehung der Phlebopathien. Zentralblatt fur Phlebol. 5: 75–82.

Montagna, W. 1962. The Structure and Function of Skin. 2nd ed. Academic Press, New York. pp. 127–37.

Mooney. M. 1940. A theory of large elastic deformation. J. Appl. Phys. 11: 582–92.

Morrison, J. B. 1968. Bioengineering analysis of force actions transmitted by the knee joint. Bio. Med. Eng. 3, 4: 164–70.

Morrison, J. B. 1970. The mechanics of the knee joint in relation to normal walking. J. Biomech. 3: 51–61.

Mucri, F. J., and Brubaker, R. F. 1969. Methodology of eye pressure measurement. Biorhel. 6(1): 37–45.

Parot, S., and Bourlière, F. 1967. Une nouvelle technique de measure de la compressibilité de la peau et du tissu sous-cutané—Influence du sexe, de l'âge et du site de mesure sur les resultats. Gerontologia 13: 95–110.

Paul, J. P. 1966. The biomechanics of the hip joint and its clinical relevance. Proc. Roy. Soc. Med. 59(10): 943–48.

Paul, J. P. 1967. Forces transmitted by joints in the human body. Proc. Inst. Mech. Eng. 181(3): 8–15.

Ridge, M. D., and Wright, V. 1964. A rheology study of skin, In R. M. Keoedi, (ed.) Biomechanics and Related Bioengineering Topics. Pergamon, Oxford. pp. 165–75.

Ridge, M. D., and Wright, V. 1965. The rheology of skin. A bioengineering study of the mechanical properties of human skin in relation to its structure. Brit. J. Dermatol. 77: 639–49.

Ridge, M. D., and Wright V. 1966a. The ageing of skin. A bioengineering approach. Gerontol. 12: 174–92.

Ridge, M. D., and Wright, V. 1966b. Rheological analysis of connective tissue. A bioengineering analysis of the skin. Ann. Rheum. Diseases 25(6): 509–15.

Ridge, M. D., and Wright, V. 1966c. The directional effects of skin—a bioengineering study of skin with particular reference to Langer's lines. J. Invest. Dermatol. 46(4): 341–46.

Ridge, M. D., and Wright, V. 1966d. An extensometer for skin—its construction and application. Med. Biol. Eng. 4: 533–42.

Rigby, B. J. 1964. Effect of cylic extension on the physical properties of tendon collagen and its possible relation to biological ageing of collagen. Nature 202, B-2: 1072–1074.

Rigby, B. J., N., Spikes, J. D., and Eyring, H. 1959. The mechanical properties of rat tail tendon. J Hirai. Gen Physiol. 43: 265–83.

Rothman, S. 1954. Physiology and Biochemistry of the Skin. Univ. Chicago Press, Chicago, Ill. Chap. I, pp. 1–7.
Saunders, D. W. 1964, Large deformations in amorphous polymers, *In* Biomechanics and Related Engineering Topics, R. M. Kenedi (ed.) Pergamon, Oxford. pp. 301–19.
Sokoloff, L. 1966. Elasticity of aging cartilage. Fed. Proc. 25(3): 1089–1095.
Sulzberger, M. B., Cortese, T. A., Fishman, L., and Wiley, H. S. 1966. Studies of blisters produced by rubbing. I. Result of linear rubbing and twisting techniques. J. Invest. Derm. 47: 456–65.
Tregear, R. T. 1958. The Physics of Rubber Elasticity. 2nd ed. Clarendon Press, Oxford.
Tregear, R. T. 1965. Interpretation of skin impedance measurements. Nature 205: 600–601.
Tregear, R. T. 1966. Physical Functions of Skin. Academic Press, London. pp. 73–75.
Tylicki, M., and Zbrodowski, A. 1966. Badania zrywalnosci krezek scienga zginacza dlugieco kciuka. [Investigations of the resistance of the mesotendium of the flexor pollicis longus to tearing.] In Polish. Polski Przeglad Chirurg. 38(2): 123–27.
Valanis, K. C., and Landel, R. F. 1967a. Large multiaxial deformation behavior of a filled rubber. Trans Soc. Rheol. 11(2): 243–56.
Valanis, K. C., and Landel, R. F. 1967b. The strain–energy function of a hyperelastic material in terms of the extension ratios. J. Appl. Phys. 38: 2997–3002.
Van Brocklin, J. D., and Ellis, D. G. 1965. A study of the mechanical behavior of toe extensor tendons under applied stress. Arch. Physiol. Med & Rehab. 46: 369–73.
Veronda, D. R., and Westmann, R. A. 1970. Mechanical characteristics of skin—finite deformations. J. Biomech. 3(1): 111–24.
Verzár, 1963. The aging of collagen. Sci. Amer. 208(4): 104–14.
Viidik, A. 1968. A rheological model for uncalcified parallel-fibered collagenous tissue. J. Biomech. 1: 3–11.
Viidik, A., and Mägi, M. 1967. Viscoelastic properties of ligaments. Dig. 7th Intl. Conf. on Med. Biol. Eng. Int. Fed. Med. Biol. Eng., Stockholm. p. 507.
Vlasblom, D. C. 1967. Skin elasticity. *In vivo* measurements at small deformations. Dig. 7th Intl. Conf. on Med. Biol. Eng. Int. Fed. Med. Biol. Eng., Stockholm. p. 369.
Walker, L. B., Harris, E. H., and Benedicts, J. V. 1964. Stress-strain relationships in human cadaveric plantaris tendon: a preliminary study. Med. Electron. Biol. Eng. 2: 31–38.
Williams, J. F., and Svensson, N. L. 1968. A force analysis of the hip joint. Bio. Med. Eng. 3(8): 365–70.
Wright, V., Dowson, D., and Lonfield, M. D. 1969. Joint stiffness—its characterization and significance. Bio. Med. Eng. 4: 8–14.

7

Stress-Strain-History Relations of Soft Tissues In Simple Elongation

YUAN-CHENG B. FUNG

I. INTRODUCTION

A history-dependent element exists in the mechanical action of all biological materials. The stress at a given point at any instant of time depends not only on the strain at that time, but also on the strain history. Such a dependence is revealed for passive materials such as connective tissues, collagen, elastin, and so on, in the phenomena of stress relaxation at constant strain, creep at constant stress, hysteresis under cyclic loading, dependence of the elastic moduli on the strain rate, and cyclic stress fatigue. For excitable tissues such as muscles, the stress-strain relation obviously depends on the course of events following the stimulation.

A mathematical description of the stress-strain-history law of tissues is required before a theoretical approach can be made on any problem in biomechanics of organs, such as the dynamics of the heart, the elasticity of the lung, the peristalsis of the ureter, and the autoregulation of the blood flow. Unfortunately, at the present time, the constitutive equation is not completely known for any biological material, mainly because a reasonably extensive two- and three-dimensional testing program is difficult to carry out on small, flimsy specimens that must be maintained in living condition. Experiments are usually limited to a one-dimensional stress field.

There are two different objectives for studying the constitutive equations of living tissues. One is to obtain a simple, general law for mathematical analysis of boundary–value problems, the other is to learn as many specific details as possible about specific tissues and organs. In engineering, Hookean solids, ideal plastic bodies, and Newtonian fluids are examples of the former; whereas detailed rheological models are examples of the latter. In this article we are more interested in the former. We hope to obtain, in a suitably restricted range of variables, a relatively simple and general expression for the stress-strain relationship.

As in engineering mechanics, generally, simplified constitutive equations are applicable only for limited ranges of stress and strain. For the description of living tissues, this attention to the restricted ranges of variables must be observed.

In the following, we shall restrict our discussion to soft tissues such as the blood vessels, connective tissues, muscles, skin, and ureter, and to simple elongation.

II. GENERAL FEATURES OF MECHANICAL RESPONSE

That the stress-strain relationship of animal tissues deviates from Hooke's law was known to Wertheim (1847) who showed that the stress increases much faster with increasing strain than Hooke's law predicts. It is also known that tissues in the physiological state are usually not unstressed. If an artery is cut, it will shrink away from the cut. A broken tendon retracts away; the lung tissue is in tension at all times.

If a segment of an artery is excised and tested in a tensile testing machine by imposing a cyclically varying strain, the stress response will show a hysteresis loop with each cycle, but the loop decreases with succeeding cycles, rapidly at first, tending to a steady state after a number of cycles (see Fig. 1).

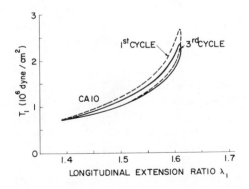

Fig. 1: Preconditioning. Cyclic stress response of a dog's carotid artery which was maintained in cylindrical configuration (by appropriate inflation or deflation) when stretched longitudinally. λ_1 is the stretch ratio referred to the relaxed length of the segment. 37°, 0.21 cycles/min. Physiological length $L_P = 4.22$ cm. $L_P/L_0 = 1.67$. Diameter, physiol. = 0.32 cm. $A_0 = 0.056$ cm^2. Dog wt. 18 kg.

The existence of such an initial period of adjustment after a large disturbance seems common to all tissues. From the point of view of mechanical testing, the process is called *preconditioning*. Generally, only mechanical data of preconditioned specimens are presented. However, since the exact meaning of preconditioning is not very clear (it is not easy to tell exactly what happened to the tissue, nor is it easy to follow a rigid criterion for repeatability of cylic responses), it is prudent in the laboratory protocol to save the preconditioning records for future references.

The hysteresis curves of a rabbit papillary muscle (unstimulated) is shown in Fig. 2. Simple tension was imposed by loading and unloading at the rates

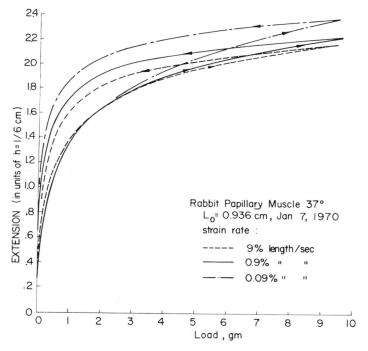

Fig. 2: The length-tension curve of a resting papillary muscle from the right ventricle of the rabbit. (a) Hysteresis curves at strain rates 0.09% length/sec.; 0.9% length/sec.; and 9% length/sec. Length at 9 mg. = 0.936. 37°C.

indicated in the figure. It is seen that the hysteresis loops did not depend very much on the rate of strain. In general, this insensitivity with respect to the strain rate holds within at least a 10^3 fold change in strain rate.

Figure 3 shows a stress-relaxation curve for the mesentery of the rabbit. The specimen was strained at a constant rate until a tension T_1 was obtained. The length of the specimen was then held fixed and the change of tension

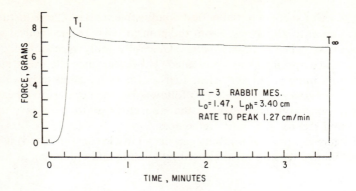

Fig. 3: Relaxation curve of a rabbit mesentery. The specimen was stressed at a strain rate of 1.27 cm/min to the peak. Then the moving head of the testing machine was suddenly stopped so that the strain remained constant. The subsequent relaxation of stress is shown.

with time was plotted. In a linear scale of time only the initial portion of the relaxation curve is seen. Relaxation in a long period of time is shown in Fig. 4, in which the abscissa is log t. It is seen that in seventeen hours a large portion of the initial stress was relaxed. If the initial stress is sufficiently high, the relaxation curve does not level off even at $t = 10^5$ sec (see the solid curve in Fig. 4). If, however, the initial stress is lower than a certain value, the stress levels off at the elapse of a long time, as shown by the dashed curve in Fig. 4. Apparently, at the higher stress levels the relaxation has not stopped even at 10^5 sec. Note that the term stress is used here in the Lagrangian sense: tension divided by the initial cross-sectional area.

Figure 5 illustrates the creep characteristics of the papillary muscles of the rabbit loaded in the resting state under a constant weight. The change of length was recorded and replotted in a logarithmic scale of time. The creep characteristics depend very much on the temperature and the load.

These features of hysteresis, relaxation, and creep at lower stress ranges ("physiological") are seen on all materials tested in our laboratory: the mesentery of the rabbit and dog, the femoral and carotid arteries and the veins of the dog, cat, and rabbit, the ureter of several species of animals, including human, and the papillary muscles at the resting state. The major differences among these tissues are the degree of distensibility. In the physiological range, the mesentery can be extended 100–200% from the relaxed (unstressed) length, the ureter can be stretched about 60%, the resting heart muscle about 15%, the arteries and veins about 60%. Data from other laboratories on skins within about 40% stretch from the relaxed lengths, tendons within 2–3%, fall into the same category. Much beyond these ranges the tissues would yield and fail, the characteristics of which are more specific with different tissues.

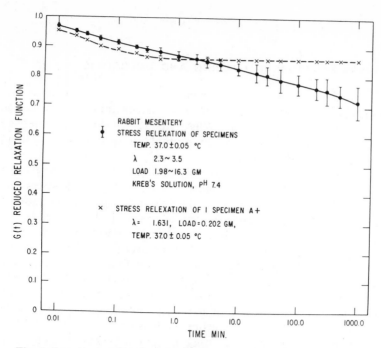

Fig. 4: Long-term relaxation of rabbit mesentery. Solid curve shows the mean reduced relaxation function for 16 different initial stress values. The dashed curve refers to one test at much lower stress level. By H. Chen.

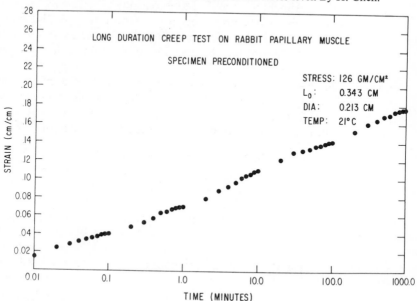

Fig. 5: Creep characteristics of the papillary muscle of the rabbit. By J. Pinto.

III. STRESS RESPONSE IN LOADING AND UNLOADING

In spite of the tremendous effort spent in the investigation of the stress-strain-history relationship for living tissues, a satisfactory, fully verified equation has eluded the investigators. What is better known is the stress-strain relationship in loading and unloading in simple elongation at a constant rate of strain. To this subject we now turn.

In the following we speak of stress and strain in the Lagrangian sense. For a one-dimensional specimen loaded in tension, the tensile stress T is the load divided by the "reference" area; the "stretch ratio" λ is the ratio of the length to the "reference" length; the tensile strain ϵ is defined as $(\lambda^2 - 1)/2$. The "reference" state can be arbitrarily chosen, subject to the requirement that it be physiological and definitive. In engineering mechanics, we always choose the "natural" state as the reference state, that is, one that represents the relaxed, unstressed material. In biological specimens it turns out that one of the most difficult practical problems is to determine the dimensions of the test specimen at the natural state. The reason is that at this state the material is usually very soft, its geometry is generally ill-defined (think of a small blood vessel or a ureter floating in a physiological solution), and it is difficult to handle when measuring the cross-sectional area and length. As a result, the "natural" length is usually not a reliable number, often the most inaccurate of the experimental data. For this reason, experimental results on biological tissues are better presented with respect to a well-defined reference state. The natural state—still very useful in any analysis—should be determined by extrapolation; but the question of accuracy should always be remembered.

For an incompressible material, the cross-sectional area of a cylindrical specimen is reduced by a factor $1/\lambda$ when the length of the specimen is increased by a factor λ. Hence the Eulerian stress σ (which is referred to in the crosssection of the deformed specimen and is sometimes called the "true" stress) is λ times T:

$$\text{(1)} \qquad \sigma = \frac{P}{A} = \frac{P}{A_{\text{ref}}}\lambda = T\lambda$$

Consider first the relationship between load and deflection in the *loading* process. If the slope of the T versus λ curve as shown in Fig. 2 is plotted against T, the result as shown in Fig. 6 is obtained. As a first approximation, we may fit the experimental curve by a straight line in the range of T exhibited and by the equation:

$$\text{(2)} \qquad \frac{dT}{d\lambda} = \alpha(T + \beta)$$

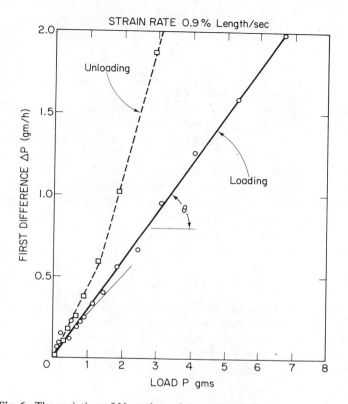

Fig. 6: The variation of Young's modulus with load at strain rate 0.9% length/sec., illustrating the method of determining the constants $\hat{\alpha}_1, \hat{\beta}_1$. Near the origin, a different straight line segment is required to fit the experimental data.

Then, an integration gives:

$$(3) \qquad T + \beta = ce^{\alpha\lambda}$$

The integration constant can be determined by finding one point on the curve, say $T = T^*$ when $\lambda = \lambda^*$. Then

$$(4) \qquad T = (T^* + \beta)e^{\alpha(\lambda - \lambda^*)} - \beta$$

Note that if λ is referred to the natural state, we must have, by definition, $T = 0$ when $\lambda = 1$; which is possible only if:

$$(5) \qquad \beta = \frac{T^* e^{-\alpha(\lambda^* - 1)}}{1 - e^{-\alpha(\lambda^* - 1)}}$$

Equation (2) includes Hookean materials, for which $\alpha = 0$ and

(6) $$\frac{dT}{d\lambda} = \text{constant}$$

A more refined representation of the experimental data shown in Fig. 6 can be made by several straightline segments. The reason for such a choice is a practical one: in the analysis of the function of the organs (such as the contraction of the heart, the peristalsis of the ureter), the function $dT/d\lambda$ appears frequently, and it is desirable to have it represented by as simple a form as possible (in the linear elasticity theory $dT/d\lambda$ is a constant). The form given in Eq. (2) is about as simple as can be, although in the case represented in Figure 6 it is necessary to specify two straight line segments:

(7)
$$\frac{dT}{d\lambda} = \alpha_1(T + \beta_1), \quad \text{for } 0 \leq T \leq T_1$$
$$\frac{dT}{d\lambda} = \alpha_2(T + \beta_2), \quad \text{for } T_1 \leq T \leq T_2$$

In this case, the integrated curve must be represented by two expressions in the form of Eq. (4), matched at the juncture $T = T_1$.

An alternative to Eq. (6) is to represent the modulus $dT/d\lambda$ as a polynomial in T. For example, the mesentery data shown in Fig. 2 can be reasonably represented by a parabola

(8) $$\frac{dT}{d\lambda} = aT(1 - bT)$$

An integration gives the following formula which is considerably more cumbersome to use than Eq. (4):

(9) $$T = \frac{e^{a\lambda}}{c + be^{a\lambda}}$$

(10) $$c = \left(\frac{1}{T^*} - b\right)e^{a\lambda^*}$$

λ^*, T^* being an arbitrary point on the curve. For the rabbit mesentery (Fung, 1967) we have $a = 12.4$; $b = 7.29 \times 10^{-4}$ cm²/dyne.

Unloading at the same strain rate results in similar straight lines with much larger slopes, as shown by the dotted curve in Fig. 6.

The ureters of the dog, rabbit, cat, and human show features similar to those presented in Fig. 6 (Yin, 1970). The same types of stress-strain curves at lower stress ranges can be found in reports on ligamenta flava (Nunley, 1958), ligamenta nuchae (Wood, 1954), plantar fascia (Wright and Rennels,

1964), flavum (Nachemson and Evans, 1968), fascia lata (Gratz, 1931), anterior cruciate (Smith, 1954), tendons (Stromberg and Wiederhielm, 1970; Takigawa, 1953; Elliott, 1965; Viidik, 1966; Walker et al, 1964; Stucke, 1950; Abrahams, 1967), cervix of rat (Harkness and Harkness, 1959), collagen (Hall, 1951; Morgan, 1960; Laban, 1962), elastin (Hoeve and Florey, 1958), skin (Ridge and Wright, 1964). Tissues which are high in collagen are less extensible, and have a fairly large initial slope at the origin (natural state). Elliott (1965) presents an interesting review on the tendon and its relation to the muscle and bone; a large bibliography can be found in his paper. Hall (1951) showed a large variation of the elasticity of collagen fibers with pH. The shrinkage of collagen at higher temperature is discussed by Theis and Steinhardt, Jr. (1950) and Rigby (1964).

We should remark that collagenous tissues show an exponential stress response only if the strain is restricted to be less than about 2%. Within this limit Stromberg and Wiederhielm (1969) obtained a value of $\alpha = 450$ for the tendon of mice tail. Creep was shown to be small for collagenous tendon, less than 0.1% over several hundred seconds.

IV. VARIOUS OTHER EXPRESSIONS IN THE LITERATURE FOR STRESS RESPONSE TO LOADING IN UNIAXIAL TENSION

The equations (2)–(8) are essentially those of Fung (1967), except that we now speak of loading and unloading curve separately instead of the vague "elastic curve" of the previous publication. In a later section (V) we shall discuss the so-called *elastic response*, which requires a further definition to be meaningful. Presently let us consider other theoretical expressions of stress-strain relationship.

One of the best known approaches to elasticity of bodies capable of finite deformation is to postulate the form of an elastic potential, or strain energy function, W (see Green and Adkins, 1960). For example, if a body is elastically isotropic and homogeneous, the strain energy function must be a function of the strain invariants. Well-known examples of strain-energy functions are those of Mooney (1940), Rivlin (1948), Rivlin and Saunders (1951), Gent and Thomas (1958), Klingbell and Shield (1964), and Hart-Smith (1966). See Chap. 6 by Crisp and the treatise of Green and Adkins (1960) where references are given.

Valanis and Landel (1967) presented a very useful strain-energy function

(11) $$W = \sum_{i=1}^{3} f(\ln \lambda_i)$$

for which we have

(12) $$\sigma_i \lambda_i = f'(\ln \lambda_i) + \bar{k}$$

where $\bar{k}$ is a Lagrange multiplier introduced to account for incompressibility and the prime denotes differentiation with respect to the argument.

Specific application of Valanis' form to soft tissues was made by Blatz, Chu and Wayland (1969), who proposed the following:

(13)   (A) $f(\ln \lambda_i) = C(\lambda_i^\alpha - 1)$

When applied to rabbit's mesentery, Blatz et al (1969) found $\alpha = 18$. For skeletal muscle at resting state they found $\alpha = 8$, for latex rubber, $\alpha = 1.5$.

(14)   (B) $f(\ln \lambda_i) = C[e^{\alpha(\lambda_i^2 - 1)} - 1]$

for which the uniaxial tension case gives

$$\sigma = \left(\frac{G}{\alpha + 1}\right)\left\{\lambda e^{\alpha(\lambda^2 - 1)} - \frac{1}{\lambda^2} e^{\alpha[(1/\lambda) - 1]}\right\}$$

(15)   (C) $W = \dfrac{G}{2\alpha}[e^{\alpha(I_1 - 3)} - 1]$.

Veronda and Westmann (1970) proposed the following form for the stress potential

(16) $W = C_1[e^{\beta(I_1 - 3)} - 1] + C_2(I_2 - 3) + g(I_3)$

where $C_1, C_2, \beta,$ are constants and $g(1) = 0$. For cat's skin, they found the following constants:

$$C_1 = 0.00394, \;\; \beta = 5.03, \;\; C_2 = -0.01985.$$

These expressions presuppose isotropy. However, it is almost axiomatic that biological tissues are not isotropic.

Other expressions proposed, not reduced to the form of strain energy, and not pretending to be generally valid for three-dimensional stress states, are the following:

Wertheim, (1847)	$\epsilon^2 = a\sigma^2 + b\sigma$	(17a)
Morgan, (1960)	$\epsilon = a\sigma^n$	(17b)
Kenedi et al, (1964)	$\sigma = k\epsilon^d$	
	$\sigma = B[e^{m\epsilon} - 1]$	(17c)
Ridge and Wright, (1964)	$\epsilon = C + k\sigma^b$	
	$\epsilon = x + y \log \sigma$	(17d)

A long time ago, in 1880, Charles Roy found that animal tissues (arteries, veins, skin, nerve, tendon, muscle) differ from the vast majority of other substances in their thermoelastic properties. They contract on being warmed and expand when their temperature is lowered. Corresponding with this he found that on being stretched their temperature rises, then declines upon their being relaxed. These features are similar to those of rubber, whose thermoelastic properties were known earlier to William Thomson (Lord Kelvin), when he studied these materials while writing an article on thermodynamics for the Encyclopedia Britanica (see Proc. Roy. Soc. May 1865, and Encyclop. Brit., 9th Ed., Art., Elasticity.) This common feature suggests that the theory of rubber elasticity may be applicable to living tissues.

Flory, Treloar, and others have theoretically derived the strain-energy function for polymers (see Treloar, 1967). King and Lawton (1950) adopted this theory to body tissues. They emphasized that when a body deforms, molecular chains may be reoriented without altering either intermolecular distances or valence angles: only the degree of randomness is decreased when the specimen is stretched. This is the entropy source for elastic stress. To set the argument more formally, consider a stretched specimen in thermodynamic equilibrium. During a reversible process the infinitesimal change in internal energy dU may be written

$$dU = TdS - pdV + \sigma dL$$

where T is the absolute temperature, S the entropy, p the external pressure, V the volume of the specimen, σ the tensile elastic force, and L the length of the specimen. For an incompressible material $dV = 0$, hence

$$\sigma = \frac{\partial U}{\partial L}\bigg|_{V,T} - T\frac{\partial S}{\partial L}\bigg|_{V,T}.$$

The last term associates the stress with entrophy changes. By the well-known identity

$$\frac{\partial S}{\partial L}\bigg|_{T} = -\frac{\partial \sigma}{\partial T}\bigg|_{L}$$

there results

$$\sigma = \frac{\partial U}{\partial L}\bigg|_{V,T} + T\frac{\partial \sigma}{\partial T}\bigg|_{V,L}.$$

From the slopes of isometric force-temperature curves, $\partial \sigma/\partial T$ and, therefore, the entropy term may be evaluated. Examinations of experimental results on the dependence of elastic force on temperature have lead Meyer and Ferri (1935) and others to identify rubber elasticity with the entropy mechanism. The results of Roy point to the same mechanism for living tissues.

Meyer and Picken (1934), and King and Lawton (1950) showed that results derived from a simple statistical mechanics theory agree well with data on living tissues. However, as Treolar (1967) said about rubber, one cannot expect the simple statistical theory to account for the mechanical property of the material in detail. For example, King and Lawton's formula for tension:

$$\sigma = C\{L^{-1}(q\lambda) - \lambda^{-3/2}L^{-1}(q\lambda^{-1/2})\} \tag{18}$$

where C and q are constants, λ is the stretch ratio, and $L^{-1}(x)$ is the inverse of the Langevin function,

$$L(x) = \coth x - \frac{1}{x} \tag{19}$$

does not fit the stress-strain curves illustrated in the previous section. However, Eq. (18) does fit the load-extension curves of a tetanically contracted muscle fiber and of a fresh myosin filament.

V. VISCOELASTICITY OF SOFT TISSUES

The experimental results illustrated in Figs. 1 through 6 show that biological tissues are not elastic. The history of strain affects the stress. In particular, there is a considerable difference in stress response to loading and unloading. This suggests an analog with the conventional theory of plasticity. However, the tissue behavior is more complex than that of an ideal plastic body because (1) the stresses in both loading and unloading processes are highly nonlinear with respect to strain, (2) pronounced stress relaxation and creep exist. In view of these complications, it seems better to regard soft tissues as viscoelastic bodies rather than plastic bodies. In any case, the concept of stress potential or strain-energy function must be questioned, because it cannot distinguish loading from unloading, and because it is easy to conceive a situation in which some components of stresses are being loaded, while others are being unloaded.

Roy's 1880 paper followed Lord Kelvin's (1865) methods and treated viscoelasticity of the arteries in essentially modern terms. Critical reviews of modern works on blood vessel are given by Wetterer and Kenner (1968, §34, p. 224 ff.), Bergel (Chap. 5), Westerhof and Noordergraaf (1970), Reuterwall, (1921) and Apter (1966, 1967, 1968, Chap. 9). The principal attention of these studies was the damping characteristics of tissues with respect to small oscillations superposed on a steady state.

Buchthal and Kaiser (1951) presented a thorough study on the creep and relaxation of the skeletal muscle at the resting state. They concluded that in the first few milliseconds of loading the muscle was thixotropic; then there followed a steady creep or relaxation. (I feel, however, that this conclusion might be affected by the dynamics and wave propagation immediately following the loading, which was not sufficiently clarified.) They give the following empirical formulas:

(20) $\qquad$ Creep: $L(t) = L(1) + C_l \log t$, (5 msec $< t < 1$ sec).
(21) $\qquad$ Relaxation: $P(t) = P(1) - C_p \log t$, (5 msec $< t < 0.1$ sec).

Galford and McElhaney (1970) found that similar empirical relations hold for the scalp, brain, and dura.

Most authors discuss their experimental results in the framework of the linear theory of viscoelasticity relating stress and strain, generally on the basis of the Voigt, Maxwell, and Kelvin models. Buchthal and Kaiser (1951) formulated a continuous relaxation spectrum which corresponds to a combination of an infinite number of Voigt and Maxwell elements. A nonlinear theory of the Kelvin type was proposed by Viidik (1968) on the basis of a sequence of springs of different natural length, with the number of participating springs increasing with increasing strain.

It is reasonable to expect that for oscillations of small amplitude about an equilibrium state, the theory of linear viscoelasticity should apply. For finite deformations, however, the nonlinear stress-strain characteristics of the living tissues must be accounted for.

Instead of developing a constitutive equation by gradual specialization of a general formulation, I shall go at once to the special hypotheses which I introduced several years ago[1] and which seem to work well. Let us consider a cylindrical specimen subjected to tensile load. If a step increase in elongation (from $\lambda = 1$ to λ) is imposed on the specimen, the stress developed will be a function of time as well as of the stretch λ. The history of the stress response, called the *relaxation function*, and denoted by $K(\lambda, t)$, is assumed to be of the form

(22) $\qquad K(\lambda, t) = G(t)T^{(e)}(\lambda), \qquad G(0) = 1,$

in which $G(t)$, a normalized function of time, is called the *reduced relaxation function*, and $T^{(e)}(\lambda)$, a function of λ alone, is called the *elastic response*. We then assume that the stress response to an infinitesimal change in stretch,

---

[1] This theory was first presented to the 5th Annual Meeting of the Rheological Society at Mission Bay, San Diego, on February 5, 1968; and to the 12th International Congress of Applied Mechanics at Stanford, California, on August 27, 1968. However, the curves in Fig. 5 were obtained by my student, H. Chen, recently. Guth, Wack, and Anthony (1946) have shown that the viscoelasticity of rubber follows Eq. (22).

$\delta\lambda(\tau)$, superposed on a specimen in a state of stretch λ at an instant of time τ, is, for $t > \tau$:

$$(23) \qquad G(t - \tau)\frac{\partial T^{(e)}[\lambda(\tau)]}{\partial \lambda} \delta\lambda(\tau)$$

Finally, we assume that the superposition principle applies so that

$$(24) \qquad T(t) = \int_{-\infty}^{t} G(t - \tau)\frac{\partial T^{(e)}[\lambda(\tau)]}{\partial \lambda}\frac{\partial \lambda(\tau)}{\partial \tau} d\tau$$

that is, the tensile stress at time t is the sum of contributions of all the past changes each governed by the same reduced relaxation function.

Rewriting Eq. (24) in the form

$$(25) \qquad T(t) = \int_{-\infty}^{t} G(t - \tau)\dot{T}^{(e)}(\tau) d\tau$$

where a dot denotes the rate of change with time, we see that the stress response is described by a linear law relating the stress T with the elastic response $T^{(e)}$. The function $T^{(e)}(\lambda)$ plays the role assumed by the strain ϵ in the conventional theory of viscoelasticity. Therefore, the machinery of the well-known theory of linear viscoelasticity (see, for example, the author's book *Foundations of Solid Mechanics*, Chaps. 1 and 15) can be applied to this hypothetical material.

The inverse of Eq. (25) may be written as

$$(26) \qquad T^{(e)}[\lambda(t)] = \int_{-\infty}^{t} J(t - \tau)\dot{T}(\tau) d\tau$$

which defines the *reduced creep function* $J(t)$. Let $T^{(e)}(\lambda) = F(\lambda)$ and $\lambda = F^{-1}(T^{(e)})$, then for a step change of the tensile stress T at $t = 0$, the time history of the extension is

$$(27) \qquad \lambda(t) = F^{-1}[J(t)]$$

The lower limits of integration in Eqs. (24), (25), and (26) are written as $-\infty$ to mean that the integration is to be taken before the very beginning of the motion. If the motion starts at time $t = 0$, and $\sigma_{ij} = e_{ij} = 0$ for $t < 0$, Eq. (24) reduces to

$$(28) \qquad T(t) = T^{(e)}(0+)G(t) + \int_{0}^{t} G(t - \tau)\frac{\partial T^{(e)}[\lambda(\tau)]}{\partial \tau} d\tau$$

If $\partial T^{(e)}/\partial t$, $\partial G/\partial t$ are continuous in $0 \leq t < \infty$, the equation above is equivalent to

(29a) $$T(t) = G(0)T^{(e)}(t) + \int_0^t T^{(e)}(t-\tau)\frac{\partial G(\tau)}{\partial \tau}d\tau$$

(29b) $$= \frac{\partial}{\partial t}\int_0^t T^{(e)}(t-\tau)G(\tau)\,d\tau.$$

Equation (29a) is suitable for a simple interpretation. Since, by definition, $G(0) = 1$, we have

(30) $$T(t) = T^{(e)}[\lambda(t)] + \int_0^t T^{(e)}[\lambda(t-\tau)]\frac{\partial G(\tau)}{\partial \tau}d\tau$$

Thus the tensile stress at any time t is equal to the instantaneous stress response $T^{(e)}[\lambda(t)]$ decreased by an amount depending on the past history, because $\partial G(\tau)/\partial \tau$ is generally of negative-value.

The question of experimental determination of $T^{(e)}(\lambda)$ and $G(t)$ will be discussed in the following sections.

VI. THE ELASTIC RESPONSE $T^{(e)}(\lambda)$

By definition, $T^{(e)}(\lambda)$ is the tensile stress instantaneously generated in the tissue when a step function of stretching λ is imposed on the specimen. Strict laboratory measurement of $T^{(e)}(\lambda)$ according to this definition is difficult, because at an infinite rate of loading, transient stress waves will be bouncing back and forth in the specimen and a recording of the stress response will be confused by these elastic waves. Fortunately, we have seen in the previous section that the stress response in a loading process is insensitive to the rate of loading over a wide range. Hence $T^{(e)}(\lambda)$ may be approximated by the tensile stress response in a loading experiment with a sufficiently high rate of loading. In other words, we may take the $T(\lambda)$ obtained in Sect. III as $T^{(e)}$.

A justification of this procedure is the following. The relaxation function $G(t)$ is a continuously varying decreasing function as shown in Fig. 4 (normalized to 1 at $t = 0$). Now, if by some monotonic process λ is increased from 0 to λ in a time interval ϵ, then at the time $t = \epsilon$ we have, according to Eq. (30),

(31) $$T(\epsilon) = T^{(e)}(\lambda) + \int_0^\epsilon T^{(e)}[\lambda(\epsilon-\tau)]\frac{\partial G(\tau)}{\partial \tau}d\tau.$$

But, as τ increases from 0 to ϵ, the integrand never changes sign, hence

(32) $$T(\epsilon) = T^{(e)}(\lambda)\left[1 - \epsilon\frac{\partial G}{\partial \tau}(c)\right]$$

where $0 \leq c \leq \epsilon$. Since $\partial G/\partial \tau$ is finite, the second term tends to 0 with ϵ. Therefore, if ϵ is so small that $\epsilon |\partial G/\partial \tau| \ll 1$, then

(33) $$T^{(e)}(\lambda) \doteq T(\epsilon)$$

VII. THE REDUCED RELAXATION FUNCTION G(t)

It is customary to analyze the relaxation function into the sum of exponential functions and identify each exponent with the rate constant of a relaxation mechanism, thus

(34) $$G(t) = \frac{\sum C_i e^{-v_i t}}{\sum C_i}$$

Two of ten important points are missed:

(1) If an experiment is cut off prematurely, one may mistakenly arrive at an erroneous limiting value $G(\infty)$ which corresponds to $v_0 = 0$ in Eq. (34).
(2) The exponents v_i should not be interpreted literally without realizing that the separation of empirical data into a sum of exponentials is a highly unstable process.

An example of the first point: Buchthal and Kaiser (1951) relaxation curves terminate at 100 msec. However, the data in Fig. 4 show that relaxation goes on beyond 1000 min! Examples of the second kind have lead one to observe that a measured characteristic time of a relaxation experiment often turned out to be the length of the experiment. Lanczos (1956, p. 276) gives an example in which a certain set of twenty-four decay observations was analyzed and found that it could be fit equally well by three different expressions for x between 0 and 1:

(35) $$\begin{aligned} f(x) &= 2.202 e^{-4.45x} + 0.305 e^{-1.58x} \\ f(x) &= 0.0951 e^{-x} + 0.8607 e^{-3x} + 1.5576 e^{-5x} \\ f(x) &= 0.041 e^{-0.5x} + 0.79 e^{-2.73x} + 1.68 e^{-4.96x} \end{aligned}$$

Lanczos comments, "It would be idle to hope that some other modified mathematical procedure could give better results, since the difficulty lies not with the manner of evaluation but with the extraordinary sensitivity of the exponents and amplitudes to very small changes of the data, which no amount of least-square or other form of statistics could remedy."

Realizing these difficulties, we conclude first that for a living tissue, a viscoelasticity law based on the fully relaxed elastic response ($G(t)$ as $t \to \infty$)

is unreliable. In fact, a formulation based on $G(\infty)$ may run into a true difficulty because often it seems that $G(t) \to 0$ when $t \to \infty$. Secondly, one should look into other experiments: creep, hysteresis, oscillation, and so forth, in order to determine the relaxation function.

In this regard the salient feature of the hysteresis curve comes to our mind: the hysteresis loop is almost independent of the strain rate within several decades of the rate variation. This insensitivity is incompatible with any viscoelastic model which consist of finite number of springs and dashpots. Such a model will have discrete relaxation rate constants. If the specimen is strained at a variable rate, the hysteresis loop will reach a maximum at a rate corresponding to a relaxation constant. A discrete model, therefore, corresponds to a discrete hysteresis spectrum, in opposition to the feature of living tissues that we have just described. This suggests at once that one should consider a continuous distribution of the exponents v_i—thus passing from a discrete spectrum α_i associated with v_i to a continuous spectrum $\alpha(v)$ associated with a continuous variable v between 0 and ∞.

To make these remarks clear, let us consider a *standard linear solid* (Kelvin model) for which the governing differential equation is (see Fung, Solid Mechanics, 1965, p. 22 ff.)

(36) $$T + \tau_\epsilon \dot{T} = E_R[T^{(e)} + \tau_\sigma \dot{T}^{(e)}]$$

where τ_ϵ, τ_σ, E_R are constants. Equation (36) is subjected to the initial condition

(37) $$\tau_\epsilon T(0) = E_R \tau_\sigma T^{(e)}(0).$$

However, our special definition of the reduced relaxation function requires that

(38) $$T^{(e)}(0) = T(0), \qquad G(0) = 1$$

hence

(39) $$E_R = \frac{\tau_\epsilon}{\tau_\sigma}.$$

By integrating Eq. (36) with the initial condition specified above, we obtain the response $T^{(e)}$ to a unit step increase in stress $T(t) = \mathbf{1}(t)$: the creep function

(40) $$J(t) = \frac{1}{E_R}\left[1 - \left(1 - \frac{\tau_\epsilon}{\tau_\sigma}\right)e^{-t/\tau_\sigma}\right]\mathbf{1}(t).$$

Conversely, the relaxation function is obtained by integrating the differential equation for stress T for a unit step increase in the elastic response:

$$G(t) = E_R \left[1 + \left(\frac{\tau_\sigma}{\tau_\epsilon} - 1 \right) e^{-t/\tau_\epsilon} \right] 1(t). \tag{41}$$

With these expressions the meaning of the constants τ_σ, τ_ϵ, E_R are clear. τ_σ is the time constant for creep at constant stress, τ_ϵ is the time constant for relaxation at constant strain, E_R is the "residual"—the fraction of the elastic response that is left in the specimen after a long relaxation, $t \to \infty$. These physical constants are related through Eq. (39).

If a sinusoidal oscillation is considered, so that

$$T(t) = T_0 e^{i\omega t}, \qquad T^{(e)}(t) = T_0^{(e)} e^{i\omega t} \tag{42}$$

then

$$T_0/T_0^{(e)} = i\omega \int_{-\infty}^{\infty} G(t) e^{-i\omega t} \, dt = \mathscr{M} = |\mathscr{M}| e^{-i\delta} \tag{43}$$

where $\mathscr{M}$ is the complex modulus:

$$\mathscr{M} = \frac{1 + i\omega \tau_\sigma}{1 + i\omega \tau_\epsilon} E_R, \qquad |\mathscr{M}| = \left(\frac{1 + \omega^2 \tau_\sigma^2}{1 + \omega^2 \tau_\epsilon^2} \right)^{1/2} E_R \tag{44}$$

δ is the phase shift, and $\tan \delta$ is a measure of "internal damping"

$$\tan \delta = \frac{\omega(\tau_\sigma - \tau_\epsilon)}{1 + \omega^2(\tau_\sigma \tau_\epsilon)}. \tag{45}$$

When the modulus $|\mathscr{M}|$ and the internal damping $\tan \delta$ are plotted against the logarithm of ω, curves as shown in Fig. 7 are obtained. The damping reaches peak when the frequency ω is equal to $1/\sqrt{\tau_\sigma \tau_\epsilon}$. Correspondingly, the elastic modulus $|\mathscr{M}|$ has the fastest rise for frequencies in the neighborhood of $1/\sqrt{\tau_\sigma \tau_\epsilon}$. If the internal friction (hysteresis loop) is insensitive to frequency ω, we must spread out the peak. This can be done by superposing a large number of Kelvin models. This is the basic reason for introducing a continuous spectrum of relaxation time into our problem.

To implement this idea, let us first rewrite the relaxation function in a different notation. Let

$$S = \frac{\tau_\sigma}{\tau_\epsilon} - 1, \qquad E_R = \frac{1}{1 + S} \tag{46}$$

then substituting into Eqs. (41) and (44), we obtain

$$G(t) = \frac{1}{1 + S} [1 + S e^{-t/\tau_\epsilon}] \tag{47}$$

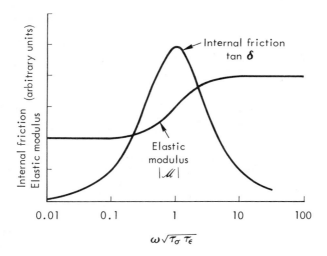

Fig. 7: The dynamic modulus of elasticity $|\mathcal{M}|$ and the internal damping $\tan \delta$ plotted as a function of the logarithm of frequency ω, for a standard linear solid.

$$\mathcal{M} = \frac{1}{1+S}\left(1 + S\frac{\omega\tau_\epsilon}{\omega\tau_\epsilon + \frac{1}{\omega\tau_\epsilon}} + iS\frac{1}{\omega\tau_\epsilon + \frac{1}{\omega\tau_\epsilon}}\right). \tag{48}$$

Now, let τ_ϵ be replaced by a continuous variable τ, and $S(\tau)$ be a function of τ. For a system with a continuous spectrum we have the following generalized reduced relaxation function, reduced creep function, and complex modulus:

$$G(t) = \left[1 + \int_0^\infty S(\tau)e^{-t/\tau}\,d\tau\right]\left[1 + \int_0^\infty S(\tau)\,d\tau\right]^{-1} \tag{49}$$

$$J(t) = \left[1 - \int_0^\infty \frac{S(\tau)}{1+S(\tau)}e^{-t/[(1+S(\tau))]\tau}\,d\tau\right]\bigg/\left[1 - \int_0^\infty \frac{S(\tau)\,d\tau}{1+S(\tau)}\right] \tag{50}$$

$$\mathcal{M}(\omega) = \left[1 + \int S(\tau)\frac{\omega\tau\,d\tau}{\omega\tau + \frac{1}{\omega\tau}} + iS(\tau)\frac{d\tau}{\omega\tau + \frac{1}{\omega\tau}}\right]\bigg/\left[1 + \int_0^\infty S(\tau)\,d\tau\right] \tag{51}$$

Our task is to find the function $S(\tau)$ which will make $G(t)$, $J(t)$ and $\mathcal{M}(\omega)$ match the experimental results. In particular, we want $\mathcal{M}(\omega)$ to be nearly constant for a wide range of frequency.

A specific proposal is to consider

$$S(\tau) = \frac{c}{\tau} \quad \text{for} \quad \tau_1 \leq \tau \leq \tau_2$$
$$= 0 \quad \text{for} \quad \tau < \tau_1,\ \tau > \tau_2 \tag{52}$$

where c is a constant. Then

$$\mathcal{M}(\omega) = \left\{1 + \int_{\tau_1}^{\tau_2}\left[c\frac{\omega\tau}{1+(\omega\tau)^2} + \frac{ic}{1+(\omega\tau)^2}\right]d(\omega\tau)\right\}\bigg/\left\{1 + \int_{\tau_1}^{\tau_2}\frac{c}{\tau}d\tau\right\}$$

$$= \left\{1 + \frac{c}{2}[\ln(1+\omega^2\tau_2^2) - \ln(1+\omega^2\tau_1^2)]\right.$$

(53) $\qquad\left. + ic[\tan^{-1}(\omega\tau_2) - \tan^{-1}(\omega\tau_1)]\right\}\bigg/\left\{1 + c\ln\frac{\tau_2}{\tau_1}\right\}$

This results in a rather constant damping for $\tau_1 \leq 1/\omega \leq \tau_2$, as can be seen from Fig. 8 for an example in which $\tau_1 = 10^{-2}$, $\tau_2 = 10^2$. The variable part of the stiffness (the real part of $\mathcal{M}(\omega)$) is also plotted in Fig. 8. The maximum damping occurs when the frequency is $\omega_n = 1/\sqrt{(\tau_1\tau_2)}$. The stifiness rises

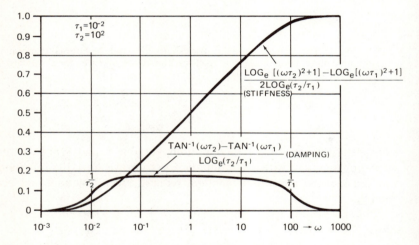

Fig. 8: The stiffness (real part of the complex modulus $\mathcal{M}$) and the damping plotted as functions of the logarithm of the frequency ω; corresponding to a continuous relaxation spectrum $S(\tau) = c/\tau$ for $\tau_1 \leq \tau \leq \tau_2$ and zero elsewhere. $\tau_1 = 10^{-2}$, $\tau_2 = 10^2$.

the fastest in the neighborhood of ω_n. At ω_n, the maximum damping is proportional to

(54) $\qquad\qquad \tan^{-1}\{\tfrac{1}{2}[\sqrt{(\tau_2/\tau_1)} - \sqrt{(\tau_1/\tau_2)}]\}$

which varies rather slowly with the τ_2/τ_1 ratio, as can be seen from Fig. 9, in which the above quantity divided by $\ln(\tau_2/\tau_1)$ is plotted.

The corresponding reduced relaxation function can be evaluated in terms of the exponential integral function which is tabulated. Fig. 10 shows an example in which the fraction of the relaxation to the residual stress is

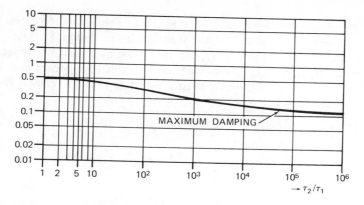

Fig. 9: The maximum damping as a function of the ratio τ_2/τ_1 for a solid corresponding to a continuous relaxation spectrum. $S(\tau) = c/\tau$ for $\tau_1 \leq \tau \leq \tau_2$, and zero elsewhere.

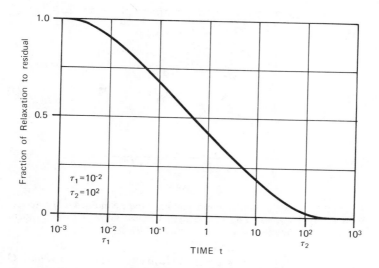

Fig. 10. The reduced relaxation function $G(t)$ of a solid with a continuous relaxation spectrum $S(\tau) = c/\tau$ for $\tau_1 \leq \tau \leq \tau_2$, and zero elsewhere.

plotted as a function of time. The portion in the middle, $\tau_1 \ll t \leq \tau_2$, is nearly a straight line; therefore the stresses decreases with $\ln t$ in that segment.

This specific spectrum, therefore, gives us the features desired. We are left with three parameters, c, τ_1, τ_2, to adjust with respect to the experimental data.

By proposing the specific spectrum (52), we do not mean to exclude other possibilities. Any fairly flat and broad continuous spectrum will serve to render hysteresis less dependent on frequency. The well-known results of Hardung, Bergel, and others on the insensitivity of damping to frequency in the arteries, as shown in Fig. 11, can be treated in a similar manner.

Historically, hysteresis of this type was described by Becker and Föppl (1928) in his study of electromagnetism in metals (see Becker and Döring,

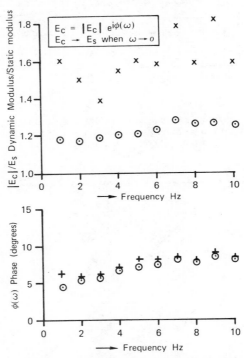

Fig. 11: The dynamic modulus and damping (angle δ in Eq. 43) of arteries from the data of Hardung and Bergel. Circles: aorta. Crosses: other arteries. From Westerhof and Noordergraf, (1970).

(1939)). Wagner (1913) described it in dielectricity. Theodorsen and Garrick (1940) introduced it into the theory of airplane flutter, and it is called "structural" damping by engineers. Knopoff (1965, Chap. 7, Vol. III B) showed that the earth's crust has an internal friction Q^{-1} which is independent of the frequency. Routbart and Sack (1966) showed that the internal friction for nonmagnetic materials is either constant or slightly decreasing with frequency in the range 1–40 k Hz. Mason (1969) attributed the reason to the existence of a kind of "kink" in the dislocation line and the associated kink energy barrier. Bodner (1968) formulated a special plasticity theory to account for the phenomena.

The concept of continuous relaxation spectrum was considered by Wagner and Becker. Wagner (1913) investigated the function

(55) $$S(\tau)\,d\tau = \frac{kb}{\sqrt{\pi}} e^{-b^2 z^2}\,dz$$

where

$$z = \log(\tau/t_0)$$

and k, b, t_0 are constants. Becker (1928) introduced the spectrum given in (52). Neubert (1963) developed the Becker theory thoroughly.

VIII. THE QUASI-LINEAR STRESS-STRAIN-HISTORY RELATIONSHIP

It is important to note that although a linear viscoelasticity law is posed in Equation (24) to relate the stress T with the elastic response $T^{(e)}$, the relation between $T^{(e)}$ and λ is nonlinear. Accordingly, the relationship between the stress T and the stretch λ may be called quasi-linear.

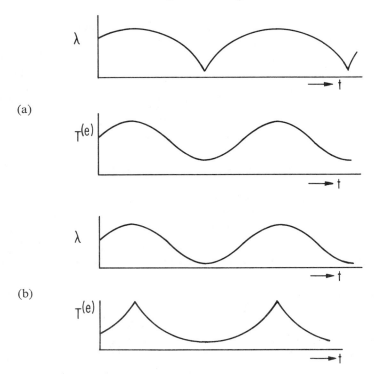

Fig. 12: The quasi-linear stress-strain-history relationship proposed in Eq. 30. (*a*) The time course of the extension λ when the elastic response $T^{(e)}$ oscillates sinusoidally. (*b*) The time course of the elastic response $T^{(e)}$ when the extension λ varies sinusoidally.

In developing the theory of continuous spectrum considered above the elastic response $T^{(e)}$ is assumed to oscillate harmonically. The corresponding oscillation of the extension is anharmonic. Figure 12a shows the time course of $T^{(e)}$ and λ when $T^{(e)}$ oscillates harmonically. If λ oscillates sinusoidally $T^{(e)}$ would oscillate anharmonically as shown in Fig. 12b. If the amplitude of oscillation is small, approximate linear relationship holds, and both stress and strain will oscillate harmonically. Patel, Tucker and Janicki, (1970) show that linear viscoelasticity applies as long as the superimposed sinusoidal strain remained below 4% (on top of $\lambda = 1.6$), and that data were reproducible up to sixteen hours following removal of tissue. Many other workers reached similar conclusions.

The mathematical details for viscoelastic stress in response to a specified history $\lambda(t)$ have been worked out for various types of processes used in our laboratory.

IX. CONCLUDING REMARKS

The extremely interesting question of correlating the mechanical properties with the fine structure of the tissue, and of determining systematically the variation of mechanical properties with physical and chemical factors (such as pH, temperature, pressure), animal species, pathology (such as arteriosclerosis), location in the body (such as blood vessels along the arterial tree), function of the organ (such as uterine cervix during pregnancy and birth of the young), resting and exercise, aging, preservation and storage (for future use or for transplantation), artificial simulation (man-made materials for medical and surgical use) will constitute a large field of research for which we can offer only a sampling of bibliography.

Our objective, the determination of a stress-strain-history law for the general use of biomechanics, has not been reached. How to go from simple elongation to a three-dimensional stress field requires further theoretical and experimental work. At the present time, we cannot but be scorned as "one-dimensional men" like most rheologists. Very little testing has been done in stress fields other than simple elongation. Fung, Zweifach, and Intaglietta (1966)'s work on simple shear lead to a stress-strain relationship similar to that in simple elongation. This is reassuring, but it does not suffice to verify a three-dimensional law. Yet, if we do not know one, how can we know three? We hope that the path from one to three will not take too long to travel.

We emphasize again that the intention of the present article is to glean from the field of published data some simple rules that seem to apply in the lower, physiological range of stresses. In a detailed comparison, each tissue, indeed, each specimen, is different, and one cannot expect a simple formula to account exactly for all the experimental data. In particular, although most original papers make some effort to justify the data as comparable to *in situ* characteristics in normal living condition, one cannot be absolutely sure. We feel, however, that these needs for further refinements should not deter us from offering a simplified general stress-strain-history law that will help biomechanics to progress beyond elementary discussion and to come to grips with significant boundary-value problems of the function of the organs.

REFERENCES

Note: Papers referred to both in this article and in Crisp's article are not duplicated here. Please see Crisp, Chap. 6, pp. 141.

Apter, J. T. et al (1966, '67, '68). Correlation of viscoelastic properties of large arteries with microscopic structure:
  I. Methods used and their justification. Apter, J. T. Circ. Res. 19: 104–21. 1966.
  II. Collagen, elastin and muscle determined chemically, histologically, and physiologically. Apter, J. T., Rabinowitz, M. and Cummings, D. H., ibid.
  III. Circumferential viscous and elastic constants measured in vitro. (Apter) ibid.
  IV. Thermal responses of elastin collagen, smooth muscle, and intact arteries. Apter, J. T. 1968. Circ. Res. 21: 901–18.
  V. Effects of sinusoidal forcing at low and at resonance frequencies. Apter, J. T., and Marquez, E. 1968. Circ. Res. 22: 393–404.

Bader, H. 1967. Dependence of wall stress in the human thoracic aorta on age and pressure. Circ. Res. 20: 354–61.

Balasubramaniam, E., and Whiteley, K. J. 1968. Stress relaxation of animal fibres. Biorheol. 5: 215–25.

Becker, E., and Föppl, O. 1928. Dauerversuche zur Bestimmung der Festigkeitseigenschaften, Beziehungen zwischen Baustoffdämpfung und Verformungsgeschwindigkeit. Forschung aus dem Gebiete des Ingenieurwesens, V.D.I., No. 304.

Becker, R., and Döring, W. 1939: Ferromagnetismus. Springer, Berlin. Chap. 19.

Bergel, D. H. 1961. The static elastic properties of the arterial wall. J. Physiol. 156: 445–57.

Bergel, D. H. 1961. The dynamic elastic properties of the arterial wall. J. Physiol. 156: 458–69.

Blatz, P. J., Chu, B. M., and Wayland, H. 1969. On the mechanical behavior of elastic animal tissue. Trans. Soc. Rheol. 13: 83–102.

Bodner, S. R. 1968. Constitutive equations for dynamic material behavior. In U. S. Lindholm, (ed.) Mechanical Behavior of Materials under Dynamic Loads. Springer-Verlag, New York, Inc. pp. 176–90.

Buchthal, F., and Kaiser, E. 1951. The Rheology of the Cross Striated Muscle Fibre with Particular Reference to Isotonic Conditions. Copenhagen: Det Kongelige Danske Videnskabernes Selskab, Dan. Biol. Medd. 21, No. 7, 318 pp.

Carew, T. E., Vaishnav, R. N., and Patel, D. J. 1968. Compressibility of the arterial wall. Circ. Res. 23: 61–68.

Elden, H. R. 1965. Biophysical aspects of aging in connective tissue. Advances in Biology of Skin 6: 229–43.

Elden, H. R. 1968. Physical properties of collagen fibers. Internat. Rev. of Connective Tissue Res. 4: 283–348.

Ellis, D. G. 1970. Temperature effects on the dynamic and transient mechanical behavior of tendon, Tech. Report. No. 11, Univ. of Michigan, Medical School, Dept. Phys. Med. & Rehabilitation.

Fitton Jackson, S., Harkness, R. D., Patridge, S. M., and Tristam, G. R. (eds.) 1964. Structure and Function of Connective and Skeletal Tissue. Proc. of Meeting, Butterworths. London 537 pp.

Flory, P. J., and Garrett, R. R. 1958. Phase transitions in collagen and gelatin systems. J. Am. Chem. Soc. 80: 4836–845.

Frasher, W. G. 1966. What is known about the physiology of large blood vessels. *In* Y. C. Fung, (ed.) Biomechanics. Amer. Soc. Mech. Engrs. pp. 1–19.

Frauke, E. K. 1951. Mechanical impedance of the surface of the human body. J. Appl. Physiol. 3: 582–90.

Frey-Wyssling, A. 1952. Deformation and Flow in Biological Systems. Interscience Pub., New York.

Fry, P., Harkness, M. L. R., and Harkness, R. D. 1964. Mechanical properties of the collagenous framework of skin in rats of different ages. Am. J. Physiol. 206: 1425–429.

Fukada, E., and Date, M. 1963. Viscoelastic properties of collagen solutions in dilute hydrochloric acid. Biorheol. 1: 101–109.

Fung, Y. C. 1965. Foundations of Solid Mechanics. Prentice-Hall, Inc., Englewood Cliffs, New Jersey.

Fung, Y. C., Zweifach, B. W., and Intaglietta, M. 1966. Elastic environment of the capillary bed. Circ. Res. 19: 441–61.

Fung, Y. C. 1967. Elasticity of soft tissues in simple elongation. Am. J. Physiol. 213: 1532–544.

Fung, Y. C. 1968. Biomechanics: its scope, history, and some problems of continuum mechanics in physiology. Appl. Mech. Rev. 21: 1–20.

Fung, Y. C. 1970. Mathematical representation of the mechanical properties of the heart muscle. J. Biomechanics 3: 381–404.

Galford, J. E., and McEllhaney, J. H. 1970. A viscoelastic study of scalp, brain, and dura. J. Biomechanics 3: 211–21.

Gratz, C. M. 1931. Tensile strength and elasticity tests on human fascia lata. J. Bone & Joint Surg. 13: 334–40.

Green, A. E., and Adkins, J. E. 1960. Large Elastic Deformations. Oxford Univ. Press, Inc., New York.

Guth, E., Wack, P. E., and Anthony, R. L. 1946. Significance of the equation of state for rubber. J. Appl. Physics 17: 347–51.

Hall, R. H. 1951. Variations with pH of the tensile properties of collagen fibres. J. of the Soc. Leather Trades Chems. 35: 195–210.

Hardung, V. 1952. Über eine Methode zur Messung der Dynamischen Elastizität und Viskosität Kautschukähnlicher Körper, insbesondere von Blutgefäszen und anderen Elastischen Gewebeteilen. Helv. Physiol Pharm. Acta 10: 482–98.

Harkness, M. L. R., and Harkness, R. D. 1959. Changes in the physical properties of the uterine cervix of the rat during pregnancy. J. Physiol. 148: 524–47.

Harkness, R. D. 1961. Biological functions of collagen. Biol. Rev. 36: 399–463.

Hill, A. V. 1939. Heat of shortening and dynamic constants of muscle. Proc. Roy Soc. London, Ser. B, 126: 136–95.

Hoeve, C. A. J., and Flory, P. J. 1958. The elastic properties of elastin. J. Am. Chem. Soc. 80: 6523–526.

Huxley, A. F. 1957. Muscle structure and theories of contraction. *In* J. A. V. Butler and B. Katz (eds.) Progress in Biophysics and Biophysical Chemistry. Pergamon Press, London, Vol. 7, pp. 255–318.

Huxley, H. E. 1969. The mechanism of muscular contraction. Science 164: 1356–366.

Kenedi, R. M., Gibson, T., and Daly, C. H. 1964. Bioengineering studies of the human skin, the effects of unidirectional tension. *In* S. F. Jackson, S. M. Harkness, and G. R. Tristram, (eds) Structure and Function of Connective and Skeletal Tissue. Scientific Committee, St. Andrews, Scotland. pp. 388–95.

King, A. L., and Lawton, R. W. 1950. Elasticity of body tissues. *In* O. Glasser, (ed.) Medical Physics. Year Book Pub. Inc., Chicago. Vol. II, pp. 303–16.

Kitamura, N. 1923. Der Einfluss der Temperatur und Spannung auf die Elastizität des Bindegewebes, insbesondere seine Härte. Pflüger's Archiv fur die gesamte Physiologie 200: 313–26.

Knopoff, L. 1965. On the independence of internal friction with frequency in earth's crust. *In* Warren P. Mason, (ed.) Physical Acoustics. Academic Press Inc., New York. Vol. III B, Chap. 7.

Laban, M. M. 1962. Collagen tissue: Implications of its response to stress in vitro. Arch. Phys. Med. 43: 461–66.
Lanczos, C. 1956. Applied Analysis. Prentice-Hall Inc., Englewood Cliffs, New Jersey.
Lawton, R. W. 1954. The thermoelastic behavior of isolated aortic strips of the dog. Circ. Res. 2: 344–53.
Lawton, R. W. 1955. Measurement of the elasticity and damping of isolated aortic strips of the dog. Circ. Res. 3: 403–408.
Learoyd, B. M., and Taylor, M. G. 1966. Alterations with age in the viscoelastic properties of human arterial walls. Circ. Res. 18: 278–91.
Lee, J. S., Frasher, W. G., and Fung, Y. C. 1967. Two-dimensional finite-deformation on experiments on dog's arteries and veins. Tech. Rept. No. AFOSR 67-1980, Univ. of Calif., San Diego.
Lee, J. S., Frasher, W. G., and Fung, Y. C. 1968. Comparison of the elasticity of an artery *in vivo* and in excision. J. Appl. Physiol. 25: 799–801.
Lindholm, U. S. 1965. Dynamic deformation of metals. In N. J. Huffington, Jr., (eds) Behavior of Materials under Dynamic Loading Amer. Soc. Mech. Engineers, New York.
Mason, W. P. 1969. A source of dissipation that produces an internal friction independent of the frequency, Rept. No. 61 on contract NONR266(91), Dept, of Civil Eng. & Eng. Mech., Columbia Univ.
Meyer, K. H. and Ferri, C. 1935. Sur l'élasticité du Caoutchouc Helv. Chim. Acta, 18: 570–589.
Meyer, K. H., and Picken, L. E. R. 1934. Thermoelastic properties of muscle and their molecular interpretation. Proc. Roy Soc. London, S. B., 124: 29.
Morgan, F. R. 1960. The mechanical properties of collagen fibres: stress-strain curves. J. Soc. Leather Trades Chems. 44: 171–82.
Nachemson, A. L., and Evans, J. H. 1968. Some mechanical properties of the third human lumbar interlaminar ligament (ligamentum flavum). J. Biomechanics 1: 211–20.
Neubert, H. K. P. 1963. A simple model representing internal damping in solid materials. Aeronautical Quarterly 14: 187–97.
Nunley, R. L. 1958. The ligamenta flava of the dog. A study of tensile and physical properties. Am. J. Phys. Med. 37: 256–68.
Patel, D. J., de Freitas, F. M., Greenfield, J. C. Jr., and Fry, D. L. 1963. Relationship of radius to pressure along the aorta in living dogs. J. Appl. Physiol. 18: 1111.
Patel, D. J., Carew, T. E., and Vaishnav, R. N. 1968. Compressibility of the arterial wall. Circ. Res. 23: 61–68.
Patel, D. J., Janicki, J. S., and Carew, T. E. 1969. Static anisotropic elastic properties of the aorta in living dogs. Circ. Res. 25: 765–79.
Patel, D. J., Tucker, W. K., and Janicki, J. S. 1970. Dynamic elastic properties of the aorta in radial direction. J. Appl. Physiol. 28: 578–82.
Peterson, L. H., Jensen, R. E., and Parnell, J. 1960. Mechanical properties of arteries *in vivo*. Circ. Res. 8: 622–39.
Peterson, L. H. 1962. Properties and behavior of the living vascular wall. Physiol. Rev. 42 (Suppl. 5, Part 2): 309–27.
Reuterwall, O. P. 1921. Über die Elastizität der Gefässwände und die Methoden ihrer nähren Prüfung. Acta. Med. Scand., Suppl. 11, 175 pp.
Ridge, M. D., and Wright, V. 1964. The description of skin stiffness. Biorheology 2: 67–74.
Routbart, J. L., and Sack, H. S. 1966. Background internal friction of some pure metals at low frequencies. J. Appl. Phys. 37: 4803–805.
Roy, Charles S. (1880–1882). The elastic properties of the arterial wall. J. Physiol. 3, 125–59.
Smith, J. W. 1954. The elastic properties of the anterior cruciate ligament of the rabbit. J. Anat. 88: 369–80.
Stacy, R. W., Zatzman, M Randall, J., and Eberstein, A. 1954. Time course of stress relaxation in isolated arterial segments. Am. J. Physiol. 177: 299–302.
Stacy, R. W. 1957. Reaction rate kinetics and some tissue mechanical properties. In J. W. Remington, (ed.) Tissue Elasticity. Amer. Physiol. Society, Washington, D. C. pp. 131–37.

Stromberg, D. D., and Wiederhielm, C. A. 1969. Viscoelastic description of a collagenous tissue in simple elongation. J. Appl. Physiol. 26: 857–62.

Stuke, K. 1950. Über das elastische Verhalten der Achillessehne im Belastungsversuch. Langenbek's Arch. u. Dtsch. Z. Chir. 265:579–99.

Takigawa, M. 1953. Study upon strength of human and animal tendons. J. Kyoto Pref. Med. Univ. 53: 915–33.

Theis, E. R., and Steinhardt, R. G. Jr. 1950. Protein axial movement studies II. Data obtained through use of dynamic and strain gage shrinkage meters. J. Am. Leather Chems. Assoc. 45: 591–610.

Theodorsen T. and Garrick E. 1940. Mechanism of flutter, Rept. 685, U. S. Nat. Adv. Comm. for Ameronautics.

Treloar, L. R. G. 1949. The Physics of Rubber Elasticity. Oxford Univ. Press, New York 2nd ed., 1967.

Valanis, K. C., and Landel, R. I. 1967. The strain-energy function of hyperelastic material in terms of the extension ratios. J. Appl. Phys. 38: 2997–3002.

Veronda, D. R., and Westmann, R. A. 1970. Mechanical characterization of skin—finite deformations. J. Biomechanics 3: 111–24.

Viidik, A. 1966. Biomechanics and functional adaption of tendons and joint ligaments. In F. G. Evans (ed.) Studies on the Anatomy and Function of Bone and Joints. Springer-Verlag. New York. pp. 17–39.

Wagner, K. W. 1913. Zur Theorie der unvollkommener Dielektrika. Ann. der Physik 40: 817–55.

Wertheim, M. G. (1847). Mémoire sur l'élasticité et la cohésion des principaux tissus du corps humain. Annales de Chimie et de Physique, Paris, Ser 3, Vol. 21, pp. 385–414.

Westerhof, N., and Noordergraaf, A. 1970. Arterial viscoelasticity: a generalized model. J. Biomechanics 3: 357–79.

Wetterer, E., and Kenner, T. 1968. Grundlagen der Dynamik des Arterien-pulses. Springer-Verlag, Berlin.

Wiederhorn, N. M., and Reardon, G. V. 1952. Studies concerned with the structure of collagen. II. Stress-strain behavior of thermally contracted collagen. J. Polymer Sci. 9, 315–25.

Wood, G. C. 1954. Some tensile properties of elastic tissues. Biochim. Biophys. Acta., 15: 311–24.

Wright, D. G., and Rennels, D. C. 1964. A study of the elastic properties of planter fascia. J. Bone & Joint Surg. 46A: 482–92.

Yin, F. C. P. 1970. Theoretical and experimental investigations of peristalsis. Ph. D. Thesis, Univ. Calif., San Diego.

Zatzman, M., Stacy, R., Randall, J., and Eberstein, A. 1954. Time course of stress-relaxation in isolated arterial segment. Am. J. Physiol. 177: 299–302.

8

Muscle Mechanics

BERNARD C. ABBOTT

The challenge to describe the mechanical characteristics of contractile tissue is enhanced by the range of elastic properties exhibited by each muscle. This variety in muscle has confounded many attempts to establish a satisfactory classification of types. The differences lie not only in gross morphological differences, but also in the ultimate ultrastructural profiles. The vertebrate skeletal muscle varies between the fast phasic system which responds with an explosive all or none twitch, and the tonic muscle with no propagated action potential but a contraction graded according to the incoming nerve barrage and characterized by slow relaxation. By contrast the myogenically rhythmic cardiac muscle discharges maximally at each beat but the mobilization of contractility can be altered by extrinsic and intrinsic factors. The majority of the smooth muscles of vertebrates occurs in the boundary walls of hollow organs and maintenance of muscle length provides steady control over the size and shape of the organ. Perhaps the ultimate extrapolation of this control is seen in the molluscan "catch" muscles in which large tensions must be actively generated but can then be maintained almost without cost.

The majority of experimental investigations have been made on the fast phasic twitch type of muscle. Such muscles exist either as resting or actively contracting systems (except for the special case of rigor mortis). The resting muscle is primarily an elastic body in which tension appears at about the maximum length assumed by the muscle in situ. Extensive studies on the tension characteristics suggest the presence of components both from the contractile element itself and from supporting elastic materials. The tension-length curve displays a discontinuity as if two elements exist. Thermoelastic heat measurements as well as the dependency of tension on temperature support the concept that as length is increased the origin of tension moves from a predominantly "rubber-like" elasticity to a "normal" elasticity. However, in other types of muscles the phenomena differ. In the relaxed "catch" muscle of the molluscan shellfish, tension appears to derive from only a single component but hysteresis is evident. When a muscle is extended and held at the longer length then the resting tension slowly drops. This moves the tension length curve along the length axis, but in these molluscan muscles the curve can be restored by an episode of excitation in which tension is actively generated in the contractile element.

Evidence of a viscous element can also be obtained. If a muscle is activated and allowed to shorten freely, with care taken not to extend it again when stimulation ends, then a transient resistance to stretch can be demonstrated. If this relaxed muscle is quickly extended (final length still being less than that for a resting tension), then a transient tension rise is seen during stretch followed by a tension decay with a rate constant closely related to the intrinsic maximum speed with which the muscle can shorten when tetanized.

In the state of rigor mortis, however, this extensibility and transience of tension disappears. The muscle is very stiff and forced extension produces a liberation of considerable heat as the bonds are forcibly broken. Plasticity of the contractile proteins demands the presence of ATP and virtual absence of free Ca^{++} ions. When ATP levels drop, the system becomes crosslinked and rigid. But if Ca^{++} level increases to 10^{-5} M or above, then the machinery becomes activated. Tension is developed and external work can be done at the expense of metabolic energy. The skeletal muscles of vertebrates are activated in this way and develop tension, but function phasically both to do external work in shortening and also to absorb work as they are forcibly stretched. The mechanical characteristics under these two conditions are quite different with a discontinuity at zero speed: in the shortening mode force falls hyperbolically with speed, but in extension force rises high above isometric value however here the unique force-speed relationship is replaced by a time-dependent force profile.

The so-called "catch" muscles differ. Tension is actively developed following the release of Ca^{++} ions and with a metabolic cost. When the tension has built up, however, internal changes occur such that the maintenance of tension is very inexpensive; but with changed conditions such that while extension is still resisted, no active shortening can occur.

Muscle tension can thus occur in supporting tissues and in contractile proteins and it can be passive or active. Other mechanical parameters such as elasticity, viscosity, and hysteresis are also demonstrated in a variety of characteristics in the muscles of the animal kingdom.

* * *

The functional characteristics of muscle are of obvious interest to many scientists who approach biology from an initial training in mechanical engineering or physics. The production and control of force and movement in the living animal challenges the model-oriented individual, while the energy mobilization challenges the thermodynamicist and the molecular coupling challenges the chemist. All of these approaches are essential but I fear that bioengineers maybe misled by biological generalizations into the formulation of naive and oversimplified models. Perhaps our goals should be defined more clearly. If models are proposed as empirical approximations to aid practical clinical diagnoses within well-defined organ systems, then the models need only be descriptive. In order to be used as tools to unravel basic mechanisms and organization, however, models must account critically for all of the biological parameters.

Most of the quantitatively definitive studies on the excitation, contraction and metabolism of muscle have been carried out on what is generally regarded as typical vertebrate striated muscle. In fact, the muscles used have been the fast, phasic all-or-none twitch-type muscle excited maximally so that all the fibers contract synchronously in a twitch or tetanus. Furthermore, the studies have been mainly on the frog sartorius for mechanics and energetics or on the rabbit psoas for biochemistry.

When model studies are extended to cover muscle systems other than skeletal, attention must be paid to the fact that these muscles are probably the most atypical of muscles, for they are highly specialized in many ways. Embryonically they have been formed by the fusion of many simpler myotubes to give multinucleate giant fibers up to 200μ diameter. These fibers are surrounded by a single continuous excitable membrane. Within the muscle cell the highly organized filament system has arrays of thick and thin filaments positioned longitudinally in sarcomeres and transversely in hexagonal orientation. The interdigitating movement of filaments is synchronized and orderly. The filaments are organized into fibrils which consist of well-defined bundles of filaments never more than 2μ diameter. These are enclosed within the fenestrated longitudinal portion of sarcotubular reticulum which in turn leads to large sacs, lateral cysternae and, through functional relays, are connected to invaginations of the outer membrane.

We know that this system is organized for release of Ca^{++} for rapid and full activation of contractile material and is equally specialized for rapid relaxation. During the action potential, the potential difference across the membrane decreases as it sweeps through the threshold for electrical excitation (which ensures that the full action potential is triggered) and on through the threshold for contraction coupling. By a still unknown mechanism it is assumed that calcium is released at a rate which increases very rapidly with further depolarization. The membrane potential overshoots the zero level before it returns and the calcium release mechanism briefly reaches its saturation level. The calcium release is assumed to be arrested as repolarization occurs. It is also assumed that the slug of calcium released around the fibril parameter diffuses the short distance to the myofilaments at a concentration such that the contractile system reaches its maximum level of active state and maintains it briefly. The active state is believed to decay as the processes of relaxation occurs, presumably by calcium reabsorption. The normal control of this system is through the motor nerve which meets this muscle in a large endplate. The interface of nerve and muscle has a surface area exaggerated by extensive infoldings (palisades) in order to produce a large depolarization and to give a propagated AP in the excitable membrane and an "all-or-none" twitch.

Contraction in the living animal occurs neither sychronously nor maximally. The muscle fibers are organized into motor units each controlled by one nerve fiber. As tension is developed motor units begin to fire asynchronously. Each unit begins to discharge at a certain muscle tension. The discharge frequency increases with tension but the dynamic range for any unit is very narrow and the frequency plateaus at around 30/sec for a very small increase in tension. Thus, at any given level of force a definite number of motor units is firing asynchronously but at the plateau level of frequency which is below mechanical fusion. This is for an isometric contraction. When movement occurs, the force developed by any individual fiber must decrease as velocity rises. Therefore, for a given force to be maintained in the whole muscle the nervous system must mobilize more motor units at the higher velocity. In the construction of a model describing the control and operation of intact muscle and limb systems, recognition must therefore be made of the patterning of activity within the muscle and of the resulting asynchronous twitch activations of the fibers.

The most obvious contrast to this concept of muscle control is seen immediately in the mammalian heart. Although the heart is a hydrodynamic pump, its operation is assured by the cumulative action of muscle fibers in the cardiac walls. The muscle fibers differ in organization and control from the skeletal system. The heart beats at a regular pace, and in each beat every muscle cell in the heart contracts maximally as defined by the conditions which exist at the instant. The heart beats spontaneously so that it does not depend on an initiating nerve impulse, although there are cardiac nerves which control the level of activity. The muscle cell of the mammalian ventricle is a single cell with a single nucleus. It is striated, with filaments, fibrils and a well-developed sarcoplasmic reticulum plus T-tubule system. However, the electrical and mechanical events differ considerably from those in the skeletal muscle. The action potential is prolonged with a plateau and delayed repolarization phase, and an early repeat excitation to give a tetanus is not possible. The sequence of calcium release giving contraction and recovery resembles that of the twitch muscle in that the time course of contraction is considerably independent of the membrane sequence. A fundamental difference exists, however, in that the slug of calcium released does not reach a saturation level at the myofilament. Instead, the calcium is poised at an intermediate level depending on the immediate previous history of the cell, and this level determines the strength of the contraction. Preceding beat intervals, the neurohumoral atmosphere outside the cell and the previous mechanical conditions determine the distribution of calcium entering the outer membrane and released from the reticulum.

A further difference from the skeletal muscle lies in the continuous dynamic control imposed by length in heart muscle. The return of blood in diastole determines the size of the ventricle and hence the length of the muscle fiber. The isometric tension developed by heart muscle varies sharply with length, so that the pressure and stroke volume of the ventricle is continuously under the control of the venuous return although the actual contribution of this to total control may vary.

The differences in operation are even more visible in smooth muscle which can develop significant tensions and can shorten proportionally more than the skeletal muscle. The fibers of the muscles, however, are single cells with very small diameters (often $5\,\mu$) joined to the others by membrane contact for transmission of tension. Fibrils are certainly absent and even the existence of two sizes of filaments is only now beginning to be demonstrated. Any sarcoplasmic reticulum is minimal and poorly organized and the motor nerve junction when present is very small. The muscle membrane is normally inexcitable and excitation occurs only by local depolarization near the terminals.

The resultant contraction is controlled in its intensity by the extent of the depolarization and it is characterized by very slow relaxation which enables tensions to be maintained efficiently and at low metabolic cost for long periods of time. Between the two extremes lies a whole range of intermediate systems and I wish to take examples over the whole range to assess our knowledge of biomechanics. Our knowledge is still meager and a discussion of the characteristics of tension production will illustrate our ignorance.

Resting tension, in skeletal muscle at least, has been assumed to be the expression of stress in passive supporting tissue (collagen) as expressed in the resting length-tension diagram. There is now a significant body of evidence that this is an oversimplification and that the resting relaxed contractile system can support tension, both of a transient and a maintained nature. Evidence for this came first from thermodynamic studies. The structural supporting materials of muscle and tendon are all long-chain polymers and the elasticity of a polymer chain can range from a pure rubber like elasticity of kinetic (entropic) origin (due to thermal motions around the monomer junctions) to a normal (potential energy elasticity depending on the extent of crossbonding between and within the chains). By a simple thermodynamic derivation Weygand and Snyder showed that the elastic characteristics can be deduced from the way in which tension in a strip of material at constant length changes with temperature. Wohlish showed that resting muscle is rubber-like in that resting tension increases as temperature rises. This is in contrast to the normal elasticity of collagen in which tension drops as temperature rises.

A.V. Hill in the mid 1950's went much further with dynamic studies of heat changes during the stretch and release of resting muscle. He showed clearly that frog muscle is rubber-like over a length range from lo to 1.2 lo (cools on release) and that over this range the tension increases exponentially with length. Above 1.25 lo, the tissue is much stiffer and the elasticity changes to a normal type (heats on release). Weiss, Fogh, and Buchthal went further and developed, in arthropod striated muscle, a technique that severed the myoplasm while leaving the membrane intact. They showed that the first (exponential) portion of the tension is in the myoplasm and that the stiffer portion is due to supporting tissue but appears only at longer lengths. Lowy extended this still further in the case of some of the molluscan smooth muscles and demonstrated the interaction of resting and active tensions under a number of conditions.

The adductor muscles of the shells of some lamellibranch molluscs display a resting tension when stretched which is an approximately exponential function of length. Tension developed activity by stimulation is maximal at about that length where resting tension appears. However, when the resting muscle is stretched the tension which appears is not maintained and drops slowly, reaching a lower steady value after some minutes. Measurement of the active tension shows that the peak value is reached at a new longer length corresponding to the length where the resting tension again first appears. If during stimulation the muscle is allowed to shorten freely, then both resting and active tension curves move back to their original length relations. This is another suggestive hint that resting tension may be within the contractile material and that tension at rest may reflect some activity.

D. K. Hill has very recently extended this concept by studying the changes in resting tension of frog muscle as the temperature is suddenly changed. From a series of studies he argues for an active origin for resting tension as a result of the presence of an activator substance whose concentration balance depends on the stretch.

In cardiac muscle, resting tension is an essential parameter in every beat. It is quite possible that this tension lies also in the contractile system for in the normal heart operating range the extensibility is high. If this is true, then any models proposed must allow for this characteristic.

Even at lengths below that for the appearance of steady resting tension many muscles display viscous properties: if shortened in an unloaded contraction and allowed to relax at short length a transient tension subsequently appears when the muscle is pulled out at rest. In fast phasic muscles the tension decay is exponential and rapid, with time constants of a few milliseconds. The slower the muscle the slower the decay, with a continuous curve relating rate constant to maximal muscle speed.

When active tension is considered, the fast phasic muscle again displays some special characteristics. Contraction is induced by excitation of the plasma membrane (identified as sarcolemma by some authors) and activity disappears when the action potentials cease. Depolarization is essential for the onset of contraction, but if steady depolarization is maintained then tension does not stay up and decays within seconds or less. Mammalian ventricular muscle with its rich reticulum behaves similarly but in most other muscles (including frog heart) the tension persists with the depolarization.

In many smooth muscles the separation of active and passive tensions is difficult. The best-known example is in the molluscan lamellibranch "catch" muscle. When the anterior byssus retractor muscle (ABRM) is exposed to Ach or to direct current stimulation, tension rises in a metabolically active process. Heat is liberated, ATP is utilized and tension redevelops after an imposed quick release of length (quick-release-recovery). When the plateau of tension is reached, internal changes produce a state of "catch," in which heat production is minimal, ATP breakdown is low and no quick-release-recovery occurs. Even when the membrane is repolarized the decay of tension is extremely slow so that the catch state persists. Relaxation can be induced in these muscles by a brief burst of tetanic stimulation or by the application of a small amount of 5-hydroxy-tryptamine.

This type of muscle represents an extreme. But there are many examples of smooth muscles in gut, bladder, and circulatory system which can exist at a wide range of levels of tone. In such muscles a given tension can be established over a wide range of lengths and active responses can be obtained by an imposed stretch. Almost nothing is known of the associated mechanochemical demands.

These comments have been made in order to describe something of the range of mechanical responses of muscle. The classical models assume that elastic components are passive, that contractile elements can display no tension at rest but when activated obey a standard force-velocity curve. The models have been extremely valuable and will certainly continue to be used, but the limitations of the assumptions must be always recognized.

9

Influence of Composition On Thermal Properties of Tissues

JULIA T. APTER

I. INTRODUCTION

Biological soft tissues (exclusive of bone and cartilage) depend for their mechanical properties on their fibrillar components: muscle, elastin, collagen and reticulin. The detailed macromolecular (polymeric) structure of these fibrils has been clarified by recent investigations (Huxley, 1957; Hanson and Lowry, 1961; Hoeve and Flory, 1958; Ramachandran, 1967) but is still far from settled. The present discussion will be limited to consideration of the possible influence of arrangements of the macromolecules on fibrillar behavior and of arrangements of fibrils on tissue behavior, with an emphasis on distinctive thermal responses. The problem is challenging since the macromolecules are found in a variety of arrangements, and rewarding because thermal responses differ markedly from one fibril type to another. A muscle fibril contains thick "filaments" of the proteins myosin and tropomyosin or paramyosin, and thin filaments of actin (Huxley, 1957) (Fig. 1). These filaments are highly ordered (crystalline) in fibrils of striated and cardiac muscle; less ordered in smooth muscle cells. Collagen also occurs as fibrils with a highly ordered protein structure (Mason, 1965) and therefore is also a crystalline polymer (Ramachandran, 1967). Elastin is an amorphous polymer which does not crystallize even when highly strained (in contrast to rubber); this amorphous state being responsible for special thermal properties of elastin (Hoeve and Flory, 1958).

The orientation of the macromolecules in the fibrils is an important factor determining the viscoelastic behavior of the fibrils. In turn, the

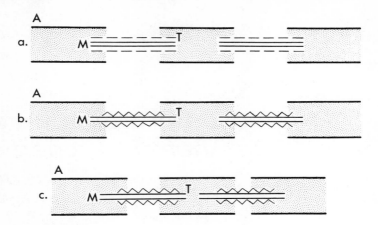

Fig. 1: Schematic of thick filaments of myosin M plus tropomyosin T and thin filaments of actin A as they might be arranged in a muscle fibril when it is *a* relaxed, *b* in isometric contraction, *c* in isotonic contraction. The filaments are surrounded in real muscle by endoplasmic reticulum, a system of membranes through which ions may pass by diffusion or by active transport to reach the filaments or to be drawn away from them.

orientation depends on the immediate environmental molecules (or ions) and temperature (Apter and Grassley, 1970a, b). In addition, the arrangement of fibrils in a fiber or in a tissue influences its overall viscoelastic behavior (Hoeve and Flory, 1958; Apter et al, 1966). By analogy with inert polymeric materials, a number of factors which determine mechanical behavior are influenced by temperature: the concentration of ions around the macromolecules; the activity of these ions and of the macromolecules themselves; and the crystalline state of the macromolecules, to name a few (Tobolsky, 1960). The measurable viscoelastic effects of thermal changes have been used in our laboratories to identify a particular fibrillar component and to ascertain the viability (and therefore transplantability) of living tissues like muscle (Apter and Najafi, 1968). The basis for this identification has been derived from extensive *in vitro* experimental studies in a number of laboratories (Weiner and Pearson, 1966) and from several physical models used to obtain tissue parameter values from experimental data. The present discussion will outline the thermal methods used in this laboratory to identify and evaluate individual fibril types (Apter and Marquez, 1968) and a model which is consistent with a wide variety of tissue responses to thermal changes and physical strains (Apter and Graessley, 1970a, b). The work of Hoeve and Flory (1958) has been very useful in developing these techniques but the present analysis uses a physical model even to develop thermodynamic considerations.

II. METHODS

A. Experimental Procedures for Measuring Effects of Temperature on Viscoelastic Properties

All tissues were studied *in vitro*; that is, after removal from a living, anesthetized animal (cat, dog, guinea pig, human, monkey, and rabbit). All specimens were prepared as rings and kept immersed in a medium suitable for living tissues, except when the effect of elastin was to be minimized. Then 30% ethylene glycol in water was used or isolated elastin (Hass, 1942) was prepared. During measurements, the rings were supported horizontally on two hooks passed through the lumen (Fig. 2). The ring specimens were stretched to six or seven strain levels (5–100%) by separating the hooks to register an initial steady force. One hook could then be manipulated laterally to superimpose small strains (2.3%) of a variety of forms: sinusoidal, ramp, step, but all within a region of linear viscoelasticity of the material. The movement of this hook was monitored with a photosensor strain transducer whose output was led to an electronic ink recorder to give strain as a function of time and also to the X direction of a storage screen oscilloscope. The other hook was attached to a force transducer whose output was also led to the electronic ink recorder to give force as a function of time and to the Y of the storage screen oscilloscope. The oscilloscope then gave force as a function of strain (stress-strain loops). The measurements of viscoelasticity were carried

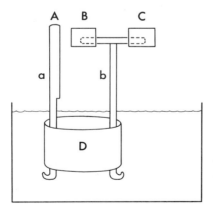

Fig. 2: Ring segment D supported in solution by two hooks, *a* and *b*. Hook *b* can be manipulated laterally via magnetic coils at B and/or C to produce a step-function strain or it can be oscillated sinusoidally. Hook *a* is attached to a stress transducer at A.

out by methods used routinely in polymer physics laboratories (Staverman and Schwarzl, 1956). These methods have proved useful in quantifying behavior, even of striated muscle, provided the proposed model was used to analyze the data.

Specimens were kept immersed in a small bath at any of ten constant temperature levels within the range of 0–75°C and changed from one temperature to another within 0.2 minutes by changing the temperature in a jacket surrounding the immersing bath. This essentially constitutes a step-function change in temperature since the transient of the response to a quick change in temperature takes several minutes to reach completion (Apter and Marquez, 1968). Transient changes in unstrained length l_0 and in elastic moduli E_1 and E_2 (Fig. 3) were measured from the relaxation curve. Steady state viscoelastic constants as functions of frequency of oscillation were obtained at ten temperature levels in the available thermal range from oscillating strains. Both the transient and the steady state data were necessary to identify and evaluate a tissue.

Single fibril types were: pure collagen in the form of aortic adventitia; pure elastin as ligamentum nuchae or formic acid-treated aorta (Hass, 1942); pure muscle as urinary bladder; heart; extraocular muscles; and iris sphincter. The aorta at various distances from the aortic valve provided good examples of composite tissue of known composition (Apter et al, 1966). The thermal response of each pure fibril was unique, consistent with our knowledge of macromolecular structure and dependent on viability. The

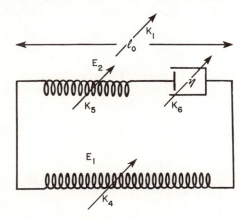

Fig. 3: Kelvin element with all parameters l_0, E_1, E_2, and η capable of varying with time, concentration of environment or strain. Unstressed length is l_0, E_1 and E_2 are force constants of springs, and η is viscosity of dashpot. When the model is extended to viscoelastic materials other than muscle, k_2 and k_3 of Eq. 3 are set to zero. Then $\dot{n} = 0$, so that l_0, E_1, E_2, and η are all constant with time. The model is still describable by Eq. 1 and is the "standard linear solid" of Kolsky (1960).

behavior of specific interest was the response to a step-function change in temperature and the steady state temperature-parameter relationship. The behavior of composite tissues sometimes matched muscle or elastin or collagen, depending on the strain, strongly suggesting that each individual fiber type could be identified by its already determined thermal response (Apter and Marquez, 1968).

B. Physical Model to Serve as Basis for Analysis of Data and Quantifying Tissue Parameters

1. Viscoelastic properties. Tissues, including muscle at any contractile state, are always viscoelastic. A model for this behavior has been developed in detail elsewhere (Apter and Graessley, 1970a, b). In brief, all these tissues show some kind of stress relaxation. Therefore, a three-parameter viscoelastic model (η, E_1, E_2) is the simplest possible description available (Fig. 3) (Apter et al, 1966). Note that these parameters, including the unstressed length l_0, can be taken to be variable rather than constant. The mechanical response of the model is given by the following equation derivable from basic principles:

(1) $$\sigma + \frac{\eta}{E_2}\dot{\sigma} = E_1\epsilon + (E_1 + E_2)\frac{\eta}{E_2}\dot{\epsilon}$$

where σ is stress and ϵ is strain, defined as

(2) $$\epsilon = \frac{l - l_0}{l_0}$$

with l being the existing length and l_0 the *instantaneous* unstretched length of the tissue. Variability in l_0 incorporates the fact familiar to physiologists that muscle may have a range of unstretched lengths, depending on its contractile state, and the elastin and collagen change length in response to the ionic strength of the environment so that $\dot{n} = 0$, but $n \neq 0$. Hoeve and Flory (1958) describe the influence of ethylene glycol on elastin and the effect of tanning on collagen is well-known. [See equation (3) below.]

Now suppose that the values of E_1, E_2, η, and l_0 depend on the macromolecular arrangement, which in turn is influenced by the temperature and by substances whose entry rates into the system depend to some extent on the strain and on the temperature. Specifically for muscle, assume that the state of contraction depends on some substance N near intracellular macromolecules. Assume also that n, the concentration of N, varies with time according to

(3) $$\dot{n} = k_2\epsilon - k_3 n$$

where k_2 and k_3 are rate constants. This model would be appropriate to elastin and collagen if N were molecules of solvent, and for unstimulated muscle if N were molecules of solute.

The instantaneous rest length and the viscoelastic parameters of muscle are taken to be the following simple functions of concentration n:

$$l_0 = l_0^\infty + \frac{l_0^0 - l_0^\infty}{1 + k_1 n} \tag{4}$$

$$E_1 = E_1^\infty - \frac{E_1^\infty - E_1^0}{1 + k_4 n} \tag{5}$$

$$E_2 = E_2^\infty - \frac{E_2^\infty - E_2^0}{1 + k_5 n} \tag{6}$$

$$\eta = \eta^\infty - \frac{\eta^\infty - \eta^0}{1 + k_6 n} \tag{7}$$

with the superscripts 0 and ∞ referring to completely relaxed ($n = 0$) and completely contracted ($n = \infty$) states, respectively.

In summary, this model represents tissues, even muscle whatever its level of contraction, as a three-parameter viscoelastic solid (Staverman and Schwarzl, 1956). The unique energy-producing (or converting) characteristic of muscle has been embodied by allowing these parameters to depend on the level of some chemical in the environs of the macromolecules responsible for the viscoelastic properties. Thus, fibrillar shortening including muscular contraction is represented simply by variations in viscosity, moduli, and rest length of the muscle. The model is faithful to the observation that muscle recontracts when it is stretched and is also capable of reproducing a large range of muscular behavior (Apter and Graessley, 1970).

2. *Influence of temperature changes on properties.*
a. *Independent of environmental ions.* Suppose the activatable macromolecule, A, can exist in a relaxed form and in a contracted form. Let $\dot{n} = 0$ and

(8)
A_r = concentration of the "relaxed form"
A_c = concentration of the contracted form
f_c = contraction force per mole of A_c
f_r = contraction force per mole of A_r
$A_0 = A_c + A_r$ = total of "contracted" and "relaxed" forms.

This last assumption follows from the reversibility of the temperature-force relationships. Then postulate a reaction

$$A_{\text{contracted}} \xrightleftharpoons{K'} A_{\text{relaxed}} \tag{9}$$

with an equilibrium constant K'. Then have

(10) $$\frac{A_c}{A_r} = k' = K'^{-1}$$

so that

(11) $$A_c = \frac{k'A_0}{1+k'}$$

and

(12) $$A_r = \frac{A_0}{1+k'}$$

Using Eq. (8), the total force f developed at any time is given by (assuming additivity)

(13) $$f = f_c A_c + f_r A_r = \frac{f_c k' A_0}{1+k'} + \frac{f_r A_0}{1+k'}$$

The thermal dependence of k' can be postulated from the thermal dependence of f if $\dot{n} = 0$. Consider this situation first, and use data relating steady state force developed in smooth muscle to temperature, keeping within the reversible range. If smooth muscle is taken below 0°C or above 41°C, the reversible thermal-force relationship found between 0–39°C is lost. Within that temperature range, however, the relationship is a reversible negative sigmoid curve with highest steady state forces f_{max} developed from 0–4°C and lowest forces f_{min} from 37–39° (Apter, 1961). Then at any temperature T from 0–39°C, the corresponding force f will be, from (13)

(14) $$f = k'A_c + f_{min}$$

and

(15) $$f = f_{max} - k'A_r$$

since

(16) $$f_{max} = f_c A_0$$

and

(17) $$f_{min} = f_r A_0.$$

This gives

(18) $$f = \frac{k' f_{max} + f_{min}}{1+k'}$$

in terms of measurable forces and the one unknown k'. From reaction rate theory,

$$(19) \qquad k' = \exp\left[\frac{\Delta F°}{R}\left(\frac{1}{T} - \frac{1}{T°}\right)\right]$$

at T_0, $k = 1$, $A_c = A_r$. If

$$(20) \qquad B = \frac{\Delta F°}{RT_0 T} \simeq \frac{\Delta F°}{RT_0^2} \quad \text{a constant,}$$

then Eq. (18) becomes

$$(21) \qquad f = \frac{f_{max} \exp[B(T_0 - T)] + f_{min}}{\exp[B(T - T_0)] + 1}$$

The parameters f_{max}, f_{min} and T_0 can be obtained from experimental data while the f for each corresponding T can be used to give

$$(22) \qquad k' = \frac{f - f_{min}}{f_{max} - f}$$

as a function of temperature. A plot of log k' versus the reciprocal of the absolute temperature would be a straight line of slope B, from the Van't Hoff equation, if (21) is reasonable. This follows from the integral form of (21)

$$(23) \qquad \int_{f_1}^{f_2} \frac{d}{dT} \ln \frac{f - f_{min}}{f_{max} - f} = 2.3 \, B \int_{T_1}^{T_2} \frac{1}{T^2}$$

What is more, both ΔH, the enthalpy, and ΔS, the entropy, changes can also be computed by finding T_0 at the place where $A_c = A_r$. How this is done will be illustrated in the Results section.

b. Dependent on environmental ions. Now consider the case where the ion N can influence the forces developed at any given temperature; that is $\dot{n} \neq 0$. The N effect can be considered as increasing the amount of A_c, which then proceeds in the reaction

$$A_c \rightleftharpoons A \quad \text{as before, or}$$
$$(24) \qquad N + A_r \stackrel{K}{\rightleftharpoons} NA_c$$
$$\quad (A'_c) \qquad (A''_c)$$

Since N reacts with A_r to produce A'_c, but A_c in the reaction

$$A_r \stackrel{k'}{\rightleftharpoons} A_c$$

is not changed. Then the total A_c is

(25) $$A_c = A'_c + A''_c$$

where A'_c is the free A_c or activated A_r and where A''_c is identical with NA_c, which is that bound to N. From equation (24), with $n = (N)$ and $k = K^{-1}$

(26) $$A'_c = \frac{A_c}{1 + nk}; \qquad A''_c = \frac{nA_c k}{1 + nk}$$

Let the elastic parameters E (E_1 or E_2 or l_0 or η) be a linear combination of the activities A'_c of the free and A''_c of the bound A_c; that is,

(27) $$E_1 = \alpha_4 A'_c + \beta_4 A''_c = \frac{\alpha_4 A_c}{1 + kn} + \frac{\beta_4 kn A_c}{1 + kn}$$

or

(28) $$E_1 = \beta_4 A_c + \frac{(\alpha_4 - \beta_4) A_c}{1 + kn}$$

Comparing equation (28) with equation (5), it is apparent that

(29) $$E_1^0 = \alpha_4 A_c; \qquad E_1^\infty = \beta_4 A_c$$

The simplest model could have

(30) $$k_1 = k_4 = k_5 = k_6 = K^{-1}$$

and therefore all the k_i would be related to the equilibrium constant of equation (24).

Under some circumstances, the equilibrium constant K might be relatively independent of temperature and then the thermal dependence of the equilibrium constant K' would be determined from force-temperature curves as already shown. However, K might not be thermally independent, for muscle especially, and its effect could be estimated by proper solution of the constitutive equation for the model with $\dot{n} \neq 0$.

Equations (1)–(7) describe the stress in the tissue as a function of the strain history and, therefore, together comprise the constitutive relations for the model. They have been used successfully to estimate parameters from stress-relaxation curves and with these to predict the responses to many other types of forcings (Apter and Graessley, 1970) for a variety of muscles.

For this purpose, these equations were solved on an analog computer to give the response of a muscle to any forcing including a step-function change in temperature. A step-function increase in temperature was simulated by imposing a step-function increase in all k_i ($i = 1, \ldots 6$) such that $\Delta k_3 >$

$\Delta k_2 > \Delta k_1 = \Delta k_4 = \Delta k_5 = \Delta k_6$. It would be reasonable for the change in k_2 of Eq. (3) to be less than the change in k_3 if k_2 reflected diffusion alone while other rate constants reflected enzyme activity.

Polymer physicists have been using the steady state temperature-modulus relationship in order to gain insight into macromolecular arrangements in a viscoelastic polymer. However, studies of the interactions between collagenous tendons and salt solutions at various concentrations suggest (Puett et al, 1965) that the muscle model may be satisfactory for the crystalline polymer collagen and for amorphous elastin as well, with N being solvent molecules instead of the solute ions used here for muscle.

In Flory's (1956; Flory and Spurr, 1961; Flory and Weaver, 1960) thermodynamic theory of elastic mechanisms in fibrous proteins, a thermodynamic relation between force, temperature and length was developed. This relationship was based on the assumption of an equilibrium between proteins in a melted (short) state and a crystalline (long) state, corresponding to A_c and A_r. The present theory is not in conflict with Flory's, but no attempt will be made here to show further dovetailing that may exist between the present analysis directed toward this specific problem and his more general one.

III. RESULTS

A. Experimental Data

Only reversible, reproducible relationships will be discussed here (Fig. 4). Pure collagen has a small negative temperature-modulus relationship, higher modulus appearing at low temperatures. Collagen responds to a step-function increase in temperature with a step-function drop in modulus and a drop in mean stress—probably signifying an increase in unstrained length l_0. Pure elastin has a small positive temperature-modulus relationship and negative l_0-temperature relationship, and responds promptly to a step-function thermal change. Thus the crystalline polymer collagen behaves quite differently from the amorphous polymer elastin.

Viable muscle, a crystalline polymer, like collagen has a negative temperature-modulus relationship which is unique in being sigmoid in form (Fig. 4). The characteristics have already been outlined in the Methods section. The response to a step-function thermal change of muscle is also unique; it is biphasic and is delayed behind the thermal change. Figure 5 shows the transient of the responses of various tissues: of essentially pure collagen (D) or elastin (B) or muscle (A). Also shown is the time course of the response of a material having both elastin and collagen simultaneously under stretch (C). In that instance, it behaves like a two-phase system, that is the mean tension rises (like elastin), but the elastic modulus drops (like collagen) when the temperature is raised abruptly. This behavior of composite

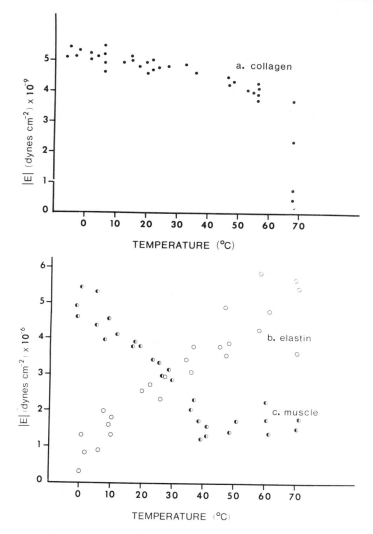

Fig. 4: Temperature-elastic modulus $T-|E|$ relationships of collagen A, elastin B and arterial smooth muscle C. Each point is a single determination.

materials is not noticed in a 30% glycerol-water mixture which prevents the reduction in l_0 of elastin at high temperatures (Hoeve and Flory, 1958). Then only the drop in the modulus characteristic of collagen (D) appears. The steady state thermal-modulus relationships of elastin and collagen are in Fig. 4. Both the transient and the steady state relationships can be used with the model to obtain tissue (or model) parameters as shown below.

228 INFLUENCE OF COMPOSITION ON THERMAL PROPERTIES

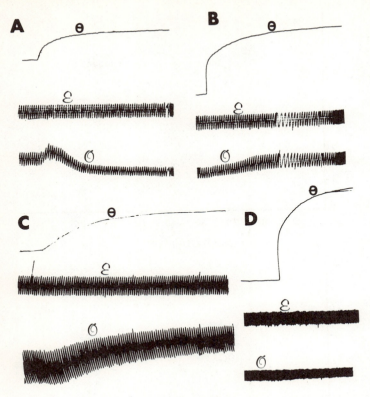

Fig. 5: Time course of response of tissues to a step-function increase in temperature (θ). Oscillating strain ϵ of constant amplitude was superimposed on an initial strain. Amplitude of oscillating stress σ is indicative of elastic modulus ($|E| = \sigma/\epsilon$) and changes in mean stress level reveal changes in l_0 and in elastic modulus.
A is smooth muscle showing biphasic change in mean stress and elastic modulus.
B is elastin with higher modulus and mean stress at higher temperature.
C is elastin plus collagen showing higher mean stress of elastin and lower modulus of collagen at higher temperature.
D is collagen showing small drop in modulus with rise in temperature (see Apter and Marquez, 1968).

The response of muscle to a thermal change is strongly dependent on its viability. This was of especial importance in the case of cardiac muscle because it could serve as a test for transplantability. Only viable muscle (viability being checked by high ATP levels) retains a reversible negative temperature-modulus relationship (Fig. 6) while muscle with low ATP levels had an irreversible relationship (Fig. 7). This study was also pertinent to the quantification of the properties of excised ureters which must be normally muscular after transplantation of the attached kidney. The behavior of ureters

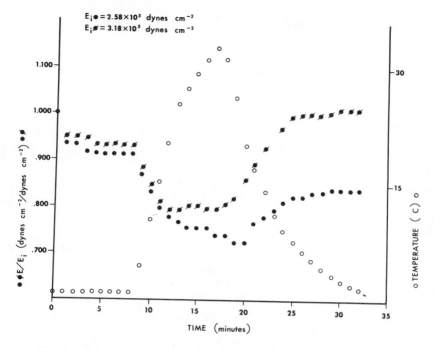

Fig. 6: Reversible temperature–modulus (T–$|E|$) relationship of portion of canine heart muscle removed from two dogs a few minutes before testing. ATP levels were 2.5 and 3.4 mgm P/gm muscle. E is existing and E_i (dynes cm^{-2}) is initial modulus.

in vivo has been attributed to nervous control. However, excised and transplanted ureters appeared uniformly responsive to all forcings, including thermal, despite isolation from a nervous supply. Thus, as will be shown, the model was as reliable for analyzing the behavior of excised ureteral muscle as of other smooth muscle. Ureteral muscle recontracts after a stretch just like other smooth muscle. This behavior suggests that the model promises to match the peristalsis of excised ureters. If so, it would contribute importantly to current thought about ureteral behavior.

The thermal response of excised composite materials containing muscle is consistent with the hypothesis that contracted muscle is always the first element put under strain. Other fibrillar elements are coiled up and therefore not under strain under those conditions, as shown by Luft (1964) for aorta. When smooth muscle is excessively stretched, it, like striated muscle, cannot exert its usual active tension because actin no longer sufficiently overlaps the myosin. Thus at high strains sufficient to stretch muscle excessively, the elastin and collagen contribute more than muscle to the elastic modulus. These fibers probably comprise part of the "series elastic" elements of many

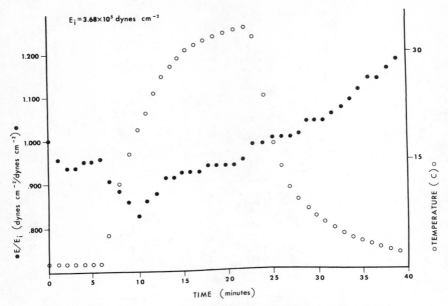

Fig. 7: Irreversible temperature-modulus (T–$|E|$) relationship of canine cardiac muscle tested 45 minutes after removal from beating in situ heart. After a quick rise in temperature, the modulus drops, as in Figure 6, but instead of remaining low, it rises even to exceed its previous level. A subsequent drop in temperature does not reverse the modulus as it did in Figure 6. ATP level was 1.5 mgm P/gm muscle.

analysts of muscular behavior. The fibers can be identified by their distinctive thermal response. However, the number of these fibrils under stretch will determine the elastic modulus: the more, the higher the modulus. Hoeve and Flory (1958) have strong evidence for this concept. Collagen appears to have longer fibers than does elastin of ligamentum nuchae, so that collagen comes under stretch at highest strains. This is not to say, however, that elastin is not also under strain then also; it probably is, as Fig. 4c suggests.

Organs more highly muscular than the thoracic aorta (like the ureter, the bladder, and the abdominal aorta) seem to have more collagen than elastin (Apter et al, 1966). The thermal behavior characteristic of muscle changes to that of collagen when the organ is stretched. None of the positive thermal-modulus relationship of elastin could be found in these tissues.

B. Behavior of Model

The model behaves like elastin, collagen and muscle to a wide variety of forms of forcing functions under isothermal conditions. These have been published in detail and will not be derived here.

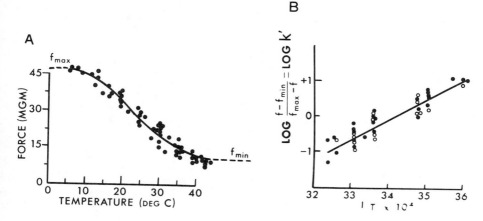

Fig. 8: A. Steady state temperature-force relationship of cat iris sphincter, a smooth muscle (see Figure 4c).
B. Same data plotted as $\log k' = \log (f - f_{min})/(f_{max} - f)$ versus reciprocal of absolute temperature. Slope of line gives $\Delta F°$ and ΔH of the postulated reactions $A_{contracted} \rightleftharpoons A_{relaxed}$ (see Apter, 1961 for similar behavior of striated muscle).

From equation (18) the thermal dependence of k' can be computed if $\dot{n} = 0$. A plot of log k' versus the reciprocal of the absolute temperature T is shown in Fig. 8. Although a variety of muscles were used, both striated and smooth, the essential linearity and correlations of the data are apparent (Apter, 1961). The slope of the line gives $\Delta H/R$; and $\Delta F°$ at any temperature is

(31) $$\Delta F° = -RT \log k'$$

Having ΔF and ΔH, we can calculate the entropy change ΔS using the Gibbs equation. For this group of muscles $\Delta H = -29,900$ cal/mole of muscle, indicating that the transformation from $A_{relaxed}$ to $A_{contracted}$ is exothermic. Indeed, heat production measurements on other muscles indicate that heat is produced during the phase of contraction (Bozler, 1930; Bugnard, 1934; Hill, 1938). By analogy, collagen also would be exothermic in this phase transition, and elastin endothermic, in agreement with Flory. $\Delta S = -108$ cal/mole/degree for all smooth muscles tested. We hesitate at this time to compute similar thermodynamic quantities for elastin and collagen since the reversible thermal range has not yet been worked out.

The response of the model to a step-function change in temperature with $\dot{n} \neq 0$ is shown in Fig. 9 along with the transient of the response of smooth muscle. It shows the time course of a biphasic change in tension—both in viable smooth muscle and the model. The resemblance is satisfying although

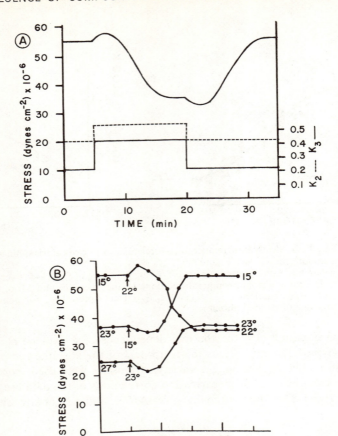

Fig. 9: A. Uppermost curve shows biphasic time course of stress response to a step function rise and a step function drop in temperature. Temperature rise is formalized as an increase in k_2 and k_3, a drop as a decrease in k_2 and k_3 in lower curves.

B. Time course of stress response of dog iris sphincter to a step function drop in temperature from 27° to 23°, from 23° to 15° and a step function rise in temperature from 15° to 22°. Biphasic stress transient is apparent. Resemblance of response of real muscle to model is satisfactory, even quantitatively.

we have not yet developed a physically valid basis for the assumption that $\Delta k_3(T) > \Delta k_2(T)$, which was our formulation for the effect of a change in temperature. Only viable smooth muscle responds this way. Thus it becomes a method for distinguishing viable material muscle from nonviable. It was this test which made it possible (Apter and Mason, Amer. J. Physiol. in press) to identify viable ureter and thereby ascertain the general uniformity of ureters and to quantify its ability to recontract.

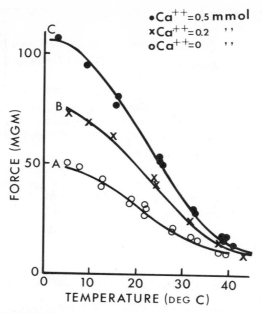

Fig. 10: Steady state temperature-force relationship of cat iris sphincter at 3 levels of calcium ion concentration in environmental solution.

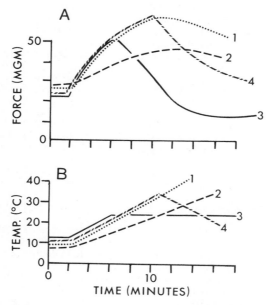

Fig. 11: Superposition principle exemplified showing response of model A to various thermal forcings B. Rise in temperature was formalized as an increase in k_2 and k_3. Data on real muscle are not yet available to check this model.

The response of the model to other forcings is shown in Figs. 10 and 11. Figure 10 indicates that Ca^{++} could be the ion N since f_{max} increases with increasing doses of Ca^{++}. The fact that f_{min} does not increase suggests that the model outlined in a previous publication (Apter, 1961) is applicable. Figure 11 exemplifies the superposition principle with regard to the response of the model to various thermal forcings: ramp, zigzag, and so forth. When experimental data of this kind become available, this behavior can be examined to help fortify or revise the model.

Acknowledgment. *This work is supported in part by grants No. GM-14659, CA 06475 and HE 05808.*

REFERENCES

Apter, J. T. 1961. Thermodynamics of the intraocular and extraocular muscles. Amer. J. Ophthal. 51: 1141/269–1146/274.

Apter, J. T., and W. W. Graessley 1969. Mathematical development of a physical model of muscular contractile element. Proc. 7th Ann. Symp. on Biomath. pp. 32–33.

Apter, J. T., and W. W. Graessley 1970a. A physical model for muscular behavior. Biophys. J. 10: 539–55.

Apter, J. T., and W. W. Graessley 1970b. Progress in quantification of muscular action. Sixth Ann. NASA-Univ. Conf. on Manual Control, pp. 165–76.

Apter, J. T., and E. Marquez 1968. A relation between hysteresis and other viscoelastic properties of some biomaterials. Biorheol. 5: 285–301.

Apter, J. T., and P. Mason. Dynamic mechanical properties of mammalian ureteral muscle. Amer. J. Physiol. In press.

Apter, J. T., P. Mason and Lang, G. Urinary bladder wall dynamics (to be published).

Apter, J. T., and H. Najafi 1968. A test for transplantability of animal hearts. J. Amer. Med. Assoc. 206(13): 2881–882.

Apter, J. T., M. Rabinowitz and D. H. Cummings 1966. Correlation of viscoelastic properties of large arteries with microscopic structure: I. Methods used and their justification. II. Collagen elastin, and muscle determined chemically, histologically, and physiologically. III. Circumferential viscous and elastic constants measured *in vitro*. Circ. Res. 19: 104–21.

Bozler, E. 1930. The heat of contraction of smooth muscle. J. Physiol. 69: 442–62.

Bugnard, L. 1934. The relation between total and initial heat in single muscle twitches. J. Physiol. 82: 502–19.

Flory, P. J. 1956. Theory of elastic mechanisms in fibrous proteins. J. Am. Chem. Soc. 78: 5222–235.

Flory, P. J., and O. K. Spurr, Jr. 1961. Melting equilibrium for collagen fibers under stress. Elasticity in the amorphous state. J. Am. Chem. Soc. 83: 1308–316.

Flory, P. J., and E. S. Weaver 1960. Helix–coil transitions in dilute aqueous collagen solutions. J. Am. Chem. Soc. 82: 4518–525.

Hanson, J., and J. Lowry 1961. The structure of the muscle fibres in the translucent part of the adductor of the oyster, Crassostrea angulata. Proc. Roy. Soc. London. Ser. B. 154: 173–96.

Hass, G. M. 1942. Elastic tissue: 1. Description of method for the isolation of elastic tissue. Arch. Pathol. 34: 807–11.

Hill, A. V. 1938. Heat of shortening and dynamic constants of muscle. Proc. Roy. Soc. London. Ser. B. 126: 136.

Hoeve, J., and P. J. Flory 1958. Elastic properties of elastin. J. Am. Chem. Soc. 80: 6523–6526.

Huxley, H. E. 1957. The double array of filaments in cross-striated muscle. J. Biophys. Biochem. Cytol. 3: 631–48.

Kolsky, H. 1960. Viscoelastic waves, In N. Davies (ed.), Intern. Symp. Stress Wave Propagation Materials. Interscience Publishers, New York. pp. 59–89.

Luft, J. H. 1964. Fine structure of the vascular wall, In R. J. Jones (ed.), Evolution of the Atherosclerotic Plaque. University of Chicago Press, Chicago. pp. 3–14.

Mason, P. 1965. Viscoelasticity and structure of keratin and collagen. Kolloid-Z. 202: 139–47.

Puett, D., A. Ciferri, and L. V. Rajagh 1965. Interaction between proteins and salt solutions. II. Elasticity of collagen tendons. Biopoly. 3: 439–59.

Ramachandran, G. N. 1967. Treatise on Collagen. Academic Press, New York. Three volumes.

Staverman, A. J., and F. R. Schwarzl 1956. Linear deformation behavior of high polymers, In H. A. Stuart (ed.), Die Phyzik der Hochpolymeren. IV 1. Springer-Verlag, Berlin.

Taylor, R. L., and T. H. Brown 1965. The use of non-linear estimation techniques in simple molecular orbital calculations. J. Chem. Phys. 69: 23-16–319.

Tobolsky, A. V. 1960. Properties and Structure of Polymers. John Wiley & Sons, Inc., New York.

Weiner, P. O., and A. M. Pearson 1966. Inhibition of rigor mortis by ethylene diamine tetraacetic acid. Proc. Soc. Exp. Biol. Med. 123,: 185–87.

10

Biomechanics of Bone: Mechanical Properties, Functional Structure, Functional Adaptation

BENNO K. F. KUMMER

Bone is neither morphologically nor mechanically homogeneous, but the inhomogeneities do not disturb the functional adaptation; on the contrary, they contribute to it. The adaptation to the mechanical function concerns the gross shape: axis form, material distribution over the length and over the cross section, as well as the cross section, and the inner structure: the trajectorial pattern of substantia spongiosa. It is very likely that bone is a body of uniform strength in the technical sense, and that it is constructed optimally. The adaptational processes are similar to a feed-back mechanism which is controlled by the mechanical stress. They concern the remodeling of the tissue as well as the exchange of mineral salts.

* * *

I. INTRODUCTION

Biomechanics deals with all kinds of interactions between the tissues and organs of the body and their mechanical stresses. This definition includes problems concerned with non-biological materials implanted into the human body.

Biomechanical relationship between the bone and its mechanical function is expressed by the concept of functional adaptation (Roux, 1895; Pauwels, 1965), in which we distinguish the adapted state from the event of adaptation

II. BONE AS A MATERIAL OF CONSTRUCTION

A. GENERAL CONSIDERATIONS

The main *mechanical properties* of interest to an engineer are: deformation (strain), modulus of elasticity, and strength. For a cylindrical bar in simple elongation, the extensional *strain* is defined as the alterations in length per unit length of the bar. When a bar is pulled, its longitudinal elongation ($+\epsilon$) or shortening ($-\epsilon$) is accompanied by transversal shortening ($+\epsilon_q$) and transversal elongation ($-\epsilon_q$) respectively. These are related to the *stresses* in the material, which has the unit of force per unit area of the cross section.

For a simple elongation of a bar of an elastic material, the strain (ϵ) and stress (σ) are connected with each other by Hooke's law:

(1) $$\epsilon = \alpha \cdot \sigma$$

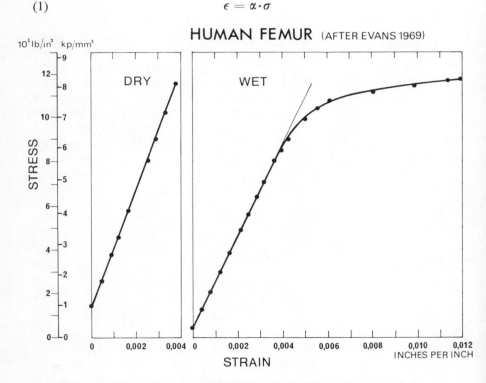

Fig. 1: Stress-strain curves of human femoral bone (adapted from Evans, 1969). At the straight parts of the curves, Hooke's law is applicable.

where α is a characteristic constant for the material. From this formulation the *modulus of elasticity* is derived:

(2) $$E = \frac{1}{\alpha} = \frac{\sigma}{\epsilon}$$

This relationship is valid only within certain limits. Beyond the boundaries of linear behavior, Hooke's law is not valid. (See Fig. 1.)

At certain stress or strain the material breaks. The term *strength* is understood to mean the breaking strength, the maximal stress at the moment of breakdown of the material.

As it is revealed under the microscope, the bone tissue is neither homogeneous nor isotropic morphologically. Whether it can be regarded as homogeneous with respect to its mechanical properties or not remains to be seen. This must be remembered when the mechanical properties of the bone are discussed.

For the explanation of the mechanical behavior of bone substance, several theories have been put forward. These will be discussed below.

B. THE TWO-PHASE HYPOTHESIS

An attempt was made to compare bone tissue with a two-phase material like fibreglass. Currey (1964) discusses this hypothesis. He points out that apatite crystals of differing sizes, with an average cross section of 2500 Å^2 (about 50 by 50 Å) are orientated along the length of the collagen fibrils which are built up by 2800 Å long tropocollagen molecules. The organic bone matrix has a low modulus of elasticity whereas the included crystals have a much higher modulus of elasticity. The volumes occupied by both collagen and apatite are about equal.

In a two-phase material there should be found a modulus of elasticity intermediate between that of the two components, but a strength higher than either of the individual components.

Currey (1964) refers to Bhimasenachar (1945) and reports a modulus $E = 24 \cdot 10^6$ lb in^{-2} for fluorapatite along the axis, the value for hydroxyapatite apparently has not been determined. On the other hand, collagen does not obey Hooke's law exactly; its tangent modulus of elasticity seems to be about 180,000 lb in^{-2}. Currey (1964) concludes that "two-phase materials can function efficiently only if there is very firm bonding between the fibres and the matrix. But the nature of the bonding between the collagen and the apatite is uncertain." (loc. cit., p. 8).

C. THE COMPOUND BAR HYPOTHESIS

According to Young (1957) the tensile stresses in bone are resisted by the collagen, and the compressive stresses by the apatite. Further, Currey (1964) reports that bone as a brittle substance has a very high tensile strength in relation to its compressive strength, namely, 15,000 lb- in^{-2} and 25,000 lb-in^{-2} respectively. Currey tries to calculate the loads borne by the apatite and the collagen with the standard formulae

(3)
$$W_a = \frac{P \cdot A_a \cdot E_a}{E_a \cdot A_a + E_c \cdot A_c}$$

and

(4)
$$W_c = \frac{P \cdot A_c \cdot E_c}{E_c \cdot A_c + E_a \cdot A_a}$$

where the meaning of the symbols is as follows:

A_a : crosssectional area of apatite
A_c : crosssectional area of collagen
E_a : modulus of elasticity of apatite
E_c : modulus of elasticity of collagen
P : load borne by the whole bone
W_a: load borne by the apatite
W_c: load borne by the collagen

The relationship between the stress borne by the collagen and that borne by the apatite in the bone at a stress of 15,000 lb in^{-2} is shown in Fig. 2 following Currey's calculations. Comparing these theoretical results with the petrotympanic bone of fin whale with 13% of collagen (against 30% in nor-

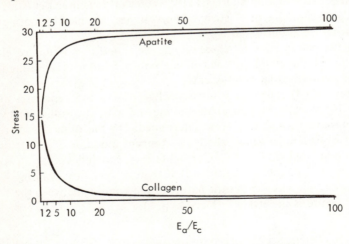

Fig. 2: Stresses in apatite and collagen for different values of E_a/E_c, at a net stress in bone of 15,000 lb in^{-2} (from Currey, 1964).

mal bone), Currey finds that the apatite should be suffering less stress than in normal bone, in fact the bone is weaker. He concludes "that the collagen and the apatite must have a more complex relationship than the simple one we have considered so far." (Currey, 1964, p. 3).

D. THE PRESTRESS HYPOTHESIS

Knese (1958) defends the opinion that the collagen fibrils in bone are prestressed with the following arguments: (1) In loaded bone the collagen fibrils are gently curved, but not undulated as expected if they were not under initial tensile stress. (2) The fibrils are arranged in the form of a catenary. Like the chain of a catenary, they are stressed along their length even where there is no external stress acting. (3) The difference between tensile and compressive strengths is not as great as might be expected from such brittle material as bone.

Currey (1964) rejects all three hypotheses of Knese. He states that undulations of fibrils are seen. Further he maintains that it is impossible to assert that a particular fibril is describing a catenary rather than a parabola on the basis of visual inspection. He finds it easy to construct theoretical models with nearly catenary curved fibrils, yet entirely uninfluenced by stress. Finally, Currey calculates a prestress of 20,000 lb in^{-2} for diminishing the tensile stress in the apatite to about 5000 to 10,000 lb in^{-2} at an applied tension of 15,000 lb in^{-2} on the whole bone. For an applied compression of 25,000 lb in^{-2} the same prestress would raise the compressive stress in the apatite up to 67,000 lb in^{-2}, an unbearable amount. Furthermore, it is hard to imagine how such a great prestress of the collagen fibrils could be produced in bone.

E. THE STRENGTH AND DENSITY OF BONE

Under these circumstances of more or less plausible hypotheses on the mechanical behaviour of the bone substance, it is not surprising that the investigations of the physical properties give a very complex pattern of results. The obtained values depend apparently on several factors, such as the state of the specimen (dry or wet, fresh or embalmed), the kind of element (long or short bones; even among the long bones there are differences), the location of the particular sample, and the structural factors of the bone tissue.

1. Compressive strength of dry bone. The easiest way of determining bone strength is the compression of dry bone. As is shown in Table I, the results of these examinations differ from 7160 lb in^{-2} (tibia; Jouck, 1970) up to 48,500 lb in^{-2} (femur; Schmitt, 1968). It is important to compare the results of different investigators, for the strength depends on the load as well

TABLE I
Compressive strength of dry bone

Author	Material		Strength in 10^3 lb in^{-2}
HÜLSEN (1896)	humerus	∥ to axis	18.650–20.970
		⊥ to axis	14.220–18.050
RAUBER (1876)	femur	∥ to axis	21.330–32.990
		⊥ to axis	23.660–26.160
CALABRISI and SMITH (1951)	femur	∥ to axis	30.150
SCHMITT (1968)	femur	∥ to axis	17.100–48.500
JOUCK (1970)	tibia	∥ to axis	7.160–22.300

as on the loading speed, as has recently been discussed by Schmitt (1968). He loaded with a speed of 90 kp mm^{-2} min^{-1} and obtained a compressive strength about 50% higher than Rauber (1876) who loaded with 0.5 kp mm^{-2} min^{-1}.

2. Tensile strength of dry bone. As is demonstrated in Table II, the tensile strength of dry bone is about half of its compressive strength, varying from 425 lb in^{-2} (parietal bone; Evans and Lissner, 1957) to 17, 480 lb in^{-2} (Humerus; Hülsen, 1896).

TABLE II
Tensile strength of dry bone

Author	Material	Strength in 10^3 lb in^{-2}
HÜLSEN (1896)	humerus	14.850–17.480
EVANS (1957)	femur	14.400–16.150
EVANS and LISSNER (1957)	parietal	0.425– 1.112

3. Strength differences between "dry" and "wet" bone. Since drying of bone is quite an unphysiological procedure, the effect of this alteration was studied by several authors. Evans (1957b) compared the tensile strength of wet and dry bone of 242 specimens from the femora of six white males (see Fig. 3). He found a considerable increase in strength after drying. Sedlin and Hirsch (1966) also report a clear effect of strength increase after one hour of drying (investigation on femora of twenty adult subjects).

4. The effect of embalming on the strength of bone. The comparison of the tensile strength of femoral bone, as obtained by Wertheim (1847) and Rauber (1876), with the results of Evans and Lebow (1951) on embalmed femora shows no significant difference (Fig. 4). The same is reported by Sedlin and Hirsch (1966). After fixation in formalin no significant change of the modulus of elasticity was seen.

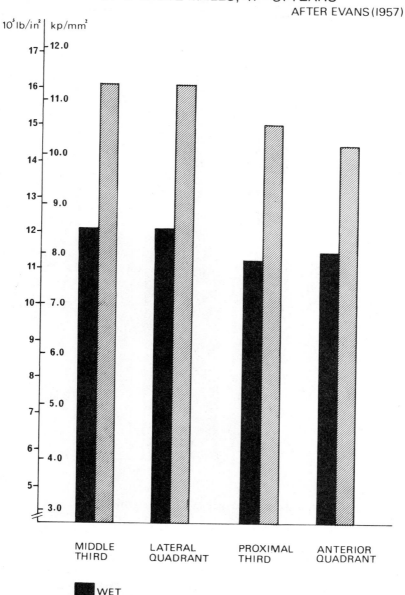

Fig. 3: Differences in tensile strength of wet and dry bone (after data from Evans, 1957). For further explanation see text.

244 BIOMECHANICS OF BONE

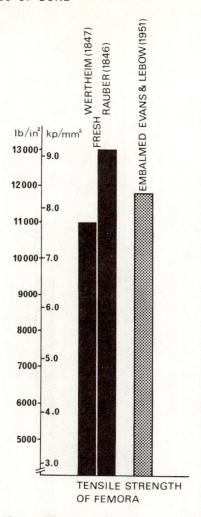

Fig. 4: No significant difference in the tensile strength of fresh and embalmed bone is seen (after data from Evans and Lebow, 1951). For further explanation see text.

5. *The strength in different skeletal elements.* It seems that the strength of bone may differ from one skeletal element to the other. This is true at least for the tensile strength. In the direction parallel to the long axis of bone tensile strength is highest in the humerus, lower and very similar in femur and fibula. The weakest bone against tensile stress seems to be the parietal (Table III).

With respect to compression, the strength differences are not significant (Table IV). The values reported by the authors vary for the humerus between

TABLE III
Tensile strength (in 10^3 lb in^{-2}) || to the axis

	Author	Humerus	Femur	Fibula	Parietal
fresh	RAUBER (1876)	14.560	12.840		
	HÜLSEN (1896)	15.000			
embalmed	WERTHEIM (1847)		11.040	11.700	
	EVANS and LEBOW (1951)		11.840		
	EVANS and LISSNER (1957)				10.230

TABLE IV
Compressive strength (in 10^3 lb in^{-2})

Author	Humerus	Femur	Fibula	Tibia	Parietal
|| to the long axis fresh:					
RAUBER (1876)	19.820	20.000		23.880	
HÜLSEN (1896)	29.030			29.760	
CALABRISI and SMITH (1951)		30.150		25.280	
dry, unembalmed:					
HÜLSEN (1896)	19.410				
RAUBER (1876)		26.990			
embalmed:					
EVANS and LISSNER (1957)					22.080
SCHMITT (1968)		32.800			
⊥ to the long axis dry, unembalmed:					
HÜLSEN (1896)	16.060				
RAUBER (1876)		25.310			

19,410 lb in^{-2} and 29,030 lb in^{-2}; for the femur between 20,000 lb in^{-2} and 30,150 lb in^{-2}; for the tibia between 23,880 lb in^{-2} and 29,760 lb in^{-2}; averages are reported as 22,080 lb in^{-2} for the parietal bone and 22,730 lb in^{-2} for vertebrae.

6. Relationship between the strength and the direction of compression.
Rauber (1876), Hülsen (1896), Evans and Lissner (1957) have shown that there are differences in compressive strength if the bone is stressed in the direction of its long axis or perpendicular to it (Table V). In the long tubular

TABLE V
Differences of compressive strength, depending on the direction

Author	Material	strength in 10^3 lb in^{-2}		
		|| to the long axis		⊥ to the long axis
		Humerus		
HÜLSEN (1896)	fresh	29.030	>	23.460
	dry	19.410	>	16.060
		Femur		
RAUBER (1876)	dry	26.990	>	25.310
		Parietal		
EVANS and LISSNER (1951)	embalmed	22.080	<	24.280

bones the tissue is strongest against compression when stressed parallel to the axis; in the parietal it is the contrary. A relationship to the presumed function of these elements seems to be evident.

7. *Distribution of compressive strength within the same bone.* It has been shown by Evans (1957a, b) that the strength of bone changes at different locations within the same skeletal element. But for the analysis of functional adaptation we need a more complete pattern of the strength distribution. Amtmann (1968) calculated the distribution of average compressive strength in the femur from the data procured by Schmitt (1968). The results were

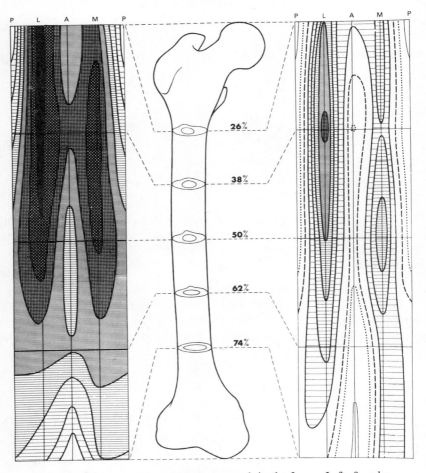

Fig. 5: Levels of equal compressive strength in the femur. Left: 9 males from 56–66 years. Right: 10 males and females from 74–81 years. Explanation of symbols: P: posterior; L: lateral; A: anterior; M: medial (from Amtmann, 1968)

obtained on dry embalmed bones. As we have seen in Sec. II, E, 4, embalming does not alter the physical properties, but drying does (see II, E,3). Nevertheless, Amtmann shows by comparing the data found in the literature that the relative distribution of the strength is the same in dry and wet bone, respectively (see Evans, 1957a).

For the study of the functional adaptation it is sufficient to know the relative strength distribution (see Sec. IV). The pattern of compressive-strength distribution is shown in Fig 5. The strongest regions are in the upper third of the femur lateral and medial, with a predomination of the lateral site. Weakest are the sites proximal posterior and distal anterior.

8. The distribution of density within the same bone. In the living subject we cannot determine bone strength in a direct way, but we can measure the density of the tissue by radiography. As we have seen, the strength of the bone depends at least partially on its mineral content (Sec. II, B), and

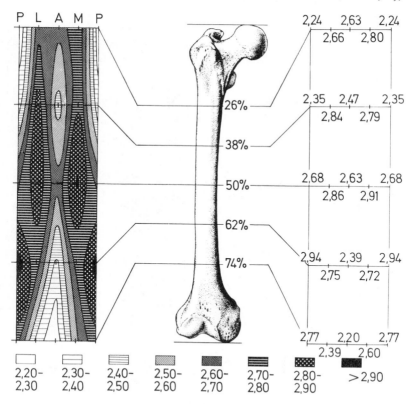

Fig. 6: Distribution of average density in femora of 8 individuals from 56–80 years. Situation of crosssections is indicated in percent of the total length. Same symbols as in Fig. 8 (from Amtmann and Schmitt, 1968).

248 BIOMECHANICS OF BONE

the mineral salts accounts for more than 90% of the radiographic density of the bone. For this reason it may be of interest to know the distribution of density.

Figure 6 shows the distribution of average densities in human femora as determined by Amtmann and Schmitt (1968). Concentrations are seen in the middle of the shaft medially and laterally, with a predomination of the medial side. The maximum is to be found posterior in the distal third. Minima of density are of the sites proximal posterior and distal anterior.

9. Relationship between the compressive strength and the density. Comparison of the distributions of compressive strength and density demonstrates immediately that there is no full harmony of both patterns. Amtmann calculated the correlation between them and found that only about 40–42% of the differences in strength may be explained by differences in density (see Amtmann and Schmitt, 1968, p. 33ff).

10. Compressive strength and structure of bone. We conclude from these calculations that another explanation has to be found for the amount of strength not related to density. Amtmann (1968a, b; 1971) puts forward the hypothesis that this difference might be due to the structure; he speaks

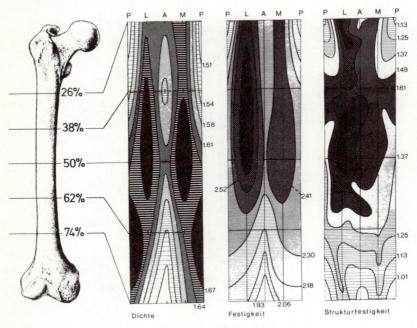

Fig. 7: Relationship between density, compressive strength and "structural strength" in the human femur. Same symbols as in Figs. 8, 9 (from Amtmann, 1971).

of "structural strength." Fig. 7 shows the distribution of this part of strength unrelated to mineral content.

This theory is supported by the results of Evans and Bang (1966, 1967), who found significant differences in strength as well as in structure between human femora, tibiae and fibulae.

11. Influence of sex and age on compressive strength and density of bone. Recently, Amtmann (1971) demonstrated significant differences in the average breaking strength of bone in men and women. Fig. 8 shows these sex differences for two density groups and for right and left femora. Besides these sex-related differences there is a decrease of the compressive strength with the age. The differences between the younger and the older group of men are as great as the difference of strength between the younger group of men and women respectively (Fig. 9). Even the "structural strength" is different for different sexes (Amtmann, 1971; Fig. 10). Amtmann (1971) further reports that there is also a significant difference in density with respect to the age (Fig. 11). Finally, he finds that the losses of density and strength are disproportionate in a somewhat paradoxial way: the greatest weakening has been just at the weakest regions (Fig. 12).

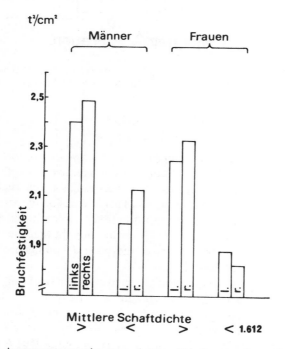

Fig. 8: Average compressive strength (t cm^{-2}) in femora of men (left) and women (right) for two density classes. 1: left femur; r: right femur (from Amtmann, 1971).

250 BIOMECHANICS OF BONE

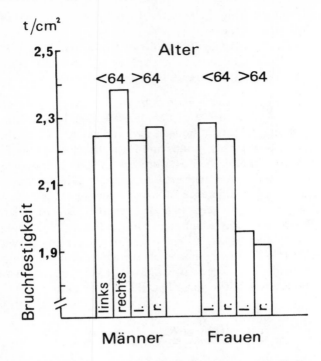

Fig. 9: Average compressive strength of right (r) and left (l) femora of men (left hand side) and women (right hand side) in two classes of age (from Amtmann, 1971).

12. Discussion of the distribution and change of strength and density in bone. A survey of the distribution of compressive strength and density in human femora shows clearly that the bone, even as a constructive material, cannot be considered as homogeneous. This is true for the strength as well as for the density and the relationship between strength and density. This disproves the opinion of Petersen (1930), who thought that the inhomogeneities in the microscopic structure gave the bone a homogeneous behaviour with respect to its mechanical properties.

From these results the question is raised whether the bone is exclusively adapted to mechanical needs. Amtmann and Schmitt (1968) discuss the possibility of the existence of two kinds of bone tissue: "supporting bone" and "metabolic bone." With this they refer to a hypothesis maintained by earlier authors (Lacroix, 1953; Vincent et Haumont, 1960; Marotti, 1963). Nevertheless, we should keep in mind that this is nothing more than a hypothesis. The presumption that the bone is adapted to its functional stress is still not disproved. On the other hand we do not have sufficient knowledge about the exact local stresses and their relationship to the structure to

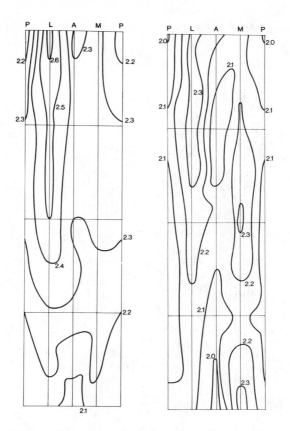

Fig. 10: Levels of equal "structural strength" ($t\,cm^{-2}$) in femora of men (left) and women (right) (from Amtmann, 1971).

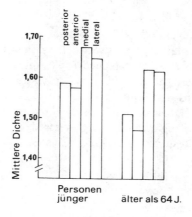

Fig. 11: Average density in the femur shaft of younger (left) and older (right) individuals (from Amtmann, 1971).

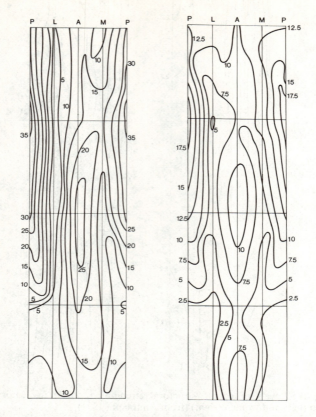

Fig. 12: Changes in femora of women over 64 years of age. Left: loss in strength (in percent of the local strength in younger individuals). Right: loss in density (in percent of the local density in younger individuals). Same symbols as in Fig. 8 (from Amtmann, 1971).

maintain the exclusiveness of the functional adaptation. But since we cannot reject functional adaptation, we have to study this phenomenon in further detail. From this viewpoint it is indeed irrelevant to know the compressive strength on sites where the bone tissue is only stressed by tension and vice versa. At the locations where pure tensile stress acts we should expect structural adaptation to tension and consequently higher tensile strength in comparison to other regions normally stressed by compression. For this consideration it is presumed that local stresses in bone generally do not change their quality. But if they did, it would be of eminent importance for our concept of functional structure.

III. FUNCTIONAL CONSTRUCTION OF BONE

A. THE GROSS SHAPE OF LONG BONES

In this section the problem whether or not the general bone form is adapted to its mechanical function will be discussed. For this purpose we will consider first the principles of bone stressing.

1. General considerations on the stressing of bone. For the following considerations it is important to state that bone is stressed via joints. It follows that in static situations the movements on the joints must be in equilibrium. Hence the resultant of forces must go through the center of rotation, which in a spherical joint is identical with the geometric center. In monoaxial joints there is only the condition that the resultant of forces must intersect rectangularly with the joint axis. In biaxial joints the rectangular intersection has to take place with both axes. For irregular shaped joints one can generally construct an "instantaneous axis of movement." In the following, for

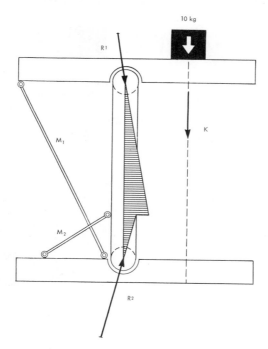

Fig. 13: Simplified model of a long bone, stressed by a weight and two muscles. For explanation see text.

sake of brevity the projections of these axes into the drawing plane will be called "rotational centers."

Figure 13 illustrates the stressing of a vertical rod, bedded in joints at its upper and lower end. It articulates with two horizontal bars. When the upper bar is loaded excentrically, it tilts to the side of the load. The bar can be brought into equilibrium by shifting the load exactly over the joint or by a counterweight. In the body the counterweight is given by muscles. Since it is highly improbable that the directions of the load (K) and of the muscle (M_1) are parallel to the axis of the rod, the resultant R_1 for the upper articulation will not go through the rotational center of the lower articulation, and the latter has to be balanced at least by a second muscle, M_2. The resultant R_2 for the lower articulation originates from the forces R_1 and M_2. Since the two resultants R_1 and R_2 are diverging from the rod's axis, the rod itself is stressed by bending.

The bending stress can be prevented if the rod is bent in a way so that its axis coincides with the lines of action of R_1 and R_2 (Fig. 14). This fact has important consequences for the distribution of the bone material.

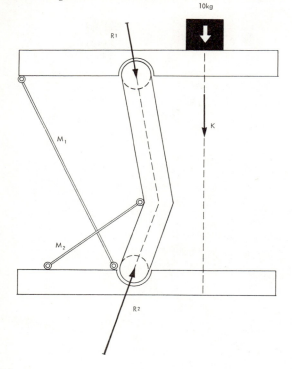

Fig. 14: Adaptation of the bone axis to the direction of stressing forces. The rod in the model is free of bending. For further explanation see text.

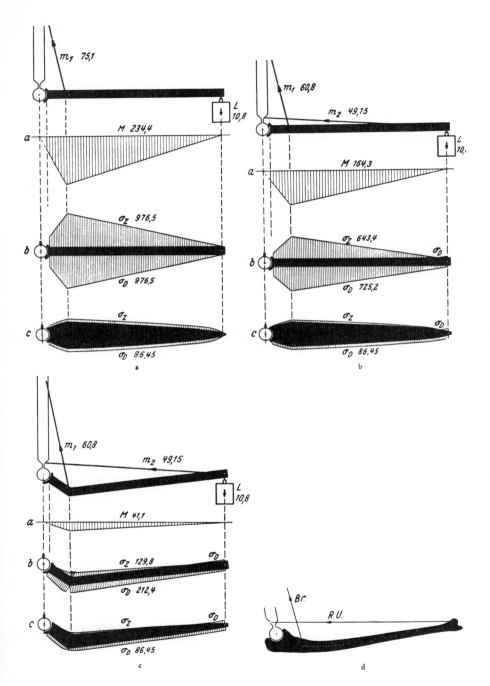

Fig. 15: From (a) to (d) better approximation of form and distribution of material in a bar to the adapted form of the human ulna. The goal is to make the stresses equal everywhere in the bone (from Pauwels, 1965).

256 BIOMECHANICS OF BONE

2. The axial form of long bones and distribution of over the length. Pauwels (1965) defends the idea that the skeleton is adapted to its bending stress by the bone form and by the muscles, which play the role of tensile strings. Here we will discuss only the bone form.

The problem can be illustrated best by two cases, studied by Pauwels (1948b, 1950; 1965). He demonstrated first how a bar, stressed like the human ulna, can be adapted on its bending stress by a specially curved axis form (against the sense of bending) and by a peculiar distribution of its material over the length of the bone (Fig. 15). The second example is taken from the lower limb. Here Pauwels again demonstrates how the bending stress of the leg skeleton can be reduced by the adapted form of the axis.

But, as we shall see, bending moments cannot be eliminated entirely.

3. Distribution of material in the cross section. According to Pauwels (1965, 1968), the bone material is arranged in the cross section in such a manner that its amount is proportional to the local values of stresses. This theory presumes bending stress and a material of homogeneous strength.

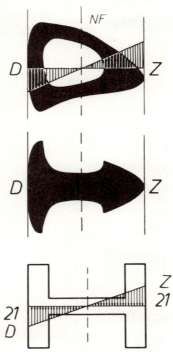

Fig. 16: Upper diagram: (cross section of the tibia with stress distribution. Middle diagram: contracted area of the cross section. Lower diagram: stress distribution in a Warren's truss under bending. (Courtesy of Prof. Pauwels.)

Under these conditions the mass distribution is to be compared with the profile of the stress diagram. This can best be done if the cross-sectional area has been contracted in the direction of the bending plane. The material of construction is used in the most economical way, if the contraction profile and the profile of the stress diagram are similar.

Pauwels (1965) tried to prove that these considerations may be applied to the human long bones, especially to the femur. He showed the parallel between the stress diagram and the local thickness of the diaphyseal compacta using as examples the femur and tibia (Fig. 16).

Amtmann (1971) states that under a constant bending load the bone tissue should be reduced to zero in the neutral zone. According to him, the actual shape of the contracted area is more compatible with bending stresses in various directions; and the maximal and average strains are better paralleled to the local amount of bone tissue.

These measurements and theoretical derivations permit us to consider the bony element as a whole as a body of homogeneous strength in the technical sense, however it cannot be proved definitely.

4. *Mineral content and strain distribution.* Pauwels (1954; 1965) first compared the radiological density with the strain distribution in a plastic

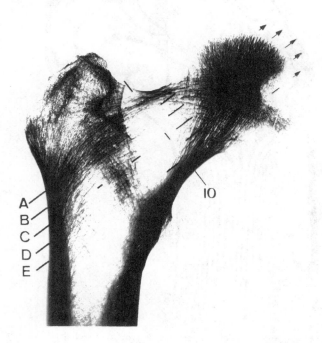

Fig. 17: Radiography of the upper femur end (from Knief, 1967).

model of the upper femur end. Again the mechanical homogeneity of bone tissue and a constant relation of mineral content to strength were presumed.

Later, Knief (1967a, b) gave a quantitative comparison of x-ray densitometry and photoelastic experiments (Figs. 17, 18). In these investigations, some minor differences between the distributions of bone minerals and strains showed up. They could not be explained fully by Knief.

Recently Konermann (1970, 71) published an easy photographical method for the determination of the density distribution in x-ray pictures. Zones of "aequidenses" can be marked by cross-hatching or with colours. According to the newer results, a still better harmony between mineral distribution and local stress appears highly probable (Fig. 19).

B. The functional architecture of spongy bone

1. The problem of minimal amount of material. Roux in 1895 described the so-called "maximum-minimum law," which states that a maximum of strength is reached with a minimum of constructive material. Roux calls

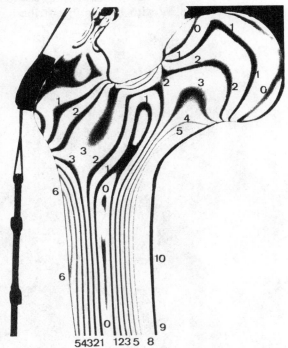

Fig. 18: Isochromes = levels of equal strain in a model of the upper femur end (from Knief, 1967).

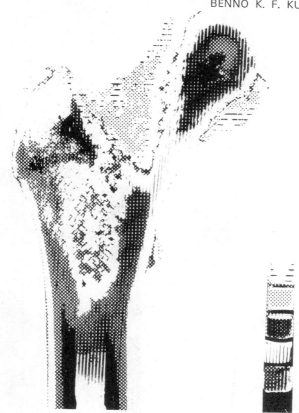

Fig. 19: Aequidenses from a radiogram of the upper femur end (from Konermann, 1970).

these structures "minimum structures." This expression must not be understood in an absolute sense, because in bones, as in technical material, a safety factor plays an important role. Consequently, a minimum structure in the sense of Roux is given if a minimum of material is used to resist the actual stresses including a certain safety factor. This safety factor may be higher or lower; in each case the whole element built up this way will represent a body of homogeneous strength.

2. *The theory of trajectorial architecture.* Roux (1895) further maintains the theory that the substantia spongiosa represents a trajectorial architecture in the sense of Meyer (1867) and Wolff (1884), being for that reason a minimum structure. Roux and his successors failed to prove this presumption. Only in 1948 did Pauwels demonstrate the trajectorial nature of the architectures of the substantia spongiosa in normal and pathological bones.

A photographical method in photoelasticity gives not only an impression of the trajectorial pattern, but also of the locally different density (Fig. 20; Kummer, 1956). The objection by Knese (1956) that two-dimensional models were not representative since the bone is a three-dimensional body and the three-dimensional stress pattern must differ widely from the two-dimensional one has been refuted (Kummer, 1959). It was possible to demonstrate the similarity between a geometrically constructed three-dimensional trajectorial pattern (Fig. 21) and the spongious architecture (Fig. 22).

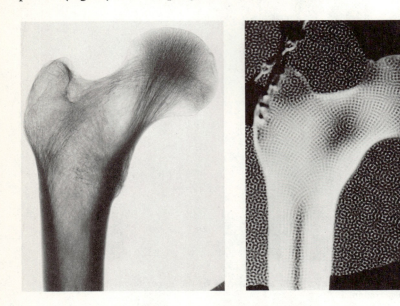

Fig. 20: Comparison of x-ray picture (left) and trajectorial pattern as revealed in the photoelastical experiment (right) Not only the architecture in the model but also the repartition of density is similar to the bone.

IV. FUNCTIONAL ADAPTATION OF BONE

According to Roux (1895, p. 157) functional adaptation means: "adaptation of an organ to its function by practising the latter." Roux presumes that functional adaptation works by the means of hypertrophy and atrophy.

A. Pauwels' Hypothesis on the Interactions between Stress and Bone Formation

1. Functional hypertrophy and atrophy. Pauwels' hypothesis is based on the observation, known to Roux (1895), that the use of an organ leads

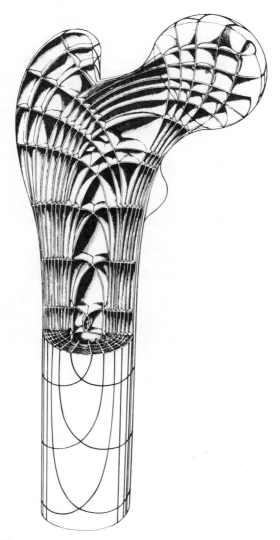

Fig. 21: Theoretical construction of three-dimensional trajectorial system in a femur model (from Kummer, 1966).

to a hypertrophy if the forces remain within certain physiological limits, and that non-use is followed by atrophy. In X-ray pictures there can be seen a densification of the bone at the sites of augmented stresses, whereas the bone becomes less dense where stresses are lower.

2. *Bone as a system, controlled by feed-back.* From these observations Pauwels concluded that bone formation is influenced by mechanical

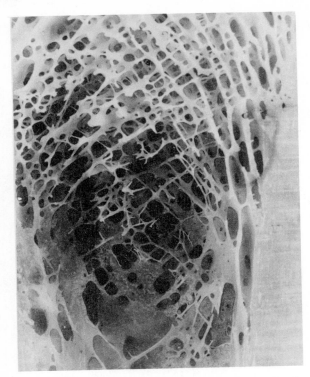

Fig. 22: Surrounding of the attractive singular point in the substantia spongiosa of the upper femur shaft. This pattern is to be compared with the corresponding region of Fig. 21 (from Kummer, 1966).

stresses. At an optimal stress value σ_s the constant remodeling of bone tissue is balanced, that is, in a time unit as much bone tissue is removed by resorption as is built up by apposition. Within the tolerance limits σ_u and σ_o, the apposition will predominate when the actual stress σ_i is greater than σ_s; resorption predominates when σ_i is below σ_s. Stresses higher than σ_o destroy the bone by pathological resorption; at stresses below the limit σ_u the resorption ceases and bone is not removed to zero.

We conclude from this that bone behaves like a system controlled by feed-back, as shown in Fig. 23. An applied force causes stresses and strains in bone. Strains are the stimuli for the remodeling process. In the balanced steady state (optimal strain), activity of osteoblasts and osteoclasts are equal. Strains greater than the optimum cause bone hypertrophy by a relative predominance of the osteoblast activity and the increase of the load-bearing cross section is followed by lowering of stresses. This may lead to strains below the optimum, the predominance of osteoclasts causes atrophy, and the stresses rise again.

Fig. 23: Bone as a system, controlled by feed-back. For further explanation see text.

3. A simple mathematical model. Pauwels' hypothesis can be illustrated by a simple mathematical model. A qualitative approximation to the observed reactions is represented by a cubic function (see Fig. 24). The meaning of the symbols is as follows:

U: bone remodeling, positive values mean predomination of apposition, negative values indicate predomination of resorption
a : factor of proportionality, related to the speed of remodeling
σ_i: actual stress
σ_s: optimal stress; its relationship to the breaking stress controls the safety factor
σ_u: lower limit of the tolerance
σ_o: upper limit of the tolerance.

By means of the formula represented in Fig. 24, the behaviour of bone can be simulated by a computer. The value of the factor a defines this kind

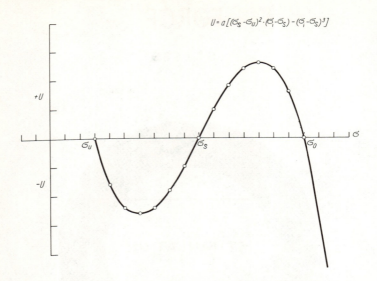

Fig. 24: Mathematical approximation of the remodeling process in bone and its relationship to the stress. For further explanation see text.

of adaptive process: if a is small, the stress approximates the optimum asymptotically (Fig. 25), at a greater a the stress changes in form to a damped oscillation, approximating the optimum again. A certain value of a leads to a resonance of remodeling processes, so that the stress oscillates undamped between two values beyond and below the optimum.

Until now it had been difficult to say which of these three types of remodeling processes best approached the behavior of the living tissue, because our observations were but qualitative ones. The need for exact quantitative measurements of the adaptive remodeling in bone is obvious.

4. Functional remodeling of the cross-sectional shape of long bones. Pauwels (1968) was able to show that these remodeling processes discussed above can, in fact, explain the observed adaptation of the form of the cross-section to abnormal stresses. With the example of three femora, deformed by ricketts, he demonstrated a shape of the cross section quite similar to the calculated theoretical form (Fig. 26).

5. Functional remodeling of the substantia spongiosa. The same rule that governs the adaptational process in the cross section also controls the formation of the trajectorial pattern in the spongious bone. According to Pauwels (1960, 1965) it is again the play of hypertrophy and atrophy that finally changes the direction of a trabecula, which becomes oblique to the direction of a stressing force until its axis coincides with position and direc-

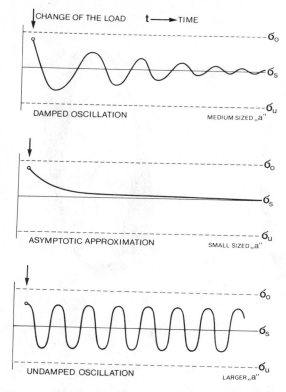

Fig. 25: Three types of computer remodeling processes. For explanation see text.

tion of the force. If an oblique bar representing the bony trabecula is stressed by bending, material will be added where the stresses exceed the optimal value (*cf* IV, A, 2), and removed where they are below, so finally the bar will change its shape and direction. Pauwels demonstrated that ultimately the bar is stressed axially and consequently built up with a minimum of material. If all the trabeculae of the substantia spongiosa are orientated by these remodeling forces, they form a rigid framework and a trajectorial pattern.

B. CRITICAL DISCUSSION

The models of bone remodeling and functional adaptation developed by Pauwels are based on the presumption that the material is homogeneous with respect to its mechanical properties. But this is not true as we have seen in Sec. II, E, 12. Nevertheless, the adaptation is perhaps better than Pauwels

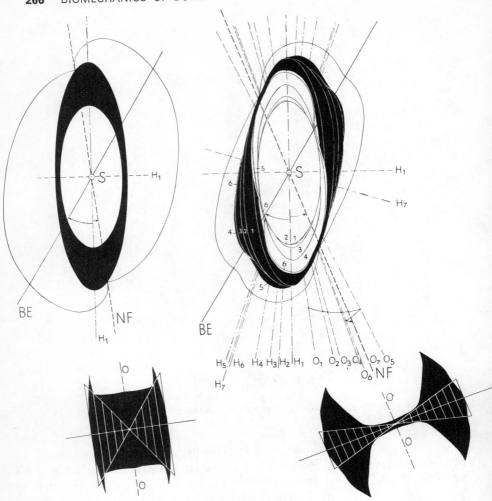

Fig. 26: Remodeling of an elliptical cross section of bone by calculation. The result is most similar to observed forms (from Pauwels, 1968).

knew. The results obtained by Amtmann (1971) show that the bone is relatively stronger just at those sites where the maximal stresses occur. In this way the bone disposes of two separate possibilities of adaptational mechanisms (Fig. 27): the long-time mechanism of apposition and resorption of bone tissue depending on the activity of osteoblasts and osteoclasts, and a short-time mechanism of uptake and output of mineral salts. Normally the exchange of calcium in bone is balanced and a steady state is maintained. As the observations on astronauts demonstrate, the first reaction of bone to a decrease of stresses is a decalcification, going on so fast that it is obviously

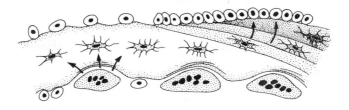
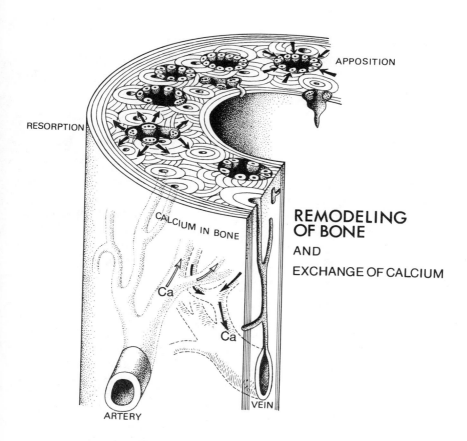

Fig. 27: Remodeling of bone by apposition and resorption and the role of calcium exchange.

a loss of mineral salts rather than a resorption of bone tissue. Consequently, recalcification occurs within a short time. On the other hand, the remodeling of bone tissue needs more time; for this reason the normalization of an atrophic bone goes on over a period of months, perhaps years.

C. Conclusions

1. Safety factor and body of homogeneous strength. It has been shown that the adaptational processes are controlled by the mechanical stresses. This control is expressed by the optimal stress σ_s and the tolerance limits σ_u and σ_o. Theoretically it is not unlikely that the values for σ_s, σ_u, and σ_o may vary

(1) from one individual to the other
(2) with age in the same individual
(3) from one skeletal element to another within the same individual
(4) from one region to another within the same skeletal element

The factor of safety S can be understood in relationship to the breaking stress σ_b and the maximal stress σ_m under normal conditions:

$$(5) \qquad S = \frac{\sigma_b}{\sigma_m}$$

The maximal stress σ_m depends on the optimal stress σ_s. The lower σ_s, the lower is σ_m also. Consequently, since σ_m is a function of σ_s, the safety factor S depends finally on σ_s.

A body of homogeneous strength demands an equal factor of safety in every region of the element. Therefore, the problem of a body of homogeneous strength is connected with the homogeneous repartition of the optimal stress σ_s within the bone. In fact, a body of homogeneous strength is to be expected only if the optimal stress σ_s is equal everywhere within it.

2. Allometrical consequences

a. On earth. The problem of scales is familiar to the engineer and was known to Galilei (1638). A reduced scale model, made of the same material as the prototype, can be relatively stronger. For example, a column supporting its own weight becomes relatively weaker by linear growth. For such a column, in order to keep strength while growing, has to increase more in breadth than in length. If we call the factor of increase in length n, then the diameter D has to grow by $n^{3/2}$.

From this we conclude that as animals become taller, their bones must grow allometrically and become thicker, if the bone substance is to maintain the same strength.

b. Under various gravitational circumstances. The weight (G) of the body is the product of its mass (m) and the acceleration (g):

(6) $$G = m \cdot g$$

Under the gravitational force on the surface of the earth, a body of the mass 1 kg exerts a force of 1 kp on the supporting ground. In absence of a gravitational force, the body becomes weightless and need not be supported. With an increase of the gravitational acceleration (for example, at the surface of a tall celestial body) the weight exerted on the support increases also. If the weight is supported by a tubular column then at equal stresses it can be shown that a disproportionately large increase of the cross-sectional area is required when the gravitational acceleration is increased.

It is to be expected that the bone will be adapted to changing gravitational forces by means of the processes discussed in Sec. III,B. That means, under decreasing gravitational acceleration or weightlessness, the bone will first lose calcium and the tissue will later be resorbed by osteoclasts. From this reaction a danger may occur: Theoretically it seems possible that bone could be reduced and decalcified so far that a return to terrestrial conditions and normal body weight would increase the stresses in the ground substance so much, that they would go beyond the upper tolerance limit σ_o (see IV, A, 2). In this case, the system would be out of control and the feed-back mechanism would break down.

REFERENCES

Amtmann, E. 1968*a*. Bruchfestigkeitsverteilung im menschlichen Femurschaft. Nat. wiss. 55: 392.

Amtmann, E. 1968*b*. The distribution of breaking strength in the human femur shaft. J. Biomechanics 1: 271–277.

Amtmann, E. 1971. Mechanical stress, functional adaptation and variation structure of the human femur diaphysis. Ergebn. Anat. (in press).

Amtmann, E. and H. P. Schmitt 1968. Über die Verteilung der Corticalisdichte im menschlichen Femurschaft und ihre Bedeutung für die Bestimmung der Knochenfestigkeit. Z. Anat. 127: 25–41.

Bell, G. H. 1956. Bone as a mechanical engineering problem. *In:* G. H. Bourne: "The biochemistry and physiology of bone." Acad. Press Inc., New York, pp. 27–52.

Bhimasenachar, J. 1945. Proc. Indian Acad. Sci. 22a: 209, Cited in Currey, 1964.

Calabrisi, P. and F. C. Smith 1951. The effect of embalming on the compressive strength of a few specimens of compact bone. U. S. Naval Med. Research Inst. MR-51-2,000, 018.07.02.

Currey, J. D. 1964. Three analogies to explain the mechanical properties of bone. Biorheology 2: 1–10.

Evans, F. G. 1957*a*. Studies on the biomechanics and structure of bone. C. R. Ass. Anat. 44. Réun. Leyden. pp. 272–276.

Evans, F. G. 1957*b*. Stress and Strain in Bones. Their Relation to Fractures and Osteogenesis. C. C. Thomas Publisher, Springfield, Ill.

Evans, F. G. 1969. The mechanical properties of bone. Artificial Limbs, 13: 37–48.

Evans, F. G. and S. Bang 1966. Physical and histological differences between human fibular and femoral compact bone. *In:* Studies on the Anatomy and Function of Bone and Joints. Springer, Berlin, Heidelberg, New York. pp. 142–155.

Evans, F. G. and S. Bang 1967. Differences and relationships between the physical properties and the microscopic structure of human femoral, tibial and fibular cortical bone. Am. J. Anat. 120: 79–88.

Evans, F. G. and M. Lebow 1951. Regional differences in some of the physical properties of the human femur. J. Appl. Physiol. 3: 563–572.

Evans, F. G. and H. R. Lissner 1957. Tensile and compressive strength of human parietal bone. J. Appl. Physiol. 10: 493–497.

Galileo, G. 1638. Discorsi e dimostrazioni matematiche intorno à due nuove scienze.... Elsevirii, Leida.

Hülsen, K. K. 1896. Specific gravity, resilience and strength of bone. Bull. Lab. Biol. St. Petersburg 1: 7–35, [In Russian.]

Jouck, T. 1970. Die Festigkeit von Tibiae nach Spanentnahme an verschiedenen Orten. Z. Anat. Entw. Gesch. 130: 345–364.

Knese, K.-H. 1956. Belastungsuntersuchungen des Oberschenkels unter der Annahme des Knickens. Morph. Jb. 97: 405–452.

Knese, K.-H. 1958. Knochenstruktur als Verbundbau. G. Thieme, Stuttgart.

Knief, J.-J. 1967a. Materialverteilung und Beanspruchungsverteilung im coxalen Femurende. Densitometrische und spannungsoptische Untersuchungen. Z. Anat. 126: 81–116.

Knief, J.-J. 1967b. Quantitative Untersuchung der Verteilung der Hartsubstanzen im Knochen in ihrer Beziehung zur lokalen mechanischen Beanspruchung. Methodik und Biomechanische Problematik, dargestellt am Beispiel des coxalen Femurendes. Z. Anat. 126: 55–80.

Konermann, H. 1970. Dichteverteilung im Röntgenbild des Skeletts. Die Naturwiss. 57: 255.

Konermann, H. 1971. Funktionelle Analyse der Knorpelstruktur des Talonaviculargelenks. Z. Anat. (in press).

Kummer, B. 1956. Eine vereinfachte Methode zur Darstellung von Spannungstrajektorien, gleichzeitig ein Modellversuch für die Ausrichtung und Dichteverteilung der Spongiosa in den Gelenkenden der Röhrenknochen. Achter Beitrag zur funktionellen Anatomie und kausalen Morphologie des Stützapparates von Friedrich Pauwels. Z. Anat. 119: 223–234.

Kummer, B. 1959. Bauprinzipien des Säugerskeletes. G. Thieme, Stuttgart.

Kummer, B. 1963. Principles of the biomechanics of the human supporting and locomotor system. IX. Congr. Internat. Chir. orthop. et de Traumatol. II. Vienna. pp. 60–81.

Kummer, B. 1966. Photoelastic studies on the functional structure of bone. Fol. biotheoret. 6: 31–40.

Lacroix, P. 1953. Sur le métabolisme du calcium dans l'os compact du chien adulte. Bull. Acad. Méd. Belg., Sér VIa 18: 489.

Marotti, G. 1963. Quantitative studies on bone reconstruction 1. The reconstruction in homotypic shaft bones. Acta anat (Basel). 52: 291–333.

Meyer, H. 1867. Die Architektur der Spongiosa. Reichert u. Dubois-Reymond's Arch. 1867: 615–628.

Pauwels, Fr. 1948a. Bedeutung und kausale Erklärung der Spongiosaarchitektur in neuer Auffassung. Ärztl. Wschr. 3: 379.

Pauwels, Fr. 1948b. Die Bedeutung der Bauprinzipien des Stütz- und Bewegungsapparates für die Beanspruchung der Röhrenknochen. I. Beitrag zur funktionellen Anatomie und kausalen Morphologie des Stützapparates. Z. Anat. 114: 129–166.

Pauwels, Fr. 1950. Die Bedeutung der Muskelkräfte für die Regelung der Beanspruchung des Röhrenknochens während der Bewegung der Glieder. Dritter Beitrag zur funktionellen Anatomie und kausalen Morphologie des Stützaparates. Z. Anat. 115: 327–351.

Pauwels, Fr. 1954. Über die Verteilung der Spongiosadichte im coxalen Femurende und ihre Bedeutung für die Lehre vom funktionellen Bau des Knochens. Siebenter Beitrag zur funktionellen Anatomie und kausalen Morphologie des Stützapparates. Morph. Jb. 95: 35–54.

Pauwels, Fr. 1965. Gesammelte Abhandlungen zur funktionellen Anatomie des Bewegungsapparates. Springer Verlag, Berlin, Heidelberg, New York.
Pauwels, Fr. 1968. Beitrag zur funktionellen Anpassung der Corticalis der Röhrenknochen. Untersuchungen an drei rachitisch deformierten Femora. 12. Beitrag zur funktionellen Anatomie und kausalen Morphologie des Stützapparates. Z. Anat. Entwickl.-Gesch. 127: 121–137.
Petersen, H. 1930. Die Organe des Skeletsystems. In, Handb. d. mikr. Anat. d. Menschen. II/3, pp. 521–678.
Rauber, A. A. 1876. Elasticität und Festigkeit der Knochen. Engelmann, Leipzig.
Roux, W. 1895. Gesammelte Abhandlungen über Entwicklungsmechanik der Organismen. I u. II. Engelmann, Leipzig.
Schmitt, H. P. 1968. Über die Beziehungen zwischen Dichte und Festigkeit des Knochens am Beispiel des menschlichen Femur. Z. Anat. 127: 1–24.
Sedlin, D. E. and C. Hirsch 1966. Factors affecting the determination of the physical properties of femoral cortical bone. Acta orthop. Scand. 37: 29–48.
Vincent, J. et G. Haumont 1960. Identification autoradiographique des ostéones métaboliques après administration de Ca^{45}. Rev. franc étud. clin. biol. 5: 348.
Wertheim, M. G. 1847. Mémoire sur l'elasticité et la cohésion des principaux tissus du corps humain. Ann. Chim. et de Phys. 21: 385–414.
Wolff, J. 1884. Das Gesetz der Transformation der inneren Architektur der Knochen bei pathologischen Veränderungen der äußeren Knochenform. Sitz. Ber. Preuss. Akad. d. Wiss., 22. Sitzg., physik. -math. Kl.
Young, J. Z. 1957. The Life of Mammals. Oxford University Press, London.
For further literature see Amtmann, 1971 and Kummer, 1959.

11
The Interstitial Space

CURT A. WIEDERHIELM

I. INTRODUCTION

The human body is composed of approximately 70% water and 30% solids. The body fluids are distributed in what may be described as three compartments—the intracellular fluid compartment, the interstitial fluid compartment, and blood. The intracellular fluid, that is the fluid contained within the cell membranes of all cells in the body, comprises approximately 50% of the body weight. Blood plasma represents approximately 5% of the total body weight. The remaining 15% is distributed in the interstitial space—the space which envelops the cells and blood vessels in the tissues. Fluid exchanges continuously between the plasma and interstitial space as well as between the intracellular space and interstitial space. Despite the complexity of these fluid fluxes among different compartments, their volume remains surprisingly constant. Regulation of this fluid distribution is of prime importance since it serves to maintain physical-chemical stability in the immediate environment of the cells as required for optimal cell function. The objective of this report is to describe the internal structure of the interstitial space and some of its physical-chemical characteristics.

II. STRUCTURE OF THE INTERSTITIAL SPACE

The interstitial space has often been considered as a single well-stirred compartment. Seemingly this assumption is valid when one considers the dispersion of small radioactive tracer molecules passing through a capillary bed. However, diffusion of moderate size molecules (MW greater than 500) is drastically impeded, indicating that the interstitial space may be a more highly organized structure.

274 THE INTERSTITIAL SPACE

The interstitium of the skin is composed of collagen and elastic fibers, forming a loose network. Within the meshes of this network is found a class of chemicals, the so-called mucopolysaccharides, which have an unusually high molecular weight (10^6). In contrast to the well-mixed pool concept, an alternative viewpoint has been advanced. In this concept the interstitial space is organized such that the space between the cells is composed entirely of a gel-like matrix.

However, when dye is injected either into blood or lymphatic capillaries, it tends both to escape into tissues in fine filaments and slowly to spread uniformly along the vessels. This observation is compatible with the concept that the interstitial space is a two-phase system with a colloid-rich gel-like phase and a free-fluid (colloid poor) phase (Fig. 1a). The free-fluid phase may take the form of vesicles or pathways through which bulk flow occurs at rates faster than diffusion fluxes. Several morphologic studies support this two-phase concept of the interstitial space (Bondareff, 1957; Chase, 1959).

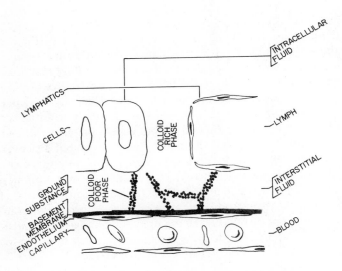

Fig. 1(a): Two-phase organization of interstitial space.

Organization of the interstitium at the molecular level is represented in Fig. 1b which shows the space between endothelial cells of a blood capillary (bottom) and a lymphatic capillary (top). Fluid and protein enter the interstitium through a large pore, or from a vesicle in the capillary endothelial cells. Similarly, fluid and protein freely enter the large space between the endothelial cells in the lymphatic capillary. A condensation of mucopolysaccharides and fine reticular collagen fibers, the basement membrane, surrounds the blood capillary. Between the two capillaries is a three-dimensional

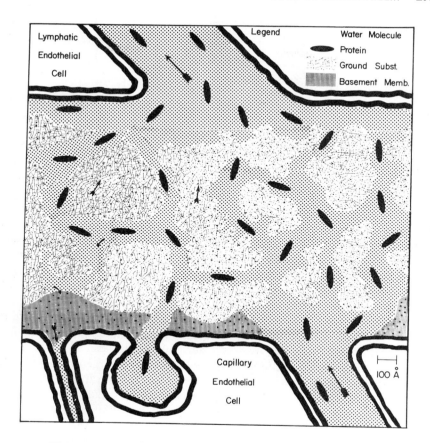

Fig. 1(b): Interstitium at molecular level (for description see text).

mesh work of mucopolysaccharide complexes, collagen and elastin fibers. As discussed later, the mucopolysaccharides tend to exclude proteins, which are confined to the free-fluid spaces between the mucopolysaccharide complexes. Small molecules such as water and electrolytes may diffuse through the mucopolysaccharide complexes albeit at reduced rates.

III. PHYSIOLOGIC SIGNIFICANCE OF INTERSTITIAL MUCOPOLYSACCHARIDES

A major component of the ground substance, mucopolysaccharides, is hyaluronate which has an extremely high molecular weight (14,000,000). Its molecular configuration is an unbranched, stiff random chain longer than 1 micron, and about 8 AU in diameter. Due to its shape, the molecule

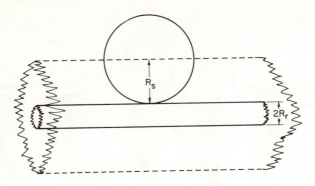

Fig. 2: Exclusion of a sphere by a rod (Laurent, 1968).

possesses a high surface to volume ratio as compared with a sphere and consequently will exclude a large solvent volume from other macromolecules, that is, two molecules cannot occupy the same space at the same time. Figure 2 illustrates volume exclusion of a sphere by a rod. In this analogy hyaluronate is considered as a rod and a globular protein as a sphere. The center of the sphere can never approach the rod closer than its radius. Thus the rod is surrounded by a cylinder with a cross-sectional area of $\pi(R_s + R_r)^2$ from which the protein is excluded. Much of this cylindrical volume is available, however, to small molecules such as water and crystalloids.

Since long molecules such as hyaluronate occupy a large domain, much interaction would be expected even at very low concentrations. The concentration of hyaluronate in the connective tissue interstitial space has been estimated to be 1%. At this concentration, hyaluronate molecules overlap so extensively that their mobility is reduced and they form a tangled network. A theoretical derivation for the volume exclusion under these conditions has been presented by Ogston and Phelps (1961). A high degree of molecular interaction is reflected in a marked deviation from Van't Hoff's Law. This can be seen in Fig. 3, which shows the relationship between the osmotic pressure and concentrations of hyaluronic acid and albumin. Note that despite its much higher molecular weight, the osmotic pressures of hyaluronate are in general considerably higher than those of albumin. The curves are markedly nonlinear and may be fitted to a power series:

(1) $$\pi_p = 2.8C_p + .18C_p^2 + .012C_p^3$$
(2) $$\pi_h = .014C_h + 4.25C_h^2 + .14C_h^3$$

where π and C refer to osmotic pressure and concentration and p and h to protein and hyaluronate respectively.

These equations relate the osmotic pressure of solutions of varying concentrations of albumin and hyaluronate to their respective concentrations.

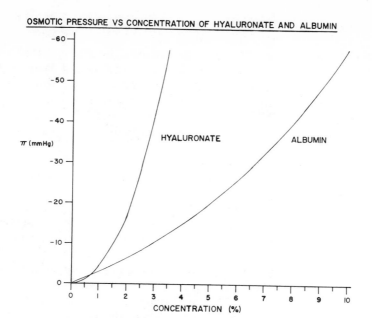

Fig. 3: Relationship of hyaluronate and albumin concentration and osmotic pressure.

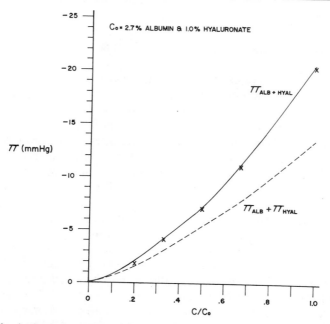

Fig. 4: Osmotic pressure of mixed solutions of hyaluronate and albumin.

277

For albumin the coefficient for the C term is highest, indicating that at moderately low concentrations a linear approximation may be used. With hyaluronate, however, the coefficient for the C^2 term dominates, indicating that over a wide range of concentrations the osmotic pressure is proportional to the square or the concentration. This difference in the behavior of albumin and hyaluronate may be attributed to the difference in their molecular conformation.

The osmotic pressure of mixtures of hyaluronate and albumin is higher than the algebraic sum of the osmotic pressure of each substance taken separately as shown in Fig. 4 (Wiederhielm, et al, in preparation). The data points indicated by the X's represent measurements of the osmotic pressure of a stock solution of 2.5% albumin and 1% hyaluronate at varying degress of dilution. The experiment was performed by placing 0.2 cc of the stock solution in the sample chamber of a membrane osmometer, and diluting it with Ringer solution by factors of 2/3, 1/2, 1/3 and 1/5. For comparison the dashed line indicates the algebraic sum of the osmotic pressure for the corresponding concentrations. The increase in osmotic pressure of the mixture can be attributed to the volume exclusion by the hyaluronate. Computer tabulations predict that the osmotic pressure of mixtures of mucopolysaccharides and proteins in concentrations normally found in the interstitial space of skin would be about 11 mmHg. Experiments were designed to test this prediction.

IV. SWELLING PRESSURE OF SKIN SPECIMEN

The swelling pressure of excised rabbit skin was measured by placing tissue specimens on a membrane osmometer impermeable to plasma proteins (Fig. 5). The membrane osmometer is connected to a pressure transducer (P1), which allows continuous recording of osmotic pressure. The pressure transducer is also connected through a valve to a reservoir of Ringer solution maintained at atmospheric pressure. When this valve is open the tissues withdraw Ringer solution through the membrane, thus allowing the sample to swell to a new volume. The sample chamber is enclosed and connected to a metal bellows which is expanded by a solenoid to accommodate any volume changes in the sample. A second pressure transducer (P2) connected to the sample chamber is balanced so that the atmospheric pressure is always maintained at the membrane level through an integrating servosystem. Swelling of the tissue sample increases the pressure in the sample chamber which is sensed by the pressure transducer and translated into a movement of the bellows sufficient to cancel out the pressure change. Displacement of the bellows is measured by a linear variable differential transformer and can be calibrated directly in terms of volume changes.

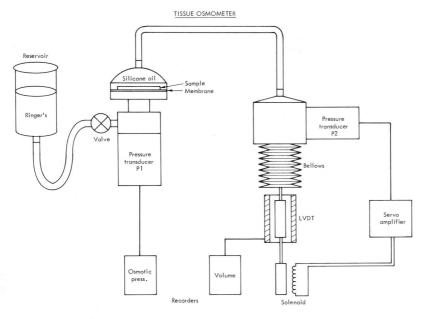

Fig. 5: Diagram of tissue osmometer (for description see text).

Skin samples approximately 3/4 inch in diameter from the flank of the rabbit were excised as rapidly as possible under conditions which minimized dehydration by evaporation. The samples were weighed and placed on the osmometer membrane, which previously had been dehydrated by applying negative pressure behind the membrane. After the tissue sample was placed on the membrane it was coated with silicone oil; the chamber was assembled, filled with silicone oil and connected to the bellows. Initial readings of the swelling pressure of the skin samples were obtained. Following this initial measurement, the valve connecting the pressure transducer to the fluid reservoir was opened, and the sample was allowed to swell for a time. The swelling pressure at the new volume was then measured. Repetition of this process allowed reconstruction of curves relating swelling pressures to tissue volume. Normal (initial) values from ten experiments are shown in Table 1. In eight experiments the swelling pressure ranged from 9.0 mmHg–14.0 mmHg. In experiments 4 and 8 the pressure was abnormally high, possibly due to dehydration during preparation of the tissue samples. The average of all experiments was 11.9 ± 2.3 mmHg. By permitting the tissue samples to swell as described previously and measuring the osmotic pressure at the new volume, swelling pressure versus tissue volume curves could be plotted (Fig. 6). The relationship between the swelling pressure and the normalized

280 THE INTERSTITIAL SPACE

TABLE I
Normal Swelling Pressure in Rabbit Skin

EXPERIMENT NUMBER	PRESSURE mmHg
3	11.5
4	16.0
5	10.4
6	10.7
7	9.0
8	16.0
10	12.2
11	12.0
11	12.0
16	10.4
Average	11.9
S.D.	±2.3

TISSUE OSMOTIC PRESSURE VS SWELLING

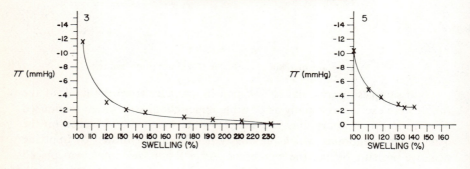

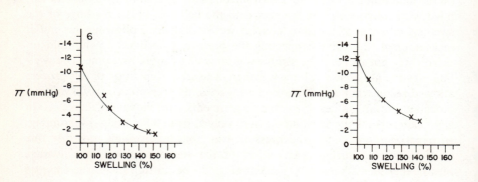

Fig. 6: Swelling pressure vs. volume of excised rabbit skin specimen. The tissue volume has been normalized to allow for variation in initial volume of different skin samples. 100% is considered the initial volume of sample.

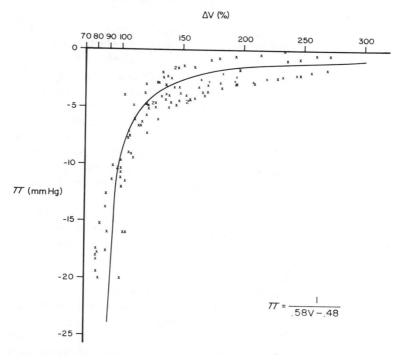

Fig. 7: Swelling pressure vs. volume; summary of 10 experiments. Volume changes are referred to a normalized initial volume of 100%. Data points in the range of 70–100% were obtained by dehydrating the skin samples by application of −20 mm Hg behind the osmometer membrane.

volume of the tissue samples is markedly nonlinear. The data of ten experiments are summarized in Fig. 7. These data may be fitted to a quadratic hyperbola with an asymptote of 83% on the volume axis as shown by the regression equation. The data points between 10 mmHg and 20 mmHg were obtained by dehydrating the sample with a negative pressure applied behind the membrane for 12 hours.

V. OSMOTIC AND HYDROSTATIC TISSUE PRESSURES

The average swelling pressure of normal tissue samples (11.9 mmHg) agrees well with the values predicted in the *in vitro* studies of mixtures of hyaluronate and albumin described previously. They also agree well with values predicted in an earlier computer simulation study on capillary fluid balance (Wiederhielm, 1968). The swelling pressure of the tissue measured in this manner represents the algebraic sum of the hydrostatic and osmotic pressure of the tissue specimen of the skin.

The question of hydrostatic tissue pressure has been quite controversial. McMaster (1946) used a direct puncture technique in which 30 gauge needles were inserted into the connective tissue of the mouse ear, and in some instances human skin, under microscopic observation. Pressure recorded in this manner averaged 1–2 mmHg. This technique has been criticized on the grounds that 30 gauge needles induce tissue trauma, leading to increased capillary permeability and altered hydrostatic tissue pressure. In 1968 we repeated McMaster's experiments using a servopressure recording system in which micropipettes of $.3\mu$ tip diameter were used as pressure transducers (Wiederhielm et al, 1964). Results obtained in this manner agreed with those of McMaster's. The average of forty-six pressure measurements was $+1.6$ mmHg (Wiederhielm, 1969).

Guyton (1963) devised an alternative method of measuring tissue pressure. In these studies perforated plastic capsules were implanted subcutaneously in the experimental animals. After 3–4 weeks scar tissue had grown through the perforations, forming a complete internal lining of each capsule. In the center of the capsule was a free-fluid space, containing protein with a concentration 2–2.5%. When a hypodermic needle, connected to a pressure transducer, was inserted into the free-fluid space, subatmospheric pressures of 6–7 mmHg were recorded. Guyton felt that the fluid within the capsule communicated freely with the surrounding interstitium and that the pressure recorded in the capsule represented an average value of the hydrostatic pressure in the surrounding connective tissue.

Scholander (1968) studied tissue pressure in reptiles by inserting polyethylene catheters into the tissues through a large bore hypodermic needle. Prior to insertion the catheter was plugged with a cotton wick and filled with Ringer solution. In these studies somewhat smaller negative pressures than those obtained by Guyton were recorded. Thus a distinct difference appears to exist between tissue pressure measurements obtained by direct puncture methods and those obtained by other techniques.

To evaluate the relative magnitude of the osmotic versus the hydrostatic components of the swelling tissue in excised skin specimens, the tissue osmometer technique described earlier was modified slightly. In these experiments the tissue specimens were placed on a membrane with an average pore diameter of 2000 AU. Initial subatmospheric pressures were considerably smaller (Table II) than those obtained with a membrane impermeable to protein (Table I). Initially the swelling pressure averaged 2.7 mmHg with a standard deviation of ± 1.6 mmHg. Over approximately 18 hours the recorded pressures declined and reached an equilibrium of -1.9 mmHg with a standard deviation of ± 1.2 mmHg.

Table II
Swelling Pressure on Large Pore Membranes

EXPERIMENT NUMBER	INITIAL PRESSURE mmHg	EQUILIBRIUM PRESSURE mmHg
12	1.50	1.2
13	1.25	1.2
18	3.50	2.0
20	1.90	1.0
21	4.30	3.0
22	4.90	4.0
23	3.80	2.4
24	.80	.3
Average	2.7	1.9
S.D.	±1.6	±1.2

The membranes used in these experiments were freely permeable to plasma proteins, whereas the mucopolysaccharides were still retained in the skin sample. The time required for reaching equilibrium pressure reflects the time required for protein to diffuse through the tissues and establish equilibrium distribution on both sides of the membrane. Molecules as large as albumin would be expected to diffuse extremely slowly through the spaces between the mucopolysaccharide complexes in these tissues. These results indicate that the major component of the recorded swelling pressure is osmotic in nature. Were the hydrostatic pressures within the tissue negative, one would expect a rapid equilibration time, and equilibrium pressures of -6 to -7 mmHg should have been recorded.

The slightly subatmospheric pressure found in these experiments may be explained on the basis of unequal distribution of protein and mucopolysaccharides across the membrane. The proteins are free to distribute themselves on both sides of the membrane, whereas the mucopolysaccharides are retained within the tissue specimen. In essence this experiment is equivalent to an equilibrium dialysis procedure, in which the membrane is permeable to only one species of macromolecules, while other macromolecules are retained on one side of the membrane.

Under these conditions the hydrostatic pressure can be predicted to be slightly subatmospheric. In fact, the negative pressure found in Guyton's capsule can be simulated with a simple dual membrane model (Fig. 8) similar to that originally proposed by Curran and McIntosh (1962). They used such a model to demonstrate that bulk fluid flow can be induced against an osmotic gradient in a system where two membranes with different reflection coefficients separate three compartments. In the system shown in Fig. 8A volume flow will occur from the left compartment into the center compartment and finally into the compartment to the right despite the fact that the sucrose

MODEL OF STEADY STATE CONDITIONS IN SUBCUTANEOUS CAPSULES

A. CURRAN DUAL MEMBRANE MODEL (REF: CURRAN et al., NATURE, 193, 4813, (1962))

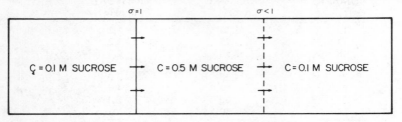

B. CAPSULE MODEL

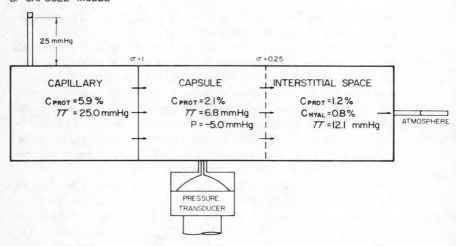

Fig. 8: Simulation of subcutaneous capsules by a modified Curran-McIntosh dual membrane.

concentration is higher in the center compartment. The theory of this system has been analyzed by Patlak and his co-workers (1963).

In the simulation of the subcutaneous capsule (Fig. 8B), the compartment to the left represents the capillaries. It is filled with 5.9% albumin solution exerting an osmotic pressure of 25 mmHg, which approximates normal total plasma osmotic pressure for man. In addition, a standpipe generates a hydrostatic pressure of 25 mmHg, which is representative of the average capillary pressure in man. The left-hand compartment is separated from the center compartment by a membrane which is totally impermeable to protein; that is, its reflection coefficient for albumin is one. The center compartment, representing the subcutaneous capsules, contains an albumin concentration of 2.1%. The right-hand compartment, representing the interstitial space, is separated from the center compartment by a membrane which has a reflection coefficient* for albumin of 0.25. This compartment is filled with a mixture of 0.8% hyaluronate and 1.2% albumin. It is maintained at atmospheric pressure by a horizontal standpipe. When the center compartment is connected to a pressure transducer, the pressures are found to be subatmospheric (about -5 mmHg). The fluid fluxes occur in one direction only, namely, from the left-hand compartment. This is an example of bulk fluid flux occurring against hydrostatic pressure gradients in the presence of osmotically active macromolecules.

The presence of mucopolysaccharides in the interstitial space thus appears to affect profoundly the osmotic forces exerted by the tissues on capillary walls. The interactions of mucopolysaccharides with plasma proteins in the tissues give rise to unexpectedly high tissue osmotic pressures. Finally, we have demonstrated that negative pressures recorded within Guyton subcutaneous capsules may also be a result of mucopolysaccharides in the scar tissue surrounding the capsule.

REFERENCES

Bondareff, W. 1957. Submicroscopic morphology of connective tissue ground substance with particular regard to fibrillogenesis and aging. Gerontologia 1: 222–33.
Chase, W. H. 1959. Extracellular distribution of ferrocyanide in muscle. A. M. A. Arch. Pathol. 67: 525–32.
Curran, P. F. and J. F. McIntosh 1962. A model system for biological water transport. Nature 193: 347–49.
Guyton, A. C. 1963. A concept of negative interstitial pressure based on pressures in implanted perforated capsules. Circul. Res. 12: 399–414.
Laurent, T. C. 1968. The exclusion of macromolecules from polysaccharide media. In G. Quintarelli (ed.), The Chemical Physiology of Mucopolysaccharides. Little-Brown and Company, Boston. pp. 153–70.
McMaster, P.D. 1946. The pressure and interstitial resistance prevailing in the normal and edematous skin of animals and man. J. Exp. Med. 84: 473–94.
Ogston, A. G. and C. F. Phelps 1961. The partition of solutes between buffer solutions and solutions containing hyaluronic acid. Biochem. J. 78: 827–33.
Patlak, C. S., D. A. Goldstein and J. F. Hoffman 1963. The flow of solute and solvent across a two-membrane system. J. Theoret. Biol. 5: 426–42.
Scholander, P. F., A. Hargens and S. L. Miller 1968. Negative pressure in the interstitial fluid of animals. Science 161: 321–28.
Wiederhielm, C. A., J. W. Woodbury, S. Kirk and R. F. Rushmer 1964. Pulsatile pressures in the microcirculation of the frog's mesentery. Amer. J. Physiol. 207: 173–76.

*The reflection coefficient is defined as the ratio of solvent flux through a leaky semi-permeable membrane induced by osmotic driving forces (J_π) to that induced by a hydrostatic driving force (J_p). For an ideal semi-permeable membrane the ratio J_π/J_p will equal one. For membranes with pore dimensions sufficiently large to allow leakage of solute molecules, the reflection coefficient will range from 0–1. The reflection coefficient thus constitutes a qualitative measure of the degree of membrane permeability to a given molecular species.

Wiederhielm, C. A. 1966. Transcapillary and interstitial transport phenomena in the mesentery. Fed. Proc. 25: 1789–798.
Wiederhielm, C. A. 1968. Dynamics of transcapillary fluid exchange. J. Gen. Physiol. 52: 29–63.
Wiederhielm, C. A. 1969. The interstitial space and lymphatic pressures in the bat wing. *In* A. P. Fishman and H. H. Hecht (eds.), The pulmonary circulation and interstitial space. University of Chicago Press, Chicago. pp. 29–41.
Wiederhielm, C. A., J. R. Fox, D. R. Lee and D. D. Stromberg, 1971. The role of mucopolysaccharides and interstitial plasma protein on capillary water balance. In preparation.

PART III

Biomechanics in Physiology, Medicine, and Surgery

12
Mechanics of Contraction in the Intact Heart

JAMES W. COVELL

The hemodynamic factors that influence the performance of the heart have been the subject of intensive investigation for many years. This classical body of knowledge describing the hemodynamic determinates of cardiac performance forms the basis for modern cardiovascular physiology (Wiggers, 1928, Mountcastle, 1968). During recent years, attention has been focused on the muscle that makes up the heart, its function, energy requirements, and the coupling of this muscle to the circulation. The heart itself is a complex structure consisting of chambers, valves, specialized conducting tissue, and cardiac muscle arranged in a continuum of spiraling fibers. The object of the current report is to review the information currently available on the mechanics of contraction in the intact heart and to present this information for review by the student of either Bioengineering or Physiology in such a fashion that access to the classic and current literature on the subject is available. An attempt will be made to review current knowledge on structure-function relationships, strain, strain rate, and stress changes within the myocardium. Since the majority of the detailed knowledge of stress-strain relationships in the intact heart concerns the left ventricle, the scope of this report will be limited to that chamber and will review structure, structure function relationships, dimensional changes during contraction, techniques for estimating wall stress and finally stress-strain relationships.

I. STRUCTURE-FUNCTION RELATIONSHIPS

Cardiac and skeletal muscle provide one of the clearest examples of precise structure-function relationships at the ultrastructural level. Each muscle cell contains many myofibrils which contain a series of sarcomere or contractile units and each sarcomere is made up of interdigitating thick and thin filaments. There is now evidence that the average sarcomere length increases proportionally to increases in muscle length over the ascending portion of the length tension curve and that within this range the increase is directly related to the force developed by the muscle (Spiro, 1966). These concepts have been investigated extensively in isolated muscle and are reviewed in several recent texts (Sonnenblick, 1966). Although changes in sarcomere dimensions have been observed during different phases of contraction (Sonnenblick et al, 1967) and at different levels of ventricular filling (Spiro et al 1964, Sonnenblick and Ross, 1967), because of the complex structure of the intact myocardium the direct relationship of these changes in ventricular

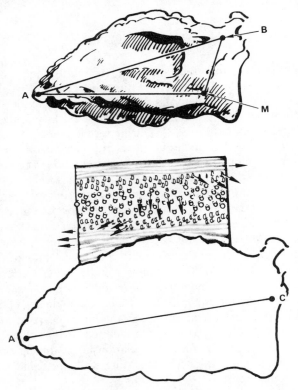

Fig. 1: (Top) Schematic representation of a cast of the internal cavity of the left ventricle arrested during systole. A = apex; B = center of the aortic valve; M = center of the mitral valve. (Below) Schematic outline of the left ventricular cavity shown with a section of left ventricular myocardium. Arrows indicate fiber bundle direction as shown above. A = apex; C = mid-point of the junction of the aortic and mitral valve rings; fiber angle orientation is described relative to the A to C axis.

dimensions to alterations at the ultrastructural level has not been extensively investigated and will not be considered in detail in this report.

Substantially more information has recently become available on static and dynamic changes in muscle fiber orientation during systole (Streeter et al, 1966, 1969, Sallin, 1969). Figure 1 (top) shows the stippled outline of a cast of the cavity of the left ventricle. The rough trabeculated surface and large indentations caused by the papillary muscles are apparent. If one examines the direction of fibers in hearts arrested during systole and diastole relative to the major axis of the left ventricle (Fig. 1, lower), it is evident that there is a continuum of fiber direction from the endocardium to the epicardium, where the fibers are parallel to the major axis, but they become parallel to the plane of the minor axes near the mid-wall. At the level of the greatest hoop dimension of the left ventricle the majority of fibers (60%) are oriented in the direction of the minor axis (Streeter et al, 1969). This proportion of fibers oriented close to 90° to the major axis decreases progressively towards the apex of the left ventricle. Moreover, fiber orientation is not significantly changed during active contraction.

II. DIMENSIONAL CHANGES DURING CARDIAC CONTRACTION

The fibers comprising the ventricular wall shorten markedly during systole and an accurate description of these changes in cardiac dimensions during rest and contraction has been the subject of intensive investigation. The results of these studies were the subject of a recent symposium (1968), and were recently reviewed (Sandler, 1970).

The major dimensional changes in the left ventricle during systole are summarized in Fig. 2. This example of dimenstional changes was obtained from the analysis of the movement of endocardial and epicardial beads implanted at the minor equator, apex and base of the left ventricle using a technique similar to that employed by Mitchell et al (1969). During ejection there is a 10–35% decrease in the minor axis dimension at the endocardium (4.5–4.0 cm, Fig. 2) while the apex to the aortic valve distance (outflow tract) decreased by an average of 1–5%. (See Ross et al, 1967, Mitchell et al, 1967, 1969, Rushmer et al, 1956, Hawthorne, 1961, Chapman et al, 1959.)

There also appears to be functional significance to the anatomic separation of inflow and outflow tract in the left ventricle and a distinct series of dimensional changes during contraction has been observed (Mitchell et al, 1969). Isovolumic contraction is initiated by an abrupt shortening of the boundaries of the inflow tract (Fig. 1, AM), and an expansion of outflow tract length (Fig. 1, AB) and the circumference (Hinds et al, 1969). During ejection, there is very little change in the inflow tract dimension and a decrease in the

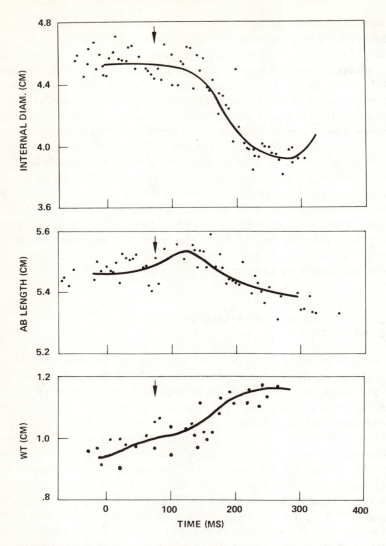

Fig. 2: Representation of internal diameter (ID) apex to base length (AB length) and wall thickness (WT) determined at 5–20 ms intervals during contraction (See text.) (Dog.)

outflow tract length and circumference (Hinds et al, 1969, 1965, Kluemper and Puff, 1960) suggesting that the ventricle contracts asymmetrically during ejection with the center of mass changing from the inflow tract and to the outflow tract (Hawthorne, 1966, Hinds et al, 1967, 1969) and presumably aiding in the directional flow of blood. During these changes the left ventricular wall thickens by an average of 25.5% (Ross et al, 1967, Cothran et al, 1967,

Feigel and Fry, 1964) and these rearrangements are accompanied by significant shear in the left ventricular wall (Feigel and Fry, 1964).

III. VENTRICULAR WALL STRESS

A variety of techniques have been employed to calculate left ventricular wall stress using geometric reference figures based on the dimensional measurements described above. These reference figures have varied from a thin walled sphere (Burton 1957, Ross et al, 1966), an ellipsoid of revolution (Sandler and Dodge, 1963, Fry et al, 1964) or treatments of the thick walled ventricle as a passive elastic body (Mirsky, 1969; Ghista and Sandler, 1969). More recently these reference figures have been extended to include a theoretical analysis of fiber angle distribution within the myocardium (Streeter et al, 1970; Wong and Rautaharju, 1968). Although most of the evidence now available tends to favor the thick walled ellipsoid as the most reasonable reference figure for representing left ventricular wall stress, there is little direct confirmatory evidence that this is the case (Sandler and Ghista 1969, Hood et al, 1969; Falsetti et al, 1970). Figures 3 and 4 summarize the results of recent studies (Burns et al, 1971, Feigel et al, 1967) in which wall stress was measured directly with auxotonic force gauges (Feigel et al, 1967) and mean stress was estimated by measuring or calculating wall thickness. Figure 3 illustrates the relationship between left ventricular pressure, aortic flow, directly measured minor axis wall stress and calculated ellipsoidal and spherical stress. Wall thickness at the auxotonic gauge was determined at the end of each experiment and corrections were made for dynamic changes in wall thickness using the data of Feigl and Fry (1964). Wall stress was then calculated from directly measured force and wall thickness. In all but two of the sixteen measurements, peak left ventricular wall stress calculated using either a spherical (PSσ) or ellipsoidal model (PEσ) was greater than peak measured stress (PMσ_1) assuming a constant wall thickness. In all experiments PMσ_1 was 84 ± 9 gm/cm², whereas PSσ averaged 79 ± 10 gm/cm², and PEσ was 119 ± 30 gm/cm² for a left ventricular end-diastolic pressure ranging from 1–3 mmHg; and 168 ± 28, 211 ± 21, and 286 ± 24 for left ventricular end-diastolic pressures ranging from 6–20 mmHg. Following correction of the directly measured wall forces for dynamic wall thickness changes and coupling of the myocardium to the pins of the gauge, measured wall stress (PMσ_2) increased to 192 ± 29 gm/cm². In all experiments, the PMσ_2 was greater than PSσ and in seven of the sixteen determinations, it was greater than PEσ. Figure 4 summarizes these results and it is evident that an ellipsoidal reference figure appears to provide a better estimate of measured wall stress. Moreover, the average ratio of measured major to minor axis wall stress of $43.0 \pm 3.3\%$ compared favorably with that calculated from the ratio of the numbers of fibers distributed in these two axes and their radii of curvature (Streeter et al, 1970) or with calculated major axis stress (Hood et al, 1969).

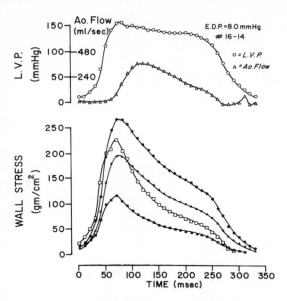

Fig. 3: The relationship between left ventricular pressure (LVP), aortic flow rate (Ao. Flow) and directly measured wall stress in a single contraction originating from a left ventricular end-diastolic pressure (EDP) of 8 mm Hg. Wall stress symbols: closed squares (■) directly measured wall stress calculated assuming a constant wall thickness $M\sigma_1$ (see text); open squares (□) directly measured wall stress corrected for dynamic variations in wall thickness; closed triangles (▲) wall stress calculated with a spherical model ($S\sigma_2$); closed circles (●) wall stress calculated with an ellipsoidal reference figure ($E\sigma$). (Dog.)

The applicability of the many formulations for mean wall stress as a descriptor of ventricular myocardial function is dependent on the assumption that there are no large local variations in wall force (Fung, 1968). If this were the case, more complex systems for the calculation of myocardial wall force (Mirsky 1969, Ghista et al, 1969, Wong et al, 1968) would be necessary. In the study mentioned above, at least in the minor equatorial plane, only small variations (avg. 15.3%) in wall stress were observed, thus supporting the utilization of mean wall stress as a descriptor of myocardial function in the circumferential plane.

Despite wide variations in techniques and assumptions, it is interesting that very little quantitative difference appears to exist between myocardial wall stresses calculated from either thick or thin walled ellipsoidal reference figures (Hood et al, 1969). The results of the data reviewed here reinforce these conclusions and, in addition, have shown that there are relatively minor differences between directly measured myocardial wall stresses ($M\sigma_2$) at the minor equator and wall stress calculated assuming relatively simple geometric reference figures, a sphere or ellipsoid and thin wall theory (the Laplace

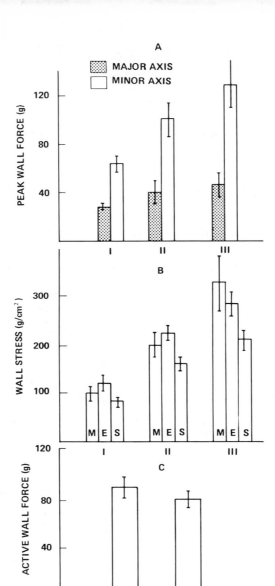

Fig. 4: Directly measured wall forces determined at three different levels of left ventricular end-diastolic pressure (I = 3.0 ± .6 mmHg; II = 6.4 ± .8 mmHg; III − 12.5 ± 2.7 mmHg). Panel A: The relationship between wall force determined in the major and minor axes of the left ventricle at the three levels of filling. Panel B: The relationship between wall stress corrected for dynamic variations in wall thickness (M) and that calculated from ellipsoidal (E) and spherical (S) reference figures for the left ventricle. Panel C: The relationship between wall force determined in the plane of the minor axis of the left ventricle and on the septal (S) and lateral (L) walls. (Dog.)

relationship) for the basis of the calculation. Although both the sphere and ellipse gave a good approximation of measured wall stress these results indicate that an ellipsoidal reference figure provides a better overall estimate of mean wall stress.

IV. STRESS-STRAIN RELATIONSHIPS OF THE INTACT HEART

The viscoelastic properties of the intact heart have been the subject of investigation for many years (Kabat and Visscher, 1939). However, the recent emphasis on examining the performance of the left ventricle in terms of muscle mechanics has stimulated renewed interest in the viscoelastic behavior of the *in situ* heart during contraction (Levine and Britman, 1964; Forward et al, 1966; Ghista et al, 1968, 1969; Covell and Taylor, 1971; Fry et al, 1964). Several studies have indirectly examined the "active stiffness" of ventricular muscle (Abbott and Mommaerts, 1959; Sonnenblick, 1962a, 1962b, 1964, Brady, 1967; Parmley and Sonnenblick, 1967; Hefner and Bowen, 1967). These techniques have all involved an examination of *in vitro* tissue or the transition from isovolumic to auxotonis contraction in the intact heart and are of necessity associated with detectable amounts of contractile element shortening during this early phase of contraction. Despite this difficulty, both Forwand et al (1966) and Fry et al (1964) were able to obtain estimates for the stiffness of the series elasticity in the intact canine left ventricle that were quite similar to values obtained from isolated cardiac muscle. In addition, Ghista and Sandler (1968) and Ghista and Vayo (1969), employing an elastic-viscoelastic model for the shape and forces in the left ventricle, have examined the elastic properties of left ventricular muscle during contraction by analyzing the stress-strain relationships of the intact, contracting left ventricle. These studies have provided useful information about cardiac function and have been used to examine the time dependent elastic characteristics of the myocardium. However, since the examinations were performed in the intact heart the properties determined also reflect those of the systemic arterial tree during systole and these characteristics do not necessarily reflect specific changes within the myocardium. Recently we have taken a different approach to the direct examination of the stress-strain properties of the myocardium of the intact heart (Covell and Taylor, 1967).

The contractile properties of active muscle may be described by a mechanical analog consisting of an active contractile element capable of developing force and shortening arranged in series with a passive elastic component (Hill, 1939). This analog has been extremely useful in the investigation of the contraction of isolated cardiac muscle and in the examination of the relationship between contraction and energy utilization by the contractile element (Coleman, 1968; Pool et al, 1968; Britman and Levine, 1964;

Rodbard et al, 1959; McDonald et al, 1966; Graham et al, 1967; Monroe and French, 1961). The evidence for the application of this analog to cardiac muscle stems primarily from quick release studies in isolated cardiac muscle. Although there is some controversy as to the particular arrangement of parallel and series elastic components that best describes the function of cardiac muscle, there is general agreement about the applicability of this type of model to both skeletal and cardiac muscle. (Sonnenblick, 1962, 1964; Brady, 1967; Hefner and Bowen, 1967). These concepts have recently been applied to the analysis of myocardial function and energetics in the intact heart. However, in the intact heart, estimates of the relationship between the force of contraction and the velocity of contractile element shortening must of necessity be based on the assumption that the series elastic component in the intact heart is essentially similar to that observed in the isolated papillary muscle. Utilizing a quick-release technique in which volume was rapidly withdrawn from the heart during systole, we have explored the possibility that the function of the intact heart may be described by a similar mechanical analog containing an undamped series elastic element and we have described its load-extension curve (Covell and Taylor, 1967). The load-extension curves were fitted to the relationship by least squares approximation

(1) $$S = a[e^{b(l/lo)-c} - 1]$$

where a, b are constants and c represents l/lo at $S = 0$. $S = S_2$ (final stress) $- S_1$ (initial stress), $l =$ length (midwall circumference) at S_{l_o}, $-$ length at S_1.

Since both the expressions $PR_i^2/(2h)$ and $PR_i^2/(R_o^2 - R_i^2)$ have been used to represent wall stress, assuming a spherical model for the left ventricle (Ross et

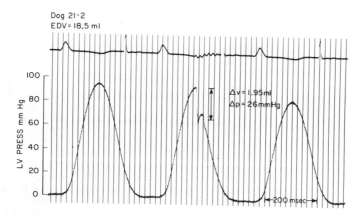

Fig. 5: Recordings obtained during quick release of the left ventricle. $\Delta V =$ volume withdrawal; $\Delta P =$ the resultant pressure fall; LV Press = left ventricular intracavity pressure.

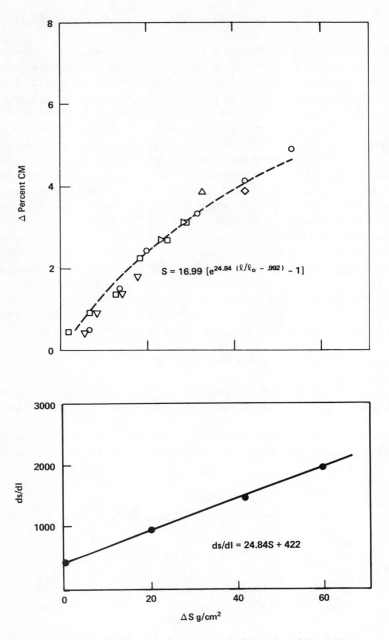

Fig. 6: Relationship between the percentage change in mid-wall circumference (from initial circumference) and the change in wall stress in a single experiment in which the time of quick release was varied throughout systole (circles = release at peak pressure; squares = 100 ms following peak pressure; triangles = 40 ms prior to peak pressure); ds/dl = the inverse slope of the above relationship. See text for definition of symbols.

al, 1966, Covell et al, 1969), both these equations were utilized. The results of this study did not allow a precise examination of the effects of preload on *SE* extension and consequently it was not possible to quantitatively examine the applicability of different arrangements of the *SE* and *PE* about the contractile element. However, active stress [Maxwell model (Sonnenblick, 1964)] was examined. Figure 5 shows the contour of the left ventricular pressure during quick release. For each series of releases (from 0.5 ml to that required to drop pressure to zero) a stress-strain relationship was calculated (Fig. 6). This series of points was then fitted to the exponential relationships shown above, and the load-extension relationship determined. The average constants for this relationship using $PR_i^2/(P_o^2-R_i^2)$ were: $a = 17.8 \pm 2.2$, $b = 27.2 \pm 2.1$, and $c = .997 \pm .001$.

These studies have also shown that active stiffness in the left ventricle does not vary greatly during active contraction. Thus, quick releases determined at several points during systole show approximately the same relationship. Moreover, in an individual heart this relationship was not influenced by changes in inotropic state induced by norepinephrine infusion.

In summary, the wide variety of information available about dimensional alterations in the normal and abnormal left ventricle during systole, the direct measurement or calculation of wall stress within the left ventricular myocardium, and the indication that the intact heart may be described as if it contained an undamped series elastic component, has made it possible to describe quantitatively myocardial behavior and to examine the relationships between force, velocity, length and time. Thus, as discussed by others in this symposium, with the knowledge currently available it has been possible to characterize the inotropic state of the left ventricular myocardium under experimental conditions and clinically in the cardiac catherization laboratory.

Acknowledgment. *This work is supported by USPHS NIH grant No. HE-12373.*

REFERENCES

Abbott, B. C. and W. F. H. M. Mommaerts 1959. A study of inotropic mechanisms in the papillary muscle preparation. J. Gen. Physiol. 42: 533.

Brady, A. J. 1967. The three element model of muscle mechanics. Its applicability to cardiac muscle. The Physiologist. 2: 75.

Britman, N. A. and H. J. Levine 1964. Contractile element work: A major determinant of myocardial oxygen consumption. J. Clin. Invest. 43: 1397.

Burns, J. W., J. W. Covell, R. Meyers and J. Ross, Jr. 1971. A comparison of directly measured left ventricular wall stress and stress calculated from geometric reference figures. Circulation Research. In press.

Burton A. C. 1957. The importance of the shape and size of the heart. Am. Heart J. 54: 801.

Chapman, C. B., O. Baker and J. H. Mitchell 1959. Left ventricular function at rest and during exercise. J. Clin. Invest. 38: 1202.

Coleman, H. N. 1968. Effect of alterations in shortening and external work on oxygen consumption of cat papillary muscle. Am. J. Physiol. 214: 100.

Cothran, L. N., W. C. Bowie, J. E. Hinds and E. W. Hawthorne 1967. Left ventricular wall thickness changes in unanesthetized horses. In R. D. Tanz, F. Kavaler, and J. Roberts (eds), Factors Influencing Myocardial Contractility. Academic Press Inc., New York. p. 163.

Covell, J. W., R. R. Taylor, and J. Ross, Jr. 1967. Series elasticity in the intact left ventricle determined by a quick release method. Fed. Proc. 26: 382.

Covell, J. W., J. S. Fuhrer, R. C. Boerth and J. Ross, Jr. 1969. The production of isotonic contractions in the intact canine left ventricle. J. Appl. Physiol. 27: 577.

Falsetti, H. L., R. E. Mates, C. Grant, D. G. Greene and L. L. Bunnell 1970. Left ventricular wall stress calculated from one-plane cineangiography. Circ. Res. 26: 71.

Feigl, E. O. and D. L. Fry 1964a. Myocardial mural thickness during the cardiac cycle. Circ. Res. 14: 541.

Feigl, E. O. and D. L. Fry 1964b. Intramural myocardial shear during the cardiac cycle. Circ. Res. 14: 536.

Feigl, E. O., G. A. Simon and D. L. Fry 1967. Auxotonic and isometric cardiac force transducers. J. Appl. Physiol. 23: 597.

Forwand, S. A., K. M. McIntyre, Lipana, J. G. and Lefine, H. J. 1966. Active stiffness of the intact canine left ventricle. Circ. Res. 19: 970.

Fry, D. L., D. M. Griggs, Jr. and J. C. Greenfield, Jr. 1964a. Myocardial mechanics: tension velocity-length relationships of heart muscle. Circ. Res. 14: 73.

Fung, Y. C. B. 1968. Biomechanics: Its scope, history and some problems of continuum mechanics in physiology. Appl. Mechs. Reviews. 21: 1.

Ghista, D. N. and H. W. Vayo 1969. The time-varying elastic properties of the left ventricular muscle. Bull. Math. Biophysics. 31: 75.

Ghista, D. N. and H. Sandler 1969. An analytic elastic-viscoelastic model for the shape and the forces in the left ventricle. J. Biomech. 2: 35.

Graham, T. P., Jr., J. Ross, Jr., J. W. Covell, E. H. Sonnenblick and R. L. Clancy 1967. Myocardial oxygen consumption in acute experimental cardiac depression. Circ. Res. 21: 123.

Hawthorne, E. W. 1961. Instantaneous dimensional changes of the left ventricle in dogs. Circ. Res. 9: 110.

Hawthorne, E. W. 1966. Dynamic geometry of the left ventricle. Am. J. Cardiol. 18: 566.

Hefner, L. L. and T. E. Bowen, Jr. 1967. Elastic components of cat papillary muscle. Am. J. Physiol. 212: 1221.

Hill, A. V. 1939. The heat of shortening and the dynamic constants of muscle. Proc. Roy. Soc. London, Ser. B. 126: 136.

Hinds, J. E. and E. W. Hawthorne 1965. The eccentricity of motion of left ventricular circumference during the cardiac cycle in awake animals. Fed. Proc. 24: 146.

Hinds, J. E., E. I. Franklin, E. A. Rhode and E. W. Hawthorne 1967. Left ventricular aorta to apex length changes in awake dogs. Fed. Proc. 26: 718.

Hinds, J. E., E. W. Hawthorne, C. B. Mullins and J. H. Mitchell 1969. Instantaneous changes in the left ventricular lengths occurring in dogs during the cardiac cycle. Fed. Proc. 28: 1351.

Hood, W. P., Jr., W. J. Thomson, C. E. Rackley and E. L. Rolett 1969. Comparison of calculations of left ventricular wall stress in man from thin-walled and thick-walled ellipsoidal models. Circ. Res. 24:575.

Kabat, H. and M. B. Visscher 1939. The elastic properties of the beating tortoise ventricle with particular reference to hysteresis. Am. J. Physiol. 125: 437.

Kluemper, A. and A. Puff 1960. On the course of contraction in the cardiac left ventricle. Z. Kreislaufforsch. 49: 500.

Levine, H. J. and N. A. Britman 1964. Force-velocity relations in the intact dog heart. J. Clin. Invest. 43:1383.

McDonald, R. H., Jr., R. R. Taylor and H. E. Cingolani 1966. Measurement of myocardial devloped tension and its relation to oxygen consumption. Am. J. Physiol. 211: 667.

Mirsky, I. 1969. Left ventricular stresses in the intact human heart. Biophys. J. 9: 189.

Mitchell, J. H. and C. B. Mullins 1967. Dimensional analysis of left ventricular function. *In* R. D. Tanz, F. Kavaler, and J. Roberts (eds.), Factors Influencing Myocardial Contractility. Academic Press Inc., New York. p. 177.

Mitchell, J. H., K. Wildenthal and C. B. Mullins 1969. Geometrical studies of the left ventricle utilizing biplane cinefluorography. Fed. Proc. 28: 1334.

Monroe, R. G. and G. N. French 1961. Left ventricular pressure-volume relationships and myocardial oxygen consumption in the isolated heart. Circ. Res. 9: 362.

Mountcastle, V. B. 1968. Medical Physiology. Vol. 1, Part 1, Circulation. The C. V. Mosby Co., St. Louis, Mo.

Pool, P. E., B. M. Chandler, S. C. Seagren and E. H. Sonnenblick 1968. Mechanochemistry of cardiac muscle. II. Isotonic contraction. Circ. Res. 22: 465.

Parmley, W. W. and E. H. Sonnenblick 1967. Series elasticity in heart muscle. Its relation to contractile element velocity and proposed muscle models. Circ. Res. 20:112.

Rodbard, S., F. Williams and G. Williams 1959. The spherical dynamics of the heart (myocardial tension, oxygen consumption, coronary blood flow, and efficiency). Am. Heart J. 57: 348.

Ross, J., Jr., J. W. Covell, Sonnenblick, E. H. and Braunwald, E. 1966. Contractile state of the heart characterized by force-velocity relations in variably afterloaded and isovolumic beats. Circ. Res. 18: 149.

Ross, J., Jr., E. H. Sonnenblick, J. W. Covell, G. A. Kaiser and D. Spiro, 1967. The architecture of the heart in systole and diastole. Technique of rapid fixation and analysis of left ventricular geometry. Circ Res. 21: 409.

Rushmer, R. F., D. L. Franklin and R. N. Ellis 1956. Left ventricular dimensions recorded by sonocardiography. Circ. Res. 4: 684.

Sallin, E. A. 1969. Fiber orientation and ejection fraction in the human left ventricle. Biophys. J. 9: 954.

Sandler, H. and H. T. Dodge 1963. Left ventricular tension and stress in man. Circ. Res. 13: 91.

Sandler, H. and D. N. Ghista 1969. Mechanical and dynamic implications of dimensional measurements of the left ventricle. Fed. Proc. 28: 1344.

Sandler, H. 1970. Dimensional analysis of the heart—a review. Amer. J. of Med. Sci. 260: 56.

Sanmarco, M. E. and J. C. Davila 1967. Continuous measurement of left ventricular volume in dogs: Estimation of volume-dependent variables using the ellipsoid model. *In* R. D. Tanz, F. Kavaler, and J. Roberts (eds.), Factors Influencing Myocardial Contractility. Academic Press Inc., New York. p. 199.

Sonnenblick, E. H. 1962*a*. Force-velocity relations in mammalian heart muscle. Am. J. Physiol. 202: 931.

Sonnenblick, E. H. 1962*b*. Implications of muscle mechanics in the heart. Fed. Proc. 21: 975.

Sonnenblick, E. H. 1964. Series elastic and contractile elements in heart muscle: Changes in muscle length. Am. J. Physiol. 207: 1330.

Sonnenblick, E. H. 1966. The mechanics of myocardial contraction. *In* S. A. Briller and H. L. Conn, Jr. (eds.), The Myocardial Cell. Univ. of Penna. Press. Phila. Penna., p. 173.

Sonnenblick, E. H., J. Ross, Jr., J. W. Covell, H. M. Spotnitz and D. Spiro, 1967. The ultrastructure of the heart in systole and diastole. Changes in sarocomere length. Circ. Res. 21: 423.

Sonnenblick, E. H. and J. Ross, Jr. 1967. Some ultrastructural considerations in myocardial failure: Sarcomere overextension and length dispersion. *In* R. D. Tanz, F. Kavaler, and J. Roberts (eds.), Factors Influencing Myocardial Contractility. Academic Press Inc., New York. p. 43.

Spiro, D. and E. H. Sonnenblick 1964. Comparison of the ultrastructural basis of the contractile process in heart and skeletal muscle. Circ. Res. 15 Suppl. II: p. II–14.

Spiro, D. 1966. The fine structure and contractile mechanism of heart muscle. *In* S. A. Briller and H. L. Conn, Jr. (eds.), The Myocardial Cell. Univ. of Penn. Press. Phila. Penn. p. 13.

Streeter, D. D., Jr. and D. L. Bassett 1966. An engineering analysis of myocardial fiber orientation in pig's left ventricle in systole. Anat. Rec. 155: 503.

Streeter, D. D., Jr., H. M. Spotnitz, D. P. Patel, J. Ross, Jr. and Sonnenblick, E. H. 1969. Fiber orientation in the canine left ventricle during diastole and systole. Circ. Res. 24: 339.

Streeter, D. D., Jr. R. N. Vaishnav, D. J. Patel, H. M. Spotnitz, J. Ross, Jr. and E. H. Sonnenblick 1970. Stress distribution in the canine left ventricle during diastole and systole. Biophs. J. 10: 345.

Symposium: Dynamic Geometry of the Left Ventricle. 1969. Presented at the 52nd Annual Meeting of the Fed. of Am. Soc. for Exp. Biol., Atlantic City, N. J. April 16, 1968. Fed. Proc. 28: 1323.

Wiggers, C. J. 1928. The Pressure Pulses in the Cardiovascular System. Longmans, Green and Co. Ltd. London, New York.

Wong, A. Y. K. and P. M. Rautaharju 1968. Stress distribution within the left ventricular wall approximated as a thick ellipsodial shell. Am. Heart J. 75: 649.

13

Determinants of Cardiac Performance

CHARLES W. URSCHEL
and
EDMUND H. SONNENBLICK

I. DETERMINANTS OF CARDIAC PERFORMANCE

Any attempt to quantitate the determinants of cardiac performance is oriented toward three primary questions. First, is the performance of the heart as a pump adequate to meet the needs of the circulation—both at rest and during stress? Second, what is the intrinsic capability of the myocardial muscle; and, third, is it possible to predict the response of the heart to a given hypothetical stress? A comprehensive clinical or experimental evaluation of the heart requires consideration of all three aspects. Although a complete quantitative description in these terms is not yet possible, it is feasible to describe cardiac performance in terms of basic muscle capability, and the lumped performance of the ventricle as a response to input and output loading. Three primary parameters therefore, describe the behavior of the heart—each of which may be considered as relatively independent determinants of performance. The first is ventricular diastolic volume or preload, as reflected by the filling pressure of the ventricle. Second is the load faced by the heart after ejection begins, the afterload. The third determinant of performance is cardiac contractility which refers to the intrinsic capability of the myocardial muscle to shorten and generate force. It is clear that any attempt to evaluate the performance of the heart without taking into account the effects of each of these three factors will be incomplete. The remainder of this paper will be concerned with a detailed discussion of each of these three parameters.

I PRELOAD

Pioneering studies by Frank (1895) and Starling (1918) described the basic dependence of pressure generation and stroke output on the diastolic or filling pressure of the left ventricle (Fig. 1). These workers demonstrated clearly that as the filling pressure or preload was increased, both the peak isovolumic pressure that the ventricle could generate, as well as the stroke volume that the heart could eject, were proportional to the filling pressure. More recently (Sarnoff and Mitchell, 1962), this dependence of performance on preload has been determined with considerable precision (Fig. 2). It was shown that where both the contractile state of the muscle and the mean afterload against which the heart ejected were held constant, the stroke

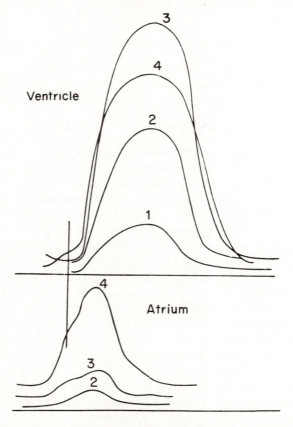

Fig. 1: Effect of preload on developed pressure. This data from Otto Frank's original publication illustrates the dependence of systolic isovolumic pressure on the filling pressure or preload of the ventricle. Where preload is excessive, the pressure begins to fall (tracing #4).

volume and stroke work of the ventricle could be described as a ventricular function curve (Fig. 2). A change in the inotropic state of the heart by pharmacologic interventions, for example, would result in a different curve. Thus, at any given end-diastolic volume, an increase in the contractile state of the heart would result in an increase in the stroke output work; and a decrease in the contractile state would result in a corresponding decrease. It is evident that the heart at a lower contractile state would need a substantially higher filling pressure in order to maintain the same stroke output as a more healthy heart. Conversely, a heart under marked inotropic stimulation could maintain a normal stroke output at a substantially reduced filling pressure.

It has been possible to make similar measurements in the clinical catheterization laboratory and findings paralleling the work in animals have been noted (Ross and Braunwald, 1964). Thus, those patients with severe myocardial disease require a higher filling pressure to maintain the same stroke output as normal persons; while at any given filling pressure the output can be significantly augmented by the use of inotropic agents. Thus, it is clear that the Frank–Starling mechanism plays a role in governing the cardiac output both in the experimental and clinical situations. This mechanism pro-

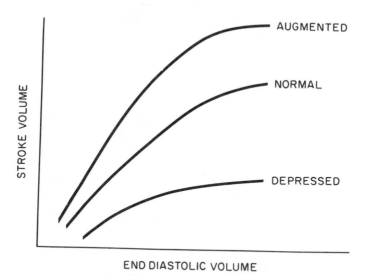

Fig. 2: Ventricular function curve. The stroke volume of the heart is dependent on the end-diastolic volume or preload of the ventricle. At any given end-diastolic volume, stroke volume can be augmented or depressed with changes in the inotropic state of the heart muscle. Normally the heart functions on that portion of the curve which is relatively steep. At high preloads very little additional increment in stroke volume can be achieved by a further increase in preload.

vides a means both for steady state and prompt beat-to-beat adjustment of stroke volume as a function of venous return.

II AFTERLOAD

The second major determinant of cardiac performance is the afterload or impedance faced by the ejecting ventricle (Urschel et al, 1968; Imperial et al, 1961). If an abrupt change in arterial pressure is made during a period when preload and contractility are constant, the stroke output of the heart changes inversely as a function of the average impedance to ejection. Thus, as demonstrated in Fig. 3, a decrease in impedance will cause an increase in stroke volume, while an increase in impedance has an opposite effect. It is evident that if the impedance is decreased the same stroke volume may be maintained despite a subnormal filling pressure while a normal filling pressure could sustain a heightened stroke volume. Similarly if the impedance to ejection is increased, stroke volume can be maintained only by an increased filling pressure. This mechanism plays a role, in the presence of certain physiological stresses, in helping to enhance cardiac output. During exercise or hyperthermia, for example, the impedance falls significantly. Similarly, certain disease states are characterized by a low impedance against which the heart must eject a large stroke volume—a load which is usually well maintained (Urschel et al, 1968). Conversely, conditions which place a high

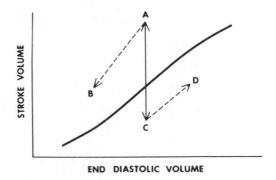

Fig. 3: Effects of altered impedance on the ventricular function curve. Stroke volume can be altered by an isolated change in the impedance to ejection. Thus, at constant end-diastolic volume (vertical solid arrow), stroke volume increases from the control point on the function curve to point. A following a decrease in impedance. Similarly, the same stroke volume as in the control state could be maintained at a subnormal end-diastolic volume, point B. In the opposite direction, an increase in afterload decreases stroke volume (point C), or if stroke volume is to be maintained in the face of the increased afterload, end-diastolic volume must be increased (point D).

impedance load on the ventricle are characterized by the development of substantial myocardial hypertrophy in order to maintain effective cardiac function in the face of the abnormal afterload.

A major problem has been to quantitize this dependence of stroke volume on the impedance to ejection in order to predict the response of the heart to different loads. It would appear that the diseased heart is substantially more sensitive to changes in impedance than is the healthy heart. One approach has utilized a model for the heart in which the ventricle is depicted as an energy (flow) source with an internal or self-impedance coupled to an external load, the aortic input impedance (Fig. 4). The source flow and the source impedance represent theoretical quantities which describe internal energy loss and which include myocardial elastance and ventricular shear or wall frictional forces operative during ejection. The load impedance includes valvular resistance to flow and aortic compliance. If the source impedance is substantially greater than the load impedance, there will be relatively little change in flow through the load as a function of changes in load. On the other hand, if the source resistance is substantially less than the load resistance the system operates essentially as a constant pressure source. Any change in the load resistance now will have a substantial effect on the flow through the external load but relatively little effect on pressure. If the heart can be adequately described by such a model, it would be possible to predict the response to a given change in impedance. Figure 5 illustrates

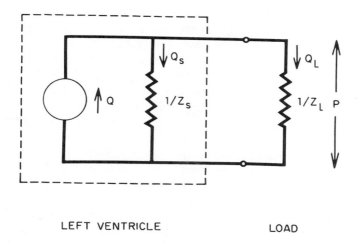

LEFT VENTRICLE   LOAD

Fig. 4: Electrical analogue for the ventricle. The left ventricle has been modeled as an energy source with an internal or source impedance. Flow from the source is, therefore, divided between the internal flow Q_s and flow through the external bad, Q_L.

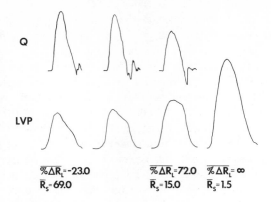

Fig. 5: Data permitting calculation of source and load impedances. Aortic arch flow (Q) and left ventricular pressure (LVP) are illustrated at four different afterloads created by occlusion or abrupt alteration of central aortic pressure. Experimentally, the control beat (second from left) is followed by a beat at abruptly altered afterload. Since preload and the inotropic state are constant over this short interval, the changes in pressure and flow reflect changes in the impedance to ejection. Utilizing the electrical model illustrated in Fig. 4, the internal and load impedance can be calculated. It is noted that the calculated source resistance (R_s) appears to depend on the nature of the change in afterload ($\% \Delta R_L$) which is used for the calculation. Thus, with a decrease in afterload (lefthand data), the source resistance appears substantially higher than with either an increase in afterload (second from right) or an infinite afterload (far right).

flow and pressure data from an experiment designed to describe the ventricle in this manner. Abrupt changes in the load resistance (input impedance to the aorta) can be made utilizing a wide-bore tube in the aorta connected to a high or a low pressure source. Such changes can be made during the interval between beats allowing two beats to be compared at different load impedances, but with preload and contractility otherwise unchanged. Given the "control" and "intervention" flows and pressures, the source and load impedances can be easily calculated. Isovolumic beats (infinite load resistance) can also be obtained by the adrupt inflation of a balloon placed just above the aortic valve. If such a model is used to describe the system, it is seen that the calculated source impedance is a function of the change in load resistance used to calculate it, and indicates that this simple linear model is an incomplete description of the myocardium. On the other hand, it has been noted that the healthy heart is consistently characterized by a higher source impedance (it behaves as a flow source) while the depressed or sick heart has a substantially lower source impedance and has marked flow sensitivity to changes in load impedance.

III CONTRACTILITY

The third governing parameter of myocardial performance is the intrinsic contractile state or contractility of the myocardial muscle. Although the term is used by many authors in a variety of ways, most agree that the attributes of a change in the contractile state are those of speed, power, energy flux, strength of contraction and oxygen and chemical substrate utilization. Thus, the contractile or inotropic state of the heart is generally acknowledged to be a function of the intrinsic biochemical reactions controlling the rate of energy release.

Many indices have been proposed to assay the contractile state of the heart (Sarnoff and Mitchell, 1962; Rushmer, 1966; Wallace et al, 1963; Covell et al, 1966). Most simply, the speed of shortening, force generating capability and shortening during ejection increase if the load and preload are held constant. Any complete evaluation of the intact heart requires separation of those changes which result from a change in contractility from those which result from changes in preload or afterload. Thus, as can be seen from Figs. 2 and 3, if preload is held constant, an increase in stroke volume may result from either a decrease in aortic input impedance or an increase in the contractile state of the heart. Similarly, a decrease in stroke volume at a given preload may represent either an increase in impedance to ejection or a decline in myocardial contractility. Where preload is allowed to change, the evaluation is even more complex.

Ideally, an index of contractility should therefore be (1) insensitive to preload and (2) unaffected by changes in afterload, that is, it should represent contractility alone. Clearly, simple measures of stroke work or stroke volume as a function of preload do not meet this criterion. Indices of the speed of contraction or shortening such as the first derivative of the pressure tracing (dP/dt) or ejection velocity are also sensitive to preload, afterload, or both.

One approach to the quantitation of contractility has been the use of the model for muscle proposed by A. V. Hill (1938) (Fig. 6). This model conceptualizes the muscle as consisting of three discrete elements. The first of these, the contractile element, is a force generating element which has a characteristic shortening velocity which is inversely related to the load on the contractile element at any given instant. The relationship between generated force and contractile element shortening during a given contraction can be described as either an inverse hyperbolic or exponential function which can be extrapolated to zero load, V_{max}. In isolated muscle this extrapolated theoretical velocity of shortening to zero load, V_{max}, is relatively independent of preload, but can be altered by inotropic agents in a direction parallel to the inotropic effect (Sonnenblick, 1962). Thus, the use of this

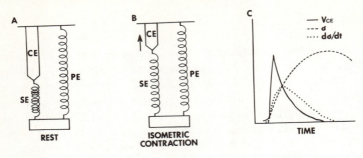

Fig. 6: Application of the Hill model for muscle to the isometric contraction. The three-element model for heart muscle proposed by A. V. Hill is shown in A. During an isometric contraction (panel B) the contractile element shortens, stretching the series elastic (*SE*) and generating force. During an isometric contraction, there is no overall muscle shortening so the parallel elastic is unchanged and supports a constant load. Panel C, redrawn from actual data, illustrates the time course of wall stress (σ), the rate of change of stress per unit time ($d\sigma/dt$), and calculated contractile element velocity (V_{ce}). It is seen that contractile element velocity rises to a peak very early and as shortening of the series elastic proceeds, force increases rapidly.

conceptual model for muscle presents a means by which the contractile state can be assayed relatively independently of preload or afterload.

In the Hill model, the contractile element is connected to the load via an undamped series elastic model, the contractile element is connected to the load via an undamped series elastic element. It has been shown experimentally that this elastic element has an exponential stress-strain relationship (Parmley and Sonnenblick, 1967). The velocity of lengthening or shortening of a spring is proportional to the rate of change of force across the spring divided by the modulus of the spring at that instant. For an exponential spring (Fig. 7), the modulus is a linear function of the instantaneous load

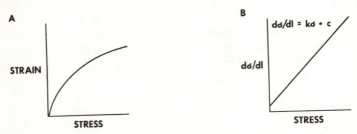

Fig. 7: Exponential stress-strain characteristic of the series elastic element. The series elastic element has an exponential stress-strain relation (Parmley and Sonnenblick, 1967). Thus, at any point on this curve, the modulus of elasticity is equal to the reciprocal of the slope of the curve, i.e., $d\sigma/dl$. If the modulus of elasticity is plotted as function of stress, it is noted (panel B) that the modulus is not constant (as for a Hookean spring), but is a linear function of the force across the spring.

on the spring. The velocity of lengthening or shortening of this spring is therefore equal to $d\sigma/dt$ (where σ is stress). If the muscle is studied during a period of time when the overall length of the muscle can be assumed to be fixed (an isometric contraction), the only contractile element shortening which occurs stretches the series elastic and generates force (Fig. 6). During this period of time, the velocity of the contractile elements is identical to the lengthening of the spring and is equal to $(d\sigma/dt)/(k\sigma + C)$ as described above. The constants k and C of the exponential spring can be determined experimentally. If the velocity thus calculated is plotted as a function of the instantaneous load, an inverse relation is obtained which, in the isolated papillary muscle, can be extrapolated to a zero load intercept (V_{max}). For accurate calculation, the third element of the Hill model must be taken into account, the parallel elastic element (*PE* in Fig. 6). This element is assumed to be an entirely passive undamped spring which supports the elastic diastolic load. During diastole the contractile element is assumed to be freely extensible and the series elastic element is not preloaded (Fig. 6). *If the muscle does not shorten*, as assumed previously, the amount of force supported by the parallel elastic will remain constant and can be subtracted throughout in order to obtain the force supported by the series elastic. When the calculations are made in this manner using developed force (that is, the total force minus preload) the extrapolation to V_{max} is rather constant despite variations in initial muscle lengths over a physiological range of preloads.

In order to apply this methodology to the intact heart, several major assumptions must be made (Fig. 8) (Taylor et al, 1967; Urschel et al, 1970). Unfortunarely, the validity of these assumptions cannot be directly checked. Validation of the technique depends on the empirical usefulness and repro-

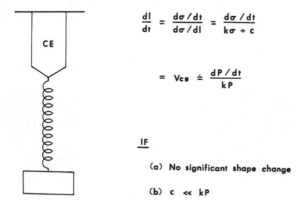

Fig. 8: Application of the Hill model in the intact heart. During an isometric contraction, contractile element velocity (dl/dt or V_{ce}) is equal to $(d\sigma/dt)/(d\sigma/dl)$. This velocity is approximately equal to $(dP/dt)/kP$ during the isovolumic period if there is no significant shape change and if the constant C is small.

ducibility of the results and the degree to which the results obtained in the intact heart parallel those from isolated muscle. The primary assumption is that there is no significant muscle fiber shortening or ventricular shape change during the period of ventricular systole prior to aortic valve opening. Inasmuch as there is some muscle rearrangement and shape change during this isovolumic portion of systole (Ninomiya and Wilson, 1965), this assumption represents only a first approximation. Calculations of actual fiber shortening made during this period show that muscle fiber rearrangement during the isovolumic portion of systole contributes only a small additional contractile element velocity to that calculated assuming no fiber shortening (Hugenholtz et al, 1970). It is, therefore, valid to assume that the ventricle during this period is isometric as well as isovolumic. The second major simplification that can be made if the ventricle is approximately isometric during this period, permits the use of pressure instead of wall stress for the calculations. Stress appears in both the numerator and denominator of the equation for contractile element velocity. Since stress is a function of pressure, radius and wall thickness, it is clear that if the radius and wall thickness are unchanged (implicit in the assumption of an isometric period) the equation can be simplified to $V_{ce} = dP/dt/(kP)$. The C constant is assumed to be relatively small. A final simplification is the use of pressure instead of stress as the horizontal axis for plotting the force velocity relation. It can be shown that if only the extrapolation to zero load is desired, the scale factor on the horizontal axis is unimportant and independent of whether the curve is a straight line, hyperbola or exponential. Thus, during an isometric period, stress is equal to pressure times a constant, that is, a scale factor.

These assumptions represent a gross simplification and in essence treat the entire ventricle as a single lumped papillary muscle. The validity and empirical usefulness of this approach have been studied extensively in a large number of acute canine experiments. Data was collected at 5 msec intervals utilizing a digital computer and analogue to digital converter. The contractile element velocity was calculated as $\Delta P/\Delta t/(28P)$ at 5 msec intervals (Fig. 9).

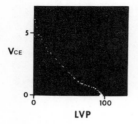 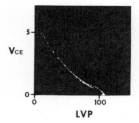

Fig. 9: The isovolumic force velocity relation in the intact heart. The calculated force velocity relation from 5 consecutive beats in the intact heart is illustrated in left panel. Mathematical description of the curve is achieved (right panel) by the use of a least squares exponential fit.

The calculated force velocity relation was then displayed on an oscilloscope, plotting contractile element velocity as a function of developed pressure. An exponential equation to describe the data of the form $V_{ce} = A + Be^{-C(LVP)}$ was obtained utilizing a least squares technique. The validity of the least squares fit was checked by plotting the data and the equation on the oscilloscope (Fig. 9).

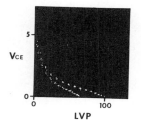 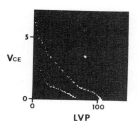

Fig. 10: Changes in the isovolumic force velocity relation with changes in preload and inotropic state. On the left are seen the force velocity relations calculated from the intact heart at 2 separate end-diastolic pressures. It is noted that the intercept of the curve at zero LVP (V_{max}) is essentially unchanged by this intervention. Conversely, in the right-hand panel, this zero load intercept is substantially increased by the addition of calcium to the heart.

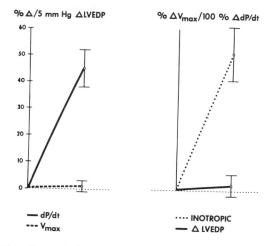

Fig. 11: The effects of changes in preload and inotropic state on V_{max}. The left-hand panel illustrates the change ($\pm$1S.E.) in V_{max} and dP/dt (the first derivative of the pressure in the left ventricle) after a 5 mm Hg change in end-diastolic pressure. dP/dt is sensitive to alterations in preload while V_{max} is not. On the right, the change in V_{max} as a function of the % change in dP/dt is plotted. Thus it is noted that with a change in LVEDP which results in an increase of 100% in dP/dt, there is no increase in V_{max} (solid line). On the other hand, following inotropic interventions (dotted line), V_{max} increases proportional to dP/dt.

314 DETERMINANTS OF CARDIAC PERFORMANCE

Typical data illustrating the calculated force velocity relations following alterations either in preload or inotropic state is illustrated in Fig. 10. Figure 11 demonstrates graphically the sensitivity of the method and its insensitivity to preload change. It can be seen (Figs. 10 and 11) that V_{max} reflects a change in inotropic state with a sensitivity slightly better than half that of the peak first derivative of the left ventricular pressure trace. Unlike the first derivative of pressure, however, which is sensitive to changes in preload as well as inotropic state, V_{max} is rather insensitive to changes in preload. Thus, if it it is necessary to quantitate the contractile state of the heart, despite changing preload, V_{max} calculated utilizing developed pressure provides a pragmatic

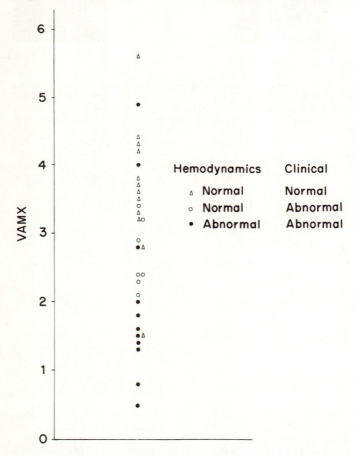

Fig. 12: Evaluation of V_{max} in children. The figure illustrates V_{max} from thirty children studied at the time of cardiac catheterization (Hugenholtz et al, 1970). It is noted that in general, V_{max} reflects quite well the hemodynamic and clinical status of the heart as determined by other methodology.

index of contractility which is sensitive and yet unaffected by changes in preload. However, if an experiment can be designed such that preload is held constant, the first derivative of the pressure (dP/dt) is more sensitive than V_{max}.

This methodology for quantification of contractility has been applied to man (16) and the results are summarized in Fig. 12. V_{max} can be calculated with reasonable facility in man during routine cardiac catheterization. Although there is considerable scatter and some overlap, V_{max} correlates rather well with other approaches used to estimate the underlying state of the myocardium. Major questions about the use of the technique remain, but this data suggests that V_{max} has potential usefulness as an index of the contractile state in man.

SUMMARY

Parameters that govern cardiac performance have been identified and discussed. The intrinsic contractility of the myocardial muscle can be measured and the dependence of ventricular performance on preload and afterload quantitized. Although the goal of a simple overall mathematical formulation that would allow precise prediction of cardiac performance is still not met, it is possible to describe in some detail the basic parameters of control. Thus, the total response of the heart reflects the interaction and summation of three basic control mechanisms and a given cardiac output may be achieved via a variety of effector pathways. Complete analysis of the behavior of the ventricle therefore requires definition of the contributing roles of each of the determinants of cardiac performance.

Acknowledgment: *This work is supported by U.S. Army contract No. DADA 17–67C7164 and U.S.P.H.S. grant No. HE–11306–02.*

REFERENCES

Covell, J. W., J. Ross Jr., E. H. Sonnenblick and E. Braunwald 1966. Comparison of the force-velocity relation and the ventricular function curve as measures of the contractile state of the intact heart. Circ. Res. 19: 364.

Frank, O. 1895. Zur Dynamik der Herzmuskels. Ztschr. f. Biol. 32: 370.

Hill, A. V. 1938. Heat of shortening and dynamic constants of muscle. Proc. Roy. Soc. London B, 126: 136.

Hugenholtz, P., R. C. Ellison, C. W. Urschel, E. H. Sonnenblick and I. Mirsky 1970. Myocardial force-velocity relationships in clinical heart disease. Circul. 51: 191.

Imperial, E. S., M. N. Levy and H. Zieske, Jr. 1961. Outflow resistance as an independent determinant of cardiac performance. Circ. Res. 9: 1148.

Ninomiya, I. and M. F. Wilson 9165. Analysis of ventricular dimension in the unanesthetized dog. Circ. Res. 16: 249.

Parmley, W. W. and E. H. Sonnenblick 1967. Series elasticity in heart muscle. Its relation to contractile element velocity and proposed muscle models. Circ. Res. 20: 112.

Ross, J., Jr. and E. Braunwald 1964. Studies on Starling's law of heart. IX. Effects of impeding venous return on performance of normal and failing human left ventricle. Circul. 30: 719.

Rushmer, R. F. 1966. Initial ventricular impulse, a potential key to cardiac evaluation. Circul. 29: 168.

Sarnoff, S. J. and J. H. Mitchell 1962. Control of function of the heart. *In* W. F. Hamilton and P. Dow (eds.), Handbook of Physiology. Sec. 2. Circulation: Vol. 1. p. 758. American Physiological Society, Washington, A.C.

Sonnenblick, E. H. 1962. Implications of muscle mechanics in the Heart. Fed. Proc. 21: 975.

Starling, E. H. 1918. The Linacre lecture on the Law of the Heart, given at Cambridge, 1914. Longmans, Green & Co. Ltd. London. 27.

Taylor, R. R., J. Ross, Jr., J. W. Covell and E. H. Sonnenblick 1967. A quantitative analysis of left ventricular myocardial function in intact sedated dog. Circ. Res. 21: 99.

Urschel, C. W., J. W. Covell, E. H. Sonnenblick, J. Ross, Jr. and E. Braunwald, 1968. Myocardial mechanics in aortic and mitral valvular regurgitation: the concept of instantaneous impedance as a determinant of the performance of the intact heart. J. Clin. Invest. 47: 867.

Urschel, C. W., A. H. Henderson and E. H. Sonnenblick, 1970. Model dependency of ventricular force-velocity relations: Importance of developed pressure. Fed. Proc. 29: 719.

Wallace, A. G., N. S. Skinner, Jr. and J. H. Mitchell 1963. Hemodynamic determinants of the maximal rate of rise of left ventricular pressure. Am. J. Physiol. 205: 30.

14

Lung Elasticity

GEORGE C. LEE
and
FREDERIC G. HOPPIN, JR.

I. INTRODUCTION

In this chapter we offer a perspective of some problems of solid and structural mechanics in the physiology of the lung. We shall present the background of these problems in terms which a non-physiologist can handle and indicate some approaches which we and others are currently taking. There will be few substantive results because the field is relatively new—in contrast to extensive work already done on the cardiovascular system and microcirculation, for example.

The plan of the chapter is as follows. First some physiologically important problems of lung elasticity will be discussed. (For more background the reader is referred to appropriate sections of physiology textbooks. Guyton (1964) is an easy start. The Handbook of Physiology (1964) provides definitive reviews of all aspects.) Next, we shall identify and characterize the components of the lung which bear tensile forces and describe the anatomy of the lung from the viewpoint of structural mechanics. Then an extremely simplified model and some more refined approaches to lung elasticity will be presented.

Various biomechanical talents could be applied to problems within the broad field of lung physiology. Consider the following general disciplines as applied to lung physiology:

A. Transport and diffusion—problems associated with exchange of respiratory gases in the lung, blood, and tissue.
B. Fluid mechanics and aerodynamics—problems of airway resistance, air flow in branching distensible channels, and so on.

C. Solid mechanics—problems associated with lung elasticity, the behavior of the chest wall, and other.
D. Control—respiratory regulation.
E. Systems—the global approach treating the intimate coupling of all the physiologic functions cited.

This chapter is limited to C, that is to problems within the purview of solid mechanics, although the physiological implications of these problems are broad.

From the viewpoint of structural mechanics, the lung can be characterized as a pretensed passive elastic meshwork. This concept will be useful in investigating two major aspects of lung function. One aspect is the gas volume and change of gas volume (ventilation) of the airspaces. Gas composition within the lung is maintained by mass transport of fresh air, each individual airspace or *alveolus* changing size like a bellows during the breathing cycle. These alveoli are simply spaces within a complex, largely continuous, passive elastic meshwork. Therefore, the volume of gas in each alveolus and the change of volume during breathing depend largely on the tensile forces developed in that particular part of the meshwork. These forces in turn, are influenced by gravity, the shape of the chest cavity and the homogeneity and integrity of the meshwork (especially as affected by disease).

A second major aspect is the mass transport of air and blood through the airways and blood vessels. These structures, which ultimately serve every airspace, pass through and are largely supported by the elastic meshwork. The diameters of the airways and blood vessels and, therefore, their aerodynamic and hemodynamic properties, depend critically on the forces with which the elastic meshwork tethers them. These mechanical relationships are of particular importance in emphysema, asthma, and atelectasis or collapsed lung. Thus, a better understanding of the solid mechanics aspects of the lung will have broad physiological and pathological import.

II. FORCE BEARING MATERIALS AND STRUCTURE

A. Introduction

The structure of the lung which bears tensile forces will be described, first in terms of its constituents, then in terms of its anatomy; that is the microscopic and macroscopic architecture of those constituents.

Tensile forces within the lung are borne by the tissues and air-fluid surfaces. The major force bearing components of the tissues are the extracellular fibrous proteins *collagen* and *elastin*.

B. Collagen

Collagen is a relatively inextensible fibrous protein. (It can be considered the most important structural element of animals; tendons and ligaments are almost pure collagen, bone and skin are largely collagen.) The tensile strength of collagen is due to the nature of its basic unit, a long, tightly packed and tightly bound triple polypeptide chain. Collagen can be identified by light or electron microscopy. It can be isolated relatively intact from tissues and quantified chemically.

Collagen molecules probably are formed within cells (fibroblasts) and extruded into the extracellular region where aggregation and bonding occur to form fibrils. Little is known of the mechanisms which control the mechanical anatomy of collagen and the rates of production and breakdown. In general, collagen can be deposited in large amounts during growth. It turns over relatively slowly in nongrowing tissues, remaining intact for months or years. In response to injury it may be rapidly deposited, forming scar tissue. This reparative process itself may be harmful as in fibrosis of the lung after infection, radiation, inhalation of particles (silicosis), malignancies, and other situations.

The mechanical properties of aggregates of collagen depend somewhat on their weave, orientation, density, and relationship with other tissue elements. For example, reticulin is an extremely fine collagen network which serves to connect and support cells, whereas tendons are massive aligned collagen bundles. The factors which determine the complex and specific geometric patterns in which collagen is laid down probably include biochemical factors and physical forces (Elden, 1968), but the process is not well understood.

Mechanical studies of collagen have generally used tendon. As its elements are overwhelmingly parallel, the tensile properties of the whole have been considered the simple sum of the properties of the constituent fibril. Tendon will extend approximately 10% in length before rupturing under a stress of about 7000 lbs per inch^2 or 4.4×10^8 dynes per cm^2 (Gratz, 1931). At high stresses, the length-force relationship is reasonably linear and Young's modulus has been calculated as high as 1×10^9 dynes per cm^2 (Burton, 1954). Some alinearity of tendon at low stresses has been attributed to geometric disposition of fibers and testing artifact. The single fiber is probably linear in extension. (See Crisp's chapter on Properties of Tendon and Skin.)

C. Elastin

Elastin is an extensible fibrous protein present in large amounts in skin, blood vessels, bladder, and other organs and is primarily responsible for the elasticity of these structures. Like collagen, elastin consists of polypeptide chains, but unlike collagen, they are tangled, elongated, and sparsely cross-linked (Ayer, 1964). Elastin has proven difficult to characterize. It can be stained for visualization by light microscopy. It is altered by the methods used to isolate it and is usually impure (Piez, 1968). It may be quantified by amino acid analysis.

Elastin is probably formed extracellularly. Its chemical similarity to collagen and the fact that it never appears without collagen suggests that their formulation may be somewhat similar. Elastin turnover is slow. In contrast to collagen, elastin is probably not formed in the reparative process. Apparent increases of elastin, as in emphysema, may be due to the loss of other tissue or microscopic misinterpretation of degenerated collagen. Definitive biochemical studies of elastin in the lung, particularly with regard to questions of changes with age and in degenerative diseases, are wanting.

Information on the mechanical behavior of elastin can be obtained from *ligamentum nuchae* of the ox which has a preponderance of elastin over collagen. It is much more extensible than collagen; at higher stresses it has a Young's modulus of the order of 3 to 8×10^6 dynes per cm^2 (Burton, 1954). Commonly, it can be extended 100% before failure. A single fiber would appear to be stiffer than the aggregate (Carton et al, 1962) suggesting either that the bonding of fibers or their geometrical relationships are important even in an apparently aligned tissue. The nature of both collagen and elastin is discussed in greater detail in other chapters of this book.

D. Air-fluid surface and surfactant

The third component of the lung which bears tensile forces is the air-fluid surface. The interior of the lung has an enormous surface, all of it wet. Any crosssection of lung, then, intersects many alveolar walls usually with an air-fluid interface on both sides. (For an idea of the magnitude involved here, in very approximate figures, consider that the surface area of the lung is 100 m^2, and there are 100 cm of wall per cm^2 of crosssection. Assuming the surface tension to be that of serum, 40 dynes per cm, and approximating alveolar shape by a sphere of radius 140 μ (La Place) the surfaces would account for roughly 5100 dynes force per cm^2 of crosssection, the equivalent of a pressure of 6 cm H$_2$O.) In fact, the surface tension of the fluid lining of the lung differs from that of other interstitial fluids due to the presence of a surface active substance called pulmonary *surfactant*.

Surface active material may be obtained from the interior of the lung by washing. The agent thought to be primarily responsible is a phospholipid, dipalmitoyl lecithin. It is probably secreted by a specific type of cell in the alveoli. The surface behavior of the lung extract can be studied with a number of methods; the height of a column in a capillary tube, the size of

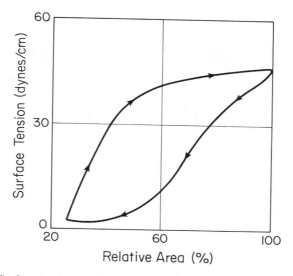

Fig. 1: Surface tension as a function of surface area measured on a Wilhelmy balance (after Clements and Tierney, 1964). By permission of the authors and the Amer. Physiol. Soc.

drops, the pressure in a gas bubble, and pull on a ring or plate (Clements and Tierney, 1964) and the shape of a bubble in a hydrostatic field (Graves, 1967). These measurements show that surface tension varies widely as surface area is changed; with expansion of the surface, surface tension may rise to nearly 50 dynes per cm and, with contraction, surface tension may fall to 2 dynes per cm (Fig. 1). This hysteresis also shows time dependence. Currently, there is widespread interest in surfactant. Its dysfunction has been implicated in respiratory distress syndrome of the newborn, pump-lung syndrome, influenzal pneumonia and many other disorders.

E. Other structural components

Several other components of the lung have been considered as possibly bearing tensile forces: smooth muscle, vessels distended with blood, and fluid in the tissue (edema). Their effects have been studied through the distensibility of the lung (see volume-pressure relationship discussed below). Smooth muscle is primarily located in the walls of airways and blood vessels and has an important role in governing the flow resistance of those structures. However, varying widely among animals, smooth muscles appear in the alveolar ducts (see below) where it might be expected to influence directly the tensile forces in the meshwork. Significant changes in the distensibility of the cat and sheep lungs have been attributed to the direct action of the smooth muscle (Colebatch and Halmagyi, 1963; Nadel et al, 1964), but it is not known how much force is ordinarily sustained in them. Likewise, there are minor changes in lung distensibility resulting from the pressure of blood in the lung vessels (von Basch, 1887). Fluid in the lung tissue apparently changes the volume-pressure behavior of the lung primarily by displacement of volume, and not by direct changes on the elasticity (Cook et al, 1959). The effect of blood pressure and fluid is reviewed in detail by Radford (1964) who concludes that their direct influence is small. Thus, the major force bearing components are collagen, elastin, and the air-fluid surface as modified by surfactant, with smooth muscle being of probable importance.

F. Lung anatomy

Anatomically, the lung is mostly air divided into approximately 4×10^8 compartments by a largely continuous elastic meshwork (Fig. 2). The mesh-

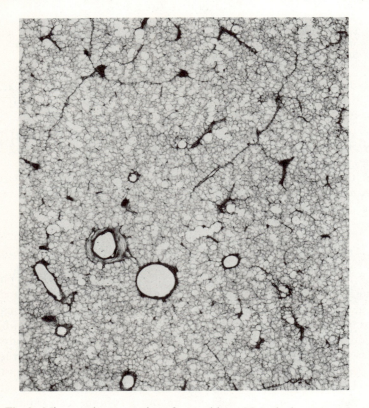

Fig. 2: Microscopic crosssection of normal lung. Note the homogeneous appearance of the meshwork of alveolar walls. (Photograph kindly supplied by Dr. Robert Wright (1961) and reprinted by permission of the Amer. J. of Path.)

work is comprised of interconnecting alveolar walls. Each wall is generally flat with air on both sides. The many small blood vessels (capillaries) in the wall have been likened to a marsh; the spaces between capillaries being potentially smaller than the capillaries themselves. Surrounding the capillaries and bridging the spaces between is a reticulin (fine collagen) network. Coarser collagen and elastin fibers pass across alveolar walls (Fig. 3). Leaving one alveolar wall, the fibers pass into contiguous alveolar walls or other structures. They are dense at the edges where the wall attaches to other walls or structures and at free edges.

The meshwork of alveolar walls is impressively homogeneous. Each alveolar wall usually intersects with two others. The alveoli as cut in the plane are usually hexagonal or pentagonal and appear to have a constant range of size throughout the entire lung (if it is submitted to an even distending pressure during preparation). In three-dimensions the alveoli are usually polygons of nearly uniform size.

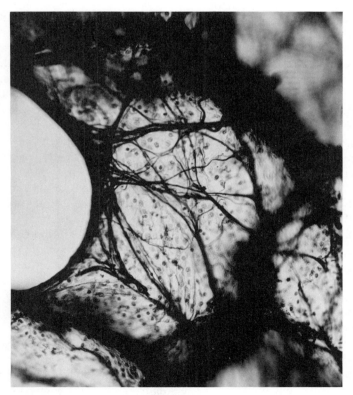

Fig. 3: Microscopic view of elastic fibers in walls of alveoli (right) and forming ring around alveolar duct (left). Note the fibers crossing alveolar walls, passing into adjacent walls, and condensing at free and joined edges of the alveolar wall. (Photograph kindly supplied by Dr. Robert Wright (1961) and reprinted by permission of the Amer. J. of Path.)

But this appearance of homogeneity at the level of the alveolus should be questioned on the basis of three anatomical facts. First, roughly 25% of the lung gas volume is not subdivided into alveoli, but is related to airways serving the alveoli. The majority of that airway volume exists in *alveolar ducts*, roughly cylindrical spaces defined by the open mouths of the alveoli and the free borders of the alveolar walls (see Figs. 2, 3). We will refer to the alveolar duct and its associated alveoli as a *respiratory unit*. Secondly, there are anatomical differences between the two ends of the respiratory unit. The free borders of the alveolar walls which define the alveolar duct show a progressive increase of elastin, collagen and smooth muscle towards the central end (von Hayek, 1960). Thirdly, these units connect with the three ramifying systems, pulmonary arteries, pulmonary veins, and airways, which are grossly different from the lung meshwork. Peripherally, the structures of the three systems are thin-walled and of small diameter. They are firmly oriented in the elastic meshwork of alveolar walls by continuous elastic and collagen fibers. Progressing centrally, with confluence of the branches in each system, the diameters become greater and the walls thicker with smooth muscle, elastin, and collagen arranged in layers and oriented variously. (In addition, the larger airways have cartilagenous plates.) Centrally the larger airways and blood vessels are independent from the surrounding lung in that they are separated from its limiting membrane by loose connective tissue, lymphatic spaces or what amounts to an invagination of the pleural space (see below).

These anatomical considerations raise questions about the apparent homogeneity and isotropy of the elastic meshwork. First, there is evidence that the alveolus changes shape during deflation, becoming relatively narrow at the mouth (Klingele and Staub, 1970). And consequently, the alveolar duct, being defined by the mouths of the alveoli, may empty more than the alveoli. Secondly, insofar as the respiratory units may be systematically oriented by radiating outwards from the central airways, the meshwork may be anisotropic. And thirdly, the ramifying systems, being structurally different from the meshwork through which they run, may influence the behavior of the meshwork. There is some suggestive evidence to the contrary. The length of airways in the excised lung varies simply as the cube root of lung volume (Hyatt, personal communication), that is, the airway does not appear to be tugging on the lung. Removal of bronchial segments from the lung does not appreciably alter the mechanical behavior (Hyatt, 1968), again suggesting that the airways and lungs are, in effect, mechanically independent.

The lung tissue of *parenchyma* is organized at several levels—lobules, lobes, and the lungs, right and left. The elastic meshwork of alveolar walls is continous within a lobule, which is a polygonal subdivision of lung on the order of 1–3 cm on a side. The lobule is partially demarcated by a limiting membrane: that is, a planar surface of tissue, which is similar to, but denser than, the alveolar wall. In man and many animals lung tissue deep within the lung is interconnected. But in many places, particularly near the surface of the lung, the limiting membranes of adjacent lobules can be slightly mutually displaced, being separated by loose tissue traversed by a few loose connecting fibers.

Ultimately the lung tissue is bounded by the *visceral pleura*. The outer surface of this membrane is smooth, moist, and slippery, while the inner surface is loosely connected to the limiting membranes of the lobules beneath. The visceral pleura consists in large part of a relatively thick layer of collagen and elastin, the collagen showing a disposition to run in parallel bands in some regions, and the elastin fibers being irregular (von Hayek, 1960). The visceral pleura virtually completely invests (and thus defines) each of the *lobes* of the lung. In man, two lobes comprise the left lung and three the right. The right and left chest cavities are separated by a vertical wall, the *mediastinum*, in which lie the heart, great vessels, and esophagus. As there is no fibrous connection on the outer surface of the visceral pleura, each lobe is largely free to move along the pleural plane relative to the inside of the chest cavity or neighboring lobes except at the central attachment to the mediastinum where the major airways and blood vessels enter the lobe. The degree of lobulation and lobation and the shape of the chest cavities vary widely among mammals. Whereas man's diaphragm is relatively transverse, many animals have sloping diaphragms. Some diving mammals can displace enormous amounts of blood into reservoirs within the chest. And the adult elephant has no pleural space at all (John West personal communication).

In concluding this anatomical section, it may be appropriate to note that our presentation differs somewhat from the traditional physiologist's viewpoint. They have tended to think of the alveolus as a functional entity attached to the end of three parallel systems of branching tubes, whereas we have emphasized the structural aspects of the network of walls.

III. VOLUME-PRESSURE BEHAVIOR AND THE SINGLE DEGREE OF FREEDOM MODEL

Despite their anatomical complexity, the overall mechanical behavior of healthy lungs has been usefully characterized as a single degree of freedom system (Mead and Milic-Emili, 1964) in which the behavior of multiple branched pathways is approximated by lumped constants:

(1) $$\Delta P_L = \frac{1}{C}(\Delta V) + R(\dot{V}) + I(\ddot{V})$$

where P_L represents transpulmonary pressure (defined as the pressure difference between the airway opening (mouth) and the pleural surface), V, $\dot{V}$, and $\ddot{V}$ represent lung volume, flow, and volume acceleration respectively,

C lung compliance, R flow resistance, and I inertance. This equation of motion has been useful despite the fact that neither C nor R are constant over much of a range of alveolar volume, and that P_L and alveolar volume vary significantly in different parts of the lung. In disease, marked inhomogeneities of C and R show up and are reflected by their marked frequency dependence.

With very slow respiratory maneuvers, both $\dot{V}$ and $\ddot{V}$ terms virtually drop out and the elastic term, dependent primarily on volume, is isolated:

$$\Delta P_L = \frac{1}{C} \Delta V \qquad (2)$$

Much of the known information regarding the elastic properties of the lungs derives from static pressure-volume (PV) curves. In the chest pleural pressure varies from place to place, but is usefully approximated by pressure in the esophagus which can be readily measured in a cooperative subject with a small balloon. When the lung is excised, it shows very similar pressure-volume behavior and continues to do so over a period of hours or days. When suspended from the hilum, it is exposed to a uniform pleural pressure and shows great uniformity of alveolar size (Glazier et al, 1967) (thus being considerably closer to the single degree of freedom system than when in the chest.) Figure 4 shows a single curve representing the PV relationship of an excised lung first inflated slowly from a minimal volume to maximal volume and then deflated slowly. The transpulmonary pressure is due to the tensile forces in the elastic meshwork including, in the air-filled lung, the force due to air-fluid surface tension. The hysteresis of the pressure-volume relationship is largely due to the area-dependent behavior of pulmonary

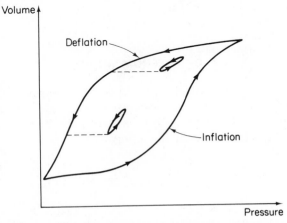

Fig. 4: A sketch indicating the hysteresis in the pressure vs. volume relationship of excised lungs.

surfactant. Some hysteresis, particularly during inflation from low lung volumes, is probably due to high opening pressures of collapsed units. The *PV* relationship is consistently located within the boundaries of the loop, and depends primarily on the volume history and to a lesser degree on the time. The *PV* relationships during normal breathing are represented as the small loops in Fig. 4. (The openness of the loop in this case is primarily attributable to the airway resistance, the $\dot{V}$ term in Eq. 1.)

The contribution of tissue components can be isolated by substituting a physiological solution (saline) for the air so that surface tension is eliminated. A typical *PV* curve for a saline filled lung, is sketched in Fig. 5 and compared with that of an air-filled lung. Note that at a maximum practical volume the pressure in the saline filled lung is approximately 12 cm H_2O, and that in the lung filled with air to the same volume the pressure is approximately 30 cm H_2O. Note also that most of the hysteresis was due to surface phenomena; there is at most a 10–15% difference in volume at a given pressure in the curve for the saline-filled lung.

The limitation of expansion at about 30 cm H_2O has an obvious protective function: that the lung tissue not be disrupted with a deep inspiration. It has long been recognized that this limitation implied a stiffening on the part of the stretched elements of the elastic meshwork, specifically the tensing of collagenous components. Setnikar (1955) proposed that the combination of collagen and elastin in the lung resulted in a length-force relationship of the following form (as later modified by Mead, 1961)

$$(3) \qquad \frac{dl}{dF} = B\frac{l_{mx} - l}{l_{mx} - l_0}$$

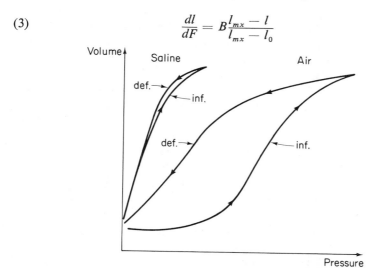

Fig. 5: A sketch compares the pressure vs. volume relationships of excised lungs when filled with air (right) and filled with saline (left). Note the significant difference in hysteresis in these two cases.

where l_0 represents the unstretched length, l_{mx} the maximal length, l the length, F the force, and B a constant representing the linear compliance due to elastin. This relationship, when applied in three dimensions, generates a curve qualitatively similar to the classical volume-pressure curves (Mead, 1961).

Fukaya (1968) has measured the force–length relationship of individual alveolar walls in uniaxial extension. Radford (1957) uniaxially stretched whole lung tissue in a saline bath and found maximum uniaxial extension ratios in the neighborhood of 1.7–1.8. It should be noted that the experimentally determined elongations are generally referred to as the "rest length" of the tissues. Our experience has indicated that "rest length" is highly variable. It should also be noted, because the elements in the lung are anything but aligned, that uniaxial stretching of whole tissue does not represent the behavior of the elastic elements alone but also reflects an unusual degree of distortion of their geometric relationships.

IV. SOME CURRENT ANALYTICAL MODELS

A. INTRODUCTION

As pointed out earlier, the single degree of freedom model has provided a useful characterization of normal lung. However, such an approach is unsatisfactory for the prediction of the behavior of diseased lungs, or for dealing with inhomogeneous aspects of lungs (ventilation in emphysema, stress distribution around airways and blood vessels, lung stress distribution in the field of gravity, respiratory unit mechanics, and so on). In order to deal with these aspects it is necessary to think of the lung as a continuous structure consisting of alveolar walls and distended at its boundaries, rather than as an airway system connecting to the millions of alveoli each, in effect, independently distended by transpulmonary pressure. It is also necessary to define the level of problems recognizing the appropriate anatomy. First, at the microscopic level there is the basic respiratory unit consisting of the alveolar duct and its associated alveoli. Second, at a larger level, one can consider a region within a lobule or a lobe containing a number of basic respiratory units which may be modelled as a homogeneous network or a highly compressible solid continuum. Thirdly, there are larger levels such as a lobe or lobule bounded by pleura or limiting membrane, being passive structures with lubricated boundaries. Finally, there are problems at the level of the lung and the chestwall as a structural system to include the driving power of respiratory muscles, the hydraulics of the abdomen, etc. We will now address ourselves to the second level, that within the lobe or lobule.

B. The Spring Network Model

The first attempt to consider lung as an interconnected network is very recent. On an analytical basis, Mead et al (1970), predicted that the "stress" (tensile forces in the elastic elements distributed over the surface of a region within the lung) applied to a region of inhomogeneity will at least vary inversely with the surface area of that region. Physiologically, this mechanism acts to preserve homogeneity. They pointed out that, in addition, stress will vary due to local variations in tensile forces in the network. Such variations were studied in a model network of Hookean springs as the size of a central unit was varied (see Fig. 6). In an extension of this approach (in collaboration with Dr. Stanley Dawson) the integrated form of Eq. (3) has been used to represent the more realistic stiffening characteristics of the spring elements.

$$(4) \qquad F = -\frac{l_0}{B}\left(\frac{l_{mx}}{l_0} - 1\right)\ln\left(\frac{\frac{l_{mx}}{l_0} - \frac{l}{l_0}}{\frac{l_{mx}}{l_0} - 1}\right)$$

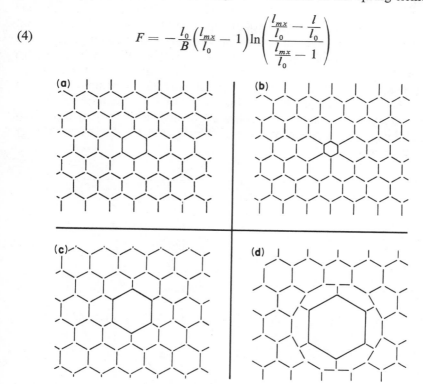

Fig. 6: Spring network model studied by Mead et al (1970). This figure shows some geometric changes of the network when the central hexagon is: (A) unchanged relative to other hexagons; (B) reduced; (C) enlarged; and (D) very much enlarged. Changes in forces in the elastic elements were determined from changes in length of the springs (after Mead et al, 1970, by permission of the authors and the Am. Physiol. Soc.).

Since displacements are finite, computer solutions are possible by a relaxation process. Preliminary results confirm those of Mead et al with Hookean springs, and indicate that stress varies nearly twice as much as predicted on the basis of change of the surface area of the region of inhomogeneity alone.

C. Continuum Formulation

In order to determine some fundamental properties, we are presently conducting experiments to obtain uniaxial and biaxial stress-strain relationships of lung tissue sheets in saline. (To date there is still no satisfactory procedure to handle testing of specimens of air-filled lung tissue.) We measure the three extension ratios of tissue slabs under uniaxial, equal, and unequal biaxial stretching. These two-dimensional experiments would provide the necessary information for a continuum formulation to predict the behavior of fluid-filled lungs. A typical experimental result of uniaxial and biaxial stretching of a saline-filled lung tissue slab obtained in this study is shown in Fig. 7. Note that in this figure the extensions are normalized with respect to the rest length and the forces normalized with respect to rest area at the undistorted degassed state. (When the air or fluid is completely drawn out of the airways and alveoli, the lung tissue volume is incompressible and, therefore, a constant value.) Predictions about air-filled lungs may be analogous, although measurement of some of the material constants may be very difficult to achieve.

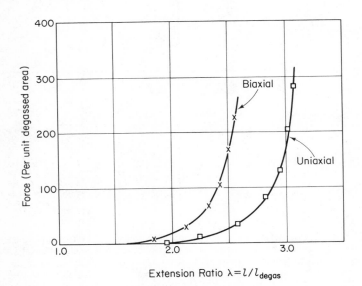

Fig. 7: Experimental force vs. elongation relationships of dog's lung tissue slabs in saline. For details see text.

Based on the above-mentioned two-dimensional experimental data on lung tissue sheets, it seems that a continuum formulation might be attempted for the "material" constitutive relationships of saline-filled, excised lungs. Let the three principal directions be 1, 2, and 3 (thickness) respectively. Then for compressible materials, the relationships between stresses and the strain energy density function, for pure homogeneous deformation, may be written as (Green and Zerna, 1968):

(5a) $$s_{11} = 2\lambda_1^2 \frac{\partial W}{\partial I_1} + 2\lambda_1^2(\lambda_2^2 + \lambda_3^2)\frac{\partial W}{\partial I_2} + 2I_3 \frac{\partial W}{\partial I_3}$$

(5b) $$s_{22} = 2\lambda_2^2 \frac{\partial W}{\partial I_1} + 2\lambda_2^2(\lambda_1^2 + \lambda_3^2)\frac{\partial W}{\partial I_2} + 2I_3 \frac{\partial W}{\partial I_3}$$

(5c) $$s_{33} = 2\lambda_3^2 \frac{\partial W}{\partial I_1} + 2\lambda_3^2(\lambda_1^2 + \lambda_2^2)\frac{\partial W}{\partial I_2} + 2I_3 \frac{\partial W}{\partial I_3}$$

$$s_{ij} = 0, \; i \neq j$$

where

$s =$ the stress per unit area of undeformed body
$s_{ij} = \sigma_{ij}\sqrt{I_3}$, σ is based on deformed body
$\lambda =$ the extension ratio
$W =$ the strain energy function per unit volume of the deformed body
$I_i =$ the strain invariants.

(6a) $$I_1 = \lambda_1^2 + \lambda_2^2 + \lambda_3^2$$
(6b) $$I_2 = \lambda_1^2\lambda_2^2 + \lambda_2^2\lambda_3^2 + \lambda_3^2\lambda_1^2$$
(6c) $$I_3 = \lambda_1^2\lambda_2^2\lambda_3^2$$

From Eq. (5c) when $s_{33} = 0$,

(7a) $$\frac{\partial W}{\partial I_3} = -\frac{1}{I_3}\left[\lambda_3^2 \frac{\partial W}{\partial I_1} + \lambda_3^2(\lambda_1^2 + \lambda_2^2)\frac{\partial W}{\partial I_2}\right]$$

Upon substitution,

(7b) $$\frac{\partial W}{\partial I_1} = \frac{1}{2(\lambda_2^2 - \lambda_1^2)}\left[\frac{-\lambda_1^2 s_{11}}{(\lambda_1^2 - \lambda_3^2)} + \frac{\lambda_2^2 s_{22}}{2(\lambda_2^2 - \lambda_3^2)}\right]$$

(7c) $$\frac{\partial W}{\partial I_2} = \frac{1}{2(\lambda_2^2 - \lambda_1^2)}\left[\frac{s_{11}}{(\lambda_1^2 - \lambda_3^2)} - \frac{s_{22}}{(\lambda_2^2 - \lambda_3^2)}\right]$$

Since the experimental data provide sets of information on s_{11}, s_{22}, ($s_{33} = 0$), λ_1, λ_2 and λ_3 Eqs. (6) and (7) provide for the direct determination of the strain energy density function derivatives and hence the strain energy function W.

Such an attempt must overcome several difficulties: (1) the adequate functional form of $W(I_1, I_2, I_3)$ to be assumed regarding whether and how to include product terms of the strain invariants and what order of strain invariants we must assume and (2) the appropriateness of using Eq. (6) since it only governs pure homogeneous deformation (Green and Zerna, 1968).

D. Discrete Element Approaches

The continuum approach described above, when successfully accomplished, can provide correlation with experimental measurements and furnish solutions to certain simple questions of inhomogeneity such as axial symmetric cases. Furthermore, any numerical or approximate solution procedures developed may be compared with the continuum solution to gain more confidence. However, in order to seek answers to a variety of questions in an inhomogeneous lung, the complexity of solutions suggests that discrete element procedures may prove to be realistic and useful. One of such an example is the spring network model presented earlier.

Currently, a finite element procedure has been adopted by John West and Frank Matthews (personal communication) for the modeling of the entire lung in the field of gravity. This effort is primarily aimed at obtaining information on the distributions of stress and strain in the lung and at the boundary (pleural surface). The discrete element procedure used in this study was proposed by Argyris (1965). In this approach, the element stiffness for tetrahedra was obtained by assuming the linear strain-displacement relationship. The large displacements were accomplished by the introduction of a geometric matrix, modifying the element stiffness matrix, to account for the effects of change of geometry on the equilibrium conditions. The nonlinear loading history, similar to that defined by Eq. (4), was handled by a step-by-step linearization incremental process. In this modeling, when the element (tetrahedron) is assumed to consist of linear elements, only two material constants are required: the modulus of elasticity and the Poisson ratio.

Using finite element procedures, we are currently developing experimental techniques to measure the coefficients for the element stiffness matrices by using both saline-filled and air-filled lung parenchyma pieces. In the meantime, calculation of the element stiffness matrices for finite deformations are being made by selecting appropriate geometries of meshwork of one-dimensional elements and two dimensional membrane elements based on a procedure similar to that proposed by Oden (1967); Oden and Sato, (1967).

At the microscopic level, a discrete model has been formulated by Jack Hildebrandt (personal communication). First, measurements of the force–elongation relationships of alveolar walls were made. A typical result is

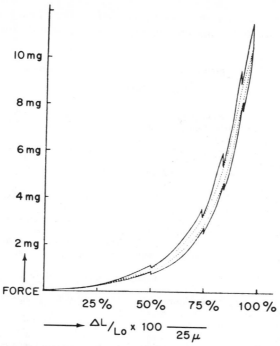

Fig. 8: Experimental force vs. elongation relationship of alveolar wall of cat's lung. The zig-zag curve represents the time-dependent study. The dotted curve represents the "static" situation in which very little hysteresis is present. The specimens were approximately $30 \times 30 \times 200\,\mu$. The rest length, L_0, is not defined clearly in the paper. (After Fukaya et al by permission of the authors and the Am. Physiol. Soc.)

given in Fig. 8 (Fukaya, et al, 1968). The response of a structure model consisting of alveolar walls and the alveolar duct (respiratory unit) has been examined.

V. SUMMARY

Obviously, there is need for development of techniques and theory, for measurement and description in every aspect mentioned above, and for correlation of solid and structural mechanics with other biomechanical, physiological, and pathological aspects of the lung. We hope that the material contained in this chapter may serve to clarify and to emphasize such a need.

Acknowledgement. *The authors wish to express their sincere appreciation to Dr. Jere Mead for his encouragement and valuable advice in our research as* well as in the preparation of this chapter, and to Dr. Stanley Dawson, for many useful ideas underlying the authors' approach described in this manuscript.

REFERENCES

Argris, J. H. 1965. Three-dimensional anisotropic and inhomogeneous elastic media matrix analysis for small and large displacements. Ingenieur-Archiv. 34: 33–55.

Ayer, J. P. 1964. Elastic tissue. Int. Rev. Conn. Tiss. Res. 2: 33–100.

Burton, A. C. 1954. Relation of structure to function of the tissues of the wall of blood vessels. Physiol. Rev. 34 (4): 619–42.

Carton, R. W., J. Dainauskas and J. W. Clark 1962. Elastic properties of single elastic fibers. J. App. Physiol. 17 (3): 547–51.

Clements, J. A. and D. F. Tierney 1964. Alveolar instability associated with altered surface tension. In Handbook of Physiology. Sec. 3, pp. 1565–583.

Colebatch, H. J. H. and D. F. J. Halmagyi 1963. Effect of vagotomy and vagal stimulation on lung mechanics and circulation. J. Appl Physiol. 18 (5): 881–87.

Cook, C. D., J. Mead, G. L. Schreiner, N. R. Frank and J. M. Craig 1959. Pulmonary mechanics during induced pulmonary edema in anesthetized dogs. J. Appl. Physiol. 14 (2): 177–86.

Elden, H. R. 1968. Physical properties of collagen fibers. Int. Rev. Conn. Tiss. Res. 4: 283–348.

Fukaya, H., C. J. Martin, A. C. Young and S. Katsura 1968. Mechanical properties of alveolar walls. J. Appl. Physiol. 25 (6): 689–95.

Glazier, J. B., Hughes, J. M. B., Maloney, J. E. and West, J. B. 1967. Vertical gradient of alveolar size in lungs of dogs frozen intact. J. Appl Physiol. 23 (5): 694–705.

Gratz, C. M. 1931. Tensile strength and elasticity tests on human fascia lata. J. Bone and Joint Surg. 13: 334.

Graves, D. J. 1967. A study of lecithin and the lung surfactant. Ph. D. Thesis, Dept. of Chem. Eng., Mass. Inst. Tech.

Green, A. E. and Zerna, W. 1968. Theoretical Elasticity. 2nd ed. Oxford at the Clarendon Press.

Guyton, A. C. 1964. Function of the Human Body. W. D. Saunders and Company, Philadelphia.

Handbook of Physiology 1964. Section 3, Respiration, Amer. Physiol. Soc. Wash., D. C. Vols. I, II.

Hyatt, R. E. 1968. Bronchial Mechanics. 9th Aspen Emphysema Conference. pp. 239–55 (PHS Publication 1717).

Hogg, J. and S. Nepszy 1969. Regional lung volume and pleural pressure gradient estimated from lung density in dogs. J. Appl. Physiol. 27: 198–203.

Klingele, T. G. and N. C. Staub 1970. Alveolar shape changes with volume in isolated air filled lobes of cat lung. J. Appl. Physiol. 28: 411–14.

Mead, J. 1961. Mechanical properties of lungs. Physiol. Rev. 41, 2: 281–330.

Mead, J. and Milic-Emili, J. 1964. Theory and methodology in respiratory mechanics with glossary of symbols. In Handbook of Physiology. Amer. Physiol. Soc. Wash., D. C. pp. 363–76.

Mead, J., T. Takishima and D. Leith 1970. Stress distribution in lungs: A model of pulmonary elasticity. J. Appl. Physiol. 28 (5): 596–608.

Nadel, J. A., H. J. H. Colebatch and C. R. Olsen 1964. Location and mechanism of airway constriction after barium sulfate microembolism. J. Appl. Physiol. 19 (3): 387–94.

Oden, J. T. 1967. Numerical formulation of nonlinear elasticity problems. J. Struct. Div. Amer. Soc. Civil Engrs. Vol. 93, No. ST3, pp. 235–55.

Oden, J. T. and T. Sato 1967. Finite strains and displacements of elastic membranes by the finite element method. Internat. J. Solids and Structures. 1: 471–88.

Piez, K. A. 1968. Crosslinking of collagen and elastin. Ann. Rev. Biochem. 37: 547–70.
Radford, E. P., Jr. 1957. Recent studies of mechanical properties of mammalian lungs. *In* Tissue Elasticity. Amer. Physiol. Soc. Wash., D. C.
Radford, E. P., Jr. 1964. Static mechanical properties of mammalian lungs. *In* Handbook of Physiology. Amer. Physiol. Soc. Wash., D. C. pp. 429–49.
Setnikar, I. 1955. Meccanica respiratoria, Estratto dal vol. N. 3 degli Aggiornamenti di fisiologia, Dott. Firenze, p. 4.
von Basch, S. 1887. Ueber eine function des carilladruckes in den lungen alveolon. Wien Med. Blatter 10: 465–67. [Quoted by Radford.]
von Hayek, H. 1960. The Human Lung. [Trans. from German by V. E. Krahl.] Hafner Pub., New York.
Wright, R. R. 1961. Elastic tissue of emphysematous and normal lungs. Am. J. Path. 39: 355.

15

Toward A Nontraumatic Study of the Circulatory System

MAX ANLIKER

The need for the development of transcutaneous or nontraumatic diagnostic methods to determine changes in the essential physiological parameters has been reemphasized by manned space flight. Man's response to new environmental conditions, of which weightlessness is only one example, cannot be predicted unless we have a thorough quantitative understanding of the physiological functions affected. The detection of gradual changes produced by environmental factors or by disease or control mechanisms calls for accurate and reliable means of quantifying physiological parameters, preferably on a noninvasive basis. It is shown that with the help of mathematical models we can utilize phenomena of sound generation and propagation to determine changes in the distensibility and hemodynamic properties of the blood vessels. The interpretation of wave transmission data is critically reviewed in terms of the mathematical models postulated. Specific attention is given to nonlinear behavior and to the overall modeling of the systemic circulation.

A series of experiments is described which demonstrates the limitations of certain mathematical models commonly used in cardiovascular dynamics. The dispersion and attenuation of distension, torsion and axial waves with frequencies between 20 and 200 Hz indicate that the blood vessel wall is viscoelastic and anisotropic. Also, its mechanical properties are shown to be strongly stress dependent and to vary with humoral neural stimuli.

If we postulate that no measurable active responses take place within a time span of about 0.1 second, the results obtained from the venae cavae of anesthetized dogs document that the stimulated vessel can be expected to be more distensible than the relaxed vessel. The diameters and effective Young's moduli for the circumferential direction may be as much as 50% higher for the relaxed vena cava than they are for the stimulated one.

To determine the overall distensibility of heart cavities and its relationship to the naturally occurring heart sounds, a vibrating piston was emplaced in the left ventricle of anesthetized dogs. By inducing intermittent sinusoidal volumetric displacements with the piston it was possible to generate artificial heart sounds whose amplitudes depend, in a direct way, on the geometry and stiffness of the ventricle. On the basis of a simple mathematical model for the mechanical behavior of the left ventricle, it is shown that the effective Young's modulus for the ventricular wall during the filling phase is of the order of 2.4×10^6 dynes/cm²; this parameter assumes a value of about $7 \times 10^6 – 15 \times 10^6$ dynes/cm² during systole.

Recent theoretical analyses of large amplitude pressure and flow pulses in major arterial conduits are reviewed and compared with experimental evidence. The results suggest that any meaningful theoretical analysis used as a basis of interpretation of the causes for small changes in the flow pulse must take nonlinearities into account as accurately as possible. Also, they indicate that the dicrotic wave is caused by reflections; however, the degree to which the dicrotic wave is due to the geometric taper and due to outflow has yet to be determined. Since the temporal and spatial profiles of the flow pulse can be obtained transcutaneously using pulsed ultrasound Doppler devices, the local hemodynamic phenomena may serve as diagnostic indicators.

* * *

I. INTRODUCTION

The capability of detecting relatively small changes in the essential physiological parameters plays a key role in the refinement of symptomatology and consequently in the advancement of preventive medicine. Early identification of manifestations of disease and reliable quantification of the gradual development of pathological patterns in the physiology of man would enable us to study many of the major illnesses as well as aging in a more systematic manner. For example, atherosclerosis can even today only be diagnosed after it has reached an advanced state, and since it evolves over a period of decades, we have some difficulties in delineating inhibitory exogenous or endogenous factors.

Aside from general medical considerations, we may have other motives to sharpen our diagnostic tools, such as the need to know man's response to new environmental conditions. We are continually changing man's environment on a massive scale without actually knowing how to fully assess the effects of these changes on his health. From a more specific point of view, the need for accurate and reliable methods of measuring changes of certain physiological functions and parameters appears to be especially acute when we are addressing ourselves to the biomedical problems of manned space flight (NASA, 1966; Nat. Acad. Sci., 1966, 1968; Anliker and Billingham, 1971). Subjecting men to unusual conditions such as high accelerations and prolonged weightlessness, we have to anticipate potential hazards and devise preventive measures and the means to determine the effectiveness of these measures. Yet, to accomplish this successfully, we must either have sufficiently accurate mathematical models of the physiological phenomena affected, which allow us to make reliable predictions as to the various possible response patterns of individual astronauts, or have an extensive backlog of empirical data from which we can infer the effects of prolonged exposure to the space environment.

An additional reason to pursue the development of improved techniques of measuring gradual changes in the parameters defining vital physiological phenomena is the need for better data pertaining to the complex regulatory mechanisms of these phenomena. Generally speaking, our knowledge of the control mechanisms of physiological functions is quite fragmentary (Iberall, 1969), mainly because we are lacking precise methods of distinguishing between active and passive responses of the physiological systems and of quantifying changes in the controlling and controlled parameters.

Whether we are aiming to establish new and more precise diagnostic methods to monitor the gradual development of a disease and of environmental effects or whether we are primarily interested in studying control mechanisms, it is, in either case, desirable that we can measure the physiological parameters in a nontraumatic or noninvasive manner without significantly disturbing the subject. Much of our comprehension of human physiology is based on animal experiments in which the normal behavior and the control mechanisms may have been drastically altered by anesthesia and trauma. It is obvious that such alterations must be avoided if we are to succeed in reliably measuring small changes of certain physiological parameters and quantities and in relating these changes to pathogenic or environmental factors or to normal regulatory mechanisms.

While most of the arguments presented so far are not restricted to a specific physiological system, henceforth we shall limit our considerations to the circulatory system, which encompasses a wide spectrum of mechanical phenomena and, as such, is an especially interesting topic of biomechanics. Considering its life-sustaining role, we also should give the cardiovascular system a high priority from the medical point of view. Besides this, we know that its mechanical behavior is significantly affected by disease, aging and environmental factors like the alterations in force fields encountered during high accelerations and weightlessness. Moreover, we have only limited understanding of its intricate control mechanisms, which regulate the flow of blood to different parts of the body, and normally ascertain adequate perfusion of the vital organs in situations of stress by changing the regional distensibility, geometry and other hemodynamic properties of the vasculature (Grodins, 1959; Guyton, 1963; Lamberti et al, 1968; Beneken, 1965). Since we also must expect both general and local changes in these properties as a result of disease and environmental factors, the challenge of measuring them transcutaneously or noninvasively is compounded by the problem of identification of the cause or causes for the measured change.

The distensibility, geometry and other hemodynamic properties of blood vessels with a diameter larger than about 0.1 cm can be determined noninvasively if we resort to indirect approaches. By utilizing phenomena of sound generation and propagation, the availability of ultrasound transducers (Franklin et al, 1966; Rushmer et al, 1966; Arndt, 1969) and mathematical models which interrelate these phenomena to the local mechanical vessel properties, we should be capable of acquiring the desired information in a nontraumatic manner. Yet, it should be emphasized that the data obtained by such an indirect method is only as accurate as the performed measurements and the mathematical models used to describe the measured quantities in terms of the geometric and mechanical features of the vessels. Ideally, the measurements should not only be sufficiently accurate to allow for a reliable interpretation of the observed changes in the cardiovascular system, but should actually yield excess data with which we can verify the adequacy of the mathematical models. It seems, therefore, appropriate to review some of the mathematical models which have been postulated and to examine their limitations on the basis of strategically designed experiments.

II. THEORETICAL AND EXPERIMENTAL STUDIES OF WAVES IN BLOOD VESSELS

The mathematical modeling of the mechanical behavior of blood vessels has been a topic of interest for more than a century. Extensive reviews of earlier work have been included in many of the recent publications and are therefore omitted here. With very few exceptions the models postulated so far assume linearity for both the vessel wall behavior and the hemodynamics of the circulatory system. Also, they can in general be separated into two categories depending on whether they treat aspects of solid mechanics or fluid mechanics more rigorously. Besides this, most of these models have not been validated by experiments.

II-1. Waves in Cylindrical Vessels Filled with an Inviscid Fluid and Having Elastic or Viscoelastic Walls

The interrelationship between the distensibility of a blood vessel and the speed with which it transmits a pressure pulse has long been recognized and applied in studies of the circulatory system. In a first approximation the speed c of a pressure signal in a cylindrical vessel is

$$(1) \qquad c^2 = \frac{A}{\rho_f \frac{dA}{dp}} = \frac{Eh}{2\rho_f a}$$

where A is the cross-sectional area of the vessel, p the transmural pressure, ρ_f the density of the fluid, E the effective Young's modulus in the circumferential direction of the vessel wall, h the wall thickness of the vessel, and a its radius. This relation does not include the effects of blood viscosity and compressibility, vessel wall mass and initial stresses in the wall. Also, it is based on the assumption that the vessel is linearly elastic, that the pressure signals have a small amplitude and that the wavelength of the perturbations is so large that the influence of the bending rigidity of the wall is negligible.

A refined treatment of the mechanical behavior of the vessel wall (Anliker and Maxwell, 1966; Maxwell and Anliker, 1968) which takes into account the effects of bending rigidity, wall mass, transmural pressure, initial axial stretch, viscoelasticity of the vessel wall and compressibility of the blood, confirms earlier linear analyses by Klip (1962, 1967) insofar as it indicates that arteries and veins should be capable of transmitting axial and torsion waves in addition to pressure waves. It also predicts that nonaxisymmetric waves are strongly dispersive (Fig. 1) and may exhibit cut-off frequencies, whereas axisymmetric waves are mildly dispersive. According to this model, we can expect the influence of the transmural pressure and initial axial stretch on the phase velocities and cut-off frequencies to be quite noticeable, even when the elastic properties of the wall remain unaltered by the wall stresses (Anliker and Maxwell, 1966). Moreover, it suggests that the effects of the compressibility of the blood are insignificant for waves with frequencies below 1000 Hz.

In unstressed vessels which are firmly tethered in the axial direction, that is, in vessels whose walls cannot move in the axial direction, the speed of axisymmetric pressure waves with frequencies approaching zero is

$$(2) \qquad c^2 = \frac{Eh}{2(1 - v^2)\rho_f a}$$

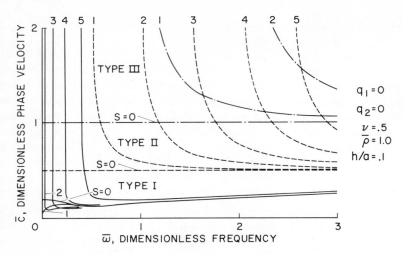

Fig. 1: Dispersion curves for waves in fluid-filled circular cylindrical shells with walls made of isotropic elastic materials. The solid lines represent pressure waves called here Type I. In these waves the radial displacement component dominates the corresponding wall displacement pattern at high frequencies. The dashed lines refer to torsion waves (Type II) in which the circumferential wall displacement is generally dominant. The dash-dot lines represent the dispersion curves for the axial waves (Type III) for which the axial wall displacement dominates the displacement pattern at the higher frequencies. $q_1 = q_2 = 0$ indicates that the axial stretch and transmural pressure have both been taken as zero in this case. The Poisson ratio v of the wall material is 0.5 and hence the vessel wall has been assumed as incompressible. The numbers attached to the curves refer to the circumferential wave number s which indicates the number of sinusoidal lobes in the displacement pattern for the circumferential direction. $\bar{\rho} = \rho_f / \rho_w$ is the ratio of the fluid density to wall density and h/a denotes the wall thickness-to-radius ratio.

if $\rho_w h / \rho_f a \ll 1$. ρ_w is the density of the wall tissue and v is the Poisson ratio. (Since the speed of sound in soft tissue is about the same as that in water, we have $v \approx 0.5$.) When we consider exclusively waves of extremely low frequencies and untethered vessels without initial stresses, this model corroborates relation (1) for the pressure waves and predicts

$$(3) \qquad c^2 = \frac{E}{\rho_w(1 - v^2)}$$

for axial waves and

$$(4) \qquad c^2 = \frac{G}{\rho_w}$$

for torsion waves, where G is the shear modulus of the wall material. We conclude from this that neither axisymmetric axial nor torsion waves should exhibit dispersion at low frequencies if the vessel wall is elastic or only slightly viscoelastic and if the blood is inviscid.

Extensive wave transmission experiments have been carried out by Klip et al, (1967) on latex rubber tubes. While their data provided for general confirmation of this theory of the transmission characteristics of the three basic types of waves in elastic fluid-filled tubes, we cannot infer from their work that blood vessels must *ipso facto* exhibit the same properties and that the theory can therefore also be applied to them. In view of our goals, the adequacy and limitations of mathematical models postulated for the mechanical behavior of live blood vessels can only be examined by studying the mechanical phenomena predicted by this model on blood vessels under *in vivo* conditions.

If we would restrict ourselves to arteries we could, in principle, utilize the pressure and flow pulses generated by the heart and attempt to compare their propagation characteristics with the theoretically anticipated ones. However, the diameter changes induced by the natural pulse are associated with strains of the order of 5% and, as such, exceed the usual limitations one must impose for a model which is based on a linear elastic behavior. Also, at wave speeds of the order 500 cm/sec the wavelength corresponding to the frequency of the natural pulse (1 Hz) would be 500 cm, which implies that within each pulse period multiple reflections are taking place at the terminal and at the beginning of any arterial conduit. These reflections compound the difficulties encountered in determining the phase velocities and attenuation for various arteries.

It seems, therefore, much more convenient to study the mechanical properties of blood vessels and the changes in these properties in terms of the propagation characteristics of waves with higher frequencies. This can be done by superimposing, on the naturally occurring pressure fluctuations, artificially induced transient signals in the form of finite trains of sine waves (Anliker et al, 1968a). For sufficiently small amplitudes the blood vessels should exhibit a linear behavior with respect to these trains of sine waves. Also, if these signals are of sufficiently short duration or if the waves are strongly attenuated, it is possible to minimize reflection interference. For high frequencies and short trains the waves can be recorded before their reflections at branching points and the terminal points of the conduit arrive at the recording sites. With strong attenuation the reflections of a wave of small amplitude can be damped below recognition before they reach the transducer locations, provided, of course, that the transducers are not in the close proximity of a major reflection site.

By selecting transient signals in the form of finite trains of sine waves we can also eliminate the need for Fourier transform computation when the medium (vessel) is mildly dispersive, that is, when the phase velocity varies only a few percent whenever we change the frequency by 10%. This can be inferred from the Fourier spectrum of such signals which is dominated by the frequency of the sine waves in increasing proportion to the length of the trains. Consequently, we can interpret the speed of the transient signal as a good approximation to the phase velocity corresponding to the frequency of the sine waves.

Guided by such considerations, we investigated the propagation of small sinusoidal pressure waves with frequencies ranging from 20–200 Hz in the aortae of anesthetized dogs (Anliker et al, 1968a). Figure 2 shows the initial experimental setup, and Fig. 3 illustrates representative tracings of recordings of the natural pulse wave modulated by artificially induced sinusoidal pressure signals. Even for pressure modulations as large as those shown in Fig. 3 we can only discern an attenuation but no distortion of the sinusoidal waves with propagation. This indicates that there are no noticeable effects of reflection and that the aorta is only mildly dispersive for the frequency range considered in this study. An accurate determination of the wave speed as a function of frequency for sine wave trains induced during diastole and at a given transmural pressure confirms the nondispersiveness of such signals.

SCHEMATIC OF EXPERIMENTAL ARRANGEMENT

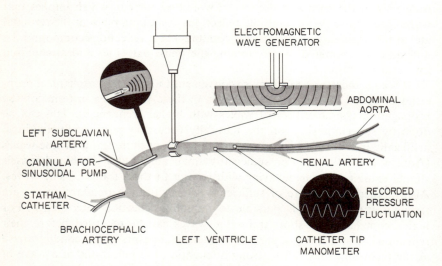

Fig. 2: Experimental setup for generating small sinusoidal pressure waves in the canine aorta by a sinusoidal pump or an electromagnetic wave generator. The transient signals are sensed by wind tunnel pressure gages which were adapted for use as catheter-tip manometers in blood vessels. The displacements of the vessel wall caused by the electrically driven impactor have axisymmetrical as well as nonaxisymmetrical components. The corresponding pressure fluctuations indeed exhibit components of both kinds but only within close proximity of the wave generator. The nonaxisymmetrical waves are completely attenuated within a distance of about 10 vessel radii (5–6 cm) from the impactor.

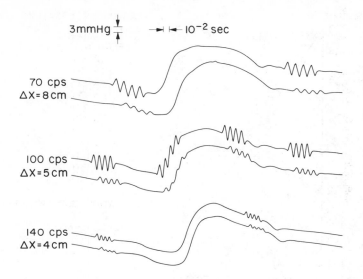

Fig. 3: Representative tracings of recordings of the natural pulse wave in the thoracic aorta of an anesthetized dog, with artificially superimposed trains of sinusoidal waves. The transient signals can be induced at any time during the cardiac cycle. Note that the sine waves are highly damped but retain their sinusoidal character during propagation. The pressure curves are drawn to the same scale but each of the curves has a different zero point for illustration purposes.

If their amplitude is plotted as a function of the distance of propagation measured in terms of wavelengths (Fig. 4), one finds the same exponential decay pattern for all frequencies as shown

(5) $$A = A_o e^{-k(\Delta x/\lambda)}$$

where A_o is the amplitude at the proximal transducer and A that at the distal one. k is the logarithmic decrement, Δx the distance between the transducers, and λ the wavelength. For waves propagating in the downstream direction (away from the heart) the values for k seem to be of the order 0.7–1.0.

As illustrated in Fig. 3, the transient signals can be induced at any instant of the cardiac cycle and can therefore be used to examine the effects of pressure and flow on the speed and attenuation of the sine waves. In the experimental arrangement devised for this investigation, the transmission time of the signal between the transducers is usually of the order of 10 milliseconds. The propagation characteristics of the sine waves should therefore yield quasi-instantaneous properties of the aorta. This aspect is especially important for control system studies since they require the quantification of the changes in the instantaneous mechanical behavior of the blood vessels. From

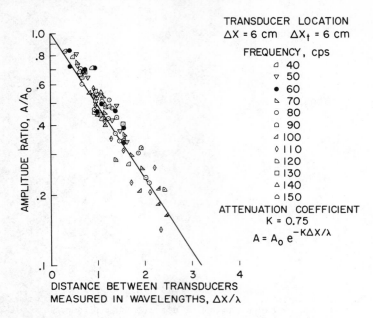

Fig. 4: Typical attenuation of sinusoidal pressure waves with A/A_o plotted on a logarithmic scale. The slope of the line approximating the variation of $\ln(A/A_o)$ with $\Delta x/\lambda$ defines the logarithmic decrement k.

the graph in Fig. 5 which gives the typical variation of the signal speed with the instantaneous pressure, we note that the wave speed can increase by as much as 50% between diastole and systole. The scatter of the data points obtained during systole exceeds by far the experimental error of the measurement and can be attributed to the fact that rapid changes in mean flow occur during each cardiac cycle which are not in phase with the pressure changes (Wetterer and Kenner, 1968). The mean flow at a given pressure during early systole is quite different from that at the same pressure during the latter part of systole. Also, the naturally occurring fluctuations in cardiac output can contribute to the scatter by causing normal variations in the mean flow. However, irrespective of the scatter it is clearly evident from Fig. 5 that the wave speed increases with pressure. This implies a stiffening of the aortic wall with pressure, that is, an increase in the effective Young's modulus with stress. According to equation (1) we would actually expect a decrease in the wave speed with pressure if E does not change since a rising pressure causes a distension of the aorta and thus a reduction in the h/a ratio.

Theoretical studies show that waves in elastic tubes containing a streaming fluid are convected by the flow onto which the waves are superimposed (Morgan and Ferrante, 1955; Wells, 1969). If $\bar{U}$ denotes the steady state flow velocity, and if $\bar{U}$ is small compared to the wave speed c one finds for

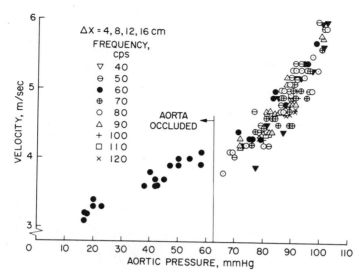

Fig. 5: Speeds of sinusoidal pressure waves of various frequencies shown as a function of the instantaneous aortic pressure. In this case the diastolic pressure varied between 65–85 mm Hg while the systolic pressure was between 85–100 mm Hg. Data points corresponding to pressures below the diastolic were obtained by occluding the aorta for about 10 seconds. The steeper rise of the wave speed with pressure between diastole and systole is attributed to a stiffening of the aorta with pressure and to an increase in mean flow.

zero flow, then the speeds of downstream and upstream moving pressure waves of small amplitude are respectively

(6) $$c^D = c + \bar{U} \quad \text{and} \quad c^U = c - \bar{U}$$

This convection phenomenon was studied by Müller (1950) on rubber tubes, who found that the change in wave speed was only about $\pm 0.83\ \bar{U}$. To examine the phenomenon on blood vessels and to separate the increase in wave speed with pressure from that due to flow, a number of dual impactor experiments were performed as schematically illustrated in Fig. 6 (Histand, 1969). Representative recordings of waves traveling upstream and downstream are shown in Fig. 7. With equations (6) we can evaluate the wave speed c at a given pressure and the instantaneous mean flow velocity from measurements of c^D and c^U. Typical results obtained from data taken over several cardiac cycles are given in Fig. 8, which shows the temporal variation of the mean flow velocity $\bar{U}$ determined from c^D and c^U. This flow pattern is in good agreement with recordings made with a Pieper flow gage (Pieper and Paul, 1968) and thus demonstrates the convection phenomenon on blood vessels.

LOCATION OF EXPERIMENTAL APPARATUS

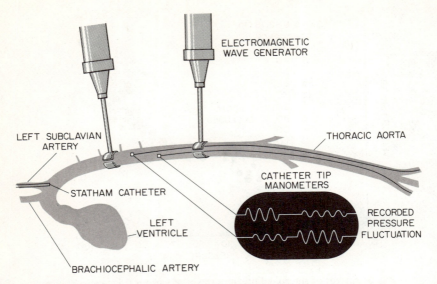

Fig. 6: Arrangement of the experimental apparatus used in dual impactor experiments. Two electromagnetic impactors produced upstream and downstream waves in a single aortic segment. The number of sine waves in each train is different for upstream and downstream waves.

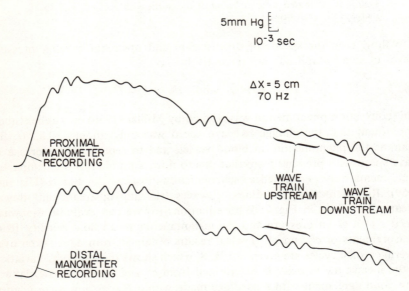

Fig. 7: Typical recordings from the two pressure manometers between the two impactors. Trains of four sine waves are propagating upstream and trains of three sine waves are traveling downstream.

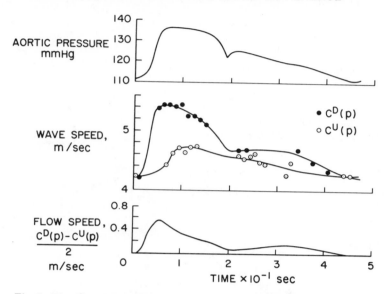

Fig. 8: Flow pattern determined from wave speed measurements. Top: Natural pulse wave defining cardiac phase at which the data shown in lower two graphs were obtained. Middle: Upstream wave speeds (open symbols) and downstream wave speeds (solid symbols) measured at different instants of the cardiac cycle. The upstream and downstream data correspond to two heart beats a few seconds apart but with matching pressure patterns. Bottom: Mean flow velocity $\bar{U}$.

Of particular significance is the dependence of the wave speed on the aortic pressure which is illustrated in Fig. 9 for high and low pressures. As will be seen later, the values for dc/dp are sufficiently large to produce strong nonlinear phenomena in the propagation of the natural pressure pulse.

A demonstration of axial and torsion waves in the aortae of anesthetized dogs is extremely difficult in view of the numerous branches and the inaccessibility of the vessel. By contrast, the carotid artery of a 20–30 kg dog has a 15–20 cm long segment which is free of branches and quite accessible, and therefore, almost ideally suited to confirm the existence of all three types of waves. The experimental arrangement with which these waves were studied in the carotid is illustrated in Fig. 10 (Anliker et al, 1968b; Moritz, 1969). The technique and apparatus used permitted the simultaneous generation and recording of axial and pressure waves of one frequency, and torsion waves of the same or another frequency. Each of the three waves was sensed in terms of its dominant feature. This allowed for the identification of the individual waves without having to resort to extensive numerical analysis of the signal recordings. Speed and attenuation data for all three

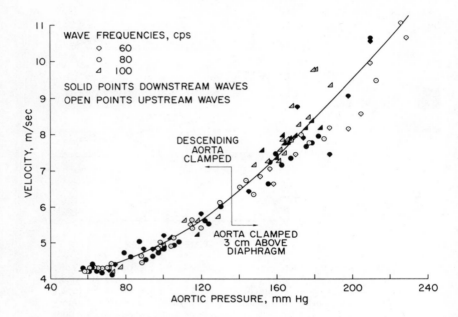

Fig. 9: Wave speed–pressure variation for a segment of the thoracic aorta in an anesthetized dog. The pressure was varied beyond its normal range by occluding the aorta above and below the segment of interest. The solid points represent waves recorded as they propagate in the downstream direction (away from the heart), while the open points were obtained from waves traveling in the upstream direction (towards the heart). To record simultaneously up- and downstream waves, two electromagnetic signal generators were used, one being placed below and one above the aortic segment studied. The sharp rise of the wave speed with pressure suggests that the peak of a large-amplitude pressure pulse travels at a higher speed than does the foot of the pulse. Accordingly such pulses change their shape with propagation.

types of waves obtained from the exposed carotid arteries of six anesthetized dogs are given in averaged form in Figs. 11 and 12 (Moritz 1969). They show that artificially induced pressure waves with frequencies ranging from 20–100 Hz are nondispersive and have a speed of about 11 m/sec at diastolic pressures between 140–160 mm Hg. Their logarithmic decrement is about 1.0. The torsion and axial waves are mildly dispersive with speeds increasing over the same frequency range from about 15–24 m/sec and 25–35 m/sec respectively. The corresponding logarithmic decrements are of the order 4–5. To allow for a comparison of the experimental wave transmission data with

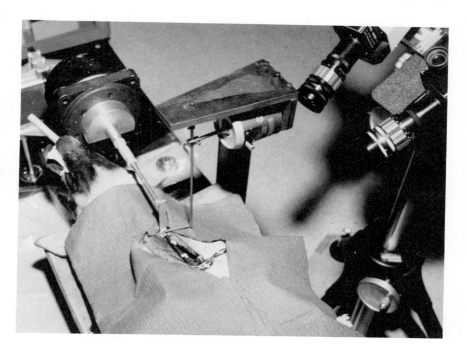

Fig. 10: Experimental arrangement to determine the dispersion and attenuation of axial, torsion and pressure waves in the exposed external carotid artery of an anesthetized dog. A collar device at the cephalic end of the exposed carotid can be displaced sinusoidally in the axial direction and simultaneously rotated about the axis of the carotid. The small constriction of the vessel caused by the collar device induces not only axial but also pressure waves as it is displaced in the axial direction. Axial and torsion waves are monitored with a pair of PhysiTech electro-optical trackers which detect the axial and circumferential displacements of the wall at two locations along the artery. Since the optical tracking system is designed to measure the motion of a line of contrast in two directions perpendicular to the optical axis, special paper targets were attached to the artery. The pressure waves are sensed by miniature catheter-tip transducers located at the sites of the paper targets.

theory, the radius to wall thickness ratio and initial axial stretch were measured in some thirty dogs on which waves of at least two kinds were studied. According to these measurements the average a/h ratio was 8.5 and the average *in vivo* strain for the normal head position of the supine dog was 40%. With these values and the measured pressure wave speeds, the theoretical results based on isotropic wall properties (Anliker and Maxwell, 1966) suggest that the observed torsion and axial wave speeds are only about half as fast as anticipated. This deviation from theory appears to indicate an anisotropic wall behavior, with an effective circumferential Young's modulus

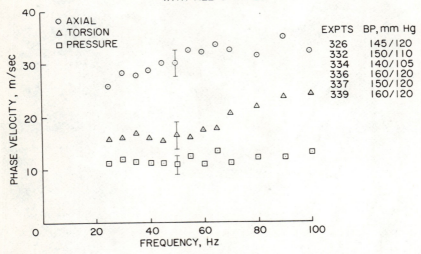

Fig. 11: Average dispersion pattern of artificially induced axial, torsion and pressure waves in the exposed carotid artery of six dogs in which all three types of waves were studied. Each data point represents the average of approximately 30 velocity determinations. Also shown are the 95% confidence intervals for each type of wave at 50 Hz. Similar intervals exist at higher and lower frequencies. The speeds of the axial and torsion waves are much lower than predicted by theoretical analysis which indicates anisotropic wall behavior. Note that the blood pressure is relatively high in all cases considered.

which is about 2–3 times higher than the axial one. The degree of anisotropy may depend on the axial stretch and the transmural pressure. As in the case of the aorta, the pressure wave speed increased markedly with transmural pressure. However, the axial and torsion wave remained essentially unchanged when the transmural pressure was altered by about ± 50 mm Hg. These findings are contrary to theoretical predictions and imply that a change in circumferential wall tension primarily affects the extensibility in that direction. The complexity of the wall behavior seems to be further compounded by the fact that a reduction in the axial stretch leads to a much greater decrease in the speeds of all three wave types than expected on the basis of the theory.

While numerous studies have been devoted to the mechanical behavior of arteries, we know very little about the wave transmission properties and distensibility of veins in a quantitative sense. This is not so surprising when we consider the difficulties in relating the wave speed to distensibility, wall

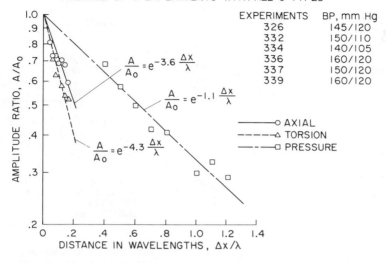

Fig. 12: Average attenuation ratio for all three types of waves. The cases documented here correspond to the dispersion data illustrated in Figure 11.

properties and geometry in the case of veins where geometric features are strongly affected by relatively small decreases in the transmural pressure which may be produced, for example, by normal muscular activity, respiration or changes in posture. Yet we expect the veins to play a more important role than arteries in regulating blood flow since the venous wall contains a higher percentage of smooth muscle than does the arterial wall. Also, in view of the high degree of flexibility of the veins, their contraction by smooth muscle activation should be much more pronounced than in the case of arteries.

To examine the validity of the various mathematical models of wave transmission in blood vessels for large veins, a series of experiments was performed in which the speed and attenuation of small and large pressure waves were measured in the abdominal vena cava of anesthetized dogs (Anliker et al, 1969; Yates, 1969). The corresponding experimental setup is schematically illustrated in Fig. 13. The recordings of large amplitude pressure waves indicated noticeable reflection phenomena and nonlinear effects such as the steepening of the wave front during propagation due to an increase in the wave speed with transmural pressure (Anliker et al, 1969). When small amplitude sine waves (signals with peak-to-peak pressure amplitudes less than 30 mm H_2O) were induced, the vena cava exhibited a linear mechanical behavior insofar as the wave transmission properties were not amplitude dependent. No reflection phenomena could be discerned for such waves,

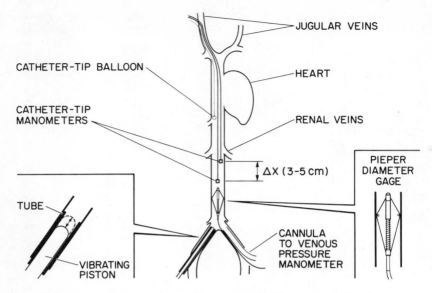

Fig. 13: General experimental preparation for the study of the transmission characteristics of small artificially induced pressure waves in the abdominal venae cavae of anesthetized dogs. Sinusoidal pressure signals are generated by the volumetric displacements of a vibrating piston in the right common iliac vein. To assess the changes of the mechanical behavior of the vessel with pressure, a catheter-tip balloon is positioned between the heart and the hepatic veins. Its inflation produces changes in venous pressure from control values of about 100 mm H_2O within 10 to 20 seconds. The internal diameter of the vena cava is monitored with a modified Pieper Diameter Gage.

and the dispersion was rather mild. (See Fig. 14.) However, when the waves were recorded over short distances (3–4 cm), their speeds changed with location, presumably due to local variations in the geometry and mechanical properties of the vessel as might be caused by branching structures or connective tissue. These local fluctuations may not be evident when the waves are monitored over longer distances, since the transmission time of a pressure signal between two points along a vessel is defined by the average of the mechanical properties of the vessel between the two points. As in the case of the aorta, the wave speed increased substantially with transmural pressure, suggesting also a stiffening of the vessel wall with increasing circumferential stress (Fig. 15). In addition, the waves were convected by the prevailing mean flow, which in some cases was found to reach velocity levels of the order 100 cm/sec during inspiration (Anliker et al, 1969). Evidence for such mean flow velocities was also obtained by other investigators using the Mills electromagnetic catheter-tip flow probe (Mills, 1966). Finally, from Fig. 16 we conclude that the attenuation pattern of the small pressure waves on the vena cava

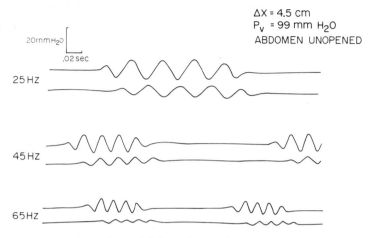

Fig. 14: Sample tracings of finite trains of sinusoidal pressure waves of different frequencies recorded in the abdominal vena cava of an anesthetized dog. The tracings have been shifted vertically for illustrative purposes. P_V is venous pressure and Δx is the distance between the catheter-tip manometers. Note that the peak-to-peak amplitude of the transient signals is generally less than 1 mm Hg (13.6 mm H_2O).

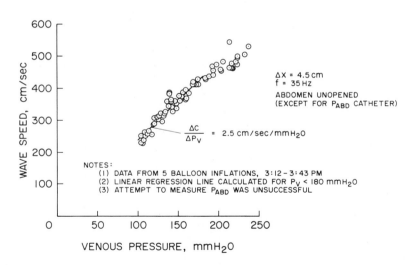

Fig. 15: Speed of pressure waves in the abdominal vena cava plotted as a function of intraluminal (venous) pressure. The increase in wave speed with venous pressure is typical and corresponds to that observed when the abdomen was opened and the viscera were displaced such as to expose the outside of the vena cava to atmospheric pressures.

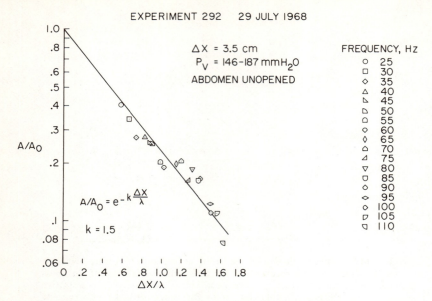

Fig. 16: Attenuation of pressure waves in the abdominal vena cava; abdomen unopened.

is the same as in the aorta but with a logarithmic decrement which can range from 0.6–3.3.

Since as yet no axial or torsion waves were studied on veins, we cannot assess the degree of anisotropy they may exhibit. However, the approximate determination of the effective Young's modulus for the circumferential direction of the vena cava from equation (1) indicates that for transmural pressures between 75–100 mm H_2O, a/h values between 30–45 and frequencies from 20–100 Hz, the values for E_{eff} range from about 5×10^5 to 5×10^6 dynes/cm². These values for E_{eff} are comparable with those of the thoracic aorta, yet the distensibility of the vena cava is much higher than that of the aorta because veins have radius to wall-thickness ratios which are almost an order of magnitude greater than those of arteries.

The experimental data described so far indicate that the various waves studied are strongly attenuated in an exponential pattern with logarithmic decrements which are essentially independent of frequency between 40–200 Hz. This damping can be attributed to three different mechanisms: viscosity of the blood, radiation of energy into the surrounding medium and viscoelasticity of the vessel wall. In the case of pressure waves in the aorta, carotid artery and vena cava, the removal of the surrounding medium did not alter noticeably the logarithmic decrement, which implies that the attenuation must primarily be due to wall viscoelasticity and blood viscosity, at least in the frequency range considered. The individual effects of wall viscoelasticity and blood viscosity on the transmission characteristics of waves in arteries and veins are examined in the next section.

II-2. Waves in Cylindrical Vessels Filled with a Viscous Fluid and Having Elastic or Viscoelastic Walls

A linearized analysis of axisymmetric waves in a cylindrical vessel filled with a viscous incompressible fluid and having a wall which is isotropically viscoelastic and subjected to elastic, independent radial and axial constraints has recently been completed as an extension of earlier studies (Womersley, 1957; Chow and Apter, 1968; Atabek and Lew, 1966). This analysis (Jones et al, 1971) also predicts axial waves which should travel faster than the pressure waves. The speeds of both waves are shown to be quite sensitive to the properties of the surrounding medium, particularly for high values of the nondimensional parameter $\alpha = a[(\omega/\eta)\rho_f]^{1/2}$ in which ω denotes the circular frequency of the wave and η the viscosity of the fluid (blood). For unconstrained elastic vessels and values of $\alpha < 10$ and $\alpha \to 0$, the speed of the pressure waves begins to deviate noticeably from that defined by equation (1) and approaches zero. Also, with $\alpha \to 0$ the attenuation due to the blood viscosity becomes increasingly pronounced, whereas for $\alpha > 10$ the attenuation appears to be dominated by the viscoelasticity of the wall. At normal heart rates ($\omega = 2\pi$ sec^{-1}) and blood viscosity ($\eta/\rho_f = 0.04$ cm^{-1} sec^{-1}) we should accordingly expect a lower pressure wave speed than predicted by equation (1) in unconstrained vessels with a radius less than 0.8 cm.

In the case of axial waves in the unconstrained elastic vessel and $\alpha < 2$ the wave speed is independent of α and given by

(7) $$c = 1.8\sqrt{\frac{Eh}{2\rho_f a}}, \quad \alpha < 2$$

For increasing $\alpha(\alpha > 2)$ the speed of axial waves approaches the limiting value

(8) $$c = \left[\frac{E}{(1-\nu^2)\rho_w}\right]^{1/2}, \quad \alpha \gg 2$$

in such vessels. Fluid viscosity causes only slight attenuation for these waves when $\alpha < 0.2$ since the velocity profile is very flat, and the wall moves axially with approximately the same velocity as the fluid. Maximum attenuation of axial waves by fluid viscosity occurs in the range $2 < \alpha < 6$. With increasing $\alpha(\alpha > 10)$ the attenuation decreases but is still larger than for pressure waves.

In the presence of radial constraints the pressure wave speed can be substantially higher than that predicted by equation (1). Yet, axial constraints do not seem to produce marked changes in the speed of such waves. Radial elastic constraints lead to a substantial increase in the attenuation of the waves due to the viscosity of the fluid. The corresponding increase induced by axial constraints is considerably smaller but still significant.

The speed of axial waves is progressively increased by strengthening either type of constraints. Interestingly, these waves are attenuated less when the radial constraint is enhanced. Conversely, an increasing axial constraint leads to a very rapid dissipation of the axial waves by fluid viscosity, particularly at large values of α.

Comparing the results of the analysis for elastic walls with the measured transmission characteristics of the high frequency pressure waves in the thoracic aorta and in the carotid artery of anesthetized dogs (Anliker et al, 1968a, 1968b), we find that the wave speed is indeed constant for large values of α; however, the observed attenuation exceeds by far that predicted on the basis of the blood viscosity. Also, the axial waves in the carotid exhibit essentially the dispersion pattern obtained from the theory, but in relation to the pressure wave speeds, the magnitude of the axial wave speeds is substantially less than anticipated. This discrepancy is most likely due to anisotropy in the vessel wall behavior, as noted earlier. Another point of agreement between theory and experiment is the independence of the axial and pressure waves for sufficiently small amplitudes.

In view of the fact that blood viscosity cannot account for the measured damping of both axial and pressure waves, it seems that the viscoelasticity of the vessel wall must be the dominant mode of dissipation. The mathematical model for the mechanical behavior of the blood vessels was therefore modified to include the effects of viscoelasticity by postulating that the wall material has a complex Young's modulus E whose imaginary part is small compared with the real part. For large values of α the predicted speeds of both pressure and axial waves are independent of the imaginary part of Young's modulus, which is compatible with experimental observations. Since the measured logarithmic decrements are independent of frequency between 20–200 Hz, we infer from the analysis that the ratio of the imaginary and real parts of E is independent of frequency in this range. Such a material behavior can not be simulated by the Voigt model applied in recent theoretical investigations (Maxwell and Anliker, 1968; Chow and Apter, 1968).

II-3. WAVES IN CURVED VESSELS

Blood vessels are not straight circular cylindrical vessels with uniform wall properties. Branches and arches together with a cross section that diminishes with increasing distance from the heart for any given arterial conduit constitute geometric complications whose effects on the transmission characteristics of waves should be examined. As part of an attempt to establish a mathematical model for the mechanical behavior of the semicircular canals and the phenomena involved in the sensation of angular motion, a study was made of waves propagating in a toroidal membranous canal filled with fluid. (Anliker and Dorfman, 1970; Anliker and Van Buskirk, 1970). The semicircular canal ducts considered in this study also serve as a model for a curved artery such as the aortic arch. By characterizing the waves in terms of average features of the fluid and wall displacements and linearizing the equations of motion which approximately interrelate the inertia, elastic and hydrodynamic forces of the system, the aortic arch is shown to transmit three basic types of waves. They are identified by their displacement patterns as extensional, breathing and flexural modes. The extensional mode corresponds to the axial wave in the straight tube, and the breathing mode is analogous to the pressure wave and as such exhibits a relatively strong pressure signal. The flexure mode corresponds to the $S = 1$ nonaxisymmetric pressure wave (Type I in Fig. 1) in a straight vessel. (Since a twisting defor-

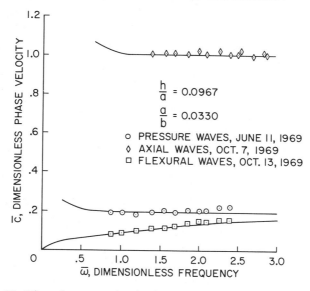

Fig. 17: Dispersion curves for the three types of waves in toroidal fluid-filled elastic shells simulating semicircular canals or the aortic arch. The experimental arrangement with which the data were obtained is shown in Figure 18. The dimensionless phase velocity $\bar{c}$ is defined as the true wave speed c divided by $c_p = \sqrt{E/\rho_w(1-\nu^2)}$ where E is Young's modulus of the wall material, ρ_w its density, and ν the Poisson ratio. The symbols a, h, and b denote the radius of the circular cross section of the toroidal shell, the wall thickness of the shell, and the ring radius of the torus, respectively. As nondimensional frequency, we have here $\bar{\omega} = b\omega/c_p$ with ω representing the circular frequency of the wave.

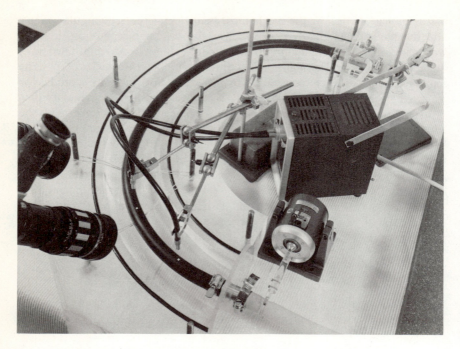

Fig. 18: Laboratory model and associated equipment assembled for the generation and monitoring of flexural waves. The membranous canal or aortic arch is simulated by a bicycle inner tube filled with fluid. The pressure fluctuations associated with the transient signals are measured by catheter-tip manometers placed inside the tube at the same stations where the targets are mounted for the determination of the wall displacements.

mation of the toroidal tube was suppressed in this analysis, no torsion mode could be predicted.) Typical dispersion curves of those waves are given in Fig. 17, together with experimental data obtained from the laboratory model shown in Fig. 18. The results suggest that, in a first approximation, the aortic arch transmits pressure and flexure waves like a straight artery.

II-4. Waves in Vessels with Elliptic Cross Sections

Another geometric complication may arise in conjunction with the shape of the vessel cross section. Arteries are generally well distended and have a near-circular cross section. This is ascertained by a wall thickness to radius ratio of the order of 1/3 to 1/10 and positive transmural pressures which, at least in the larger arteries, exceed 50 mm Hg. By contrast, the veins have considerably more flexible walls. Their wall thickness to radius ratio ranges from 1/20 to 1/100, and their transmural pressure often is near zero or even slightly negative. Accordingly, the cross section of veins is often of a near-elliptical shape and sometimes fully collapsed. An approximate analysis of the dispersion of pressure waves in cylindrical vessels with an elliptic cross section, however, shows that the wave speed remains practically unaffected by an ellipticity of the cross section as long as the eccentricity is less than 0.5 (Bailie, 1969).

III. ACTIVE CHANGES IN THE MECHANICAL BEHAVIOR OF BLOOD VESSELS

Reliable identification of the cause or causes for pathologic changes in the hemodynamic properties of blood vessels requires a thorough understanding of the naturally occurring changes, especially those induced by humoral and neural stimuli. If no measurable active responses can take place within about 50 milliseconds, we can infer from wave transmission phenomena the changes in the quasi-instantaneous mechanical properties of the circulatory system induced by the activation of the smooth muscle in the blood vessel walls. This was demonstrated for the canine vena cava with the experimental preparation schematically illustrated in Fig. 13, and by recording simultaneously its diameter and the speed of artificially induced pressure waves before, during and after application of a well-defined stimulus. To examine the effects of the stimuli over a wider range of wall stresses, the catheter-tip balloon placed in the inferior vena cava just caudal to the heart was rapidly inflated. This caused the venous pressure to rise from control values of about 100 mm H_2O to a level between 200–300 mm H_2O within 10–20 seconds. Neural and humoral stimuli are generally expected to produce a constriction of the vessel, that is, a change in the circumferential resting length. In addition, we anticipate a change in the effective circumferential Young's modulus for a given wall stress since, in the stimulated vessel, the smooth muscle presumably takes up a larger percentage of the hoop tension than in the relaxed vessel. Correspondingly, we conjecture that the elastin and collagen carry less of the hoop tension. Since it is rather difficult to determine the circumferential resting length of the vessel under *in vivo* conditions, we shall describe the circumferential elastic behavior of the wall in terms of an effective incremental Young's modulus and its variation with stress rather than strain. Knowing the transmural pressure P_T, the radius a of the vessel and its wall thickness h_0 at a reference radius a_0 and assuming incompressibility of the wall material and $h/a \ll 1$, we can give the wall stress as

$$\sigma = \frac{P_T a^2}{a_0 h_0} \tag{9}$$

Conventionally, the effective incremental Young's modulus has been determined from pressure-diameter curves (Alexander, 1962) such as that shown in Fig. 19 for the canine inferior vena cava. However, these curves do not represent the instantaneous elastic behavior of the vessel, since they were obtained over a period of 10–20 seconds during which active changes may have taken place altering the circumferential resting length. This uncertainty regarding possible active changes can be resolved by evaluating the effective modulus from the speed c of a sinusoidal pressure perturbation observed at a given radius a and transmural pressure P_T

$$(10) \qquad E = \frac{2\rho_f a c^2}{h} = \frac{2\rho_f a^2 c^2}{h_0 a_0}$$

In principle we also could determine E from the small diameter changes induced by the sinusoidal pressure perturbations, but the Pieper diameter gauge did not have a sufficiently high frequency response after it was modified for use in veins.

Extreme results as to P_T versus D and E versus σ curves for a stimulated and a relaxed vessel are given in Figs. 19–22. They show that the canine inferior vena cava can change its diameter by as much as 50% when the

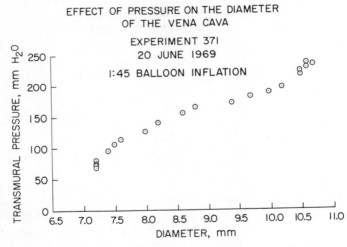

Fig. 19: Effect of transmural pressure on the diameter of the abdominal vena cava. The vessel is initially constricted or stimulated, and by inflating the catheter-tip balloon illustrated in Figure 13, the transmural pressure in the abdominal vena cava was raised from about 70–220 mm H_2O. This elevation in pressure causes the diameter to increase in typical sigmoid fashion from about 7.2–10.7 mm. We have thus a case where the effective strain is of the order of 50%. If active changes in the mechanical properties of the venous wall could be ruled out, this pressure-diameter curve would yield the incremental Young's modulus as a function of instantaneous stress.

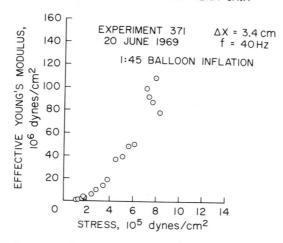

Fig. 20: Effect of transmural pressure on the effective Young's modulus of the abdominal vena cava. The data points shown were obtained from wave speed measurements performed during the balloon inflation experiments illustrated in Figure 19 and by using equations (9) and (10). Each wave speed measurement requires less than 50 msec and thus yields information on the quasi-instantaneous mechanical behavior of the vena cava. We know that the effective Young's modulus increases by two orders of magnitude while the wall stress increases by only one.

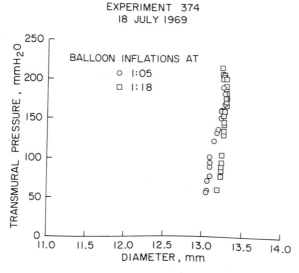

Fig. 21: Effect of transmural pressure on the diameter of the abdominal vena cava. In this case the vessel is initially dilated (relaxed) and the pressure rise caused by the balloon inflation produces only a small increase in the diameter.

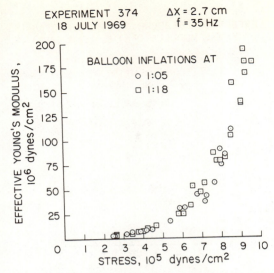

Fig. 22: Effect of transmural pressure on the effective Young's modulus of the abdominal vena cava. Data points shown are obtained from wave speed measurements performed during the balloon inflation experiments illustrated in Figure 21. As in the case of the relaxed vena cava (see Figures 19 and 20), the variation with stress of the effective Young's modulus reflects a completely different mechanical behavior of the wall than we would anticipate on the basis of pressure-diameter curves.

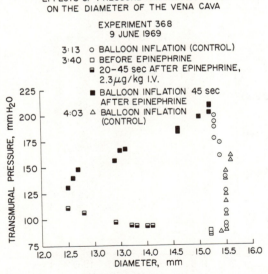

Fig. 23: Control balloon inflations (see Figure 13 for experimental arrangement) before the injection of epinephrine show that the vessel is relaxed but also stiff since it essentially does not change its diameter with pressure. In response to the epinephrine, the vessel constricts markedly while the pressure increases by about 50%. A subsequent balloon inflation shows that with rising pressure the diameter of the "stimulated" vessel increases again to its control value, indicating that the constricted vessel is more distensible than the relaxed vessel.

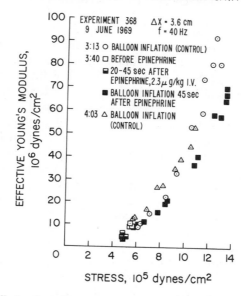

Fig. 24: Effects of transmural pressure and epinephrine on the effective Young's modulus of the abdominal vena cava. The data show that the relaxed vena cava wall exhibits a noticeably higher effective Young's modulus than does the stimulated wall at the same wall stress. The difference in the Young's modulus is as much as 100% at the higher wall stresses.

transmural pressure is elevated from about 70 to 220 mm H_2O. Also, the circumferential effective Young's modulus determined from waves with a frequency of 40 Hz can increase from 10^6 dynes/cm² at a σ of 10^5 dynes/cm² to a value as high as 100×10^6 dynes/cm² at a σ of 8×10^5 dynes/cm². For a similar increase in transmural pressure, the diameter changes of the relaxed vessel are insignificant and the values of E correspondingly higher.

The effects of epinephrine injection into a relaxed vena cava are illustrated in Figs. 23 and 24. The stimulation clearly elicits an active response during which the diameter of the vessel decreases while the pressure increases. If we were to use this part of the P_T versus D curve to determine the effective circumferential Young's modulus, we could obviously obtain a negative value. Immediately upon stimulation the balloon inflation appears to yield the familiar sigmoid-type pressure diameter curve, and 23 minutes after injection of epinephrine the vessel again exhibits its original relaxed state. From the E versus σ plot we note that at the higher wall stresses the relaxed vessel clearly has a lower E than the stimulated vessel. In this case the difference in E can be as much as 100%. Similar results were observed when active responses were induced through splanchnic stimulation (Yates, 1969).

IV. THE DISTENSIBILITY OF HEART VENTRICLES

The distensibility of a heart cavity is a dynamic variable since we have not only sizeable geometric changes occurring during the cardiac cycle, but presumably also substantial changes of the elastic properties of the ventricular wall with each phase of the cardiac cycle. Therefore, the techniques employed to determine the distensibility should yield essentially instantaneous values. We could, for example, inject into the ventricle or rapidly withdraw from it a small volume of blood and measure the resulting sound level or pressure change. Another approach would be to induce artificial heart sounds by small but rapid volumetric displacements. For a given ventricular configuration and assuming a linear elastic behavior, we can interpret the pressure changes or sound levels corresponding to small volumetric displacements as direct measures of an effective Young's modulus of the ventricular wall.

Various amounts of saline were injected into the left ventricles of anesthetized dogs with open chests by means of a spring-loaded syringe (Astleford, 1969). The pressure changes induced by the injection were measured by means of a catheter-tip manometer that was positioned in the ventricle. The injections could be timed to occur during the isometric contraction, isometric relaxation or any other phase of the cardiac cycle. But, in addition to an unsatisfactory repeatability, it was rather difficult to assess what the exact shape of the pressure curve would have been if no injection had taken place.

Aiming for a higher degree of repeatability and a better-defined perturbation, an electrically driven piston was inserted into the ventricle (Astleford, 1969). Figure 25 shows the tracings of the left ventricular and aortic pressure records obtained with the vibrating piston technique. In this case the systolic pressure was approximately 122 mm Hg, the diastolic pressure 91 mm Hg and the filling pressure less than 10 mm Hg. The curves exhibit the pressure fluctuations or sounds caused by a sinusoidal volume displacement of peak-to-peak magnitude $\Delta V = 0.062$ cm^3 and a frequency of 75 Hz. The corresponding pressure changes Δp also measured from peak to peak are found to be of the order of 2.8 mm Hg during systole. During the filling phase we cannot discern any sinusoidal pressure fluctuations on these records. This means that during diastole Δp must be less than 1 mm Hg and the effective Young's modulus of the ventricular wall considerably smaller than during systole. A particularly striking fact is that the distensibility of the ventricle does not seem to be much affected by the opening of the aortic valve. A further interesting point is that the artificial heart sounds produced by the vibrating piston are transmitted into the ascending aorta and into the aortic arch (the pressure fluctuations associated with these sounds can not be resolved in the arterial pressure record shown). We can utilize this fact and place two catheter-tip manometers at different points of the ascend-

DISTENSIBILITY OF THE LEFT VENTRICLE AT SYSTOLE
EXPERIMENT NO. 146
♂ DOG   29.5 kg $W_{HEART} = 197\,g$

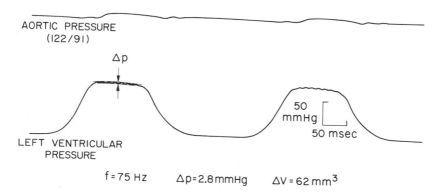

Fig. 25: Aortic and left ventricular pressure during continuous operation of sinusoidally vibrating piston inserted into the left ventricle through the left atrium. The peak-to-peak volumetric displacement of the piston ΔV is 0.062 mm³ and the piston frequency f is 75 Hz. Δp denotes the peak-to-peak pressure perturbation induced by the piston. In the recording shown here, the resolution of pressure changes was about 1 mm Hg. At autopsy, the left ventricular volume was found to be approximately 40 cm³.

ing aorta and/or the aortic arch and determine the wave transmission characteristics of the aorta in the immediate vicinity of the heart.

To deduce an order of magnitude value for an effective Young's modulus of the ventricular wall from the experimental data, the ventricle is assumed to behave mechanically like a thick-walled spherical shell with an internal radius r_i and an external radius r_o (Fig. 26). The intraventricular pressure is denoted by p_i, and the outside pressure p_o is taken as zero. We decompose p_i and r_i into unperturbed instantaneous values p_{io} and r_{io} plus the perturbations caused by the piston

(11)
$$p_i = p_{io} + \Delta p$$
$$r_i = r_{io} + \Delta r_i$$

Likewise for the internal volume

(12)
$$V_i = V_{io} + \Delta V$$

With a Poisson's ratio of $\nu = 1/2$, neglecting the inertia effects associated

DISTENSIBILITY OF INCOMPRESSIBLE THICK-WALLED SPHERICAL SHELLS

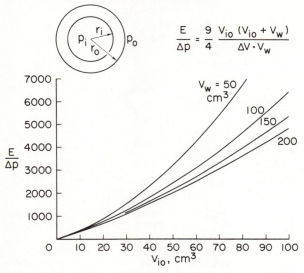

Fig. 26: Relationship between the effective Young's modulus E, the pressure perturbation Δp, the left ventricular wall volume V_w, the unperturbed intraventricular volume V_{io} and the piston displacement ΔV. The curves shown represent this relationship for various wall volumes V_w and a ΔV of 0.070 cm³; they are based on the assumption that the left ventricle behaves like a thick-walled spherical shell with respect to the small volumetric changes induced by the piston and that the inertia of the wall and the fluid can be neglected. The inner and outer radii of the spherical shell are denoted by r_i and r_o, respectively. p_i represents the intraventricular pressure and p_o the external (ambient) pressure.

with the wall and the fluid, we can express the strain at the inner surface of the sphere due to its distension as

$$(13) \qquad \epsilon_{\theta\theta} = \epsilon_{\phi\phi} = \frac{\Delta r_i}{r_i} = \frac{3\Delta p}{4E} \frac{r_o^3}{r_o^3 - r_{io}^3}$$

where $\epsilon_{\theta\theta}$ and $\epsilon_{\phi\phi}$ are the strains in the meridianal and equatorial directions. Denoting the wall volume by V_w we have $V_w = V_o - V_{io}$ and

$$(14) \qquad \frac{E}{\Delta p} = \frac{9}{4} \frac{V_{io}(V_{io} + V_w)}{\Delta V \cdot V_w}$$

as the relationship between the effective Young's modulus E, the unperturbed instantaneous volume V_{io}, the wall volume V_w, the volume increment ΔV and the pressure increment Δp for the static case. Figure 25 gives $E/\Delta p$ as a function of the instantaneous intraventricular volume V_{io} for $\Delta V = 0.070$ cm³ and representative values for the left ventricular wall volume. Assuming an internal volume of 40 cm³, we find in the case of $V_w = 100$ cm³ for $E/\Delta p$ a value of about 1800. Since our records as shown allow for a resolution of 1 mm Hg pressure change, we conclude that we can only see a pressure fluctuation produced by $\Delta V = 0.070$ cm³ when the effective Young's modulus $E = 1800 \times 1330 = 2.4 \times 10^6$ dynes/cm². At this point we should recall that for low filling pressures we did not observe a sinusoidal pressure variation during diastole. This means that theoretically Young's modulus must be smaller than 2.4×10^6 dynes/cm² during diastole, which would be in partial agreement with the values reported by others (Ghista and Sandler, 1969). When a pressure fluctuation was discernible during diastole, as was the case for filling pressures of the order 15 mm Hg, we had an effective Young's modulus of 2.4×10^6 dynes/cm². If we can rule out resonance phenomena, we may conclude from the fact that Δp was of the order 3 mm Hg during systole in the case illustrated in Fig. 26 that E_{eff} during systole is usually larger than this value and ranges from about $7-15 \times 10^6$ dynes/cm². As such, E_{eff} for systole is about one order of magnitude higher than the values obtained from two-dimensional X-ray cinematography studies in which all volumetric changes of the ventricle are interpreted as those of a shrinking or expanding ellipsoid. It is unrealistic not to allow for volumetric changes that may be due to changes in shape involving highly localized and small strains such as we may encounter, for example, in distending a paper bag. Also, we know that curved surfaces can change their shape without requiring the development of strains. (This follows (Kreyszig, 1959) from Gauss's "theorema egregium.") Therefore, if the entire volumetric change of a ventricle is interpreted as corresponding to the shrinking or expanding of an ellipsoid, its wall must be much more elastic as compared with that of a ventricle that undergoes changes in volume to a large extent by changing its shape.

V. NONLINEAR ANALYSIS OF PRESSURE AND FLOW PULSES IN ARTERIAL CONDUITS

The pressure and flow pulses propagating in arteries as a result of the intermittent ejection of blood from the left ventricle undergo characteristic changes in their shapes as they travel from the heart to the vascular bed. Some of these changes are quite sensitive to variations in certain properties of the cardiovascular system and could possibly be used as diagnostic indicators or in a study of the control mechanisms. However, a meaningful interpretation of these indicators presupposes a thorough quantitative understanding of the effects of the various cardiovascular parameters on the shape and propagation of large amplitude pressure and flow pulses in the major arterial conduits. Numerous theoretical studies have been made of the natural pulse for segments of large arteries and of the linear behavior of the system (McDonald, 1960; Rudinger, 1966; Skalak, 1966; Wetterer and Kenner, 1968; Womersley, 1957). While a few investigators included in their analysis some of the nonlinear effects, they failed to provide for realistic cardiac ejection patterns and outflow through branches compatible with an observed regional blood flow distribution, and for the variation of the distensibility or wave speed with pressure and with distance from the heart (Rudinger, 1966, 1970; Lambert, 1958; Streeter et al, 1963; Barnard et al, 1966a, b; Jones, 1969). We have therefore initiated a parametric study of both the arterial pulse in the aorta and its continuation beyond the saphenous artery (Anliker et al, 1970). This study incorporates the nonlinear phenomena observed in the experiments described in Sec. II. Also, it utilizes the method of characteristics and, therefore, can account for the effects of reflections, for the convection of the signal by the flow and the variation of the cross-sectional area with pressure and distance from the heart.

To examine the propagation of large amplitude waves in an arterial conduit, we have assumed one-dimensional motion of the blood. The conduit itself is modeled as a tapered elastic tube with a porous wall which allows for a continuously distributed seepage to simulate the outflow through branches in a manner which approximates the regional blood flow pattern. The blood is treated as a compressible fluid, and the effects of viscosity are accounted for in an approximate fashion. To define the interaction of the blood with the elastic wall we have prescribed the mechanical behavior of the conduit in terms of the speed of small pressure waves and the variation

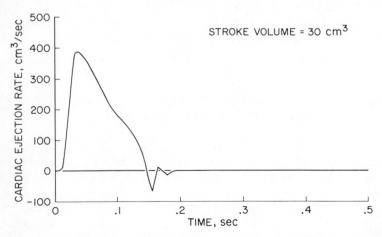

Fig. 27: Cardiac ejection rate as a function of time for the case of a heart rate of 120 beats per minute and a stroke volume of 30 cc.

of the speed with pressure and location. The equations governing the fluid flow and fluid-wall interaction are transformed according to the method of characteristics and put into a form suitable for machine computation. At the

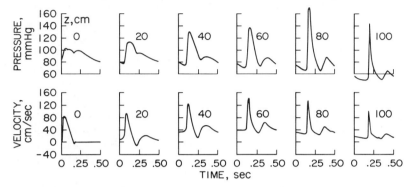

Fig. 28: Pressure and flow velocity profiles in the aorta and its continuation beyond the femoral artery for the standard case. z denotes the distance from the aortic valve. Note steepening of the wave front and peaking of the pulses with propagation. Flow patterns computed at intermediate stations along the arterial conduit show that flow reversal persists to $z = 30$ cm in this case.

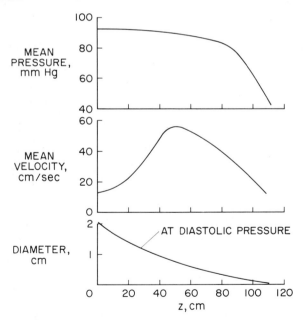

Fig. 29: Pressure and flow velocity averaged over the cardiac cycle and plotted as a function of distance from the heart for the standard case. Also internal diameter at diastolic pressure.

proximal end of the conduit, which is assumed to be connected to the left ventricle, we have specified the volume flow rate of blood as shown in Fig. 27. Recognizing that the pulsatile component of the blood pressure diminishes in the smaller arteries and arterioles and essentially vanishes in most precapillary vessels, we have assumed that the pressure at the distal end is constant.

To compare the effects of certain parameters we have computed the pressure and flow patterns in the aorta and its continuation beyond the saphenous artery of an anesthetized dog weighing 30 kg and having a heart rate of 120 beats per minute. For typical parameter values characterizing the standard case, the predicted pressure and flow velocity profiles are displayed in Fig. 28 for six different locations beginning with $z = 0$ (aortic valve) and continuing to $z = 100$ cm. Figure 29 shows the temporal mean values for pressure and flow velocity as a function of z, together with the diameter at diastolic pressure. The temporal pressure profiles plotted in Fig. 30 for various distances from the aortic valve indicate the familiar gradual change in shape as the pulse wave propagates in the aorta. We clearly note the incisura in the pressure pattern at $z = 0$. Its sharpness, however, begins to disappear with increasing distance. In our numerical studies we observed that the short interval of backflow or negative flow shown in Fig. 27 was not necessary to

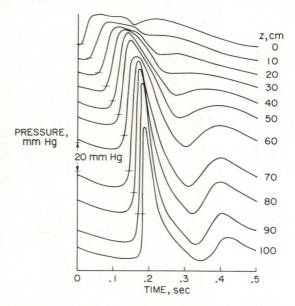

Fig. 30: Pressure-time profiles for the standard case at different distances from the aortic valve, illustrating the evolution of the natural pressure pulse. The pressure corresponding to 80 mm Hg is denoted at the beginning of each profile.

produce the incisura. Appropriate decreases in the cardiac ejection rate can cause an incisura even when there is no flow reversal. At normal pressure levels no backflow was found beyond 30 cm. As can be seen from Figs. 28 and 30, the wave front of the pulse steepens markedly with distance from the heart, which correlates well with *in vivo* observations. Linearizing the analysis, we observe prominent differences in the predicted pressure and flow patterns (see Figs. 31 and 32).

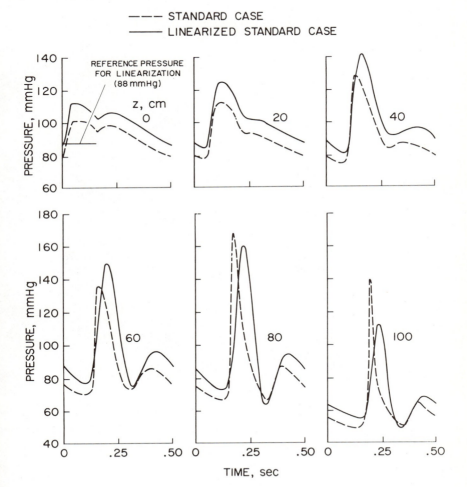

Fig. 31: Pressure patterns corresponding to the standard case in which nonlinear effects have been taken into account, and patterns computed from the linearized equations. Aside from the mean pressure, the differences in shape become more pronounced with increasing distance from the heart.

374 NONTRAUMATIC STUDY OF THE CIRCULATORY SYSTEM

The increase of the wave speed with pressure and flow suggests the possibility of the formation of shock waves in blood vessels. Under normal conditions the pressure pulse generated by the heart is not sufficiently steep or strong to produce a shock wave within any arterial conduit. But in the case

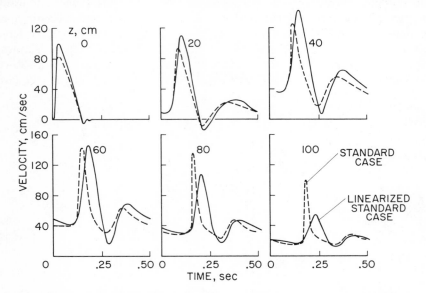

Fig. 32: Velocity patterns for the standard case computed on the basis of nonlinear and linear analyses. The corresponding pressure patterns are shown in Figure 31.

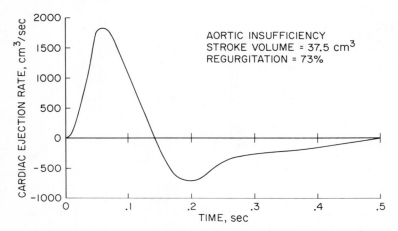

Fig. 33: Hypothetical cardiac ejection rate pattern simulating aortic insufficiency. Amount of regurgitation is 73% of the gross stroke volume. Heart rate is 120 beats per minute.

of an incompetent aortic valve (aortic insufficiency), the situation is quite different. The large amount of backflow associated with a leaking valve would reduce the normal net cardiac output per beat unless the size of the heart and the gross ejection volume were increased. A larger heart and correspondingly larger positive and negative flow rates are indeed clinically observed in patients with aortic insufficiency. Making actual recordings of cardiac ejection patterns in instances of aortic valve incompetence, we have assumed the ejection rate function given in Fig. 33. The corresponding pressure and flow patterns in the arterial conduit considered earlier are shown in Figs. 34 and 35, together with normal patterns obtained for the standard

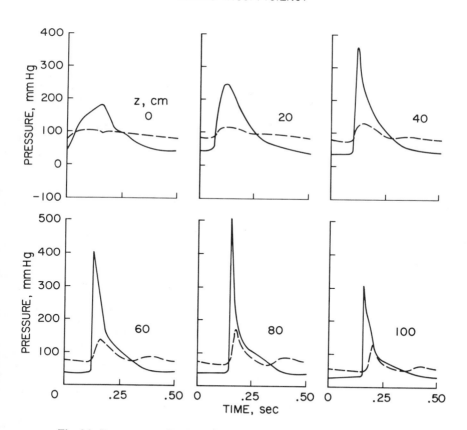

Fig. 34: Pressure profiles for hypothetical aortic insufficiency corresponding to cardiac ejection rate given in Figure 33. Other parameter values are those for the standard case. First indication of a shock wave is suggested by the steepness of the pressure front at 40 cm from the heart.

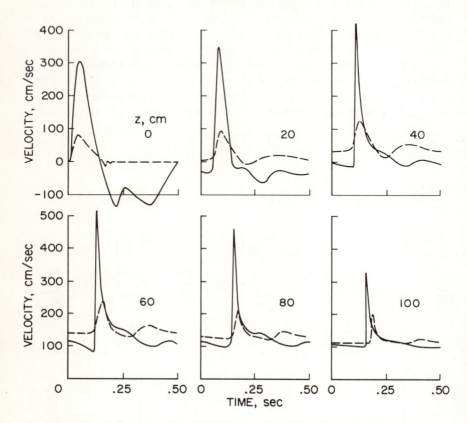

Fig. 35: Velocity profiles for hypothetical aortic insufficiency corresponding to the pressure profiles of Figure 34. The flow pattern for $z = 0$ does not faithfully reproduce the ejection pattern imposed on the system and given in Figure 33. The apparent discrepancy however vanishes when we multiply the instantaneous flow velocity by the corresponding cross-sectional area. Our model allows for the variation of the cross-sectional area at the root of the aorta with pressure.

case. A first indication of a shock wave in this particular case can be noted at the distance of 40 cm from the heart. Just as an ordinary shock wave induces audible vibrations in air, the large and rapid rise in pulse pressure can be expected to induce vibrations in the elastic arterial wall. When these vibrations are within the audible frequency range and have a sufficiently large amplitude, they represent a sound which is clinically referred to as the so-called "pistol shot phenomenon" that manifests itself with a sharp pulse which can actually deflect the palpating fingers at the radial or femoral artery in the case of man.

Acknowledgment. *The work described in this paper was carried out at the Ames Research Center of NASA with the support of NASA grant NGL 05-020-223 and of the Office of Naval Research under contract number N00014-67-A-0112-0007.*

REFERENCES

Alexander, R. S. 1962. The Peripheral Venous System. *In* W. F. Hamilton (ed.), Handbook of Physiology, Circulation. Vol. 2. Amer. Physiol. Soc. Washington, D. C.
Anliker, M. and J. A. Maxwell 1966. Dispersion of Waves in Blood Vessels. *In* Y. C. Fung (ed.), Biomechanics Symposium. ASME, New York. pp. 47–67.
Anliker, M., Histand, M. B. and Ogden, E. 1968*a*. Dispersion and Attenuation of Small Artificial Pressure Waves in the Canine Aorta. Circul. Res., 23: 539–51.
Anliker, M., W. E. Moritz and E. Ogden 1968*b*. Transmission Characteristics of Axial Waves in Blood Vessels. J. Biomech. 1: 235–46.
Anliker, M., M. K. Wells and E. Ogden 1969. The Transmission Characteristics of Large and Small Pressure Waves in the Abdominal Vena Cava. IEEE Trans. on Bio-Medical Eng. BME-16: 262–73.
Anliker, M., R. L. Rockwell and E. Ogden 1970. Theoretical Analysis of Nonlinear Phenomena Affecting the Pressure and Flow Pulse in Arteries. AGARD Conf. Proc. No. 65: Fluid Dynamics of Blood Circulation and Respiratory Flows. Naples, Italy, pp. 22-1–22-10.
Anliker, M. and M. Dorfman 1970. Theoretical Model Studies of Wave Transmission in Semicircular Canal Ducts. Ingenieur-Archiv. 39: 390–406.
Anliker, M. and W. C. Van Buskirk 1970. Experimental Model Studies of Some Dynamic Response Characteristics of the Semicircular Canals. *In* Dynamic Response of Biomechanical Systems. ASME, New York, pp. 101–132
Anliker, M. and J. Billingham 1971. Biomechanical Studies in Aerospace Physiology. IEEE Spectrum, 8 (3): 64–72.
Arndt, J. O. 1969. Über die Mechanik der intakten A. carotis communis des Menschen unter verschiedenen Kreislaufbedingungen. Archiv für Kreislaufforsch., 59: 153–97.
Astleford, W. J. 1969. Direct Determination of Distensibility of the Left Ventricle of the Heart Under *in vivo* Conditions. Engineer Thesis, Biomechanics Laboratory Report SUDAAR No. 372, Stanford University, Calif.
Atabek, H. B. and H. S. Lew 1966. Wave Propagation through a Viscous Incompressible Fluid Contained in an Initially Stressed Elastic Tube. Biophys. J. 6: 481–503.
Attinger, E. O. 1964 (ed.) Pulsatile Blood Flow, McGraw-Hill Book Co., New York.
Attinger, E. O. 1966. Hydrodynamics of Blood Flow. Adv. Hydrosci. 3: 111–53.
Attinger, E. O. 1967 Analysis of Pulsatile Blood Flow. *In* N. Levine (ed.), Advances in Biomedical Engineering and Medical Physics. Vol. 1. Wiley-Interscience, New York.
Bailie, J. A. 1969. Theoretical Studies on High Frequency Wave Propagation in Blood Vessels. Ph. D. Dissertation, Biomechanics Laboratory Report SUDAAR No. 378, Stanford University, Calif.
Barnard, A. C. L., W. A. Hunt, W. P. Timlake and E. Varley, 1966*a*. A Theory of Fluid Flow in Compliant Tubes. Biophys. J. 6: 717–24.
Barnard, A. C. L., W. A. Hunt, W. P. Timlake and E. Varley, 1966*b*. Peaking of the Pressure Pulse in Fluid-filled Tubes of Spatially Varying Compliance. Biophys. J. 6: 735–46.
Beneken, J. E. W. 1965. A Mathematical Approach to Cardio-vascular Function. Institute of Medical Physics. Da Costakade 45, Utrecht, The Netherlands.

Chow, J. C. F. and J. T. Apter 1968. Wave Propagation in a Viscous Incompressible Fluid Contained in Flexible Viscoelastic Tubes. J. Acoust. Soc. Am. 44 (2): 437–43.

Franklin, D. L., W. A. Schlegel and R. F. Rushmer 1961. Blood Flow Measured by Doppler Frequency Shift of Backscattered Ultrasound. Science 134: 564–65.

Fox, E. A. and E. Saibel 1963. Attempts in the Mathematical Analysis of Blood Flow. Trans. Soc. Rheol. 7: 25–31.

Ghista, D. N. and H. Sandler 1969. An Analytic Elastic-viscoelastic Model for the Shape and the Forces in the Left Ventricle. J. Biomech. 2: 35–47.

Grodins, F. S. 1959. Integrative Cardiovascular Physiology: A Mathematical Synthesis of Cardiac and Blood Vessel Hemodynamics. Quart. Rev. Biol. 34: 93–116.

Guyton, A. C. 1963. Cardiac Output and its Regulation. W. B. Saunders Co., Philadelphia, Penna.

Hardung, V. 1962. Propagation of Pulse Waves in Viscoelastic Tubings. In W. F. Hamilton and P. Dow (eds.), Handbook of Physiology, Circulation, Vol. 1. Amer. Physiol. Soc. Washington, D.C. pp. 107–35.

Histand, M. B. 1969. An Experimental Study of the Transmission Characteristics of Pressure Waves in the Aorta. Ph. D. Dissertation, Biomechanics Laboratory Report SUDAAR No. 369, Stanford University, Calif.

Iberall, A. S. 1969. New Thoughts on Bio Control. In C. H. Waddington (ed.), Towards a Theoretical Biology, 2. Sketches. Edinburgh University Press, pp. 166–78.

Jones, R. T. 1969. Blood Flow. Ann. Rev. Fluid Mech. 1: 223–44.

Jones, E., M. Anliker and I. D. Chang. Effects of Viscosity and Constraints on the Dispersion and Dissipation of Waves in Large Blood Vessels—Part I: Theoretical Analysis, Part II: Comparison of Analysis with Experiments. In press.

Klip, W. 1962. Velocity and Damping of the Pulse Wave. Martinus Nijhoff, The Hague.

Klip, W., P. van Loon and D. A. Klip 1967. Formulas for Phase Velocity and Damping of Longitudinal Waves in Thick-walled Viscoelastic Tubes. J. Appl. Phys. 38: 3745–755.

Kreyszig, E. 1959. Differential Geometry. University of Toronto Press, Toronto.

Lambert, J. W. 1958. On the Nonlinearities of Fluid Flow in Nonrigid Tubes. J. Franklin Institute 266: 83–102.

Lamberti, J. J., J. Urquhart and R. D. Siewers, 1968. Observations on the Regulation of Arterial Blood Pressure in Unanesthetized Dogs. Circul. Res. 23: 415–28.

McDonald, D. A. 1960. Blood Flow in Arteries. Edward Arnold, Ltd., London.

McDonald, D. A. 1968a. Hemodynamics. Annual Review of Physiology. Annual Reviews, Inc., Palo Alto, California. pp. 525–56.

McDonald, D. A. 1968b. Regional Pulse-Wave Velocity in the Arterial Tree. J. Appl. Physiol. 24: 73–78.

Maxwell, J. A. and M. Anliker 1968. Dissipation and Dispersion of Small Waves in Arteries and Veins with Viscoelastic Wall Properties. Biophys. J. 8: 920–950.

Mills, C. J. 1966. A Catheter Tip Electromagnetic Velocity Probe. Phys. Med. Biol. 11: 323–24.

Morgan, G. W. and W. R. Ferrante 1955. Wave Propagation in Elastic Tubes Filled With Streaming Liquid. J. Acoust. Soc. Am. 27: 715–25.

Moritz, W. E. 1969. Transmission Characteristics of Distension, Torsion and Axial Waves in Arteries. Ph D. Dissertation, Biomechanics Laboratory Report SUDAAR No. 373, Standford University, Calif.

Müller, A. 1950. Über die Fortpflanzungsgeschwindigkeit von Druckwellen in dehnbaren Röhren bei ruhender und strömender Flüssigkeit. Helv. Physiol. Acta, 8: 228–41.

NASA. 1966. Medical Aspects of an Orbiting Research Laboratory. Special Publ. 86, U.S. Government Printing Office, Washington, D.C.

Nat. Acad. Sci.-Nat. Res. Council. 1966. Space Research: Directions for the Future. NAS-NRC Publ. 1403, Washington, D.C.

Nat. Acad. Sci.-Nat. Res. Council. 1968. Physiology in the Space Environment. I, Circulation. NAS-NRC Publ. 1485A, Washington, D.C.

Pieper, H. and L. T. Paul 1968. Catheter-tip Gauge for Measuring Blood Flow Velocity and Vessel Diameter in Dogs. J. Appl. Physiol. 24 (2): 259–61.

Rudinger, G. 1966. Review of Current Mathematical Methods for the Analysis of Blood Flow. *In* Biomedical Fluid Mechanics Symposium. ASME, New York. pp. 1–33.

Rudinger, G. 1970. Shock Waves in Mathematical Models of the Aorta. J. Appl. Mech. (ASME Series E) 37: 34–37.

Rushmer, R. F., D. W. Baker and H. F. Stegall 1966. Transcutaneous Doppler Flow Detection as a Non-destructive Technique. J. Appl. Physiol. 21: 554–66.

Skalak, R. 1966. Wave Propagation in Blood Flow. *In* Y. C. Fung (ed.), Biomechanics. ASME Applied Mechanics Division, New York. pp. 20–40.

Streeter, V. L., W. F. Keitzer and D. F. Bohr 1963. Pulsatile Pressure and Flow Through Distensible Vessels. Circul. Res. 13: 3–20.

Wells, M. K. 1969. On the Determination of the Elastic Properties of Blood Vessels from their Wave Transmission Characteristics. Ph. D. Dissertation, Biomechanics Laboratory Report SUDAAR No. 368, Stanford University, Calif.

Wetterer, E. and T. Kenner 1968. Grundlagen der Dynamik des Arterienpulses. Springer Verlag, Berlin.

Womersley, J. R. 1957. Elastic Tube Theory of Pulse Transmission and Oscillatory Flow in Mammalian Arteries. WADC Tech. Report TR-56-614, Defense Documentation Center.

Yates, W. G. 1969. Experimental Studies of the Variations in the Mechanical Properties of the Canine Abdominal Vena Cava. Ph. D. Dissertation, Biomechanics Laboratory Report SUDAAR No. 393, Stanford University, Calif.

16

Flow and Pressure in the Arteries

THOMAS KENNER

I. INTRODUCTION

Hemodynamics comprises three general areas of problems: 1) the description of physical phenomena related to pressure and flow in the arteries, 2) the explanation of the role which these phenomena play in the living biological system, and 3) the practical application of methods for diagnostic and clinical purposes.

In no other field of biological research has modeling been emphasized so much as in hemodynamics. The way of thinking and the definition of many problems in this field cannot be fully understood without considering this fact.

The comparison of the action of the elastic distensibility of the arterial system with an air chamber (Windkessel) to that of an ancient fire pump dates back to Borelli (1681), Hales (1733) and E.H.Weber (1828). The transmission line analogy of elastic tubes was introduced in mathematical form by E. H. Weber (1850) and his brother W. Weber (1866) and has been worked out in detail by Kries (1892).

The great advent of modeling, however, came when Frank (1899) introduced the Windkessel model as a mathematical tool for the description of arterial pulse waves. This model, which is very useful as long as its limits are properly defined, has been modified, improved, and adjusted in order to simulate certain aspects of pulsatile pressure and flow (Wetterer and Kenner 1968).

The transmission line analogy reappeared in the 40's (Landes, 1942) and since then has been worked out in great detail using hydrodynamic and electrical analog simulation or digital computation. Although Euler's equations and the Navier-Stokes equations had been applied to hemodynamic problems before (Witzig 1914; Frank 1926), Womersley's work became a great stimulus for the development of modern theoretical hemodynamics (Womersley 1955a, b 1957, 1958; McDonald, 1960).

With respect to experimental research, Frank's (1903) study on manometer design principles, and the invention of the electromagnetic flowmeter by Kolin (1936) and Wetterer (1937, 1938) provided the necessary tools for the recording of pulsatile pressure and flow.

In this study it is intended to explain the physical events in the arterial system from a very general viewpoint by developing and discussing transmission line models. The motion of fluid in arteries is governed by the Navier-Stokes equations. Including the necessary boundary conditions, these equations can so far only be solved in simplified form. The usefulness of different degrees of approximation highly depends on the questions to be answered. For a basic description of overall patterns of pressure and flow, even friction may be neglected. The analogy of the linearized equations with the transmission line equations allows the application of the four-terminal network theory and the solution of problems by the use of an electrical analog model (Jager et al, 1965; Noordergraaf, 1969; Westerhof et al, 1969). One particular problem which has to be remembered in any model study, and which explains the multitude of models for the arterial system, is the fact that the parameters of nearly any model can be adjusted so that certain dependent parameters or functions can perfectly be simulated (for example, pulse contour, input impedance, see Attinger, 1968).

With regard to biological problems, the similarity in the design of the arterial system in different animals is most interesting and appears to follow slightly different rules than those an engineer would probably apply.

Finally, it is important to note that the phenomenon of pulsatility of flow and pressure in arteries is still only poorly understood from a biological viewpoint. Regarding the tremendous gain in knowledge about the physical and mathematical background of this phenomenon, this fact should be a challenge to engineers and biologists as well.

Since this study is not outlined as a literature review, the reader is referred to the following works: Attinger 1964, 1966, 1968, 1969; Fung 1966; Klip 1969; McDonald 1960; Noordergraaf 1969; Skalak 1966; Spencer and Denison 1963; Vadot 1967; Westerhof et al, 1969; Wetterer 1956; Wetterer and Kenner 1968; Whitmore 1968.

II. THE EQUATIONS OF MOTION

An external force acting on a fluid in equilibrium state generates a pressure gradient. Let F denote the external force per unit fluid volume, then

(2.1) $$\operatorname{grad} p = F$$

The hydrostatic pressure p, in this case, is the potential energy density which is necessary to keep the fluid in equilibrium. To describe fluid in a moving state, inertial resistance to the external forces has to be added. Neglecting friction thus leads to Euler's equations:

(2.2) $$\rho \frac{dv}{dt} + \operatorname{grad} p = F$$

The condition of incompressibility is contained in the equation of continuity:

(2.3) $$\operatorname{div} v = 0$$

dv/dt in Eq. 2.2 represents the total or material acceleration. If the way of a fluid particle in time dt is given by the vector (dx, dy, dz), the change of the x-component of the velocity $v_x(x, y, z, t)$ is

(2.4) $$dv_x = \frac{\partial v_x}{\partial t} dt + \frac{\partial v_x}{\partial x} dx + \frac{\partial v_x}{\partial y} dy + \frac{\partial v_x}{\partial z} dz$$

Therefore, it follows that under steady flow conditions, although $\partial v_x/\partial t = 0$, the total acceleration is not necessarily zero. This is the case, for example, in a narrowing, tapering tube where the flow accelerates due to the change in tube diameter. Eq. 2.2 can also be written in the following form:

(2.5) $$\rho \left(\frac{\partial v}{\partial t} - v \times \operatorname{curl} v \right) + \operatorname{grad} \left(\rho \frac{v^2}{2} + p + U \right) = 0$$

It is assumed that the external force F which, in most practical biological applications corresponds to gravitation, has a potential $F = -\operatorname{grad} U$. In steady irrotational flow of an ideal fluid the sum of kinetic and potential energy densities, therefore, is constant. This fact is expressed by Bernoulli's equation

(2.6) $$\rho \frac{v^2}{2} + p + U = \text{constant}$$

The hydrostatic pressure p of an incompressible fluid represents the reaction against the condition of incompressibility (Sommerfeld, 1950). Therefore, p is dependent on the magnitude of U and the kinetic energy density.

Of a different nature is the stress gradient produced by the shear stresses. In a parallel flow the shear stress on a plane surface parallel to the direction of flow is proportional to the velocity gradient perpendicular to the surface

(2.7) $$\sigma_{yx} = \eta \frac{\partial v_x}{\partial y}$$

The constant η denotes the viscosity of the (Newtonian) fluid. Introduction of the divergence of the shear stresses into the equations of motion leads to the Navier-Stokes equation:

$$\rho \frac{dv}{dt} - \eta \nabla^2 v + \text{grad } p = F \tag{2.8}$$

To examine the pulsatile flow in elastic arteries, an additional set of equations is necessary to describe the motion of the vascular wall and the proper boundary conditions for wall and fluid movement. The following dimensional relations allow simplifications of the equations: Pressure and flow waves in arteries are very long compared with the diameter of the tube; flow velocities are small with respect to the wave velocity; the radial fluid movement is very small compared with the longitudinal movement. Womersley (1957) also examined the possibility of a longitudinal movement of the arterial wall, assuming different degrees of tethering. Since the actual longitudinal tethering of most arteries *in vivo* is very strong (Patel and Fry, 1966), and since Womersley's assumptions about the homogeneity of the arterial wall material and the difficulty of establishing proper boundary conditions lead to unrealistic results (Wetterer and Kenner, 1968), a basic examination can be restricted to tubes which are fixed in the length direction.

From Eq. 2.8 follows the equation of motion of fluid in the longitudinal direction of a tube in cylindrical coordinates after omitting the external force F:

$$\frac{\partial v_x}{\partial t} + v_r \frac{\partial v_x}{\partial r} + v_x \frac{\partial v_x}{\partial x} = -\frac{1}{\rho}\left(\frac{\partial p}{\partial x}\right) + \frac{\eta}{\rho} \cdot \frac{1}{r} \cdot \frac{\partial}{\partial r}\left(r \frac{\partial v_x}{\partial r}\right) \tag{2.9}$$

As a further development, the terms of this equation may be integrated over the cross section area A of the tube:

$$\frac{\partial(A\bar{v}_x)}{\partial t} + 2\pi \frac{\partial}{\partial x}\int_0^r v_x^2 r\, dr = -\frac{A}{\rho}\frac{\partial p}{\partial x} + 2\pi \frac{\eta}{\rho}\left(r \frac{\partial v_x}{\partial r}\right)_{r=r_w} \tag{2.10}$$

The second term of this equation contains the integral of the two convective acceleration terms of Eq. 2.9, which can be shown to be equal (Lee, 1970).

The continuity equation in cylindrical coordinates is given by

$$\frac{1}{r}\frac{\partial}{\partial r}(v_r r) + \frac{\partial v_x}{\partial x} = 0 \tag{2.11}$$

Integration over the cross-sectional area and addition of a leakage term yields

$$\frac{\partial A}{\partial t} + \frac{\partial(A\bar{v}_x)}{\partial x} + pG = 0 \tag{2.12}$$

G is the admittance, per unit length of the tube, of leaks in the wall through which fluid is lost. Using the condition of continuity Eq. 2.10 can further be developed into a form which contains a generalized statement of Bernoulli's energy condition. In the following equations the subscript x for longitudinal movements will be omitted.

$$(2.13) \quad \rho \frac{\partial \bar{v}}{\partial t} = -\frac{\partial}{\partial x}\left(p + \rho \frac{\bar{v}^2}{2}\right) - \frac{1}{A}\frac{\partial}{\partial x}\rho A(\overline{v^2} - \bar{v}^2) + 2\pi \frac{\eta}{A}\left(r\frac{\partial v}{\partial r}\right)_{r=r_w}$$

The first two terms on the right side represent the gradient of the energy density. The second term contains the difference between the mean of the square velocity minus the square of the mean velocity. If the velocity profile is flat as in ideal fluid flow, this term is zero. The total energy density gradient is higher if the velocity profile is not flat as in laminar flow. In steady laminar flow $\overline{v^2}/\bar{v}^2 = 1.33$. The third term on the right side represents the pressure loss produced by the viscous drag between vessel wall and fluid. It is the shear stress acting at the circumference of the tube per unit cross-sectional area and unit length of the vessel.

For practical purposes the second and third term on the right hand side of Eq. 2.13 can be expressed by an approximate function of the averaged velocity $\bar{v}$. Various terms have been applied (Streeter et al, 1964). One possible way to rewrite Eq. 2.13 is

$$(2.14) \quad \rho \frac{\partial \bar{v}}{\partial t} + \rho \bar{v}\frac{\partial \bar{v}}{\partial x} + \frac{\partial p}{\partial x} + RA\bar{v} = 0$$

where R is the longitudinal flow resistance per unit length of the tube. The expression

$$(2.15) \quad R = f\frac{\eta Re}{8r^4\pi}$$

may be used, which has been tested for steady flow over a wide range of Reynolds numbers. $f = 64/Re$ in laminar flow and $f = 0.3164 \times Re^{1/4}$ in turbulent flow up to $Re = 8 \times 10^4$ (see Wetterer and Kenner, 1968). Eq. 2.15 shows that the resistance values in the aortae of dogs and humans are negligibly small, even at the highest physiological flow rates and Reynolds numbers. A mean pressure gradient of more than 3 mm Hg between aortic root and femoral artery in supine position under resting conditions indicates pathological changes.

Eq. 2.14 can be solved by the method of characteristics (Skalak, 1966; Streeter et al, 1964; Wylie 1966) which, however, does not consider the effect of viscoelastic wall properties. If $L_1 = 0$ is the equation of continuity (Eq.

2.12) and $L_2 = 0$ is the equation of motion (Eq. 2.14), a linear combination of the two with the unknown multiplier λ gives

(2.16) $$L_1 + \lambda L_2 = 0$$

By finding a proper expression for λ it is possible to construct the total derivatives dp/dt and $d\bar{v}/dt$ and thus to transform Eq. 2.16 into a total differential equation. It can be shown that this equation is valid only if

(2.17) $$\frac{dx}{dt} = \bar{v} \pm c$$

is satisfied, which describes the characteristics in the t, x-plane. Eq. 2.17 implies that the propagation velocity of a disturbance dx/dt is dependent on the sum of the velocity of flow $\bar{v}$ and the wave velocity. This fact was experimentally established by Müller (1951) and was recently confirmed by Anliker at al (1968).

Another possibility to attack the nonlinear Eq. 2.9 is to use the solution of the linearized equation (neglecting the convective acceleration terms) to evaluate the influence of these nonlinear terms (Womersley, 1955, 1957; Lee, 1966). Under most physiological conditions the error of neglecting the nonlinear terms is small for the description of pulsatile events. Detailed experimental data regarding this question were recently published by Cox (1970).

III. THE SIMPLIFIED WAVE EQUATIONS AND D'ALEMBERT'S SOLUTION

From the foregoing discussion it appears justified, as a first approximation, to use a linearized version of the wave equations and to neglect friction for the purpose of examining pressure and flow events in the aorta and big arteries. In spite of the simplicity, this method proves highly useful for a qualitative, and sometimes even quantitative, checking of most problems related to arterial hemodynamics (Wetterer and Kenner, 1968).

(3.1) $$i = A\bar{v}$$

is the volume flow through the cross section A of the tube.

(3.2) $$M = \frac{\rho}{A}$$

is defined as "effective mass" per unit length of the tube (Frank 1903, 1926) where ρ is the density of the fluid.

(3.3) $$C = \frac{dA}{dp}$$

is the capacitance or compliance per unit length of the tube. With these terms inserted, the equations of motion and continuity can be written in the following form:

(3.4) $$-\frac{\partial p}{\partial x} = M\frac{\partial i}{\partial t}$$

(3.5) $$-\frac{\partial i}{\partial x} = C\frac{\partial p}{\partial t}$$

By differentiating and with mutual insertion they can be transformed into

(3.6) $$\frac{\partial^2 p}{\partial t^2} = c_0^2 \frac{\partial^2 p}{\partial x^2}$$

(3.7) $$\frac{\partial^2 i}{\partial t^2} = c_0^2 \frac{\partial^2 i}{\partial x^2}$$

where

(3.8) $$c_0^2 = \frac{1}{CM} = \frac{dp}{dA}\frac{A}{\rho} = \frac{dp}{dr}\frac{r}{2\rho}$$

is the velocity by which a flow and pressure disturbance is propagated along the tube. Eq. 3.8, in terms of radial distension (right), is known as Weber's equation for the wave velocity (W. Weber, 1866). The so-called Moens' equation (Moens, 1878)

(3.9) $$c_0^2 = \frac{Eh}{2r\rho}$$

is a much used and practical approximation. E is Young's modulus of the arterial wall and h is the wall thickness. Eq. 3.9 contains the assumption

(3.10) $$Eh = \frac{dp}{dr}r$$

which, as will be shown in Sec. 7, introduces a much greater error than any other approximation.

D'Alembert's solution of Eqs. 3.6 and 3.7

(3.11) $$p(t, x) = f_1\left(t - \frac{x}{c_0}\right) + f_2\left(t + \frac{x}{c_0}\right)$$

(3.12) $$i(t, x) = \frac{1}{Z_0} \cdot \left[f_1\left(t - \frac{x}{c_0}\right) - f_2\left(t + \frac{x}{c_0}\right)\right]$$

shows that a combined pressure and flow wave is produced by any disturbance of the system.

$$(3.13) \qquad Z_0 = \frac{p}{i} = \frac{c_0 p}{A}$$

is the characteristic impedance of the tube and relates pulsatile pressure p to pulsatile flow i, as long as waves travelling in only one direction are considered ($f_2 = 0$). Related to $\bar{v}$ instead of i, Eq. 3.13 states that the momentum of the wave transmitted in unit time through a cross section of the tube is equal to the pulse pressure

$$(3.14) \qquad p = \rho c_0 \bar{v}$$

This equation is sometimes referred to as "waterhammer formula" (Wetterer and Kenner, 1968) and is equivalent to Eq. 3.13. (In the following the terms i, $\bar{v}$ and p will be used for the pulsatile components of flow, velocity and pressure.)

Eq. 3.13 or 3.14 can be found directly from Eqs. 3.4 and 3.5 by applying the method of characteristics. Eq. 2.16 then yields Eq. 3.13 or 3.14 and is valid along the characteristics $dx/dt = \pm c_0$, which is nothing else than a representation of D'Alembert's solution. A comparison with Eq. 2.17 shows that neglect of the convective acceleration term leads to a neglect of the influence of the flow velocity on the overall propagation velocity, dx/dt. The error produced hereby is highest in the ascending aorta (25%) due to the high peak flow velocity and the small wave velocity at this site.

As shown by Eqs. 3.11 and 3.12 the sign of retrograde flow waves is inverse to that of retrograde pressure waves. In a tube of finite length where all waves primarily are generated at one end, such as the aorta, retrograde waves (f_2) are produced by reflection. They repeat the pattern (f_1) of the input wave. The amplitude is attenuated by a factor K. Using Eqs. 3.11 and 3.12 it is found that any change of the characteristic impedance of the tube or a frictional resistance R at the end of the tube leads to reflection with the factor

$$(3.14) \qquad K = \frac{Z_2 - Z_1}{Z_2 + Z_1} \text{ or, respectively } K = \frac{R - Z}{R + Z}$$

Retrograde waves in the aorta are reflected with a factor near $K = +1$ at the closed aortic valves. The same reflection factor is assumed to exist during the ejection phase (Ronniger, 1955; Kenner and Ronniger, 1960).

Considering repeated reflections, the following set of equations describe the time course of pressure and flow at a point $O < x < L$ of a homogeneous uniform tube model of the aorta (Kenner and Ronniger, 1960); n is the number of the peripheral reflections

$$
\text{(3.15)} \quad p(t, x) = f\!\left(t - \frac{x}{c_2}\right) + Kf\!\left(t - \frac{2L - x}{c_0}\right) + Kf\!\left(t - \frac{2L + x}{c_0}\right) + \cdots
$$
$$
\cdots + K^n f\!\left(t - \frac{2nL - x}{c_0}\right) + K^n f\!\left(t - \frac{2nL + x}{c_0}\right)
$$

$$
\text{(3.16)} \quad i(t, x) = \frac{1}{Z_0}\bigg[f\!\left(t - \frac{x}{c_0}\right) - Kf\!\left(t - \frac{2L - x}{c_0}\right) + Kf\!\left(t - \frac{2L + x}{c_0}\right) - \cdots
$$
$$
\cdots - K^n f\!\left(t - \frac{2nL - x}{c_0}\right) + K^n f\!\left(t - \frac{2nL + x}{c_0}\right) \bigg]
$$

At the entrance of the tube ($x = 0$) the flow pattern "degenerates" to

$$
\text{(3.17)} \quad i(t, 0) = \frac{1}{Z_0} f(t)
$$

which corresponds to the cardiac ejection flow. At the end of the tube where the outflow is taking place

$$
\text{(3.18)} \quad i(t, L) = \frac{1}{R} p(t, L)
$$

An example of the pressure and flow waves at the entrance, middle, and end of a uniform tube with a positive end reflection ($K = 1/3$) according to the above equations is shown in Fig. 1. Compared with actual aortic pressure and flow pulses the pattern appears extremely exaggerated. The following facts show the qualitative agreement: The amplitude of the peripheral pressure pulse is higher than that of the central pulse; the secondary waves in the periphery are 180° out of phase compared with the central waves.*

Since the sign of the reflection for flow waves is inverse to that for pressure waves, the peripheral flow amplitudes are smaller than the central amplitude. The effect of the sign inversion can most clearly be seen in the middle of the tube. The reason for the occurrence of a "backflow," as shown in Fig. 1, was the object of a considerable controversy until the importance of wave reflections was commonly accepted (Wetterer and Kenner, 1968).

---

* Since the time Δt which a wave needs to travel from the entrance of the tube to the peripheral reflection site is equal to the time it needs for the return travel to the entrance the period between the main peak and the secondary (dicrotic) peak of a pressure wave equals 2 Δt. The peaks of the peripheral pulses appear delayed by Δt with respect to the central peaks. This is sometimes referred to as a phase shift of 180°. Since the pressure waves shift from one end of the tube to the other like a swing, the time sequence of the secondary (dicrotic) peaks is similar to the oscillations in a standing wave. A minimum of the oscillatory pressure variations can be seen in the middle of the tube (Fig. 1), corresponding to a pressure node.

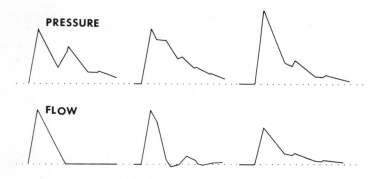

Fig. 1: Pressure and flow pulses in a uniform tube. The outflow resistance is located at the end of the tube and produces a peripheral reflection $K = +1/3$.

IV. THE EFFECTS OF TAPERING

The mammalian arterial tree consists of the main aortic tube which continues into the two iliac and femoral arteries and a number of side branches. Propagation and transformation of pressure and flow waves in this system depend on the distribution of the characteristic impedance. In the main aortic tube the characteristic impedance increases more than 10-fold from central to the periphery. The two contributing factors are the decrease of cross-sectional area (geometric tapering) and the peripheral increase of the wall stiffness (elastic tapering). The preponderance of the latter leads to a peripheral increase of the wave velocity according to Moens equation (Eq. 3.9). Average values are 400 cm/sec at the aortic arch and about 1200 to 1500 cm/sec in the femoral artery (Gauer 1936; Learoyd and Taylor, 1966).

As a first approximation, the effect of a local stepwise increase of the characteristic impedance in an elastic tube will be examined. According to Eq. 3.14 incoming waves will be reflected at this site with a factor K. Part of the wave will be transmitted. Considering the conservation of energy and momentum, or using Eqs. 3.11 and 3.12, the following amplitudes can be calculated (Kenner 1962, 1963)

	incident	reflected	transmitted
pressure	p	Kp	$(1 + K)p$
flow	i	$-Ki$	$(1 - K)i$
energy flow	pi	$K^2 pi$	$(1 - K^2)pi$

The main conclusion from this table is that a local step increase of the characteristic impedance leads to an increase in the amplitude of the transmitted pressure wave and a decrease in the amplitude of the transmitted flow wave. This agrees with what can actually be found in arteries (see Attinger, 1968; Kroeker and Wood, 1955; Spencer and Denison, 1963).

Kenner and Wetterer (1962a, b) have extensively studied the contours of pressure pulses in a two-part tube system where the peripheral part has a higher characteristic impedance than the central part. As shown in Fig. 2, pressure pulses recorded in this simple model were nearly indistinguishable from human or animal pressure pulses along the aorta. Remington and Wetterer (1963); Remington (1965); Remington and O'Brien (1970) and Bauer et al (1970) have calculated pressure pulses for aortic models assuming more than a one-step increase of the characteristic impedance. Again, the agreement with actually recorded pressure pulses is nearly perfect.

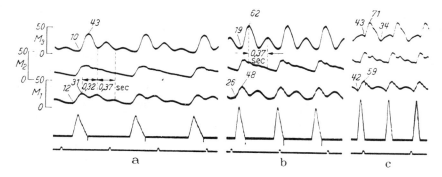

Fig. 2: Inflow and pressure pulses in a two-part tapered tube at different frequencies (Kenner and Wetterer, 1962a). Constant stroke volume 30.8 ml. Characteristic impedances: 1st tube $Z_1 = 52$ dyn. sec. cm^{-5}; 2nd tube $Z_2 = 260$ dyn. sec. cm^{-5}. Length $L_1 = 16$ cm, $L_2 = 112$ cm. Cross section: areas $A_1 = 12.1$ cm^2; $A_2 = 3.8$ cm^2. Reflection factor at the end of the second tube $K = +0.65$.
M_1 pressure at the entrance of the first tube; M_2 pressure at the site of the "pressure node" 27 cm behind the entrance of the first tube; M_3 pressure at the end of the second tube. Pressures in mm Hg.

There are two ways to examine the overall effect of continuous tapering using a simple approach (Kenner, 1963): Either to calculate the effect of a great number of very small steps of the characteristic impedance or to use the following still more simplistic way of reasoning. Assume that the characteristic impedance continuously increases from Z_1 at the entrance to Z_2 at some distance from the entrance of the tube. If the total pulsatile energy $p_1 i_1$ is transmitted from point 1 to 2 then Eq. 3.13 leads to

(4.1) $$\frac{p_1}{p_2} = \sqrt{\frac{Z_2}{Z_1}} \quad \text{and} \quad \frac{i_2}{i_1} = \sqrt{\frac{Z_1}{Z_2}}$$

Thus the pressure amplitude in a tapered tube with a matching impedance at its end increases proportionately to the square root of the characteristic impedance values. Flow amplitudes change reciprocally to the pressure amplitudes. It can be shown that the quotients of Eq. 4.1 actually are asymptotic values for high frequencies (Kenner, 1963; Taylor, 1965).

Pressure pulsations recorded in a hydrodynamic model with linear geometric taper show that the peripheral increase of amplitudes is as expected according Eq. 4.1 (Kenner and Wetterer, 1967). The pulse contours are very similar to those in a uniform tube model. A reflection factor of the same value as in a uniform tube is produced by a much higher end resistance in the tapered tube because the reflection depends on the relation of the resistance value to the value of the characteristic impedance at the end of the tube (see Sect. 9). Flow which was not recorded in these experiments (Kenner and Wetterer, 1967) ought to give somewhat unrealistic results. Since the total outflow took place at the end of the tube, the mean flow values are the same throughout the length of the tube. Due to the decrease of the flow amplitudes in the periphery of the tube, small flow pulsations are riding on a high mean flow.

An additional effect of tapering appears of great importance: it keeps the arterial hydrostatic indifference point close to the level of the heart. This point is defined as the site where, excluding active control, the pressure does not change during change of the body position from supine to upright or vice versa. Mathematically, the point is defined as the center of gravity of the total arterial compliance (Kenner and Wetterer, 1967).

V. THE EFFECTS OF TAPERING, BRANCHING AND LEAKAGE

Although during systole the heart has a certain finite internal impedance, the resistance component, under normal conditions, is high enough to consider it as a pure flow pump (Wetterer and Kenner, 1968). In contrast, the aorta acts as a pressure source towards branching arteries.

Consider a pressure and flow wave entering a side branch of the aorta, for example, the carotid artery. The wave will be transmitted towards the periphery of the branch and part of it will be reflected. Returning to the point of branching, it will be reflected nearly totally with a factor close to $K = -1$ since the characteristic impedance of the aorta is much smaller than that of the carotid artery. The ratio of the characteristic impedances is approximately inverse to the ratio of the corresponding cross-sectional areas.

$$(5.1) \qquad \frac{Z_c}{Z_A} = \frac{A_A \cdot c_c}{A_c \cdot c_A}$$

Hardung (1952) who first examined and described these reflections which keep the transmitted energy trapped in a side branch called this a "wave trap effect."

Since side branches of the aorta are relatively short, the resistive component of their input impedance prevails, in general. For this reason the outflow from the aorta can be considered as a distributed leakage. A leakage term was first introduced into the wave equations by Ronniger (1954) in analogy to the telegraph equations. As mentioned in Chap. 2, the term Gp has to be added at the right side of Eq. 3.5. (G is the admittance per unit length of the tube.)

The distributed loss of pulsatile energy into side branches of the aorta mainly accounts for the damping of the pulse waves. Neglecting wave dispersion by using an asymptotic value for the damping factor, the leakage accounts for a decline of pressure and flow amplitudes by the factor

$$(5.2) \qquad \exp\left(-\frac{1}{2}Z_0 G x\right)$$

over the distance x. A reasonable estimate for the aortae of dogs and humans for this term is $\exp(-0.01x)$, where x is given in cm. For a first approximation the leakage may be considered equally distributed along the aorta. For a more detailed examination the nonequal distribution of G along the aorta has to be taken into account, which can most efficiently be done by analog simulation (Westerhof et al, 1969).

Clamping or ligature of side branches may increase the mean pressure level and the pressure and flow amplitudes, but it influences the form of the pulsations only to a minor extent (Hardung, 1962a; Meisner and Remington, 1962). This points again to the ambiguity of some simulation experiments, where only pressure contours are considered.

Symmetric bifurcations of arteries have the same effect as if the artery were to continue in one tube with half the characteristic impedance of each of the two branches. The reflection factor at the aortic bifurcation is about $+0.05$ (Wetterer and Kenner, 1968). For this reason the aortic-femoral system can be considered as one extended leaking tapering tube.

This has been clearly shown by Kapal et al (1951) in humans. The self-oscillation period T between the main peak and the secondary dicrotic wave, which can best be recognized in the femoral pressure pulse, corresponds to the round trip time of a wave from the aortic root to the peripheral reflection site back to the aortic root. Artificially shifting the peripheral reflection site by inflating a cuff around the ankle, knee, or thigh should shorten the round-trip time and T. This actually was found in the experiments. If L is the distance from the aortic root to the site of the inflated cuffs (around both legs) and $\bar{c}$ the mean wave velocity along L, the relation

$$(5.3) \qquad L = \frac{\bar{c}T}{2}$$

was found for any value L, agreeing with the theory. The authors concluded that, under normal conditions, the main self-oscillation period of arterial pulses is determined by peripheral reflection sites located in the feet.

To some extent this may sound incredible. A rough estimate of the energetics involved leads to the following values. In the femoral pressure pulse the amplitude of the dicrotic oscillation normally amounts to about 1/4 to 1/8 of the main pressure peak. Thus, 3/4 to 7/8 of the pressure amplitude of a wave is lost during one round trip through the main aortic-femoral tube. Hence, 3/4 to 7/8 of the pulsatile energy is lost during one single trip of the wave. Approximately 1/4 to 1/8 of the pulsatile energy generated by the left ventricle is transmitted to both femoral arteries. The values agree with experimental flow and pressure measurements (Attinger et al, 1966).

With respect to the pressure pulsations, the amplifying effect of tapering counteracts the damping effect of leakage. If tapering did not exist, the femoral pulse pressure would be only 1/2 to 1/3 that in the central aorta due to the leakage loss. On the other hand, the flow amplitudes are reduced by both tapering and leakage. The femoral flow pulse amplitudes are 2–3 times smaller than they would be if the aorta were a uniform tube. Figure 3 sche-

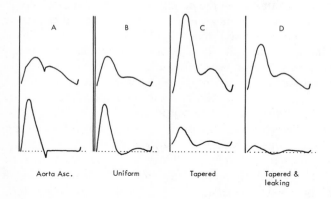

Fig. 3: Schematic drawing showing central pressure and flow (*A*), and peripheral pressure and flow in three different models (*B*, *C*, *D*), to demonstrate the effects of tapering and leakage.

matically shows central pressure and flow (A) and peripheral pressure and flow, corresponding to the site of the femoral arteries. Compared to a nonleaking uniform tube (B), the pressure amplitudes are much more increased in a nonleaking tapering tube (C) (Kenner and Wetterer, 1967). Adding leakage to the tapering tube reduces pressure as well as flow amplitudes (D). In general, leakage throughout the aorta reduces not only the flow pulse amplitudes but also the mean aortic flow values at a peripheral site. Due to the nonlinearity of the pressure flow relation in the resistance vessels (the leakage resistance), the degree of the reduction of both may not be the same. An example is the behaviour of the femoral flow during vasoconstriction. The flow pulse amplitude does not change or may even increase, whereas the mean flow may be reduced to nearly zero. This will be further discussed in Sect. 9.

In addition, it has to be noted that compliance, radius, and characteristic impedance of any artery are nonlinear functions of the distending pressure. This fact has to be considered in simulating pressure and flow pattern in different circulatory situations where remarkable changes of the mean arterial pressure are to be taken into account (Attinger et al, 1968).

The shape of pulse waves within side branches depends on the aortic

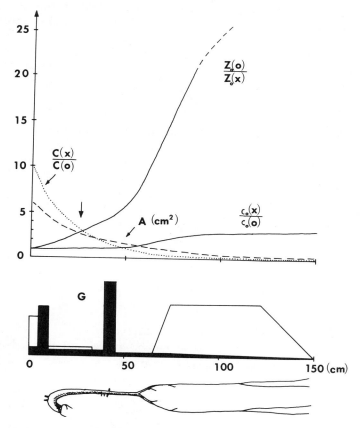

Fig. 4: Distribution of cross-sectional area (A), characteristic impedance (Z_0), wave velocity (c_0), compliance (C), and leakage (G) along a lumped model of the human aorta.
The black bars show the leakage under resting conditions; the white bars during running exercise.

pressure contour at the entrance of the branch, its length and geometric and elastic properties. In short branches, such as the renal artery, the flow contour equals the input pressure. In long side branches the input flow is distorted by reflections returning from the periphery of the branch. The contour of peripheral pressure pulses in long side branches is influenced by the fact that retrograde waves are reflected negatively. With respect to detailed studies on pressure and flow contours in different arteries the reader is referred to the literature (Spencer and Denison, 1963; Wetterer and Kenner, 1968).

To summarize the parameters discussed in this Section, Fig. 4 shows a scheme of the main thoroughfares of the human aortic femoral tube. In the upper part the lumped cross-sectional area (A), compliance per unit length (C), wave velocity (c_0) and the characteristic impedance (Z_0) are shown according to data by Gauer (1936); Patel et al (1963); Remington and Hamilton (1945); Simon and Meyer (1958); Westerhof et al (1969); and Wetterer and Kenner (1968). The vertical arrow shows the location of the arterial hydrostatic indifference point (Kenner and Wetterer, 1967). In the lower part of the figure the distribution of the leakage is shown under resting condition (black) and during running (white). Besides an increase in coronary flow (first block) and a small increase in blood flow to the respiratory muscles, the main part of the cardiac output goes to the exercising muscles (see Falls, 1968). The leakage G in Fig. 4 refers to the mean outflow values. Due to the nonlinearity of the pressure-flow relation of the resistance vessels (see above), and according to few data available (see Sect. 9), it appears probable that the pulsatile leakage values do change much less, if at all, during exercise. The problem is one of the most important in hemodynamics which is still unanswered.

VI. THE LONGITUDINAL IMPEDANCE OF A TUBE SEGMENT

The following discussion concerns the irreversible energy losses due to viscous friction in the streaming fluid and in the arterial wall. For this purpose a second approach using the terminology of the so-called "four-terminal" network theory will be adopted. (See Sect. 8 *infra*.) In particular, with Eqs. 8.1 and 8.2, the longitudinal segment impedance z_L and the transverse segment impedance z_t per wall unit length of a tube are defined. In the present section and in Sect. 7, z_L and z_t be examined closely.

So far it has been assumed that in an arterial segment without branches and leakage, z_L is determined by the effective fluid mass only

(6.1) $$z_L = j\omega M_0$$

and that z_t is due only to ideal and reversible elastic properties of the arterial wall

$$(6.2) \qquad z_t = \frac{1}{j\omega C}$$

The advantage, as well as the limitation, of an application of the algorithm of the four-terminal network theory in hemodynamics is that longitudinal and transverse resistance of an arterial segment can be treated separately (Landes, 1942; Ronniger, 1954, 1955; Taylor, 1959; Wetterer and Kenner, 1968; Womersley, 1957). It has to be assumed that an arterial segment does not change its length during pulsations. This is a reasonable assumption for most arteries in the systemic circulation which are fixed to the surrounding tissue (Patel et al, 1966). The further assumption included is that the system can be treated as linear and that any nonlinear interaction between longitudinal and transverse impedance is small. Both assumptions can be justified within the range of normal pressure and flow pulsations.

For calculation of the longitudinal impedance of a tube segment of unit length, the equation of motion (Eq. 2.9) is used, neglecting radial movements (Witzing, 1914; Womersley, 1957)

$$(6.3) \qquad \frac{\partial v_x}{\partial t} = -\frac{1}{\rho}\frac{\partial p}{\partial x} + \frac{\eta}{\rho}\frac{1}{r}\frac{\partial}{\partial r}\left(r\frac{\partial v_x}{\partial r}\right)$$

A sinusoidal pressure gradient is assumed as forcing function

$$(6.4) \qquad -\frac{\partial p}{\partial x} = P_x e^{j\omega t}$$

A solution for the longitudinal flow velocity of the following form is sought

$$(6.5) \qquad v_x = V_x e^{j\omega t}$$

For simplification the radial coordinate is transformed using

$$(6.6) \qquad y = \frac{r}{r_w}$$

where r_w is the inner radius of the wall of the tube.

It was first shown by Witzig (1914) that under these conditions, assuming laminar flow, the similarity of flow pattern is determined by the dimensionless parameter

$$(6.7) \qquad \alpha = r_w \sqrt{\omega \frac{\rho}{\eta}}$$

Introduction of Eqs. 6.4–6.7 int. Eq. 6.3 leads to the following Bessel equation

(6.8) $$\frac{d^2V_x}{dy^2} + \frac{1}{y}\frac{dV_x}{dy} - j\alpha^2 V_x = -P_x\frac{r_w^2}{\eta}$$

With the boundary condition of no slip at the wall ($V_x = 0$ at $y = 1$) the following solution is obtained

(6.9) $$V_x = \frac{P_x r_w^2}{j\eta\alpha^2}\left[1 - \frac{J_0(\alpha y\sqrt{-j})}{J_0(\alpha\sqrt{-j})}\right]$$

This equation describes the distribution of flow velocities in a rigid tube as a function of radius and time. Integration of Eq. 6.9 over the cross-sectional area of the tube yields the volume flow

(6.10) $$I = \frac{P_x r_w^4 \pi}{j\eta\alpha^2}\left[1 - \frac{2J_1(\alpha\sqrt{-j})}{\alpha\sqrt{-j}\,J_0(\alpha\sqrt{-j})}\right]$$

For zero frequency, Eq. 6.10 leads to Poiseuille's equation

(6.11) $$i_{st} = P_x\frac{r_w^4\pi}{8\eta} = P_x\frac{1}{R_{st}}$$

The index st refers to steady flow, P_x in this case is the steady pressure gradient, R_{st} is the Poiseuille resistance per unit tube length (units dyne. sec. cm^{-6}). Using an abbreviation introduced by Womersley (1957) for the bracketed term, Eq. 6.10 can be written as

(6.12) $$I = P_x\frac{1}{j\omega M_0}(1 - F_{10})$$

where M_0 is the frequency independent effective mass according to Eq. 3.2. The longitudinal impedance of the tube per unit length therefore is

(6.13) $$z_L = \frac{j\omega M_0}{1 - F_{10}}$$

Tables for the function $1 - F_{10}$ can be found in Womersley (1957) and McDonald (1960). Splitting the right side of Eq. 6.13 into its real and imaginary parts yields a form which can easily be compared with other models

(6.14a) $$z_L = R(\omega) + j\omega M(\omega)$$

Eq. 6.14a will be compared in the following with the simplified assumption

(6.14b) $$z_L = R_{st} + j\omega M_0$$

where R_{st} is defined by Eq. 6.11 and M_0 by Eq. 3.2. The effect of the frequency dependence of z_L according to Eq. 6.14a as opposed to Eq. 6.14b on pressure and flow phenomena in tubes has been called "sleeve effect" (Jager et al, 1965).

The upper part of Fig. 5 shows the relations $R(\omega)/R_{st}$ and $M(\omega)/M_0$ as functions of α. At low frequencies or in tubes with small radius or using fluids with high viscosity (see Eq. 6.7), the velocity profile approaches a parabolic form. For this reason $M(\omega)$ is higher than M_0 by a factor 1.33. This factor has been shown in Sect. 2 to be the quotient of mean square to square mean velocity and signifies the actually greater effective mass to be accelerated if the velocity profile is not flat. At increasing α, which, in a given tube and fluid is proportional to the square root of the frequency, the

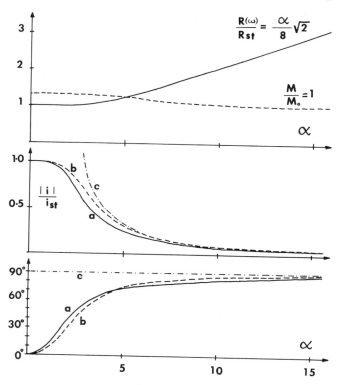

Fig. 5: Upper part: dependence of resistance R and inductance M of a rigid tube on the dimensionless parameter $\alpha = r\sqrt{\omega\eta/\rho}$. Lower part: relative flow amplitudes and phase lag at a given pressure gradient. Solid line according Eq. 6.14a; dashed line according Eq. 6.14b; dash-dotted line according Eq. 6.1.

velocity profile takes on a more and more flat shape. Therefore, $M(\omega)$ approaches M_0. At the same time the shear stress at the wall increases and also, therefore, the flow resistance $R(\omega)$. Using a series expansion of the Bessel functions in Eq. 6.10, it can be shown that $M(\omega)/M_0$ approaches 1 and $R(\omega)/R_{st}$ approaches an asymptote which increases proportionally to α. In a given tube and fluid $R(\omega)$ rises proportionally with the square root of the frequency, whereas $\omega M(\omega)$ increases proportionally with the frequency. For this reason the influence of the effective mass rises faster than that of the resistance with increasing frequency.

This is summarized in the lower part of Fig. 5. The flow amplitude at a given magnitude of the pressure gradient per unit tube length related to the corresponding steady flow value i_{st} is shown for three different models of z_L. In addition, the corresponding values for the phase lead of the pressure gradient are shown (Wetterer and Kenner, 1968). Compared are the values according to Eqs. 6.1, 6.14a and 6.14b. It can be seen that the approximation of Eq. 6.14b is fairly good above $\alpha = 5$ and improves rapidly even for the most simple approximation, Eq. 6.1 above $\alpha = 10$.

The values α for the central aorta of different animals and man at the heart rate is shown in Fig. 5 (McDonald, 1960; Wetterer and Kenner, 1968). The dependence of α on the radius of the ascending aorta can be described by the equation

(6.15)
$$\alpha = 15\sqrt{r_w}$$

which will be explained in Sect. 11.

These values of α justify the use of the simplified approach described in Sect. 3. It also shows that it is reasonable to use constant parameters to cal-

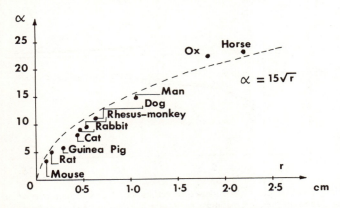

Fig. 6: Dependence of the parameter α of the first harmonic of the heart rate on the radius of the aorta ascendens of different animals. The approximate regression equation is explained in Sect. 11.

culate flow from the pressure gradient in the dog and human aortae (Fry et al, 1956, 1964) or even to neglect friction entirely (Rudewald, 1962).

Jager et al (1965) used a series expansion of the Bessel functions in Eq. 6.10 to simulate the "sleeve effect" in an electrical analog model of the human arterial system (Noordergraaf, 1969; Westerhof et al, 1969). As expected from this discussion, the influence of the "sleeve effect" was shown not to be very striking. Jager et al (1965) also simulated and examined the influence of non-Newtonian flow properties of the blood, which do not seem to play any significant role in arteries except, of course, in the microcirculation (Whitmore, 1968).

VII. THE TRANSVERSE IMPEDANCE OF AN ELASTIC TUBE SEGMENT

The problem of vessel wall elasticity is complicated by two facts: 1) the high nonlinear and anisotropic extensibility and 2) the viscoelastic properties of the wall material. The first property makes use of Young's modulus of elasticity as a rough approximation.

To describe small extensions of an elastic material from an already extended state as is necessary for the description of pulsatile vessel wall distension, incremental moduli may be used (Frank, 1906, 1920; Bergel, 1961a, b; Kenner and Wetterer, 1963; Kenner, 1967). As Kenner (1967) has shown, it is a characteristic feature of these moduli that two different definitions exist simultaneously. The following equations demonstrate this for the case of a simple extension

(7.1) $$E = \frac{dF}{A}\left(\frac{L}{dL}\right)$$

and

(7.2) $$\epsilon = d\sigma\left(\frac{L}{dL}\right)$$

dF is the increase of the extending force, L is the current length of the material, A is the current cross-sectional area of the material perpendicular to the acting force and $\sigma = F/A$ is the stress per unit cross-sectional area. Considering the components of $d\sigma$, the following relation between the two moduli can be found

(7.3) $$\epsilon = E - \sigma\frac{L}{A}\left(\frac{dA}{dL}\right)$$

Kenner (1967) has further developed these considerations for the description of two-dimensional extensions. In general, Eqs. 7.1 and 7.2 have to be replaced by matrix equations which have to obey a transformation equation similar to Eq. 7.3. The moduli as well as the corresponding Poisson ratios can be expressed in terms of the stress-energy-function of the material which, however, is unknown for vessel wall material. Since the stress-strain behaviour of vessel walls is highly dependent on the fiber structure of the tissue, the incompressibility of the material (Frank, 1920; Carew et al, 1968) is not of too much help for the calculation of the elastic behaviour.

The following equations describe the stress-strain relation in a longitudinally tethered artery in terms of the second modulus system (Eq. 7.2)

(7.4) $$\frac{dr}{r} = \frac{1}{\epsilon_t} d\sigma_t - \mu_{tL}\frac{1}{\epsilon_L} d\sigma_L$$

(7.5) $$0 = \frac{dL}{L} = -\mu_{Lt}\frac{1}{\epsilon_t} d\sigma_t + \frac{1}{\epsilon_t} d\sigma_t$$

The stress in transverse (circumferential, subscript t) and longitudinal (subscript L) direction in a thin-walled tube is given by

(7.6) $$\sigma_t = p\frac{r}{h}$$

(7.7) $$\sigma_L = p\frac{r}{2h} + \sigma_T$$

where σ_T is the stress which acts upon the wall due to the longitudinal tethering (Kenner, 1967).

Inserting these equation into Eqs. 7.4 and 7.5, solving for dr/r, and then inserting into Weber's equation (Eq. 3.8) yields the following expression for the wave velocity

(7.8) $$c_0^2 = \frac{\epsilon_t h}{2r\rho(1 - \mu_{Lt}\mu_{tL})} - p$$

A similar expression using the first module system (Eq. 7.1) contains the factor $-p/2$ instead of $-p$. The appearance of a pressure term in the wave velocity equation when incremental moduli are used was first described by Frank (1920). The omission of this factor introduces an error in the calculation of the wave velocity, or, in the calculation of the modulus from pressure of volume measurements, of about 10% at normal arterial pressure values (Wetterer and Kenner, 1968).

Another source of error is the fact that by no means do the Poisson ratios have to be 0.5 in an incompressible material, as usually is assumed for vessel walls. Due to the stretch-dependent anisotropy produced by the fiber structure of the wall, the Poisson ratios are nearly zero at small extensions and, depending on the direction of extension, may achieve values of nearly one at high extensions (Kenner, 1967). Therefore, the expression $(1 - \mu_{Lt}\mu_{tL})$ may differ considerably from the commonly assumed value 0.75. Measurements on ox carotids yield values around 0.6 at *in situ* conditions. It therefore appears that the two sources of error, the pressure factor and the Poisson ratios, may cancel each other under "normal" conditions.

Due to the lack of more accurate data on this problem, the pressure factor occurring in many calculations using incremental moduli will be neglected in the following discussion. In view of still incomplete knowledge, this omission appears justified for a basic description. It is, however, important to point out that in many cases the use of sophisticated methods may improve results from a certain aspect by a few tenths of a percent, whereas at the same time a quite appreciable error is not accounted for by neglecting some basic facts.

The viscoelastic properties of vessel walls were first extensively studied by Ranke (1934). He found that it was impossible to simulate the overall dynamic stress-strain response of vessel walls over a reasonable frequency range by using combinations of springs and dashpots. This has been confirmed since by the results of many workers (Bergel, 1961a, b; Learoyd and Taylor, 1966; Gow and Taylor, 1968). During sinusoidal extension of a vessel wall, the modulus of elasticity increases with respect to the "static" value at very low frequencies and stays nearly constant from 0.1 Hz to frequencies above 20 Hz (Lee et al, 1967). The increase of the dynamic stiffness is the greater the more smooth muscle fibers are contained in the vessel wall. In the central aorta, which contains mostly elastic fibers, the dynamic modulus is about 10% higher than the static modulus. In the carotid artery of dogs, or in other peripheral arteries like the femoral artery, the increase amounts to 70% and even more. Similar to the modulus, the phase angle between stress and extension increases at very low frequencies and stays constant at higher frequencies. Its values are relatively small, in the order of 5–15°, the higher values found in vessels containing more smooth muscle fibers.

The dynamic behaviour of the modulus of elasticity can be described by

(7.9) $$E_d = E + j\omega\eta_w$$

where E as well as the wall viscosity η_w may be functions of the frequency, and, as Lee et al (1967) showed, functions of the amplitude of the extension. The behaviour described above shows some resemblance to a magnetic hysteresis. For this reason and due to the amplitude dependence, it seems doubtful that the so-called "static" value, which corresponds to zero frequency and zero amplitude, can be used as a reference value at all.

The complex dynamic compliance of a longitudinally tethered vessel depends on the dynamic modulus of elasticity and a geometric factor:

$$(7.10) \qquad \frac{dA}{dp} = C_d = \frac{2r^3\pi(1 - \mu_{Lt}\mu_{tL})}{h(E + j\omega\eta_w)}$$

The transverse impedance of the vessel wall, therefore, can be described by

$$(7.11) \qquad z_t = \frac{1}{j\omega C_d} = R_w(\omega) + \frac{1}{j\omega C_w(\omega)}$$

The expression can be split in its real and imaginary parts. $R_w(\omega)$ then corresponds to the real wall resistance due to internal friction. $C_w(\omega)$ is the reversible elastic part of the compliance. In the simplest model of a viscoelastic wall, both R_w and C_w are constant. Actually, in arteries both parameters are frequency dependent in a hitherto not well-understood fashion.

With respect to problems of pulse propagation, the tangent of the angle ϕ, which is equal to the phase lead of pressure versus extension of an arterial segment, is the most important parameter to characterize the viscoelastic properties of a material (Taylor, 1959; Hardung, 1962a, b; Wetterer and Kenner, 1968)

$$(7.12) \qquad \tan\phi = \omega\frac{\eta_w}{E} = \omega R_w C_w$$

Leakage is an additional factor of a different nature which has to be considered in calculating the total transverse impedance if there is a distributed outflow through small leaks in the arterial wall. If G is the admittance per unit length of the tube, then the most general equation for the transverse impedance is given by (Wetterer and Kenner, 1968)

$$(7.13) \qquad z_t = \frac{1}{G + j\omega C_w/(1 + j\omega R_w C_w)}$$

The energy loss due to leakage as described by the admittance G is related only to the pulsatile component of pressure and flow and, therefore, is related to the differential value of the outflow resistance of side branches (see Sect. 9).

VIII. WAVE TRANSMISSION AND DAMPING

There are two ways to examine transmission properties and damping in a linearized tube model. In both cases it is assumed that time varying signals are sinusoidal. Thus, complex notation will be used. With the introduction of the longitudinal and transverse impedances, the equations of motion and continuity can be written in the following form

(8.1) $$\frac{-dP}{dx} = z_L I$$

(8.2) $$\frac{-dI}{dx} = \frac{1}{z_t} P$$

Since the segment impedances may be functions of x, differentiation and mutual insertion yields

(8.3) $$\frac{d^2 P}{dx^2} - \frac{1}{z_L} \frac{dz_L}{dx} \frac{dP}{dx} - \frac{z_L}{z_t} P = 0$$

(8.4) $$\frac{d^2 I}{dx^2} - z_t \frac{d(1/z_t)}{dx} \frac{dI}{dx} - \frac{z_L}{z_t} I = 0$$

After inserting the proper functions for z_L and z_t these equations can be solved by the usual methods.

A second possibility is to use the algorithm of the four-terminal network theory. If P_1, I_1 are pressure and flow at the input side and P_2, I_2 are those at the output side of an arterial segment, then the following relations exist

(8.5) $$P_1 = A_{11} P_2 + A_{12} I_2$$
(8.6) $$I_1 = A_{21} P_2 + A_{22} I_2$$

Since arteries are passive, the determinant of the coefficient matrix A has to fulfill the condition

(8.7) $$|A| = 1$$

For a small arterial segment of length dx, the segment impedances are $z_L dx$ and dx/z_t. Then the matrix A is given by

(8.8) $$A = \begin{pmatrix} 1 & z_L dx \\ dx/z_t & 1 \end{pmatrix}$$

where terms of the order of $(dx)^2$ have been neglected.

The problem of pressure and flow transmission through a chain of $n = L/dx$ segments can be solved by splitting matrix A into the product of two reciprocal and one diagonal matrix

(8.9) $$A = BDB^{-1}$$

Then the corresponding matrix A_n for the pressure and flow transmission through n segments of a uniform tube is

(8.10) $$A_n = BD^n B^{-1}$$

The solution of this problem for a uniform tube segment of length L is

(8.11) $$A_n = \begin{pmatrix} \cosh \gamma L & Z \sinh \gamma L \\ (1/Z) \sinh \gamma L & \cosh \gamma L \end{pmatrix}$$

where

(8.12) $$Z = \sqrt{z_L z_t}$$

is the characteristic impedance of the tube and

(8.13) $$\gamma = \sqrt{z_L/z_t}$$

is the transmission coefficient.

Considering an outflow impedance R at the end of the tube segment, pressure and flow at the entrance are related to the pressure at the exit of the segment by the following equation

(8.14) $$\begin{pmatrix} P_1 \\ I_1 \end{pmatrix} = \begin{pmatrix} P_2 \\ P_2/R \end{pmatrix} A_n$$

In the case of a tube with a matching impedance at the end ($R = Z$), Eq. 8.11 yields

(8.15) $$P_1 = I_1 Z$$

and

(8.16) $$P_2 = P_1 e^{-\gamma L}$$

The transmission coefficient can be split into its real and imaginary parts

(8.17) $$\gamma = \beta + jk$$

β is the damping factor which determines the spatial decrement of the wave. k is the phase delay between two points of unit distance along the tube. Therefore, the phase velocity is given by

(8.18) $$c = \frac{\omega}{k}$$

This value of the phase velocity refers to waves traveling in one direction. An apparent phase velocity can be measured in the presence of reflected waves (see Sect. 10).

The damping-free tube described in Sec. 3 will be used as reference. For this model the characteristic impedance has a real value Z_o. The damping coefficient is zero, the phase velocity is given by

(8.19) $$c_0 = \omega/k_0 \quad \text{where} \quad k_0 = \omega\sqrt{M_0 C}$$

Using the segment impedances given by Eqs. 6.14a and 7.13, the effect of fluid friction, viscoelastic wall friction and distributed leakage on the wave transmission can be calculated (Kries, 1892; Ronniger, 1954; Taylor, 1959; Hardung, 1962; Wetterer and Kenner, 1968). To examine the properties of the aorta, asymptotic solutions for large α can be used since, in general, the parameters Z, c and β converge rapidly as α rises above five. Insertion of z_L and z_t into Eq. 8.12 shows that the characteristic impedance tends towards Z_0 with increasing frequency; however, the viscoelastic wall properties introduce a small pressure phase lead which does not vanish at high frequencies.

(8.20) $$Z = Z_0\sqrt{1 + j\tan\phi}$$

Thus, the characteristic impedance of the aorta can be assumed to be equal to that of a damping-free tube and therefore, real with only a small error (Wetterer and Kenner, 1968). The phase velocity converges with increasing frequency towards

(8.21) $$c = c_o$$

The constancy of the phase velocity at high frequencies has been experimentally proven by Anliker et al (1968) in the range of 20–120 Hz. At α below three the phase velocity tends to zero if fluid friction is present, the limit value for small α being

(8.22) $$\frac{c}{c_0} = \frac{\alpha}{2}$$

In contrast to Z and c, the magnitude and frequency dependence of the damping factor β is highly dependent upon which type of energy loss is prominent.

If leakage is the only distributed energy loss, β approaches a constant asymptotic value with increasing frequency

(8.23) $$\beta = \frac{1}{2} Z_0 G$$

If fluid friction alone represents the loss, then

(8.24) $$\beta = \frac{k_0}{\alpha \sqrt{2}}$$

above $\alpha = 5$, β increases with the square root of the frequency similar to the resistance (see Sect. 6). If fluid friction and viscoelastic wall friction are both present, the latter predominates and the damping factor rises with the frequency

(8.25) $$\beta = \frac{k_0 \tan \phi}{2}$$

As is discussed in Sect. 7, the phase angle ϕ tends toward a constant value between 5–15°.

If, as in the aorta, all three types of energy losses are present, the fluid friction again is overridden in the asymptotic solution and the damping effects of leakage and viscoelastic wall friction add together

(8.26) $$\beta = \frac{1}{2} Z_0 G + \frac{1}{2} k_0 \tan \phi$$

In addition to the physiological importance of this equation with respect to the explanation of damping in the aortic tube, it is interesting to note that, Eqs. 8.25 8.26 are obtained irrespective of whether Eqs. 6.14a or 6.14b is used, that is whether or not the sleeve effect is considered.

Estimates of the frequency independent damping term of Eq. 8.26 from animal experiments range between 0.01 (Kenner, 1962) and 0.015–0.03 cm^{-1} (Hardung, 1962a, b).

The frequency dependent viscoelastic term has most accurately been determined by Anliker at al (1968) in the aortae of dogs. These authors found, in the frequency range of 40–200 Hz, the value $\beta = 0.0006\, \omega$ cm^{-1}, in close agreement with an estimate given by Hardung (1962).

Thus, for the aorta of a dog, the damping factor is given by

(8.27) $$\beta = 0.01 + 0.0006\omega (\text{cm}^{-1})$$

which also appears applicable to the human aorta.

IX. THE INPUT IMPEDANCE

The input impedance relates the the pulsatile pressure P to pulsatile flow I at a certain point of an artery including the effect of reflections. The input impedance of the ascending aorta reflects the impedance which the heart faces during the ejection phase.

From Eq. 8.14 the input impedance P/I of a uniform tube with an impedance R at $x = L$ becomes (Landes, 1942; Ronniger, 1954)

$$(9.1) \qquad R_i = Z \frac{\cosh \gamma L + \dfrac{Z}{R} \sinh \gamma L}{\sinh \gamma L + \dfrac{Z}{R} \cosh \gamma L}$$

This equation can be transformed to

$$(9.2) \qquad R_i = Z \left(\frac{2}{1 - K \exp(-2\beta L - j2kL)} - 1 \right)$$

where K is the (complex) reflection factor at $x = L$.

The Nyquist diagram of the input impedance of a uniform tube, as given by Eqs. 9.1 or 9.2 with $0 < K < 1$ and a frequency dependent damping factor (Eq. 8.26), is a clockwise narrowing spiral in the right half of the complex plane. The curve approaches Z_0 (see Eq. 8.20) with increasing frequency. Figure 7 shows an experimental example recorded in a uniform tube. The damping factor in this particular tube is of the order $10^{-6} \omega$ (cm^{-1}) and thus much smaller than in the aorta (Wetterer and Kenner, 1968).

It is interesting to note that the basic structure of Eq. 9.2 does not change if the total outflow does not take place at the end of the tube, but rather by distributed leakage throughout its length. In this case the resistance at the end

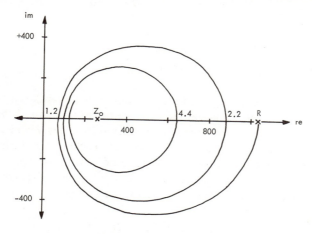

Fig. 7: Nyquist diagram of the input impedance as measured in a homogeneous elastic tube model (drawn after data by Kenner and Wetterer, 1962a). Impedance values in dyn. sec. cm^{-5}. Parameters: frequency in Hz.

may be assumed to be infinite and, therefore, $K = 1$. Using the damping factor, Eq. 8.26, the constant term $exp(-Z_0 G L)$ appears in Eq. 9.2 as "apparent reflection factor" with values of the order $+0.3$, in good agreement with values found in man and different animals (Ronniger, 1954; Kenner, 1959, 1962; McDonald, 1960). In experimental measurements, therefore, the influence of the damping term and of the peripheral impedance and reflection factor cannot be easily, if at all, distinguished. That this is true is equivalent to the fact that in modeling one parameter may be adjusted in order to simulate the effect of other parameters.

The input impedance of cascaded tubes of different length, diameter, and wall properties can be calculated by starting with the uniform segment closest to the end, inserting the resulting value R_i into Eq. 9.1 as end impedance of the next segment, and repeating this process until all segments are included. A graphic method can be used for this purpose (Gessner, 1964). In general, the input impedance of such a system follows the same pattern as shown in Fig. 7, with each segment of the tubing adding to a distortion of the spiral.

Although this method is useful when examining the influence of the tapering of the arterial system, it is easier to consider an exponentially tapering system in order to get some basic insight.

Let z_L as well as z_t depend on the distance x from the origin of the tube with the factor $exp(bx)$, where b is a constant. After inserting into Eqs. 8.3 and 8.4, and solving them, it is found that the input impedance of this leaking tapered tube which is closed at $x = L$ becomes

$$(9.3) \qquad R_i = az_{t0}\left(\frac{2}{1 - \exp(-2aL)} + \frac{b}{2a} - 1\right)$$

where

$$(9.4) \qquad a = \sqrt{\frac{b^2}{4} - \gamma^2}$$

With increasing frequency the factor a converges towards γ, az_{t0} towards Z_0 (the characteristic impedance at the entrance of the tube) and $b/2a$ towards zero. Therefore, the input impedance of a tapered tube has a similar pattern to that of the corresponding uniform tube with the characteristic impedance Z_0 throughout. There is one additional condition: with an equal input impedance pattern the total outflow resistance of the tapered tube is higher than that of the corresponding uniform tube by the factor

$$(9.5) \qquad \frac{bL}{1 - \exp(-bL)}$$

Tapering thus provides a matching between low input and high output impedance (Taylor, 1964, 1965), a fact which has been used in electrical engineering for a long time (Wagner, 1947).

The similarity of the input impedance of uniform and tapered tubes is not restricted to exponential taper, and has been experimentally confirmed by Kenner and Wetterer (1967) in a hydraulic model using a linearly tapered elastic tube.

A very marked damping effect on the pattern of the input impedance can be simulated by random branching (Taylor, 1966a, b). The damping effect in this case is produced by the random superposition of reflected waves. Taylor's randomly branching model is more closely similar to the pulmonary arterial system (Wiener et al, 1966).

The physiologic meaning of the input impedance of a vessel cannot be fully understood without including the low frequency part. There are only two studies available which have examined this low frequency spectrum in the aorta. Taylor (1966a, b) has measured the aortic input impedance in dogs down to a frequency of one cycle per 13 min using a spectral analysis method. Figure 8 shows a schematized and idealized Nyquist diagram of the aortic input impedance of a dog in which active control mechanisms have been abolished by ganglionic blockade. Below a frequency of 1/200 Hz, Taylor's results show a pressure lead which has been omitted in Fig. 8 because it may have been produced by some autoregulatory response (see Sect. 12). The diagram shows two distinct parts, one low frequency, the other high frequency. The latter resembles the one shown in Fig. 7 except that the damping influence is much more pronounced. At decreasing frequency the curve tends towards an intersection with the real axis, but then shifts towards an extended low

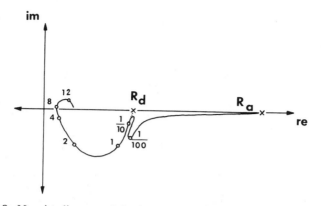

Fig. 8: Nyquist diagram of the input impedance of the aorta of a dog (using data by Taylor (1966c). R_a absolute value of the total outflow resistance, R_d differential value of the total outflow resistance. Parameters: frequency in Hz.

frequency part. The point corresponding to zero frequency (R_a) indicates the total peripheral resistance.

The distinction between low and high frequency parts is more evident from the following results. Wetterer and Pieper (1955) examined the pressure flow relation of the total arterial system of the dog at frequencies of 0.2–0.3 Hz using artificially induced arterial pressure variations. A typical set of results is shown in Fig. 9. In contrast to the well-known parabolic pressure flow curves from "quasi-static" perfusion experiments (Wezler and Sinn, 1953; Green et al, 1963), the relations shown in Fig. 9 are essentially linear within the region of induced pressure variations. These results justify the assumption that the peripheral outflow resistance can be considered linear within the range of pulsatile pressure.

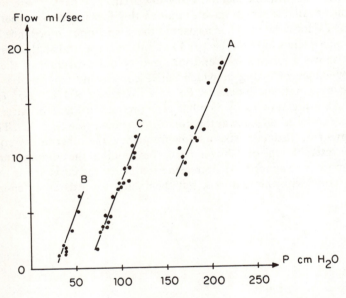

Fig. 9: Pressure flow relation of the arterial system of a dog at a frequency of 0.3 Hz (courtesy of Wetterer and Pieper, 1955). A: during adrenalin infusion; B: during ganglionic blockade; C: control state.

Frequencies of 0.2–0.3 Hz correspond to the region of the input impedance of a dog where the high frequency part of the Nyquist diagram tends towards an intersection with the real axis. According to a theory originally proposed by Ronniger (1955), this extrapolated real value corresponds to the differential value of the outflow resistance

(9.6) $$R_d = \frac{dp}{di}$$

which, as can be seen from Fig. 9, is considerably smaller than the absolute value of the outflow resistance

(9.7) $$R_a = \frac{p_m}{i_m}$$

The index m refers to mean values of pressure and flow. R_d is the resistance value which determines reflection at terminal branches and leakage G for the pulsatile region. In an idealized model the influence of R_a is restricted to zero frequency. Thus R_a would appear as a singular point on the real axis. The actual low frequency part of the Nyquist diagram, as shown in Fig. 8, is produced by the nonlinearity of the peripheral outflow resistance. As will be discussed in Sect. 12, circulatory control influences appear prominent in this region under normal conditions.

Four principal factors influence the shape of the aortic (or pulmonary) input impedance: 1) The energy losses due to fluid and wall friction, and leakage, 2) the nonlinearity of the peripheral outflow resistance, 3) the tapering of the arterial system, and 4) the influence of (random) branching. With respect to modeling, it is important to note that all four factors have a very similar effect on the shape of the input impedance, and that the omission of one or even two can be accounted for by proper adjustment of the others. For this reason the nonlinearity of the peripheral outflow resistance has been neglected in most simulations.

The pattern of the input impedance is similar in the human aorta (Gabe et al, 1964; Patel et al, 1965), the dog aorta (Attinger et al, 1966; Taylor 1966c), and in the pulmonary artery (Patel et al, 1963). It should be expected that the first maximum of the high frequency part of the input impedance appears at the resonance frequency of the system and is equal to $1/T$, where T is the period of the first natural mode. In most of the actual measurements the maximum appears at a higher frequency. Peculiar resonance phenomena in tapered tubes, which may be related to this effect, have been discussed by Kenner and Wetterer (1962, 1967).

The input impedance of the femoral artery or side branches of the aorta is similar in shape to that of the aorta, with however, a different frequency scale, depending on length and wave velocity. A resonance maximum is usually found to be more prominent in the femoral artery impedance under normal conditions, and disappears during vasodilatation (O'Rourke and Taylor, 1966).

If the pressure pattern is used as a judgment about the validity of a simulation, another factor has to be considered in addition. With respect to the generation of the input pressure pattern, the input impedance of a model, as well as *in vivo*, has to be weighted with the input flow amplitudes. Due to the somewhat smoothed triangular shape of the aortic and pulmonary input flow, the first four or five harmonics contain nearly the total pulsatile energy (Attinger et al, 1966). Therefore, higher harmonics do not play a significant role in the generation of the pressure pattern. This explains the fact that the pressure pattern in the two-part tapered tube model (Fig. 2, Kenner and Wetterer, 1962a, b) resembles closely the *in vivo* pulse pattern, although the input impedance of this model has maxima at high frequencies which do not appear *in vivo*. Models which adequately simulate the input impedance in the range of the first to the fourth or fifth harmonic of the pulse rate are sufficient to simulate "normal" pressure patterns.

A problem of great importance which is not yet sufficiently solved is the pressure flow relation in the coronary arteries. For establishing a linearized model the assumption has been made that with each systolic contraction the outflow of these arteries is impeded by a counterpressure $\bar{p}_T(t)$, which is related to an averaged value of the tissue pressure (Kenner, 1969). Using a complex notation, the matrix Eq. 8.14 in this case becomes

$$(9.8) \qquad \begin{pmatrix} P_1 \\ I_1 \end{pmatrix} = \begin{pmatrix} P_2 \\ I_2 \end{pmatrix} A_n$$

where P_1, I_1 are input pressure and flow and P_2, I_2 pressure at the periphery of the coronary artery and outflow through the peripheral resistance R, which is assumed to be constant. The outflow is given by the equation

$$(9.9) \qquad I_2 = \frac{P_2 - \bar{P}_T}{R}$$

The relation between aortic pressure P_1 and coronary inflow I_1 is influenced by the counterpressure $\bar{p}_T$ and, therefore, may be called "apparent input impedance."

$$(9.10) \qquad R_{\text{app}} = \frac{1}{\dfrac{1}{R_i} - \dfrac{\bar{P}_T/(P_1 R)}{A_{11} + A_{12}/R}}$$

where A_{11} and A_{12} are components of the matrix A_n. The actual input impedance of the "undisturbed" system R_i could only be measured in the absence of a pulsatile counterpressure. Using a friction damped uniform tube with a resistance R at the end as the simplest model, it can be shown by comparison with experimental results that $\bar{p}_T(t)$ is of the order of one-half the intraventricular pressure.

As shown in Fig. 10 the shape of the apparent input impedance reflects the active component generated by the time varying counterpressure. Although the general feature of the coronary pressure flow relation can be simulated by Eq. 9.10, a more detailed model will possibly gain more insight

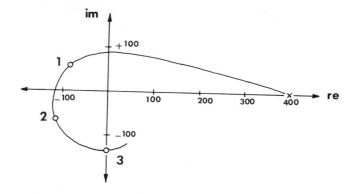

Fig. 10: Apparent input impedance of the circumflex branch of the left coronary artery of a dog. Resistance values times 10^3 in dyn. sec. cm^{-5}. 1, 2, 3: first to third harmonic of the heart rate.

into the nature and distribution of $p_T(t)$. It appears likely that direct measurements of the so-called intramyocardial tissue pressure yield too high values. In terms of Guyton's definition (Guyton, 1970) these measurements reflect the total tissue pressure, rather than the interstitial tissue pressure, the latter seemingly more closely related to p_T.

X. THE WAVE TRANSMISSION

The examination of the pressure wave transmission is an important tool used to characterize the properties of the arterial system. The pattern of the wave transmission very sensitively reflects normal and pathological changes in the arteries (Ronniger, 1954; Hardung, 1962; Kenner, 1962; Attinger, 1963; Luchsinger et al, 1964; Taylor, 1966b, Maloney et al, 1968a; O'Rourke et al, 1968).

For a basic description a uniform tube again may be used. Eq. 8.14 yields (Ronniger 1954; Hardung, 1962)

(10.1) $$\frac{P_1}{P_2} = \cosh \gamma L + \frac{Z}{R} \sinh \gamma L$$

or in another form

(10.2) $$\frac{P_1}{P_2} = \frac{1}{1+K}(e^{\gamma L} + K e^{-\gamma L})$$

P_1 is the pressure at the entrance, P_2 the pressure at $x = L$, K is the (complex) reflection factor at the end of the tube ($x = L$), R is the impedance at $x = L$.

To calculate the pressure transmission from the entrance of the tube to a point $x < L$, the following equation has to be used

(10.3) $$\frac{P_1}{P_x} = \frac{(P_1/P_2)}{(P_x/P_2)}$$

where P_x/P_2 is the transmission from point x to L.

In tapered tubes the right side of Eqs. 10.1 or 10.2 has to be multiplied with $(Z_1/Z_2)^{1/2}$ to yield an asymptotic expression for high frequencies (Kenner, 1963). This confirms that the peripheral pressure amplitudes are amplified corresponding to the square root of the characteristic impedance in tapered tubes. This can be shown to be true not only in exponentially tapering tubes but also in other types of tapering. More detailed studies have been presented by Ronniger (1954); Evans (1962); Kenner (1963); Taylor (1965), and Kenner and Wetterer (1967). The most detailed and advanced study has been presented recently by Skalak and Stathio (1966).

Figure 11 shows the typical pattern of the pressure wave transmission in the aortic system of dogs, the influence of vasodilatation and vasoconstriction, and the measurement at different points along the aorta (Kenner and Held, 1961). The basic pattern of the wave transmission from the entrance to the end of a damping-free tube with a real resistance R at the end, is an ellipse with real intercepts $+1$ and -1 and imaginary intercepts $+jZ_o/R$ and $-jZ_o/R$. Tapering shifts all intercepts, with the exception of the zero frequency point, closer to the origin. The effect of tapering can readily be seen in Fig. 11. Frequency dependent damping widens the ellipse to a counter-

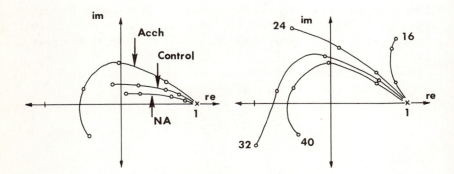

Fig. 11: Pressure transmission in the aorta of a dog. Left part: Transmission from aorta ascendens to the femoral artery during control state, infusion of noradrenalin (NA) and infusion of acetylcholin (Acch). Right: Transmission from the aorta ascendens to points at different distances (parameters in cm) during infusion of acetylcholine (Courtesy of Kenner and Held, 1961).

clockwise spiral. Vasodilatation increases damping due to increased leakage, vasoconstriction decreases this damping effect.

The pressure wave transmission from the aortic root to points at different distances leads to peaks and loops in the Nyquist diagram (Eq. 10.3). The angle at which these loops appear is typical for the relative distance $(x - L)/L$ of the distal measuring point (Lindner and Ronniger, 1955), as can be seen in Fig. 11. The pressure wave transmission has been examined in different models. Humans, dogs, cats, rabbits and rats show a remarkable similarity with respect to the position of the harmonics of the pulse rate (Wetterer and Kenner, 1968). Stenoses produce a characteristic increase of the damping effect, which may well be used for diagnostic purposes. In a randomly branching tube system the pattern of the pressure wave transmission may become more difficult to interpret. However, the general pattern in the pulmonary artery is similar to that in the aorta (Attinger, 1963; Wiener et al, 1966; Maloney et al, 1968a).

Another method which has been used to characterize arterial transmission properties is the measurement of the phase velocities (Hardung, 1962; Wetterer and Kenner, 1968). Parameters to be considered are the phase velocity of a wave traveling in one direction

$$(10.4) \qquad c = \frac{\omega}{k}$$

and the group velocity which is equal to the signal speed of a wave group in the absence of reflections

$$(10.5) \qquad c_g = \frac{d\omega}{dk}$$

At the foot of a pulse wave, where the overall wave velocity usually is measured, it can be assumed $c = c_g = c_0$ (Wetterer and Kenner, 1968)*. If, in the presence of reflections, ϕ is the phase angle of the wave transmission between two points separated by the distance Δx, then

$$(10.6) \qquad c_{ph} = \frac{\omega}{\phi}\Delta x$$

is the mean apparent phase velocity between those two points. The apparent phase velocity is defined by

$$(10.7) \qquad c_{app} = \omega \frac{dx}{d\phi}$$

* As Ranke (1934) had stated, the frequency components, which represent the sharp inflection at the foot of the pulse wave, are in the frequency region where c as well as c_g have converged toward the constant value c_0 (see Sec. 8). Kenner and Wetterer (1962a) confirmed this fact in rubber tube experiments.

and can be measured between very closely situated points at a given location x.

Due to the damping of the pulse waves, c_{ph} and c_{app} tend towards c_0 with increasing frequency. The influence of reflected waves on the measured overall pulse wave velocity, as usually is done at the foot of a pulse wave, is small.

Not much experimental data are available about the flow wave transmission I_1/I_2. An equation analogous to Eqs. 10.1 or 10.2 can be derived from Eq. 8.14. The asymptotic equation for tapering tubes contains the multiplicative factor $(Z_2/Z_1)^{1/2}$, confirming the peripheral decrease of flow amplitudes produced by tapering, and the reciprocity of flow versus pressure amplitudes.

XI. THE LAW OF SIMILARITY

The usefulness of modeling of a system depends on the possibility of proper scaling. This is true for any physical, biological or ecologic system. Galilei (1638) was the first to discuss the general problem of similitude. The law of similarity for hydrodynamic systems was first established by Reynolds (1883, 1895). A detailed description of biological similarity criteria was presented by Stahl (1963).

Two useful dimensionless parameters which describe the similarity of steady flows can be derived from the Navier-Stokes equations. The Reynolds number

$$(11.1) \qquad Re = \frac{2r\rho\bar{v}}{\eta}$$

which is the quotient of acceleration and friction force, and Eulers number

$$(11.2) \qquad S = \frac{2\Delta p}{\rho\bar{v}^2}$$

which is the ratio of the pressure drop to the kinetic energy density. Δp refers to the total pressure drop from the aorta to the right atrium.

It has been shown in Sec. 6 that the solution of the linearized Navier-Stokes equations for the case of sinusoidally varying flow contains the dimensionless parameter

$$(11.3) \qquad \alpha = r\sqrt{\frac{\omega\rho}{\eta}}$$

The similarity of pulsatile flows depends on this parameter.

It was also mentioned that α, related to the frequency of the heart beat, increases with increasing size of an animal. Similarly, as will be discussed below, the Reynolds number as related to the mean aortic flow velocity increases with the size of an animal. Only the parameter S has similar values in different animals. It appears, therefore, that the similarity of the design of the circulatory system is not related to Re or α.

Considering average resting state values of different homoiotherm animals, the following gross pattern can be seen. Density, viscosity and kinematic viscosity are similar. The radius of the aorta is size-dependent, proportional to the length dimension of the body. In discussing the pressure wave transmission it was mentioned that the shape of the Nyquist diagram was similar in different animals if the frequency scaling was done according to the harmonics of the heart rate. The same is true for the input impedance. The order of magnitude of the aortic wave velocity is the same in different animals. Looking at Moens equation (3.9) this is not so surprising, because the ratio h/r of different aortas appears to be similar, and, since the gross composition of the biomaterial of the vessel wall is likely to be similar, this is also true for the modulus of elasticity. The flow velocities in the aorta of different animals are equal (Wetterer, 1956). The aortic and pulmonary velocity contours have a somewhat rounded triangular shape with peak velocities around 100 cm/sec and a mean velocity around 20 cm/sec in animals of different size. The mean artrial blood pressure (80–100 mm Hg) and the pressure amplitudes (40–60 mm Hg) are independent of the size of an animal; although with certain exceptions, such as the giraffe, which has a considerably higher blood pressure (200–300 mm Hg). The latter fact can be "explained" by the exceptionally high hydrostatic pressure difference in this animal.

These facts, with two major exceptions which will be discussed below, fit into a very useful scheme, proposed by Lambert and Teissier (1927). Assuming a constant average density of biomaterial, mass is proportional to the cube of the length dimension. More important is the hypothesis that individual biological time periods are proportional to the length dimensions of the body.

Using λ as a nondimensional factor of proportionality, the following relations can be established by applying this hypothesis.

TABLE I

	proportionality	dimensions
length	λ	L
mass	λ^3	M
time	λ	T
surface	λ^2	L^2
velocity	1	LT^{-1}
force	λ^2	MLT^{-2}
energy	λ^3	ML^2T^{-2}
power	λ^2	ML^2T^{-3}
total peripheral resistance $R_T = p_m/i_m$	λ^{-2}	$ML^{-4}T^{-1}$
total compliance C_T	λ^3	$M^{-1}L^4T^{-2}$
characteristic impedance	λ^{-2}	$ML^{-4}T^{-1}$
overall time constant $R_T C_T$	λ	T

The overall time constant is a factor derived from the Windkessel theory. Although only an approximate measure, its proportionality with the length dimension is consistent with the fact that a constant pressure amplitude in animals of different size can only be expected if the normal resting heart period also is proportional to the length dimension of the body (Deppe, 1941). Figure 12, after data given by Lambert and Teissier (1927), gives a proof of this fact. More detailed theories to explain the same fact led to a similar conclusion (Broemser, 1935; Wetterer and Kenner, 1968).

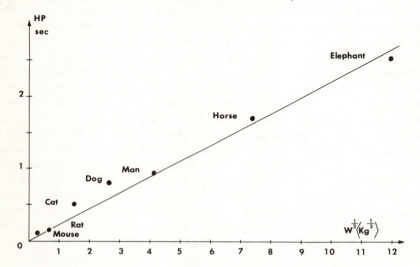

Fig. 12: Dependence of the normal resting heart period (sec) on the length dimension of the body (cube root of the body mass). Data by Lambert and Teissier, 1927.

There are two factors which do not fit into this scheme. 1) According to it the viscosity of body fluids should increase in proportion with the length dimension. Actually, it is constant. 2) The total energy consumption per unit time (basal metabolic rate, BMR) should increase in proportion to the surface area and proportional to (body mass)$^{2/3}$. Actually the relation

(11.4) $$BMR = (\text{body mass})^{3/4}$$

has been found over the range of sizes from unicellular organisms to elephants (Zeuthen, 1955). It appears likely that both deviations have a common denominator, and that the factor of proportionality of different forms of energy and energy losses is different. Thus acceleration forces follow the scheme of Table 1, but not friction forces.

As far as hemodynamic parameters are concerned, this deviation appears to have only a minor influence on the parameters listed in table 1, apparently because acceleration forces prevail over friction forces in the large arteries. However, the two dimensionless parameters Re and α, which would be expected to be size-independent according to Lambert and Teissier's scheme, increase with size. Re rises in proportion to λ, and the parameter α, which governs the similarity of phasic laminar flow, rises proportional to the square root of λ. It was shown in Fig. 6, that $\alpha = 15r^{1/2}$ gives a good approximation to the actual values of α for the first harmonic of the heart rate.

The importance of these considerations is not only of basic, but also of practical, interest. Relations as shown in Table 1 determine the body reaction of animals of different size to environmental forces and thus influence tolerance limits (Gierke 1964, 1969).

Compared with the resting state, the total oxygen consumption of the body may increase twenty-fold, cardiac output and mean aortic flow velocity may increase up to five-fold, during physical exercise (Bevegard and Shepherd, 1967; Falls, 1968). The increase of the peak velocity is somewhat less, due to the relative shortening of the diastole in tachycardia. The mean arterial pressure increases only slightly, whereas the pressure amplitudes show a more remarkable increase (Kroeker and Wood, 1955; Falls, 1968). In humans, systolic pressure values may reach levels above 200 mm Hg in maximum exercise, the total peripheral resistance and the overall time constant thus may decrease nearly to one-fifth of the resting value. It has been shown that the rise of cardiac output in exercising dogs is independent of the increase of the heart rate (Warner and Topham, 1967). The actual increase of the heart rate, thus appears to be an adjustment to the shortening of the overall time constant. Although such an interpretation would also agree with the decrease of the heart rate in well-trained persons, the high complexity of all the factors involved in the control of the heart rate demands some caution as to the generalization of this "explanation" (Stegemann and Kenner, 1971).

The normal resting state usually refers to a young adult male subject. During the phase of growth the hemodynamic parameters obey, in general, the same rules as established for the interspecies relations. However, after adolescence marked deviations may appear. The radius of the aorta continues to increase and the wall material stiffens (Learoyd and Taylor, 1966). The zero-pressure volume of the whole aorta in humans increases nearly three times from 20–70 years (Simon and Meyer, 1958). These changes are reflected in the behaviour of the pulse wave velocity. The increase of the modulus of elasticity overrides the effect of the increase of the radius (Eq. 3.9). Figure 13 shows the mean aortic wave velocity as measured between carotid artery and femoral artery, depending on mean blood pressure and age (Schimmler, 1965). The increase of aortic stiffness and of the wave velocity is reflected in a steady increase of the pulse pressure and of the systolic pressure with age.

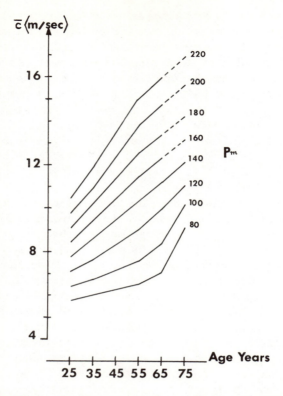

Fig. 13: Mean aortic pulse wave velocity as a function of age (abscissa), and mean arterial blood pressure (parameter, mm Hg) in humans (modified after Schimmler 1965, Courtesy of Schimmler). The dashed parts are uncertain due to selection by death.

Thus, Eulers number (S) is at a constant value and friction forces and Reynolds number (Re) play a small role, all at the cost of turbulent flow developing with higher probability and degree the bigger the size of an animal.

Two examples show that both facts may only be explainable by taking biochemical and metabolic factors into account.

The plasma of all mammals has the same protein concentration, the same viscosity, and the same colloid osmotic pressure (Stahl, 1963). In order to provide equilibrium of the body fluids, the ratio of the arterial pressure to the colloid osmotic pressure appears to be important. The ratio actually has the value 5.0 in all mammals (Stahl, 1963).

Relating vascular pressures to the work of the heart, we find the following connections. Neglecting pulsatility, the work expenditure of the heart per unit time ($\dot{W}$) amounts to

(11.5)
$$\dot{W} = i(\Delta p_1 + \Delta p_2)$$

where

(11.6)
$$i = i_c + i_p$$

is the cardiac output, which partly goes into the coronary arteries (i_c), and to the periphery of the body (i_p). Δp_1 is the pressure difference between aorta and right atrium, Δp_2 is the pressure difference between pulmonary artery and left atrium. If C_a is the arterial oxygen content, ϵ_c the fraction of oxygen which the heart muscle extracts from the perfusing blood and E_H is the thermodynamic efficiency of the heart, then

(11.7)
$$\dot{W} = kC_a\epsilon_c E_H i_c$$

where k is the energy equivalent of oxygen (approximately 5 cal/ml under normal conditions).

If the dimension cal/ml is used for the pressures, then Eq. 11.5–11.7 lead to

(11.8)
$$\frac{i_c}{i} = \frac{\Delta p_1 + \Delta p_2}{kC_a\epsilon_c E_H}$$

The dimensionless ratios on both sides of Eq. 11.8 relate the fraction of the cardiac output which provides the cardiac pump with supply via the coronary arteries (i_c/i), to the ratio between hydraulic pressures and metabolic factors. All parameters on the right side of the equation appear to be dimensional constants in all mammals under comparable resting conditions. Therefore, the relation between coronary flow and cardiac output is constant. Since the metabolic properties of tissues are similar in different animals, the relation implies that, in addition, the relation of the corresponding masses of the perfused organs is constant, a fact most strikingly documented by comparing the relation between heart weight and body weight of small mammals and whales (Black-Schaffer et al, 1965).

XII. PULSATILE PRESSURE AND FLOW

Pulsatility is the consequence of a natural design principle. Pulsatile action apparently is the most efficient way of a hollow vessel to propel a fluid in a closed circuit. The system is designed in such a way that the blood flow through the heart muscle itself is secured. Since, during each contraction of the heart the coronary flow is impeded, the main flow takes place during diastole (Gregg and Fisher, 1963). Several factors act together to keep the diastolic aortic pressure at a high level as the potential energy source for the coronary flow. The effect of these factors usually is summarized as the Windkessel action. This action, in addition, minimizes the pulsatile power expenditure of the heart and provides a smoothing of the peripheral outflow.

The same factors which produce the Windkessel action, however, not only tend to preserve but to increase the peripheral pressure amplitudes within the system. Speculations reappearing from time to time even assume active contractile pulsations of peripheral arteries, which certainly is not the case in the frequency of the heart rate. The frequency range of active contractions is below 0.3 Hz. Figure 14 shows a Nyquist diagram of the aortic input impedance of a dog which is somewhat schematically drawn after data by Taylor (1966c). It shows a clear separation between the passive high frequency region, which is similar to that shown in Fig. 8, and the low frequency part which shows the effect of active neural (baroreceptor) and peripheral (autoregulatory) control. Using a highly simplified linearized method, it is possible to describe the general feature of the low frequency part of the input impedance by the equation

(12.1)
$$R_{ic} = \frac{1}{(1/R_i)(1 + GR_i/R)}$$

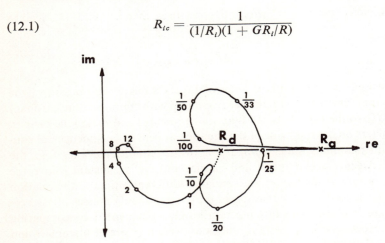

Fig. 14: Nyquist diagram of the input impedance of the aorta of a dog, using data by Taylor (1966c); the effects of circulatory control are included. See Fig. 8.

where R_{ic} is the input impedance of the controlled system, R_i the input impedance of the uncontrolled system (Fig. 9) and R is the total peripheral outflow resistance of the uncontrolled system. G is the open loop pressure gain (amplitude ratio) of the barostatic system, which is a nonlinear second order system with a time delay (Scher and Young, 1963; Levison et al, 1966; Kenner et al, 1970). Under certain conditions, the resonance maximum at 1/25 Hz may lead to pressure oscillations in this frequency range (Mayer waves).

Due to the nonlinearities of the outflow resistance of vascular beds, a passive second order rectification of mean flow may appear (Ronniger, 1955). The following expansion of $i = f(p)$ into a Taylor series explains the shift of the mean flow.

(12.2) $$i = f(p_m) + f'(p_m)(p - p_m) + \frac{1}{2}f''(p_m)(p - p_m)^2 \cdots$$

If the second order term is finite and positive, the mean flow i_m can be shown to be higher at a given mean pressure gradient p_m in pulsatile, rather than in nonpulsatile perfusion. Under certain circumstances this effect is of great importance. In an upright person the hydrostatic pressure difference from the top to the bottom of the lungs is greater than the mean pulmonary artery pressure. Therefore, only the peaks of the pulsatile pressure allow the pumping of blood through the upper part of the lungs, which otherwise would not be perfused at all (Maloney et al, 1968b).

An interesting model of this process is realized by Riva Rocci's method of blood pressure measurement (Kenner and Gauer, 1962). If the cuff around the arm is inflated to a pressure above the diastolic value, only the "upper parts" of the pulse wave are transmitted. The volume transmitted per pulse stroke is

(12.3) $$V = \frac{1}{Z} \int_{t_1}^{t_2} [(1 + K)p(t) - p_T] \, dt$$

p_T is the tissue pressure produced by the cuff, $p(t)$ is the pressure contour of the incoming pulse wave, Z is the characteristic impedance of the brachial artery, and K is the reflection factor produced at the compression site of the artery. The pulse pressure is higher than p_T from t_1 to t_2. It is interesting to note that the volume of blood being propelled into the compressed part of the artery grows with decreasing compression pressure p_T until, depending on the length of the cuff, a small pulse is transmitted through the compressed artery. The accuracy of the method, therefore, depends on the proper relation between vascular dimensions and cuff length.

A quite similar condition appears to exist during sustained contraction of muscles, where the increase of the total tissue pressure produced by the contracting muscle compresses the small vessels and increases the resistance (Anrep et al, 1933; Hirche et al, 1970).

In addition to these effects, peripheral pulsatility appears to act as an energy source to propel the lymphatic flow (Rusznyák et al, 1967). It has been shown that nonpulsatile perfusion of organs or of the whole body produces edema and may, after several hours, lead to irreversible functional and structural changes (Takeda 1960; Dalton et al, 1965; Peirce, 1969).

Besides these "passive" rectification phenomena, "active" rectification plays a role in peripheral autoregulation and barostatic control. Thus, pulsatile perfusion of the carotid sinus has a remarkable influence on the control of the arterial pressure, as compared with nonpulsatile perfusion (Kezdi, 1967; Stegemann and Tibes, 1969).

Our understanding of the interaction of all these passive and active phenomena related to the influence of pulsatility is still very poor and most discussions in the literature are somewhat contradictory.

Compared with a pump which produces steady flow, the heart has to spend an additional amount of energy to propel the same mean flow in a pulsatile manner. The power expenditure of the left ventricle is 10–20% higher than that of a corresponding steady flow pump. The extra expenditure is still higher for the right ventricle. Since most of the energy is transformed into heat during the passage of blood through the peripheral resistance, a higher "effective resistance" can be calculated for the pulsatile perfused system (Skalak et al, 1966).

The additional energy produced by the heart to generate pulsatile pressure and flow, is a source of additional information. Riva Rocci's method for noninvasive measuring of systolic and diastolic blood pressure has been mentioned above. Many methods for measuring blood flow rely on the existance of pulsations. These include, as a typical example, the so-called pulse contour methods. Ever since Broemser and Ranke (1930) and Frank (1930) published their first attempt to determine the stroke volume of the heart from the pulse contour, this possibility has attracted attention. Only recently, Broemser and Ranke's formula has been rediscovered or reinvented (Warner et al, 1953, 1968; Kouchoukos et al, 1970).

Two basic types of equations can be used to calculate the stroke volume from the pressure pulse contour (Ronniger 1955; Wetterer and Kenner, 1968). The first type is given by

$$(12.4) \qquad V_s = \frac{T p_m}{R_T}$$

where T is the pulse period and R_T the total peripheral resistance. The attempt is made to determine R_T by assuming that the central aortic pressure declines exponentially during the period of the diastole D with the overall time constant $\tau = R_T C_T$, which can be expressed by

$$(12.5) \qquad \tau = \frac{D}{\ln(p_E/p_D)} = \frac{D p_{Dm}}{(p_E - p_D)}$$

p_E is the end diastolic aortic pressure at the moment of the valve closure, and p_{Dm} is the mean pressure during diastole. Since p_{Dm} and p_m usually are very close, insertion into Eq. 12.4 yields

(12.6) $$V_s = C_T(p_E - p_D)\frac{T}{D}$$

The factor T/D or $(1 + S/D)$ is characteristic for this type of equation (S is the duration of the systole). The problem, thus, is shifted to the determination of the total arterial compliance C_T. The methods of Broemser and Ranke (1930), Frank (1930) and Wezler and Boeger (1937) take a uniform tube of length L, with the cross section area A of the ascending aorta and the mean wave velocity of the aorta as a model of the system. Thus

(12.7) $$C_T = \frac{AL}{(c^2\rho)}$$

Attempts were made to determine the length L from the autooscillation period of the arterial pulse (Frank 1930; Wezler and Boeger 1937), from the duration of the systole (Broemser and Ranke, 1930), or to relate L to the anatomical length of the aorta (Aperia, 1941). The main source of error in Eq. 12.6 lies in the calculation of the time constant. The lower the arterial pressure and the wave velocity are, the less valid is the assumption that the diastolic aortic pressure contour follows an exponential decline. For this reason the results have a greater uncertainty at low arterial pressures (Wetterer and Deppe, 1939, 1940).

The second basic type of equation is given by

(12.8) $$V_s = \frac{1}{Z_0}\int_0^S (p(t) - p_D)\, dt$$

where S is the duration of the systole (ejection phase) and $p(t)$ the central aortic pulse contour. Due to the distortion of $p(t)$ by reflections, which usually are enhanced by vasoconstriction and increased blood pressure, the error of this type of equation is expected to be greater at high blood pressure values. Additional errors are introduced by the application of noninvasive measurements of the blood pressure and the pulse contours (sphygmograms).

The empirical method of Remington and Hamilton (1945). and Hamilton and Remington (1947) can be related to both types of equations. Although the method appears to have a surprising accuracy, its application is restricted to dogs.

In conclusion, it is surprising how much our knowledge about the practical biological importance and the practical applicability of the phenomenon of pulsatility lags behind the theoretical knowledge. It is the sincere hope of the author that this study shows where some of the important problems still are hiding.

List of Symbols

(Some symbols which are used only once are defined in the text and are not listed here).

A	cross sectional area, transmission matrix
c	wave velocity
E, ϵ	modulus of elasticity
F	force
G	leakage (admittance per unit length)
h	wall thickness
i, I	volume flow (capital: complex)
j	$= \sqrt{-1}$
K	reflection factor (real or complex)
k	phase constant
L	length
M	effective mass
p, P	pressure (capital: complex)
R	resistance, impedance
r	radius
t	time
T	period
v, V	velocity (capital: complex)
$\bar{v}$	velocity averaged across cross section
V_s	stroke volume
Z	characteristic impedance
z_t, z_L	transverse and longitudinal impedances, respectively
α	dimensionless parameter
β	damping coefficient
γ	propagation coefficient
η	viscosity
λ	coefficient
μ	Poisson ratio
$\nu = \eta/\rho$	kinematic viscosity
π	$= 3.1415927$
ρ	density
σ	stress
ϕ	phase angle
$\omega = 2\pi f$	circular frequency

Subscripts

o	reference
$1, 2$	entrance and end of a tube, respectively
a	absolute
d	differential or dynamic

i input
m time average (mean)
w wall
t, L directions in a tube: transverse (tangential), longitudinal
x, y, z directions along Cartesian coordinates
$r, x, 0$ directions in cylindrical coordinates

REFERENCES

Anliker, M., M. B. Histand and E. Ogden 1968. Dispersion and attenuation of small artificial pressure waves in the canine aorta. Circulat. Res. 23: 539–51.

Anrep, G. V., A. Blalock and A. Samaan 1933. The effect of muscular contraction upon the blood flow in the skeletal muscle. Proc. Roy. Soc. Lond. B 114: 223–45.

Aperia A. 1941. A new physical stroke volume formula and its applications. Acta Physiol. Scand. 2: 64–70.

Attinger, E. O. 1963. Pressure transmission in pulmonary arteries related to frequency and geometry. Circulat. Res. 12: 623–41.

Attinger, E. O. 1964. Ed. Pulsatile blood flow. McGraw-Hill Book Company, New York.

Attinger, E. O. 1966. Hydrodynamics of blood flow. In V. T. Chow (ed.), Advances in Hydroscience. Vol. 3 Academic Press, New York, London.

Attinger, E. O. 1968. Analysis of pulsatile blood flow. In S. N.Levine (ed.), Advances in Biomedical Engineering and Medical Physics. Wiley (Interscience), New York.

Attinger, E. O. 1969. Mechanical behaviour of biological systems, In J. F. Dickson, III and J. H. U. Brown (eds.), Future goals of engineering in biology and medicine. Academic Press, New York, London.

Attinger, E. O., A. Anne, T. Mikami and H. Sugawara 1968. Modelling of pressure flow relations in arteries and veins. In A. L. Copley (ed.), Pergamon Press, New York. Hemorheology.

Attinger, E. O., H. Sugawara, A. Navarro, A. Ricetto and R. Martin 1966. Pressure flow relations in the dog arteries. Circulat. Res. 19: 230–46.

Bauer, R. D., Th. Pasch and E. Wetterer 1970. Neuere Untersuchungen zur Entstehung der Pulsformen in grossen Arterien des Menschen. Verh. Dtsch. Gas. KreislForsch. 36 (in print).

Bergel, D. H. 1961a. The static elastic properties of the arterial wall. J. Physiol. 156: 445–57.

Bergel, D. H. 1961b. The dynamic elastic properties of the arterial wall. J. Physiol 156: 458–69.

Bevegard, B. S. and J. T. Shepherd 1967. Regulation of the circulation during exercise in man. Physiol. Rev. 47: 178–213.

Black-Schaffer, B., C. E. Grinstead and J. N. Braunstein 1965. Endocardial fibroelastosis in large mammals. Circulat. Res. 26: 383–90.

Borelli, G. A., 1681. Cit. According to W. Sinn 1956.

Broemser, Ph. 1935. Ueber die optimalen Beziehungen zwischen Herztaetigkeit und physikalischen Konstanten des Gefaessystems. Zeitschr. Biol. 96: 1–10.

Broemser, Ph., and O. F. Ranke 1930. Ueber die Messung des Schlagvolumens des Herzens. auf unblutigem Weg. Zeitschr. Biol. 90: 467–507.

Carew, T. E., R. N. Vaishnar and D. J. Patel 1968. Compressibility of the arterial wall. Circulat. Res. 23: 61–68.

Cox, R. H. 1970. Blood flow and pressure propagation in the canine femoral artery. J. Biomech. 3: 131–49.

Dalton, M. L., E. C. Mosley, K. E. Woodward and Barila, T. G. 1965. The effect of pulsatile flow on renal blood flow during extracorporeal circulation. J. Surg. Res. 5: 127–31.

Deppe, B. 1941. Ueber den diastolischen arteriellen Druckverlauf im grossen Kreislanf Zeitschr. Biol. 100: 437–74.
Evans, R. L. 1962. Pulsatile flow in vessels where distensibility and size vary with site. Phys. Med. and Biol. 7: 105–16.
Falls, H. B. 1968. Ed. Exercise Physiology. Academic Press, New York, London.
Frank, O. 1899. Die Grundform des arteriellen Pulses. Zeitschr. Biol. 37: 483–526.
Frank, O. 1903. Kritik der elastischen Monometer. Zeitschr. Biol. 44: 445–613.
Frank, O. 1906. Die Analyse endlicher Dehnungen und die Elastizitaet des Kautschuks. Ann. Physik 21: 602–608.
Frank. O. 1920, 1928. Die Elastizitaet der Blutgefaesse. I: Zeitschr. Biol. 71: 255–72, II: Zeitschr. Biol. 88: 105–18.
Frank, O. 1926. Die Theorie der Pulswellen. Zeitschr. Biol. 85: 91–130.
Frank. O. 1930. Schaetzung des Schlagvolumens des menschlichen Herzens auf Grund der Wellen-und Windkesseltheorie. Zeitschr. Biol. 90: 405–409.
Fry, D. L., D. M. Griggs and J. G. Greenfield, 1964. *In vivo* studies of pulsatile blood flow: the relationship of the pressure gradient to the blood velocity. *In* E. O. Attinger (ed.), Pulsatile blood flow. McGraw-Hill Book Company, New York.
Fry, D. L., A. J. Mallos and A. G. T. Casper 1956. A catheter-tip method for measurement of the instantaneous aortic blood velocity. Circulat. Res. 4: 627–32.
Fung, Y. C. 1966. Ed. Biomechanics. Proceedings Symposium Amer. Soc. Mech. Eng. *ASME*, New York.
Gabe, I., T. J. Karnell, I. G. Porje and B. Rudewald. 1964. The measurement of input impedance and apparent phase velocity in the human aorta. Acta Physiol. Scand 61: 73–84.
Galilei, G. 1638. Discorsi e Dimostrazioni Matematiche. Cit. according to A. Sommerfeld 1950.
Gauer, O. H. 1936. Ueber Pulswellengeschwindigkeit in Aorta und Beinarterien des Menschen. Zeitschr. KreislForsch. 28: 7–16.
Gessner, U. 1964. A graphic method of computing pulsatile flow parameters. *In* E. O. Attinger (ed.), Pulsatile blood flow. McGraw-Hill Book Company, New York.
Gierke, H. E. von, 1964. Biodynamic response of the human body. Appl. Mech. Rev. 17: 951–58.
Gierke, H. E. von, 1969. Mechanical behaviour of biological systems. *In* J. F. Dickson, III and J. H. U. Brown (eds.), Future goals of engineering in biology and medicine. Academic Press, New York, London.
Gow, B. S. and M. G. Taylor 1968. Measurement of viscoelastic properties of arteries in the living dog. Circulat. Res. 23: 111–22.
Green, H. D., C. E. Rapela and M. G. Courad 1963. Resistance (Conductance) and capacitance phenomena in terminal vascular beds. *In* W. F. Hamilton and P. Dow (eds.), Handbook of Physiology. Circulation, Vol. II Washington D.C.
Gregg, D. E. and L. C. Fisher, 1963. Blood supply to the heart. *In* W. F. Hamilton and P. Dow (eds.), Handbook of Physiology. Circulation, Vol. II Washington D.C.
Guyton, A. C. 1970. Textbook of Medical Physiology, 3rd ed. W. B. Saunders, Philadelphia.
Hales, S. 1733. Statical Essays: containing Hemostaticks. W. Innys, London.
Hamilton, W. F. and J. W. Remington 1947. The measurement of the stroke volume from the pressure pulse. Amer. J. Physiol. 148: 14–24.
Hardung, V. 1952. Zur mathematischen Behandlung der Daempfung und Reflexion der Pulswellen. Archiv. KreislForsch. 18: 167–72.
Hardung, V. 1962a. Die physikalische Analyse der in der Aorta des Hundes und ihrer Hauptaeste registrierten Druckwellen. Archiv KreislForsch. 37: 87–111.
Hardung, V. 1962b. Propagation of pulse waves in viscoelastic tubings. *In*, W. F. Hamilton and P. Dow (eds.), Handbook of Physiology. Circulation Vol. I Washington D.C.
Hirche, H., W. K. Raff and D. Grün 1970. The resistance to blood flow in the gastrocnemius of the dog during sustained and rhythmical isometric and isotonic contractions. Pflügers Arch. 314: 97–112.
Jager, J. N., N. Westerhof and A. Noordergraaf 1965. Oscillatory flow impedance in electrical analog of arterial system. Circulat. Res. 16: 121–33.

Kapal, E., F. Martini, H. Reichel and E. Wetterer 1951. Ueber die Laenge der stehenden Welle bei kuenstlicher Verkuerzung des Arteriensystems. Zeitschr. Biol. 104: 429–44.

Kenner, T. 1959. Ueber die elektrokymographische Pulskurve der Arteria Pulmonalis. Archiv. KreislForsch. 29: 268–90.

Kenner, T. 1962. Experimentelle Untersuchungen zur Analyse der Umformung von Druckpulsen im arteriellen System von Kaninchen. Zeitschr. Biol. 113: 125–44.

Kenner, T. 1963. Zur Theorie der Ortsabhaengigkeit des Wellenwiderstandes im Arteriensystem und ihrer Beziehung zur peripheren Zunahme des Pulsdrucks. Zeitschr. Biol. 113: 409–21.

Kenner, T. 1967. Neue Gesichtspunkte und Experimente zur Beschreibung und Messung der Arterienelastizitaet. Archiv. KreislForsch. 54: 68–139.

Kenner, T. 1969. The dynamics of pulsatile flow in the coronary arteries. Pflügers Arch. 310: 22–34.

Kenner, T. and Gauer, O. H. 1962. Untersuchungen zur Theorie der auskultatorischen Blutdruckmessung. Pflügers Arch. 273: 23-45.

Kenner, T. and Held, D. 1961. (Results published in Wetterer, E. and Kenner, T. 1968.)

Kenner, T. and R. Ronniger 1960. Untersuchung zur Entstehung der normalen Pulsformen. Archiv KreislForsch. 32: 141–73.

Kenner, T., J. Stegemann, J. Allison, A. Anne and E. O. Attinger, 1970. Systems analysis of the open and closed loop carotid sinus baroreceptor response. Fed. Proc. 29: 515 Abstract.

Kenner, T. and E. Wetterer, 1962a. Experimentelle Untersuchungen ueber Pulsformen und Eigenschwingungen in zweiteiligen Schlauchmodellen. Pflügers Arch. 275: 594–613.

Kenner, T. and E. Wetterer 1962b. Zur Theorie der Eigenschwingungen zweiteiliger Schlauchmodelle. Pflügers Arch. 275: 614–27.

Kenner, T. and E. Wetterer 1963. Studie ueber die verschiedenen Definitionen des Elastizitaetsmoduls in der Haemodynamik. Zeitschr. Biol. 114: 11–24.

Kenner, T. and E. Wetterer 1967. Experimentelle Untersuchungen an einem Schlauchmodell, dessen Wellenwiderstand periphaerwaerts kontinuierlich zunimmt. Pflügers Arch. 295: 99–118.

Kezdi, P. 1967. Ed. Baroreceptors and hypertension. Pergamon Press, New York.

Klip, W. 1969. Theoretical foundations of medical physics: Vol. II Univ. of Alabama Press. (An introduction into medical physics.)

Kolin, A. 1936. An electromagnetic flowmeter. Principle of the method and its application to blood flow measurements. Proc. Soc. Exper. Biol. 35: 53–56.

Kouchoukos, N. T., L. C. Sheppard and D. A. McDonald 1970. Estimation of stroke volume in the dog by a pulse contour method. Circulat. Res. 26: 611–23.

Kries, J. von, 1892. Studien zur Pulslehre. Freiburg. Akademische V. B. H. von J. C. B. Mohr.

Kroeker, J. and E. H. Wood 1955. Comparison of simultaneously recorded central and peripheral arterial pressure pulses during rest, exercise, and tilted position in man. Circulat. Res. 3: 623–31.

Lambert, R. and G. Teissier 1927. Théorie de la similitude biologique. Ann. Physiol. 3: 212–46.

Landes, G. 1942. Studien zur Theorie des Blutdrucks und der Blutstroemung in den Arterien. Zeitschr. Biol. 101: 1–20.

Learoyd, B. M. and M. G. Taylor 1966. Alterations with age in the viscoelastic properties of human arterial walls. Circulat. Res. 18: 278–92.

Lee, J. S. 1966 The pressure flow relationship in long wave propagation in large arteries, In Y. C. Fung. (ed.) ASME, New York. Biomechanics.

Lee, J. S. 1970. (Personal communication.)

Lee, J. S., W. G. Frasher and Y. C. Fung 1967. Two-dimensional finite deformation experiments in dog's arteries and veins. AFOSR Report 67–1980.

Levison, W. H., G. O. Barnett and W. D. Jackson 1966. Nonlinear analysis of the baroreceptor reflex system. Circulat. Res. 18: 673–82.

Lindner, A. and R. Ronniger 1955. Zur Darstellung der Beziehung zwischen zentralen und peripheren Pulsen als Ortskurven. Arch. KreislForsch. 22: 72–80.

Luchsinger, P. C., R. E. Snell, D. J. Patel and D. L. Fry 1964. Instantaneous pressure distribution along the human aorta. Circulat. Res. 15: 503–10.
Maloney, J. E., D. H. Bergel, J. B. Glacier, J. M. B. Hughes and J. B. West 1968. Transmission of pulsatile blood pressure and flow through the isolated lung. Circulat. Res. 23: 11–24.
Maloney, J. E., D. H. Bergel, J. B. Glacier, J. M. B. Hughes and J. B. West 1968b. Effect of pulsatile pulmonary artery pressure on distribution of blood flow in isolated lung. Resp. Physiol. 4: 154–67.
McDonald, D. A. 1960. Blood flow in arteries. E. Arnold (Publishers) Ltd., London.
Meisner, J. E. and J. W. Remington 1962. Pulse contour in carotid and foreleg arterial systems. Amer. J. Physiol. 202: 527–35.
Moens, A. I. 1878. Die Pulskurve. Leiden.
Müller, A. 1951. Ueber die Fortpflanzungsgeschwindigkeit von Druckwellen in dehnbaren Roehren bei ruhender und stroemender Fluessigkeit. Helvet. Physiol. Acta 9: 162–76.
Noordergraaf, A. 1969. Hemodynamics. In H. P. Schwan (ed.), Biological Engineering. McGraw-Hill Book Company, New York.
O'Rourke, M. F. and M. G. Taylor 1966. Vascular impedance of the femoral bed. Circult. Res. 18: 126–39.
O'Rourke, M. F., J. V. Blazek, C. L. Morreels and L. J. Krowetz 1968. Pressure wave transmission along the human aorta, changes with age and inarterial degenerative disease. Circulat. Res. 23: 567–79.
Patel, D. J., F. M. De Freitas and D. L. Fry 1963. Hydraulic input impedance to aorta and pulmonary artery in dogs. J. Appl. Physiol. 18: 134–40.
Patel, D. J., F. M. DeFreitas, J. C. Greenfield and D. L. Fry 1963. Relationship of radius to pressure along the aorta of the living dog. J. Appl. Physiol. 18: 1111–117.
Patel, D. J. and D. L. Fry 1966. Longitudinal tethering of arteries in dogs. Circulat. Res. 19: 1011–1030.
Patel, D. J., J. C. Greenfield, W. G. Austen, A. G. Morrow and D. L. Fry 1965. Pressure flow relationship in the ascending aorta and femoral artery in man. J. Appl. Physiol. 20: 459–63.
Peirce, E. C. II, 1969. Extracorporeal circulation for open-heart surgery. Charles C. Thomas, Springfield, Ill.
Ranke, O. F. 1934. Die Daempfung der Pulswelle und die innere Reibung der Arterienwand. Zeitschr. Biol. 95: 179–204.
Remington, J. W. 1965. Quantitative synthesis of head and foreleg arterial pulses in the dog. Amer. J. Physiol. 208: 968–83.
Remington, J. W. and W. F. Hamilton 1945. The construction of a theoretical cardiac ejection curve from the contour of the aortic pressure pulse. Amer. J. Physiol. 144: 546–56.
Remington, J. W. and L. J. O'Brien 1970. Construction of aortic flow pulse from pressure pulse. Amer. J. Physiol. 218: 437–47.
Remington, J. W. and E. Wetterer 1963. Graphische Synthese des Arterienpulses aus Primaerwelle und reflektierten Wellen aufgrund der experimentell bestimmten Verteilung des Wellenwiderstandes im Arteriensystem. Pflügers. Arch. 278: 19 Abstract.
Reynolds, O. 1883, 1895. Cit. according to A. Sommerfeld 1950.
Ronniger, R. 1954. Ueber eine Methode zur uebersichtlichen Darstellung haemodynamischer Zusammenhaenge. Arch. KreislForsch. 21: 127–60.
Ronniger, R. 1955. Zur Theorie der physikalischen Schlagvolumenbestimmung. Arch. KreislForsch. 22: 332–73.
Rudewald, B. 1962. Hemodynamics of the human ascending aorta as studied by means of a differential pressure technique. Acta Physiol. Scand. 54, Suppl. 187: 1–64.
Ruszyák, I., M. Földi and G. Szabó 1967. Lymphatics and lymph circulation. Pergamon Press, New York.
Scher, A. M. and A. C. Young 1963. Servoanalysis of carotid sinus reflex effects on peripheral resistance. Circulat. Res. 12: 152–62.
Schimmler, W. 1965. Untersuchungen zum Elastizitaetsproblem der Aorta (Statistische Korrelation der Pulswellengeschwindigkeit zu Alter, Geschlecht und Blutdruck). Arch. KreislForsch. 47: 189–233.

Simon, E. and W. W. Meyer 1958. Das Volumen, die Volumendehnbarkeit und die Druck-Laengen-Beziehung des gesamten aortalen Windkessels in Abhaengigkeit von Alter, Blutdruck und Arteriosklerose. Klinische Woschr. 36: 424–32.

Sinn, W. 1956. Die Elastizitaet der Arterien und ihre Bedeutung fuer die Dynamik des arteriellen Systems. Akademie der Wiss. u. Lit. Mainz. Abh. Math. Naturw. Kl. Nr. 11. Wiesbaden.

Skalak, R. 1966. Wave propagation in blood flow. In Y. C. Fung (ed.), Biomechanics. ASME, New York.

Skalak, R. and T. Stathis 1966. A porous tapered elastic tube model of a vascular bed. In Y. C. Fung (ed.) Biomechanics. ASME, New. York.

Skalak, R., F. Wiener, E. Morkin and A. P. Fishman 1966. The energy distribution in the pulmonary artery. I: Theory, II: Experiments. Phys. Med. and Biol. 11: 287–94; 437–49.

Sommerfeld, A. 1950. Lectures on theoretical physics. Vol. II Mechanics of deformable bodies. Academic Press, New York.

Spencer, M. P. and A. B. Denison 1963. Pulsatile blood flow in the vascular system. In W. F. Hamilton and P. Dow (eds.), Handbook of Physiology. Circulation. Vol. II. Washington D.C.

Stahl, W. R. 1963. The analysis of biological similarity. In J. H. Lawrence and J. W. Gofman (eds.), Advances in biological and medical physics. Vol. 9. Academic Press, New York & London.

Stegemann, J. and T. Kenner 1971. A theory of heart rate control by muscular metabolic receptors. Archiv. KreislForsch. (In print.)

Stegemann, J. and U. Tibes 1969. Der Einfluss von Amplitude, Frequenz und Mittelwert sinusfoermiger Reizdrucke und den Pressorezeptoren auf den arteriellen Mitteldruck des Hundes. Pflügers Arch. 305: 219–28.

Streeter, V. L., W. F. Keitzer and D. F. Bohr 1964. Energy dissipation in pulsatile flow through distensible tapered vessels. In E. O. Attinger (ed.), Pulsatile Blood Flow. McGraw-Hill Book Company, New York.

Takeda, J. 1960. Experimental study on peripheral circulation during extracorporeal circulation, with a special reference to a comparison of pulsatile flow with nonpulsatile flow. Archiv. Japan. Chirurgie 29: 1407–430.

Taylor, M. G. 1959. An experimental determination of the propagation of fluid oscillation in a tube with a viscoelastic wall; together with an analysis of the characteristics required in an electrical analogue. Phys. Med. and Biol. 4: 63–82.

Taylor, M. G. 1964. Wave travel in arteries and the design of the cardiovascular system, In E. O. Attinger (ed.), Pulsatile blood flow. McGraw-Hill Book Company, New York.

Taylor, M. G. 1965. Wave travel in a nonuniform transmission line in relation to pulses in arteries. Phys. Med. and Biol. 10: 539–50.

Taylor, M. G. 1969a. The input impedance of an assembly of randomly branching elastic tubes. Biophys. J. 6: 29–51.

Taylor, M. G. 1966b. Wave transmission through an assembly of randomly branching elastic tubes. Biophys. J. 6: 697–716.

Taylor, M. G. 1966c. Use of random excitation and spectral analysis in the study of frequency dependent parameters of the cardiovascular system. Circulat. Res. 18: 585–95.

Vadot, L. 1967 Mécanique du coeur et des artères. L'Expansion Scientifique Francaise, Paris.

Wagner, K. W. 1947. Einfuehrung in die Lehre von den Schwingungen und Wellen, Dietrich'sche V. B. H., Wiesbaden.

Warner, H. R., R. M. Gardner and A. F. Toronto 1968. Computer-based monitoring of cardiovascular functions in postoperative patients. Circul. 37 (Suppl. II): 68–74.

Warner, H. R., H. J. C. Swan, D. C. Connolly, R. G. Tompkins and E. H. Wood 1953. Quantitation of beat to beat changes in stroke volume from the aortic pulse contour in man. J. Appl. Physiol. 5: 495–507.

Warner, H. R. and W. S. Topham 1967. Regulation of cardiac output during transition from rest to exercise. In P. Kezdi (ed.), Baroreceptors and hypertension. Pergamon Press, New York.

Weber, E. H. 1828. Annotationes Anatomicae et Physiologicae Programmata collecta

II. De utilitate parietis elastici arteriarum. (Programma de 25, Aprilis 1828.) Leipzig, 1851.
Weber, E. H. 1850. Ueber die Anwendung der Wellenlehre auf die Lehre vom Kreislauf des Blutes und insbesondere auf die Pulslehre. Ber. Verh. koenigl. Saechs. Ges. Wiss. Leipzig. p. 164.
Weber, W. 1866. Theorie der durch Wasser oder andere incompressible Fluessigkeiten in elastischen Roehren fortgepflanzten Wellen. Verh. koenigl. Saechs. Ges. Wiss. 18: 353. Leipzig.
Westerhof, N., F. Bosman, C. J. De Vries and A. Noordergraaf 1969. Analog studies of the human systemic arterial tree. J. Biomech. 2: 121–43.
Wetterer, E. 1937, 1938. Eine neue Methode zur Registrierung der Blutstroemungsgeschwindigkeit am uneroeffneten Gefaess. Zeitschr. Biol. 98: 26–36; 99–62.
Wetterer, E. 1956. Die Wirkung der Herztaetigkeit auf die Dynamik des Arteriensystems. Verh. Dtsch. Ges. KreislForsch. 22: 26–60.
Wetterer, E. and B. Deppe 1939, 1940. Vergleichende tierexperimentelle Untersuchungen zur physikalischen Schlagvolumbestimmung. I: Zeitschr. Biol. 99: 307–18; II: Zeitschr. Biol. 99: 320–37; III: Zeitschr. Biol. 100: 105–18.
Wetterer, E. and T. Kenner 1968. Grundlagen der Dynamik des Arterienpulses. Springer Verlag, Berline, Heidelberg, New York.
Wetterer, E. and H. Pieper 1955. Ein indirektes Verfahren zur Bestimmung des diastolischen Abstromes aus dem Arteriensystem und seine Anwendung zum Studium der Druck-Stromstaerke-Beziehung *in vivo*. Verh. Dtsch. Ges. KreislForsch. 21: 430–39.
Wezler, K. and A. Boeger 1937. Ueber einen neuen Weg zur Bestimmung des absoluten Schlagvolumens des Herzens beim Menschen auf Grund der Windkesseltheorie und seine experimentelle Pruefung. Naunyn Schmiedebergs Arch. exper. Path. u. Pharm. 184: 482–505.
Wezler, K. and W. Sinn 1953. Das Stroemungsgesetz des Blutkreislaufs. Cantor, Aulendorf.
Whitmore, R. L. 1968. Rheology of the circulation. Pergamon Press, New York.
Wiener, F., E. Morkin, R. Skalak and A. P. Fishman 1966. Wave propagation in the pulmonary circulation. Circulat. Res. 19: 834–50.
Witzig, K. 1914. Ueber erzwungene Wellenbewegungen zaeher, inkompressibler Fluessigkeiten in elastischen Roehren. Thesis. Bern.
Womersley, J. R. 1955a. Oscillatory motion of a viscous liquid in a thin walled elastic tube. I: The linear approximation for long waves. Phil. Mag. 46: 199–221.
Womersley, J. R. 1955b. Method for the calculation of velocity, rate of flow and viscous drag in arteries when the pressure gradient is known. J. Physiol. 127: 553–63.
Womersley, J. R. 1957. An elastic tube theory of pulse transmission and oscillatory flow in mammalian arteries. WADC Tech. Report. TR 56-614.
Womersley, J. R. 1958. Oscillatory flow in arteries. II: The reflection of the pulse wave at junctions and rigid inserts in the arterial system. Phys. Med. and Biol. 2: 313–23.
Wylie, E. B. 1966. Flow through tapered tubes with nonlinear wall properties. *In* Y. C. Fung (ed.), Biomechanics. ASME, New York.
Zeuthen, E. 1955. Comparative physiology (respiration). Ann. Rev. Physiol. 17: 459–82.

Acknowledgement. *The work described in this paper was supported by NIH Grant HE 11747.*

17

Analysis of Recent Developments in Blood Flow Measurement

ARNOST FRONEK

The multitude and variety of methods for blood flow measurement exemplify that at present not only an ideal technique is missing, but a generally accepted basic methodical approach is lacking. It is, therefore, not surprising that there are difficulties in categorizing all the varied approaches. Some possible methods of classification will be listed. It seems that none fulfills the nonoverlapping criterion. In this analysis we shall select one categorizing system: Invasive and noninvasive techniques with subdivisions: direct and indirect methods.

Invasive techniques are those which require a discontinuity either of vessels or of the skin. Direct techniques are defined as those which can be monitored directly with appropriate sensors, while indirect methods require either an additional marker or additional external energy to demonstrate changes in physical functions related to blood velocity or flow rate.

It is obvious that this categorizing scheme is overlapping, and therefore, not ideal; but it has many practical advantages. One of those is that in a number of cases the invasive techniques served as a developmental basis for future noninvasive methods and a thorough understanding of the basic physical principles enhances the chances of a future successful application in the noninvasive group.

In view of some excellent reviews in the past, only recent developments in this area will be considered and a critical analysis will be attempted.

From the variety of different techniques, only two will be discussed: the electromagnetic and the thermodilution method. Special attention will be given to advantages and disadvantages of the respective approaches and, where possible, the potential for transcutaneous applicability will be considered.

* * *

I. INTRODUCTION

The determination of blood flow represents not only one of the most important basic techniques in cardiovascular physiology but, unfortunately, one of the most difficult. At present, an ideal method is not only lacking in the diagnostic clinical area of application but is also in an undeveloped stage in the experimental field.

If we define our requirements for an ideal flow-metering system as one to be ultimately applied to the patient, we may add to the generally accepted requirements of accuracy, sensitivity, reliability and frequency response, the specific clinical requirements of least traumatization. It is obvious that at present we are very far from this goal. I would like to narrow our discussion only to those techniques which either make possible flow or velocity measurement in intact vessels using specially designed catheters or which have some potential for future noninvasive application. For lack of time we have to omit the pressure gradient and ultrasonic technique and concentrate on recent developments in the area of electromagnetic and thermodilution methods.

II. ELECTROMAGNETIC FLOW-METERING

General remarks

After it was realized that the much promising Rein's Thermostromuhr method did not fulfill expectations, the electromagnetic flow-metering technique—developed independently by Kolin (1936) and Wetterer (1937)—was anticipated by physiologists to help solve basic circulatory problems under more physiological circumstances than hitherto possible—on unopened vessels. In spite of this interest, it took almost twenty years before operational units became commercially available, indicating the amount of theoretical and technical difficulties associated with this problem. Several authoritative surveys have been published in the past (Jochim, 1962; Wetterer, 1963; Wetterer, 1963b); and we would like, therefore, to discuss only recent developments. Earlier works will be mentioned, however, in conjunction with new developments wherever necessary.

In the past decade efforts were concentrated mainly in the following directions: 1) theoretical analyses, 2) development of more adequate electronic circuitry, 3) analyses and various changes in the flow probe design, 4) introduction of the catheter-tip-electromagnetic flow or velometer, 5) extension of application to chronic experimental implantation, and clinical use during surgery.

Theoretical and Experimental Analyses of Basic Phenomena. Shercliff (1962) has published one of the most thorough and comprehensive analyses on this subject, particularly from the viewpoint of intravascular distribution of the magnetic field. The computation of the weight function W and its distribution throughout the cross section made it possible to analyze the contribution of each point to the final flow-dependent signal. Extrapolating Shercliff's conclusion, Rummel and Ketelsen (1966) have generalized the factors related to the induced electromotive force

(1) $$E = \int_v W_v \cdot B_v \cdot u_v \cdot dV$$

E = voltage induced
V = volume
B_v = component vertical to the electrode axis and also vertical to vessel axis
u_v = velocity of the dV parallel to vessel axis
W_v = weight function for the specific dV topography

Shercliff has shown that the highest W, that is the biggest contribution to the flow signal, can be observed close to the electrodes, while decreasing towards the center of the cross section as well 90° in the same plane. This shows, however, that the induced E in Eq. 1 is not only related to the velocity v but also to varying values of B_v and W_v. This explains the velocity profile dependency of E, though earlier experimental works have discounted their significance. Thürlemann (1941), for instance, has shown that the ultimately recorded E is independent of the velocity profile and there was no significant difference in induced E whether observed under laminar or turbulent circumstances. Since the effect need not be large under favorable conditions of magnetic field configuration and velocity profile, it is possible that this effect was overlooked by Thürlemann due to his understandingly higher scatter. Rummel and Ketelsen (1966) have pointed out that a real independency of the induced E requires a constant product $W_v B_v$ and therefore,

(2) $$E = W_v \cdot B_v \cdot \int_v u_v \cdot dV$$

They concluded that this can be achieved by adjusting B_v according to W_v though not necessarily in an exact reciprocity, but nevertheless a slight inhomogeneity makes the recorded potential practically velocity-profile independent even under extreme flow conditions.

Effect of the Vessel Wall. It has been shown (Einhorn, 1940; Cushing, 1958; Gessner, 1961; Shercliff, 1962; Ferguson and Landahl, 1966; Wyatt, 1968) that the specific resistance of the vessel wall has a significant effect on the amplitude of the flow signal—if the conductivity of the vessel wall is smaller than that of blood, the sensitivity increases.

The Effect of Specific Resistance of the Fluid on the Flow Signal. In view of the fact that the conductivity of the fluid in which the EMF is being induced may affect to some extent the resulting picked-up potential, it is only natural that attention has been focused on the effect of hematocrit on the sensitivity. There is a well-established relationship between hematocrit and specific resistance of blood (Schwan, 1949; Okada and Schwan, 1960), but there is also a very significant correlation between linear velocity and specific resistance of blood (Coulter and Pappenheimer, 1949; Fronek and Ganz, 1956). In effect, a number of authors have found a significant relationship between hematocrit and sensitivity, generally a decrease with an increasing hematocrit (Khouri and Gregg, 1963; Spencer and Denison, 1959; Brunsting and tenHoor, 1968; Beck et al, 1965). Other authors could not confirm these findings (Ferguson and Landahl, 1966; Bond, 1967). This is a very important question for two reasons: 1) is it necessary to use blood for flowprobe calibration or is a saline solution satisfactory? 2) Is it necessary to use some correction factors if significant changes in hematocrit are observed? A very thorough study in this respect was undertaken by Dennis & Wyatt (1969). From their findings it can be deduced that a high input impedance of the amplifier makes the sensitivity less fluid-conductivity dependent. On the other hand, even under very favorable input impedance conditions, the sensitivity has been found to be fluid-conductivity dependent. However, it must be pointed out that the authors used a system with an accuracy of 0.1 % which is about fifteen times greater than can be achieved by the commercially available systems. Since even under these extremely favorable conditions, a change in hematocrit value from 0–66% caused a decrease in sensitivity of 4%, it can be assumed that for most practical purposes the calibration using saline is justified.

Development of More Adequate Electronic Circuitry. It is surprising how little the quality and adequacy of commercially available electronic systems have changed since the basic circuit diagrams by James (1951), Dennison et al (1955), Kolin and Kado (1959) have been published. Their gating circuitry was successfully applied and has led to the commercial availability of these systems. Subsequently, various improvements have been published (Westersten et al, 1959; Shirer et al, 1959; Giori, 1960; Wyatt,

1961); ways of improving the poor signal-to-noise ratio have been considered, different carriers instead of the original sinusoidal wave have been used (Beck et al, 1965; Yanof et al, 1963), a different type of signal detection (phase-detection) (Olmstead and Aldrich, 1961), and a selective automatic gain control for a given sampling period (Spencer and Denison, 1960; Moody, 1966). Only a negligible fraction of these reports had an effect on the design of commercially available flowmeters (Hognestad, 1966, produced by Nycotron *A/S*, Oslo, Norway; Westersten et al, 1969, to be produced by Micron, Los Angeles, U.S.A.). The latter promises an electrical determination of zero by achieving a symmetry with respect to eddy currents. Testing under a variety of physiological conditions will determine the reliability of this procedure. Another system which promises a significant improvement (Wyatt, 1965) is unfortunately not commercially available in spite of a variety of attractive features as, for instance, automatic quadrature suppression, 0.1% accuracy, preset phasing, and other features. Though it is felt that the design of a proper flow probe is obscured with far more complex problems, it seems that utilization of the already published electronic innovations would bring the cardiovascular physiologist closer to his ultimate request: a flowmetering system with a reliable electric zero determination, high accuracy, sensitivity, stability and frequency response.

Analyses and Resulting Changes in the Flow Probe Design. Whatever the quality of the electronic system, an inadequate flow probe will devaluate all achievements. Additional efforts have been mobilized to eliminate, compensate or suppress the unwanted no-flow signals originating in the probe. Because the desired flow signal is about seven orders of magnitude smaller than the voltage across the coils (Wyatt, 1968) and other artifactituous no-flow signals are of the same order of magnitude as the desired signal itself, it is obvious that a good design represents a very complex task. Before discussing some of the better-known sources of spurious signals, it is worthwhile to list some constraints to be taken into consideration. The size of the probe very early became a limiting factor because, to some extent, it is in an inverse relationship to the sensitivity. However, if the linear velocity is sufficiently high, the size can be significantly reduced by using coreless coils (Kolin, 1959, 1960). The advantage of the smaller size was lost with the introduction of the Helmholtz type of coreless coils (Yanof, 1963; Kolin and Wisshaupt, 1963). The more homogeneous magnetic field produced by this coil does not outweigh its bulkiness, especially since it has been shown (Kolin and Wisshaupt, 1963; Clark and Wyatt, 1968) that the homogeneity is not crucial or even desirable (Rummel and Ketelsen, 1966).

Stability of Electrical Specifications of Insulating Material. For reasons mentioned above, high input impedance is a necessity. Any changes in this direction will cause increased instability, decreased sensitivity and phasing problems. These stressing conditions occur mainly in chronic implantation experiments, especially those extending for more than several weeks (Sellers and Dobson, 1967). Wyatt (1966) has systematically examined and analyzed the various sources of baseline errors. Besides eddy currents, leakage and multiple ground paths, the electrodes themselves are a very complex source of artifacts. The inhomogeneity of the electrical field can be improved either by using low conductivity material (Hognestad, 1966) or recessing the electrodes (Wyatt, 1966), which is, of course, less desirable for practical reasons. External interference still represents a nonnegligible source of unwanted signals, especially in the smaller-sized probes where the amplitude of the flow signal is *a priori* smaller, and efficient screening procedures may represent a technical problem. Significant progress was achieved by the so-called split-pole flow probe (Kolin and Vanyo, 1967) with its high sensitivity, relatively small size and simplicity. However, most of the mentioned sources of artifacts are still present in this and other commercially available flow probes. An interesting technique used to cancel the transformer voltage has been published recently by Folts (1970). It permits a more flexible adjustment of the magnetic flux and may also simplify the construction of the probe. The toroid flow probe described by Clark and Wyatt (1969), on the other hand, seems to be designed on a sounder theoretical basis, and it is possible that Folts' compensation procedure may not be required in this type of design.

Electromagnetic Catheter Systems. The better understanding of the principles underlying the electromagnetic flowmetering technique and improvements in electronic circuitry made it more realistic to start to miniaturize the flow head to such an extent that it could be incorporated on a tip of a catheter. Such a device is urgently needed in experimental as well as in clinical cardiology, since it is the ideal tool for mapping the velocity conditions throughout the circulatory system.

Earlier types of catheter-tip velometers were based on either unreliable principles (Kanzow's Thermostromuhr catheter, 1956, in this case a flowmeter) or on delicate differential transformer activation (Pieper 1958, 1963, 1964). Though the latter has a very adequate frequency response (up to 30 c/s), the system is mechanically very delicate and requires thorough heparinization.

With the application of the electromagnetic principle, basically two types of catheter-tip velometers can be distinguished: the self-contained system and the pick-up system with external excitation.

a. Self-contained System. Lochner and Oswald (1964) seemed to have been the first to successfully package the excitation and pick-up system on

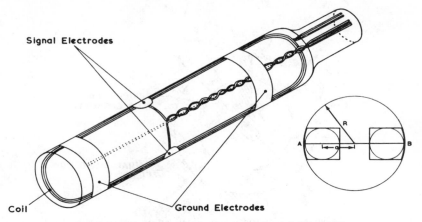

Fig. 1: Electromagnetic velocity probe (from Mills, 1966).

the tip of a catheter. An additional cuff and/or a specially designed tip orifice makes it possible to measure the flow rate either by occluding the rest of the vessel or wedging the catheter inlet into the orifice of the branching artery. Mills (1966, 1967, Fig. 1) has described an electromagnetic velocity probe with remarkably good performance specifications. Since the electrodes are immersed in blood, there is no danger of electrode-artery interface artifacts, and the input impedance of the amplifying system can be lower. The small size of the excitation and pick-up system made it possible to be used in man— in right heart catheterization (Wexler et al, 1968). A similar system, but with an iron core, was described by Bond and Barefoot (1967) and is also commercially available. Recently, Stein and Schuette (1969) have combined a miniaturized electromagnetic system with a small funnel which makes it possible to measure either velocity or flow by wedging the tip into an aortic branch, somewhat similar to Lochner's original catheter-tip system. Kolin and his co-workers (1967*a*, 1967*b*, 1968) have described a more bulky system utilizing a more conventional coil and pick-up system, while the flow is diverted through a center opening positioned over the branching orifice. A pull-wire system facilitates the proper maneuverings of the catheter.

A completely different and very interesting system was recently described by Kolin (1969*a*) using a radial magnetic field which gives rise to tangential currents (Fig. 2). However, with an introduction of a dielectric radial partition, these eddy currents are blocked and the induced EMF is no longer dissipated; adjacent electrodes can pick up velocity dependent signals. Attempting to decrease the size of the catheter, Kolin (1969*b*) has also described what is perhaps the ultimate in decreasing the size of the self-contained

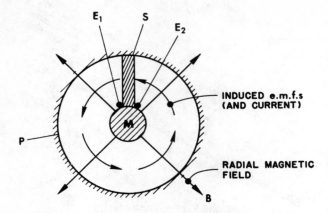

Fig. 2: Radial field electromagnetic flow sensor (from Kolin, 1969a). B = radial magnetic field; M = transducer magnet; P = pipe; S = dielectric septum; E_1 and E_2 = pick-up electrodes.

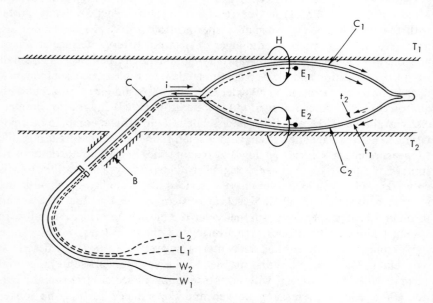

Fig. 3: Miniaturized electromagnetic flow catheter (from Kolin, 1969b). T_1 and T_2 = walls of the artery; B = art. branch; C = catheter harboring the coil leads W_1 and W_2 and electrode leads L_1 and L_2; C_1 and C_2 = legs of the coil generating the magnetic field H; i = current in the leads W_1 and W_2; t_1 and t_2 = two turns of the coil.

electromagnetic catheter system (Fig. 3). The magnetic field is generated by two parallel wires carrying equal currents in opposite directions and the electrodes are fixed to the wire producing the magnetic field. Due to the special topical arrangement, the velocity sensitivity is logarithmically related to the tube diameter.

   b. Pick-up Systems with External Excitation. All the above described systems still have significant dimensional limits (about 3 mm.), and they cannot be applied percutaneously, as for example, Seldinger's technique. This led Kolin (1970, Figs. 4 and 5) to develop an extremely small-sized electromagnetic catheter-tip system containing only the pick-up electrodes with an additional lead for quadrature voltage suppression, while the magnetic field is applied from outside. From the preliminary information available, it seems that the sensitivity depends upon the diameter of the investigated vessel, which indicates that this technique still needs a thorough theoretical and experimental analysis for absolute quantitative velocity determinations. It is, however, one of the most promising developments in this direction.

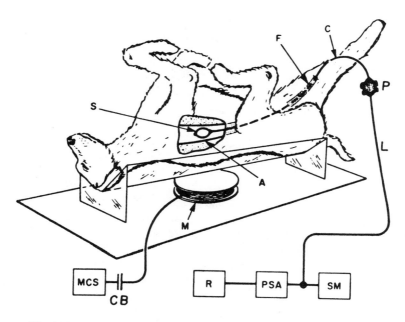

Fig. 4: External field electromagnetic catheter flow meter (from Kolin, 1970). A = aorta; S = sensor portion; M = magnet coil; F = femoral artery; C = catheter; L = leads; P = plug, MCS = magnet current supply; PSA = phase sensitive amplifier; SM = sensitivity monitor; R = recorder; CB = condenser bank resonating the coil at 400 Hz.

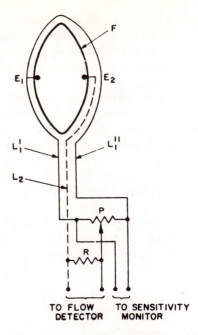

Fig. 5: Tip of flow sensor (from Kolin, 1970). Compensation of quadrature signal. F = stainless steel frame (insulated); E_1, E_2 = electrodes; L'_1, L''_1, L_2 = insulated electrode leads; P = potentiometer $10^3 \Omega$; R = resistor $10^5 \Omega$.

III. THERMODILUTION TECHNIQUES

GENERAL REMARKS

Since Fegler (1954) published his original observation indicating that heat loss between the site of injection and site of sensing seemed to be negligible, a number of authors have employed this basic modification of the indicator dilution technique with various types of application.

In comparison to the electromagnetic flowmetering technique, the thermodilution procedure is technically far more simple and less expensive, and less sophisticated instrumentation is needed. Compared with the standard dye-dilution technique, the thermodilution modification has a number of substantial advantages. In spite of these circumstances, it is surprising that the technique did not meet a broader acceptance. We shall try to identify some of the obstacles, real or imaginery, which seem to have prevented a larger and a more wide-spread use of this flowmetering technique.

There are several possibilities of sub-classification applicable to most indicator dilution techniques but those which we intend to use in this review, namely 1) total cardiac output, and 2) single vessel or regional blood flow, have special justification in the thermodilution modification due to special features given by the mobility and minute size of the sensing device.

Similarly, as in the discussion of the electromagnetic flowmetering procedure, only recent developments will be considered that were not covered by Hosie's comprehensive survey (1962). Again, earlier works will be mentioned insofar as they relate directly to recent publications.

Cardiac Output Determination. It is surprising that the big advantage of the thermodilution technique—not requiring the withdrawal of blood samples—has not attracted a larger acceptance and enthusiasm particularly in clinical studies and whenever repeated determinations are required as, for instance, in Intensive Care Units. It is conceivable that part of this hesitant approach was due to improper instrumentation design (as will be pointed out later), but mainly it was due to fear of heat loss somewhere between the site of injection and sensing. Dow's early criticism (1956) has prompted a number of authors to reinvestigate these circumstances, and it seems that the experimental results indicate those losses to be negligible (Hosie, 1962). Mohammed and co-workers (1963), extrapolating Evonuk's experiments (1961), did find slightly higher cardiac output values if sensed in the periphery (carotid or femoral artery) in contrast to values obtained in the arch of the aorta. On the other hand, the site of injection seems to be even more important than in the dye-dilution technique—cardiac output values calculated after jugular injections were higher than from right atrium injection. Goodyear's group (1959) found less scatter in the aorta than in the pulmonary artery. It seems that in this case better mixing over the longer distance outweighed the interference of heat loss in the pulmonary system. Branthwaite and Bradly (1968) have recently published good correlation results (versus the Fick technique) pointing out, however, that they had to inject from the internal jugular vein in order to obtain good results; more peripheral injection sites evidently caused some heat loss. In consideration of the importance of cardiac output determination during low output conditions where possible heat loss might be even more exaggerated, the results of Solomon and co-workers (1969) encourage the application of the thermodilution technique even under shock conditions. It seems, however, that the highest degree of accuracy and reproducibility may be expected if the site of injection is in the right atrium and the site of sensing in the pulmonary artery as described by Ganz and coauthors (1970, Fig. 6). This method obviates femoral artery puncture, and the thermistor catheter can be inserted via the cardiac injection catheter. An additional advantage is the negligible recirculation espe-

446 BLOOD FLOW MEASUREMENT

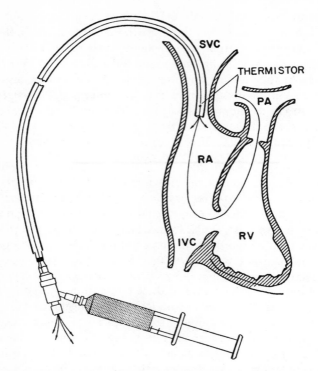

Fig. 6: Cardiac output thermodilution set-up (from Ganz et al, 1970). SVC = sup. vena cava; IVC = inferior vena cava; RA = right atrium; RV = right ventricle; PA = Pulmonary artery.

cially in the right atrium-pulmonary artery configuration. This makes it possible to use a simple integrator for on-line evaluation without requiring electronic extrapolation as is used in conjunction with the dye-dilution technique.

Summarizing, it seems that the reliability of the thermodilution technique is fairly well established provided the recent experiences regarding the injection and sensing sites are fully taken into consideration. But it has to be pointed out that these results are based on empirical findings, and a theoretical analysis of possible heat transfer, including extreme hemodynamic conditions, would be highly desirable.

Local Thermodilution. We define local thermodilution as the technique in which the sites of injection and sensing are close to one another (from several millimeters to a few centimeters, Fronek and Ganz, 1960, Fig. 7). This configuration yields a number of advantages: the possibility to measure blood flow in a single vessel, fast sequence of measurement, and no recircula-

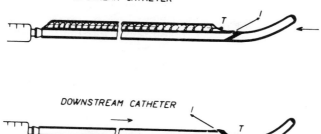

Fig. 7: Local thermodilution catheters. T = thermistor; I = injection orifice.

tion. However, it imposes a stricter constraint on the mixing since it has to be completed within a very short distance. There is an additional complicating factor: the initial part of the dilution curve is being recorded at a time when the injection is not yet completed. It was pointed out that this may be of no importance if arterial blood flow is measured because the injectate temporarily replaces the carrier. If venous blood flow is measured, the injection of the fluid may cause a transient increase in blood flow and, therefore, the equation has to be corrected accordingly (Fronek and Ganz, 1960). In our experience, however, venous flow rates above 60 ml/min do not require this correction if a slug injection technique is used.

Since Hosie's review was published (1962), more experience with the local thermodilution principle (LTD) has accumulated.

Basically two types of injection can be used: a single slug and continuous injection. The advantages of the latter are as follows: Since the steady state of the injectate mixing is recorded, the evaluation is very simple and the deflection is actually the only variable under otherwise standardized conditions. Furthermore, if some dynamics of blood flow changes are to be recorded, it is safer to have a continuous sensing rather than to rely on single dilution curves considering the inherent scatter of each value obtained in this way. However, there are some disadvantages which have to be considered. The amount of fluid injected into the system is not negligible if an average 50 ml/min infusion rate is being used for at least 20–30 seconds and if 5–15 measurements are performed. Pavek et al (1969) have shown that for cardiac output an injection rate of 1–2 ml/sec is considered a minimum for reliable mixing. Linzel (1966) has adopted the continuous infusion technique to measure mammary blood flow in the goat. He did find good correlation measuring flow rates from 200–1000 ml/min and using an injection rate from 20–50 ml/min. The injection rate naturally depends mainly on the kinetic energy of the indicator jet leaving the orifice (Andres et al, 1954; Fronek and Ganz,

1960), and, therefore, the injection rate always has to be considered in conjunction with the size of injection orifice.

The continuous infusion technique raises a question regarding the frequency response of the whole system including not only the thermistor mounted on the catheter but also that of the "mixing volume" (Lowe, 1968). In view of the complex nature of this topic, more analytical work is needed to define the "cut-off" frequency assuming the respective mixing volume under physiological conditions.

In recent years several authors have published good comparison studies using the local thermodilution principle. Some have used it to measure venous blood flow in animals and man (Shillingford et al, 1962; Kountz et al, 1964; Dowsett and Lowe, 1964; Clark and Cotton, 1966; Lowe and Dowsett, 1967; Chalmers et al, 1966); others have measured arterial blood flow mounting the thermistor and injection orifice on a Seldinger-type catheter (Ganz et al, 1964; Hlavova et al, 1965; Hlavova et al, 1966)*. (Fig. 8.) White and co-workers (1967) have been able to build a LTD catheter system applicable to the renal vein in the rabbit and measured renal outflow in nonanesthetized rabbits for one to three weeks. One of the most challenging areas of application is coronary sinus blood flow measurement. It has been shown that the LTD principle is highly applicable for this purpose, and coronary sinus outflow has been measured in this way under the influence of a variety of drugs (Ganz and Fronek, 1961; Fronek and Ganz, 1961; Ganz, 1963) in animals and also in man (Ganz et al, 1965). A useful combination of local and cardiac output measurements is shown in Fig. 9 (Ganz and Fronek, 1961) which represents two thermodilution curves—one obtained from the coronary sinus and the second recorded by a thermistor placed on a pulmonary artery catheter. One injection gives, in this case, values for coronary sinus flow and cardiac out-

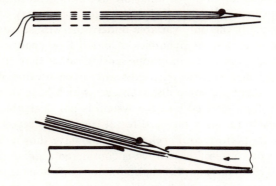

Fig. 8: Local thermodilution catheter—Seldinger type.

* By using two separate but closely positioned injection and sensing sites, the umbilical blood flow was successfully measured immediately after birth (Stembera et al, 1964).

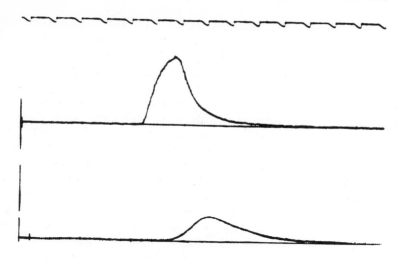

Fig. 9: Simultaneous determination of coronary sinus blood flow and cardiac output (from Ganz and Fronek, 1961). Upper tracing coronary sinus blood flow, lower tracing cardiac output.

put. Recently, Ganz and co-workers have applied the continuous infusion approach to measure coronary sinus outflow in man (1970).

In this context it is worthwhile to mention two techniques which basically operate on the LTD principle but are not usually listed in this category. Afonso (1966) has described a thermodilution flowmeter using an electrical heater as a source of energy and a thermistor as a sensing device; an electrical stirrer is provided to secure proper mixing. Khalil and co-workers (1966) have utilized a similar approach to measure cardiac output. In order to achieve proper mixing, these authors utilized high frequency excitation of a heating coil. It seems, however, that the advantage of high frequency was lost in this arrangement since the current is fed into a coil and not directly into the blood. Nevertheless, the results show good agreement with the Fick method.

It is possible that the commercial availability of suitable LTD-catheters would pave the way to a wider acceptance of the LTD-technique.

It seems appropriate to add some technical notes which seem of basic importance and have some future applicative possibilities which have not yet been fully utilized. Though the basic instrumentation is relatively very simple, it has been sometimes oversimplified by overlooking the self-heating of the thermistor. Most authors have been aware of the fact that there is a certain maximum dissipation which must not be reached if flow sensitivity of the thermistor is to be avoided. A detailed study by James and co-workers (1965) has given a general solution according to which, for instance, a thermistor of 1500 ohms must not be exposed to a voltage higher than 0.12 V.

On the other hand, the flow sensitivity of the "overheated" thermistor could be used very conveniently to measure velocity (Felix and Groll, 1953; van der Werf and Zijlstra, 1966; Ristau and Warbanow, 1968). This technique, however, has not met the expected acceptance, perhaps due to several technical disadvantages, namely, nonlinearity, low frequency response, and mainly, lack of directional sensitivity. Recently, Grahn and co-workers have described an electronic design linearizing the resistance/velocity relationship, increasing the frequency response, and, most importantly, incorporating directional sensitivity (1968). It can be expected that a combination of the LTD principle measuring blood flow, with the overheated or flow-sensitive thermistor (for instance, by increasing the voltage after completion of the injection) could combine the advantages of a single flow determination with a continuous velocity monitoring. A prerequisite, of course, would be that vessel diameter changes will be negligible during the velocity read-out.

In the above described developments we have discussed various types and modifications of the thermodilution technique. There was one common denominator: the sensing device was located in the vascular system. This, of course, is one of the main reasons why the determination of cardiac output did not become a bedside procedure, though from a pathophysiological point of view the determination of cardiac output is perhaps more important than arterial blood pressure. We are, therefore, currently exploring the possibilities of recording intravascular temperature fluctuations from the vessel wall and ultimately transcutaneously. For this purpose we are investigating, together with Mr. Silva, the possibility of measuring the heat flux instead of the temperature after injecting a cold bolus. It has become obvious, as is evidenced by the following figure (Fig. 10), that additional factors such as heat conduc-

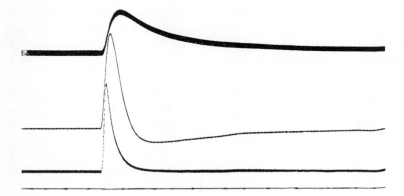

Fig. 10: Heat flux dilution curve from rabbit skin wrapped around copper tubing, flow rate 600 ml/min. From above: 1) temperature change from rabbit skin, 2) heat flux change from the same site, 3) temperature change intraluminal.

tivity, and especially heat capacitance, of the intervening biological material are of prime importance. The figure shows the difference of the intravascular temperature and heat flux recording from the vessel wall. It seems that the establishment of a suitable model will be essential for final quantitation of the results. In spite of these difficulties, it is hoped that this is a promising avenue for future transcutaneous cardiac output determination.

REFERENCES

A. Electromagnetic flowmetering

Beck, R., J. A. Morris and N. S. Assali 1965. Calibration of the pulsed-field electromagnetic flowmeter. Am. J. Med. Electronics 4: 87–91.
Bond, R. F. 1967. *In vivo* method for calibrating electromagnetic flowmeter probes. J. Appl. Physiol. 22: 358–61.
Bond, R. F. and C. A. Barefoot 1967. Evaluation of an electromagnetic catheter tip velocity-sensitive blood flow probe. J. Appl. Physiol. 23: 403–409.
Brunsting, J. R. and F. tenHoor 1968. Factors preventing accurate *in vitro* calibration of noncannulating electromagnetic flow transducers. *In* Chr. Cappelan, Jr. (ed.), New Findings in Blood Flowmetry. Universitetsforlaget, Oslo. pp. 107–11.
Clark, D. M. and D. G. Wyatt 1968. The effect of magnetic field inhomogeneity on flowmeter sensitivity, *In* Chr. Cappelan, Jr. (ed.), New Findings in Blood Flowmetry Universitetsforlaget, Oslo. pp. 49–54.
Clark, D. M. and D. G. Wyatt 1969. An improved perivascular electromagnetic flowmeter. Med. Biol. Eng. 7: 185–90.
Coulter, N. A. and J. R. Pappenheimer 1949. Development of turbulence in flowing blood. Am. J. Physiol. 150: 401–408.
Cushing, V. 1958. Induction flowmeter. Rev. Sci. Inst. 29: 692–97.
Denison, A. B., M. P. Spencer and H. D. Green 1955. A square-wave electromagnetic flowmeter for application to intact blood vessels. Circul. Res. 3: 39–46.
Dennis, J. and D. G. Wyatt 1969. Effect of hematocrit value upon electromagnetic flowmeter sensitivity. Circul. Res. 24: 875–86.
Einhorn, H. D. 1940. Electromagnetic induction in water. Trans. Roy. Soc. S. Africa 28: 143–60.
Ferguson, D. J. and H. D. Landahl 1966. Magnetic meters: effects of electrical resistance in tissues on flow measurements and an improved calibration for square-wave circuits. Circul. Res. 19: 917–29.
Folts, J. D. 1970. Electronic zero for chronic application of electromagnetic flowmeter probes. J. App. Physiol. 28: 237–41.
Fronek, A. and V. Ganz 1956. Blood velocity and the specific resistance of blood. Proc. IV Congress of the Czechosl. Physiol Soc. 23.
Gessner, U. 1961. Effects of the vessel wall on electromagnetic flow measurement. Biophysic. J. 1: 627–37.
Giori, F. A. 1960. A blood flowmeter for use in coronary heart disease research. IRE Trans. Med. Electronics 7: 211–16.
Hognestad, H. 1966. Square-wave electromagnetic flowmeter with improved baseline stability. Med. Res Engrg. 5: 28–33.
James, W. G. 1951. An induction flowmeter design suitable for radioactive liquids. Rev. Sci. Inst. 22: 989–1002.
Jochim, K. E. 1962. The development of the electromagnetic blood flowmeter. IRE Trans. Med. Electronics 9: 228–35.
Kanzow, E. 1956. Fortlaufende Messung der Koronardurchblutung im Tierexperiment bei uneröffnetem Thorax mittels Stromuhrkatheters. Dissertation, Göttingen.

Khouri, E. M. and D. E. Gregg 1963. Miniature electromagnetic flowmeter applicable to coronary arteries. J. Appl. Physiol. 18: 224–27.
Kolin, A. 1936. Electromagnetic flow meter: principle of method and its application to blood flow measurements. Proc. Soc. Exptl. Biol. Med. 35: 53–56.
Kolin, A. 1959. Electromagnetic blood flowmeters. Science 130: 1088–1097.
Kolin, A. 1960. Determination of blood flow by electromagnetic method. In O. Glasser (ed.), Medical Physics. Vol. 3, pp. 141–55.
Kolin, A. 1969a. A radial field electromagnetic intravascular flow sensor. IEEE Trans. Biomed. Engrg. BME 16: 220–21.
Kolin, A. 1969b. A new principle for electromagnetic flowmeters. Proc. Nat. Acad. Sc. 63: 357–63.
Kolin, A. 1970. A new approach to electromagnetic blood flow determination by means of catheter in an external magnetic field. Proc. Nat. Acad. Sc. 65: 521–27.
Kolin, A. 1967a. An electromagnetic intravascular blood flow sensor. Proc. Nat. Acad. Sci. 57: 1331–337.
Kolin, A. 1967b. An electromagnetic catheter flowmeter. Circul. Res. 21: 889–99.
Kolin, A. and J. Vanyo 1967. New design of miniature electromagnetic blood flow transducers suitable for semi-automatic fabrication. Cardiovasc. Res. 1: 274–86.
Kolin, A., G. Ross, J. H. Grollman and J. Archer 1968. An electromagnetic catheter flowmeter for determination of blood flow in major arteries. Proc. Nat. Acad. Sc. 59: 808–15.
Kolin, A. and R. Wisshaupt 1963. Single-coil coreless electromagnetic blood flowmeters. IEEE Trans. Biomed. Electronics 10: 60–67.
Kolin, A. and R. T. Kado 1959. Miniaturization of the elctromagnetic blood flowmeter and its use for the recording of circulatory responses of conscious animals to sensory stimuli. Pro. Nat. Acad. Sci. 45: 1312–321.
Lochner, W. 1964. Eine elektromagnetische Stromuhr für Messung des Coronarsinusausflusses. Arch. Ges. Physiol. 281: 305–308.
Mills, C. J. 1966. A catheter-tip electromagnetic velocity probe. Phys. Med. Biol. 11: 323–24.
Mills, C. J. and J. P. Shillingford 1967. A catheter-tip electromagnetic velocity probe and its evaluation. Cardiovasc. Res. 1: 263–73.
Moody, N. F. 1967. Square-wave flowmeter. Med. Res. Eng. 6: 8.
Okada, R. H. and H. Schwan 1960. An electrical method to determine hematocrits. IRE Trans. Med. Electronics ME 7: 188–92.
Olmsted, F. and F. D. Aldrich 1961. Improved electromagnetic flowmeter: phase detection, a new principle. J. Appl. Physiol. 16: 197–201.
Pieper, H. P. 1958. Registration of phasic changes of blood flow by means of a catheter type flowmeter. Rev. Sci. Instr. 29: 965–67.
Pieper, H. P. 1963. Catheter-tip blood flowmeter for measurement of pulmonary arterial blood flow in closed-chest dogs. Rev. Sci. Instr. 34: 908–10.
Pieper, H. P. 1964. Catheter-tip flowmeter for coronary arterial flow in closed-chest dogs. J. Appl. Physiol. 19: 1199–1201.
Rummel, T. L. and B. Ketelsen 1966. Inhomogenes Magnetfeld ermöglicht induktive Durchflussmessung bei allen in der Praxis vorkommenden Strömungsprofilen. Regelungstechnik 14: 262–67.
Schwan, H. 1949. Eine elektrische Methode zur Bestimmung der Erythrocytenzahl. Pflüg. Arch. 251: 550–58.
Sellers, A. F. and A. Dobson 1967. Some applications and limitations of electromagnetic blood flow measurements in chronic animal preparations. Gastroenterology 52: 374–79.
Shercliff, J. A. 1962. The theory of electromagnetic flow measurement. Cambridge University Press, Cambridge.
Shirer, H. W., R. B. Shackelford and K. E. Jochim, 1959. A magnetic flowmeter for recording cardiac output. IRE Trans. Med. Electronics. 6: 1901–912.
Spencer, M. P. and A. B. Denison Jr., 1959. Square-wave electromagnetic flowmeter: theory of operation and design of magnetic probes for clinical and experimental application. IRE Trans. Med. Electronics ME-6: 220–28.

Spencer, M. P. and A. B. Denison Jr., 1960. Square-wave electromagnetic flowmeter for surgical and experimental application, *In* H. D. Bruner (ed.), Methods in Medical Research. Year Book Publ., Chicago. pp. 321–41.
Stein, P. D. and W. H. Schuette 1969. New catheter-tip flowmeter with velocity flow and volume flow capabilities. J. Appl. Physiol. 26: 85156.
Thürlemann, B. 1941. Methode zur elektrischen Geschwindigkeitsmessung von Flüssigkeiten. Helvetica Physica Acta 14: 383–419.
Westersten, A., G. Herrold, E. Abbot and N. S. Assali 1959. Gated sine-wave electromagnetic flowmeter. IRE Trans. Med. Electronics 6: 213–16.
Westersten, A., E. Rice, C. R. Brinkman and N. S. Assali 1969. A balanced field-type electromagnetic flowmeter. J. Appl. Physiol. 26: 497–500.
Wetterer, E. 1937. Eine neue Methode zur Registrierung der Blutströmungsgeschwindigkeit am uneröffneten Gefäss. Z. Biol. 98: 26–36.
Wetterer, E. 1963a. A critical appraisal of methods of blood flow determination in animals and man. IRE Trans. Biomed. Electronics BME 9: 165–73.
Wetterer, E. 1963b. Flowmeters: their theory, construction, and operation, *In* W. F. Hamilton and P. Dow (eds.), Handbook of Physiology. Amer. Physiol. Soc., Washington. pp. 1294–324.
Wexler, L., Bergel, D. H., Gabe, Y. T., Rakin, G. S. and Mills, C. J. 1968. Velocity of blood flow in normal human venae cavae. Circul. Res. 23: 349–59.
Wyatt, D. G. 1961. A 50 c/s cannulated electromagnetic flowmeter. Electronic Eng. 33: 650–54.
Wyatt, D. G. 1965. An input transformer with low earth leakage currents. Electronic Eng. 37: 16–19.
Wyatt, D. G. 1966. Baseline errors in cuff electromagnetic flowmeters. Med. Biol. Eng. 4: 17–45.
Wyatt, D. G. 1968. The design of electromagnetic flowmeter leads, *In* Chr. Cappelen, Jr., (ed.), New Findings in Blood Flowmetry. Universitetsforlaget, Oslo. pp. 69–74.
Wyatt, D. G. 1968. Dependence of electromagnetic flowmeter sensitivity upon encircled media. Phys. Med. Biol. 13: 529–34.
Yanof, H. M., A. L. Rosen and W. C. Shoemaker 1963. Design of an implantable flowmeter transducer based on the Helmholtz coil. J. Appl. Physiol. 18: 227–32.
Yanof, H. M. P. and A. L. Rosen 1963. Improvements in trapezoidal-wave electromagnetic flowmeter. J. Appl. Physiol. 18: 230–32.

B. Thermodilution techniques

Alfonso, S. 1966. A thermodilution flowmeter. J. Appl. Physiol. 21: 1883–886.
Andres, R., K. L. Zierler, H. M. Anderson, W. N. Stainsby, G. Coder, A. S. Ghrayyib and J. L. Lilienthal, Jr. 1954. Measurement of blood flow and volume in the forearm of man; with notes on the theory of indicator-dilution and on production of turbulence, hemolysis and vasodilation by intravascular injection. J. Clin. Invest. 33: 482–504.
Branthwaite, M. B. and R. D. Bradley 1968. Measurement of cardiac output by thermal dilution in man. J. Appl. Physiol. 24: 434–38.
Chalmers, J. P., P. I. Korner and S. W. White 1966. The control of the circulation in skeletal muscle during arterial hypoxia in the rabbit. J. Physiol. (London) 184: 698–716.
Clark, C. and L. T. Cotton 1966. Venous flow measurement in man using a local thermal dilution flowmeter. Brit. J. Surg. 53: 987.
Dow, P. 1956. Estimation of cardiac output and central blood volume by dye dilution. Physiol. Rev. 36 (Suppl. 2): 77–102.
Dowsett, D. J. and R. D. Lowe 1964. Measurement of blood flow by local thermodilution. J. Physiol. (London) 172: 13P–15P.
Evonuk, E., C. J. Imig, W. Greenfield and J. W. Eckstein 1961. Cardiac output measured by thermal dilution of room temperature injectate. J. Appl. Physiol. 16: 271–75.

Fegler, G. 1954. Measurement of cardiac output in anesthetized animals by a thermodilution method. Quart. J. Exp. Physiol. 39: 153–64.
Felix, W. and H. Groll 1953. Die Messung des Blutstromes mit Thermistoren. Z. Biol. 106: 208–18.
Fronek, A. and V. Ganz 1960. Measurement of flow in single blood vessels including cardiac output by local thermodilution. Circul. Res. 8: 175–82.
Fronek, A. and V. Ganz 1961. The effect of papaverine on coronary and systemic hemodynamics and on the oxygen metabolism of the myocardium. Cor et vasa 3: 120–28.
Ganz, V. and A. Fronek 1961. The action of nitroglycerin on coronary and systemic hemodynamics and on the oxygen metabolism of the myocardium. Cor et vasa 3: 107–19.
Ganz, V. 1963. The acute effect of alcohol on the circulation and on the oxygen metabolism of the heart. Am. Heart J. 66: 494–97.
Ganz, V., A. Hlavova, A. Fronek, J. Linhart and I. Prerovsky 1964. Measurement of blood flow in the femoral artery in man at rest and during exercise by local thermodilution. Circulation 30: 86–89.
Ganz, V., K. Bergmann, K. Dejdar and A. Fronek 1965. The measurement of coronary blood flow in man using the local thermodilution method. Proc. IV European Congress of Cardiology.
Ganz, V., K. Tamura, R. Donoso, and H. J. C. Swan 1970. Measurement of coronary sinus and great cardiac vein flow in man by continuous thermodiltuion. Am. J. Cardiol. In press.
Ganz, V., R. Donoso, H. S. Marcus, J. Forrester and H. J.C. Swan 1970. A new technique for measurement of cardiac output by thermodilution. Am. J. Cardiol. 27: 392–396.
Goodyear, A. V. N., A. Huvos, W. F. Eckhardt and R. H. Ostberg 1959. Thermal dilution curves in the intact animal. Circul. Res. 7: 432–41.
Grahn, A. R., M. H. Paul and H. U. Wessel 1968. Design and evaluation of a new linear thermistor velocity probe. J. Appl. Physiol. 24: 236–46.
Hlavova, A., J. Linhart, I. Prerovsky, V. Ganz and A. Fronek 1965. Leg blood flow at rest, during and after exercise in normal subjects and in patients with femoral artery occlusion. Clin Sc. 29: 55–64.
Hlavova, A., J. Linhart, I. Prerovsky, V. Ganz and A. Fronek 1966. Leg oxygen consumption at rest and during exercise in normal subjects and in patients with femoral artery occlusion. Clin. Sc. 30: 377–87.
Hosie, K. F. 1962. Thermal dilution techniques. Circul. Res. 10: 491–504.
James, G. W., M. H. Paul and H. U. Wessel 1965. Thermal dilution: instrumentation with thermistors. J. Appl. Physiol. 20: 547–52.
Khalil, H. H., T. Q. Richardson and A. C. Guyton 1966. Measurement of cardiac output by thermal dilution and direct Fick methods in dogs. J. Appl. Physiol. 21: 1131–135.
Kountz, S. L., W. J. Dempster and J. P. Shillingford 1964. Application of a constant indicator dilution method to the measurement of local venous flow. Circul. Res. 14: 377.
Linzell, J. L. 1966. Measurement of venous flow by continuous thermodilution and its application to measurement of mammary blood flow in the goat. Circul. Res. 18: 745–54.
Lowe, R. D. and D. J. Dowsett 1967. A catheter probe for measurement of jugular venous blood flow in man by thermal dilution. J. Appl. Physiol. 23 (6): 1001–1003.
Lowe, R. D. 1968. Use of a local indicator dilution technique for the measurement of oscillatory flow. Circul. Res. 22: 49–56.
Mohammed, S., C. Imig, E. J. Greenfield and J. W. Eckstein 1963. Thermal indicator sampling and injection sites for cardiac output. J. Appl Physiol. 18: 742–45.
Pavek, K., Boska, E. and I. V. Selecky 1964. Measurement of cardiac output by thermodilution with constant rate injection of indicator. Circul. Res. 15: 311–19.
Ristau, O. and W. Warbanow 1968. Blutstrommessung mit Thermistoren. Acta Biol. Med. Germ. 20: 731–42.
Shillingford, J., T. Bruce and I. Gabe 1962. The measurement of segmental venous flow by an indicator dilution method. Brit. Heart J. 24: 157–65.

Solomon, H., M. A. San Marco, R. Y. Ellis and C. W. Lillehei, 1969. Cardiac output determination: superiority of thermal dilution. Surgical Forum 20: 28–30.

Stembera, L. K., J. Hodr, V. Ganz and A. Fronek 1964. Measurement of umbilical cord blood flow by local thermodilution. Am J. Obst. & Gynec. 90: 531–36.

van der Werf, T. and W. G. Zijlstra 1966. Über die Anwendbarkeit von Thermistorkathetern zur direkten Messung des Blutstromes in den grossen Gefässen und im Herzen. Z. Kreislfschg. 55: 53–64.

White, S. W., J. P. Chalmers, R. Hilder and P. I. Korner 1967. Local thermodilution method for measuring blood flow in the portal and renal veins of the unanesthetized rabbit. Austr. J. Exp. Biol. Med. Sci. 45: 453–68.

18

Mechanics of the Microcirculation

RICHARD SKALAK

The theory of flow of blood in small blood vessels dates back to the classic work of Poiseuille. The particulate nature of flow of blood in the microcirculation has received theoretical consideration only in recent years. The mechanical behavior of blood is principally due to properties of the red blood cells and of the plasma in which they are suspended. Plasma is known to be a Newtonian fluid. The red blood cell is considered to be a flexible membrane filled with hemoglobin which is also a Newtonian fluid of somewhat higher viscosity than plasma. A lack of quantitative information on the elastic properties of the red cell membrane impedes accurate modeling of a red cell. Sieving and other experiments indicate that the red cell membrane can tolerate very large lineal extensions while only a small increase in area will produce rupture. In capillary vessels whose diameter is equal or less than the red cell diameter, lubrication theory has been applied to show the influence of the cell's elasticity on the pressure drop. The macroscopic behavior predicted is a decreasing resistance with increasing flow velocity. A similar behavior has been shown for liquid droplets suspended in a second immiscible fluid and in large scale experiments on models of red cells. The bolus flow between red blood cells which fit closely in a capillary has been computed. It has been shown that the circulation set up in the bolus of plasma between cells is not an important factor in the mass transfer to and from the red cells. Experimental observations *in vivo* show various effects which are not incorporated in any theoretical developments. The spacing of red blood cells in small vessels is not regular. There are also pulsations of velocity which persist to vessels of relatively small size. Recent experimental work on a complete vascular bed shows an effect of inertial terms on the pressure drop at steady perfusion rates. The transition from single particle flow in the smallest vessels to continuum theory regimes in larger vessels is incomplete. Theoretical work has suggested that blood may be represented as a micropolar fluid. The extent of the validity of such models is not yet clear.

* * *

I. INTRODUCTION

The mechanics of blood flow in the microcirculation has been a comparatively neglected subject until the past several years. Neglected here means as a field of application of analytical mechanics and as compared to the much more extensive literature of wave propagation in larger arteries. There is currently a considerable amount of innovative activity in the area of mechanics of the microcirculation, and this review will attempt to survey the present state of the art despite the evidence that in a few years it should be possible

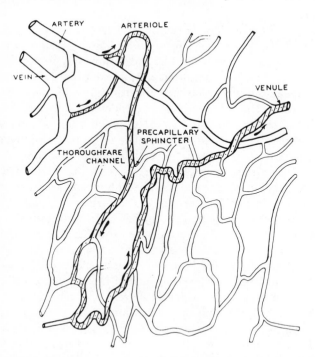

Fig. 1: Schematized tracing of the capillary bed indicating terminology for different structural components. (From B. W. Zweifach, 1961).

to write a more conclusive account. A number of recent reviews have already described the main outlines of the subject. [Bugliarello (1969); Fung (1969); Aroesty and Gross (1970).]

The term microcirculation is a recent generic term for the group of blood vessels which cannot be seen by the naked eye. The term will be used here in this sense and of any and all small blood vessels. Precise definitions of the subgroups such as arterioles, capillaries and venules have been the subject of much discussion and disagreement, which is perhaps some index of the difficulty of the subject. A recent review by Sobin (1966) gives detailed definitions based to a large extent on the structure of the vessel wall as well as the diameter. Figure 1 illustrates some of the terminology for a nutritive circulation. Figure 2 shows some of the distinctions among structures of various vessels. These distinctions are important to the discussion of physiologic function and control of blood flow. The present review is concerned with aspects of the fluid mechanics of flow in the microcirculation rather than the behavior of the wall or control mechanisms. Flow in capillaries which are of

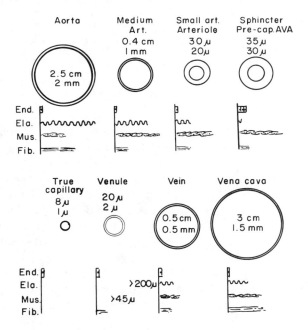

Fig. 2: Size, thickness of wall and four basic tissues in the wall of different blood vessels. The figures directly under the name of the vessel represent the diameter of the lumen; next, the thickness of the wall. End. = endothelial lining cells. Ela. = elastin fibers. Mus. = smooth muscle. Fib. = collagenous fibres. (From Burton, A. C. 1944. Relation of structure to function of tissues of the wall of blood vessels. Physiol. Rev. 34: 619–42.)

460 MECHANICS OF THE MICROCIRCULATION

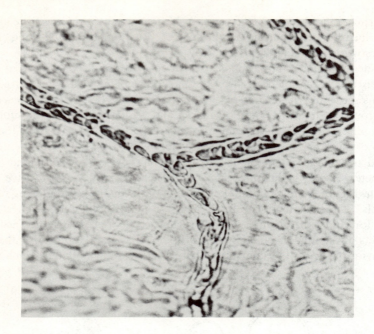

Fig. 3: Photograph of blood flow in capillary blood vessels in the mesentry of a rabbit. (Courtesy B. W. Zweifach. Reproduced from Y. C. Fung, 1969.)

the same order of diameter as that of a red cell is given primary consideration. In such vessels the red cells are severely deformed from their normal biconcave disk shape. Figure 3 shows a view of a portion of a microcirculation in which some of the characteristic deformations of the red cells can be seen.

In regard to the morphology of the microcirculation, it should be kept in mind that various organs and tissues of the body have different microvascular networks. Sobin (1966) briefly discusses the architecture and function of the microvasculature of muscle, the lung and the kidney. More extensive accounts of these and other special circulations may be found in the Handbook of Physiology (1963). The fluid mechanical effects discussed in the present review are for the most part not special to any particular vasculature, but will, in general, have bearing in any microcirculation.

II. HISTORICAL BACKGROUND

The historical development of ideas and knowledge of the circulation of blood has been elegantly recorded at length in a special volume edited by Fishman and Richards (1964). It appears that the ancient Greeks were quite familar with the anatomy of the heart and the major blood vessels, but did not realize that the blood circulated in a closed circuit. They did know that the heart valves permit flow in only one direction. Galen (131–201 A.D.), whose writing dominated medical thinking for 1500 years, taught that the heart was the source of heat and blood. There is some notion of the flow through the lungs in Galen's writings, but Michael Servatus (burned for heresy in 1553) is given credit for first properly describing the flow of blood from the arteries to veins in the pulmonary circulation (Cournand, 1964). The idea that the systemic circulation is also "in a circle" is the famous discovery of William Harvey (1578–1657). Although he demonstrated, by a variety of experiments and deductive proofs, that the blood must travel in a closed circuit, Harvey (1628) did not know the nature of the "porosities of the flesh" which he postulated. Microscopes had not yet been invented. The first direct observation of capillary vessels was reported by Marcello Malpighi (1661). Shortly after Malpighi, Leeuwenhoek (1688) reported seeing individual red blood cells flowing in clear plasma in the tail of a live tadpole. Landis (1964) gives this summary: . . . "He could distinguish single globules following each other, compressed and in single file, through the narrowest channels. Sometimes the corpuscles change into long ovals as the vessel narrowed. Both the narrowness and the number of these parallel pathways arrested his attention."

Leeuwenhoek's sketches (Fig. 4) clearly show red cells well separated, filling the capillary and in single file. It is remarkable that no attempt to solve the fluid mechanics problem of the resulting bolus flow between two cells was made until nine years ago (Prothero and Burton, 1961). The assessment of the resistance offered by the microcirculation was begun by Hales (1733) who first measured blood pressure in a "mare tied down to a gate." Hales

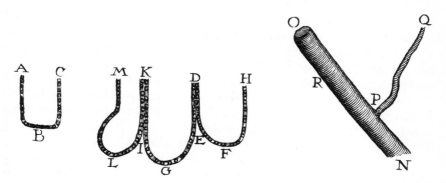

Fig. 4: Drawings from A. van Leeuwenhoek's 65th letter to the Royal Society (1688) showing capillary connections between arteries and veins and (right) junction with larger vessels in the tail of a fish. Note the representation of red blood cells as globules filling the capillaries.

realized that the principal resistance in the circulation is due to the "capillary arteries" and "capillary veins." He estimated the force driving the flow in a capillary as the product of the cross-sectional area (based on Leeuwenhoek's measurements) and the difference of arterial and venous pressures which he measured as 80 inches of blood. He remarks "whence we see both from experiment and calculation, that the force of the blood in these fine capillaries can be but very little, and the longer such capillaries are, the slower will the motion of the blood be in them."

A more sophisticated estimate of microvascular resistance was made by Thomas Young (1809) in his Croonian lecture on the functions of the heart and arteries. In a preparatory paper (Young, 1808) he derived empirical formulas for the pressure drop in pipes. He realized that in small tubes the pressure drop was proportional to the velocity, but attempted to write a single formula for water for the full range of sizes, from rivers to fine tubes, and came fairly close to the limited data available. Using this formula, he estimated the pressure drop in a mathematical model of the entire circulation made up of thirty segments from the aorta of 3/4 inch diameter and 9 inches long bifurcating twenty-nine times with a diameter ratio of 1/1.26 and a length ratio 1/1.961, "so that the diameter of the thirtieth segment may be only the eleven hundredth part of an inch, that is, nearly large enough to admit two globules of the blood to pass at once." He remarks that for this model ... "at the twenty-fifth division, where the artery does not much exceed the diameter of a human hair, the height to which water would rise, in a tube fixed laterally into the artery, is only two inches less than in the immediate neighborhood of the heart." Finally, he estimates the relative viscosity of blood and water and comes remarkably close to the typical value of 3.5:

"Hence, there can be no doubt that the resistance of the internal surface of the arteries to the motion of the blood must be much greater than would be found in the case of water: and supposing it about four times as great, instead of 20 inches, we shall have 80, for the measure of a column of which the pressure is capable of forcing the blood, in its natural course, through the smaller arteries and veins, which agrees very well with Hales's estimates."

Young also gives his own estimate of red cell diameters as 1/3000 to 1/3600 inches or about 7.1–8.5μ. This brackets the latest measurements (Canham and Burton, 1968) quite well.

The first clear definition of the capillaries is attributed (by Landis, 1964) to Marshall Hall (1831) who noted those vessels which "do not become smaller by subdivision, nor larger by conjunction; but they are characterized by continual and successive union and division, or anastomoses, whilst they retain a nearly uniform diameter."

Precise measurements of the pressure drop in fine glass capillary tubes were made by Poiseuille (1840) with the intent to understand the laws governing the pressure drop through the microcirculation. His tests were on water, alcohol and mercury. Poiseuille expressed his results by the empirical equation

(2.1) $$Q = k\,(1 + AT + A'T^2)PD^4/L$$

where P is the pressure drop over the length L of the tube, D is the diameter of the tube, T is the temperature and k, A, A' are empirical constants depending on the fluid. The analytical result which is usually called Poiseuille's law is

(2.2) $$Q = (\pi/128\mu)PD^4/L$$

where μ is the viscosity of the fluid. Equation (2.2) was not derived until 1858 and then independently by the physicists Franz Neumann and Edward Hagenbach (Schiller, 1933). Rouse and Ince (1957) point out that the tests of Poiseuille were also preceded by similar less accurate tests of Hagen (1839), and it was Hagenbach who named the resistance law after Poiseuille.

After the establishment of Poiseuille's law, there appears to be a complete gap in analytical mechanics work directed toward the microcirculation until the past decade. However, knowledge and understanding of the microcirculation continued to grow in other directions. Important among the nineteenth century advances were the histological study of the capillary wall, the elaboration of Fick's law (1855) of diffusion and its application to physiology, and Starling's hypothesis in 1896 for the transport across the capillary wall (see Landis, 1964).

Many of the aspects of the mechanics of blood flow in the microcirculation which are currently being studied were previously noted in a qualitative or rough quantitative manner. The book by Krogh (1930) which represents the foremost statement of knowledge of the microcirculation as of some forty years ago, mentions many topics on which quantitative information is being developed at present by experimental and mathematical methods. A few quotations from Krogh's first chapter illustrate the point:

"In the capillaries the pulse does not as a rule cause any variation in diameter of the vessels, but the velocity variations can often be very distinctly seen. . . ."

. . . "A special reduction in the number of corpuscles, which may even amount to their complete washing away from a certain capillary field is sometimes brought about by the process which I have termed *plasma skimming*."

. . . "When examined under a fairly high power, so that the capillary walls can be distinctly seen, some are found which allow the corpuscles to pass in a continuous current, and these generally exhibit a definite axial stream surrounded by a plasma zone through which a white corpuscle will occasionally come rolling along. Others are so narrow that the corpuscles have to pass in single file and come continuously in contact with the wall. Others again are even narrower, and the corpuscles can pass only in a deformed state. The simplest deformation is observed in capillaries down to about 4–5μ diameter (in mammals) where the edges of the flat dislike corpuscles are bent in while the length of the corpuscle measured during the passage does not exceed its diameter in the free state. In still narrower capillaries the red corpuscles are greatly deformed and compressed into a shape like sausages, the length of which may be double the normal diameter."

"The pressure necessary to bring about the deformation in narrow capillaries must be comparatively low, since the flow does not stop in a single narrow capillary, even when the same arteriole supplies several others through which the corpuscles can pass freely, but a definite estimate cannot be attained."

"It is well known that the red corpuscles will pass through filter paper, the pores of which will hold back, quantitatively, precipitates consisting of particles which are much smaller than corpuscles. There can be no doubt, though the passage has never been directly observed, that the corpuscles are greatly deformed during the passage. The pressure available for such a passage cannot exceed the height of fluid in the funnel."

"In single capillaries the flow may become retarded or accelerated from no visible cause; in capillary anastomoses the direction of flow may change from time to time."

"The most direct evidence of the wonderful plasticity and elasticity of red corpuscles is obtained when they are watched in a current, where they can be caught against a projecting edge and bent by the pressure of the current flowing past them. This happens very often in the lungs. . . . With each beat of the heart the flow in the capillaries becomes greatly accelerated, while it slows down gradually between beats."

In the last decade, as will be seen below, quantitative information has been developed on each of the items mentioned.

III. PROPERTIES OF RED CELLS AND PLASMA

The mechanical behavior of blood is principally due to properties of the red blood cells and of the plasma in which they are suspended. The plasma itself is known to behave as Newtonian fluid although it contains several types of protein molecules in suspension. Whole blood also contains platelets, which are about 2μ in diameter, but they are not present in sufficient number to affect the viscosity. The proteins are vital to the osmotic balance of fluids. Antibody mechanisms and platlets are important elements in thrombogenis but may be disregarded so far as the macroscopic mechanical behavior of the plasma under normal conditions is concerned. Gabe and Zata (1968) have shown that even under oscillatory motion plasma behaves as a Newtonian fluid. The viscosity of human plasma is about 1.2 cp (Gregersen, 1967; Chien, 1970).

Blood also contains leukocytes (white cells) which are larger and stiffer than red cells and because of these characteristics might be thought to be a controlling element in the flow through the microcirculation. However, there are about 600 times as many red cells as white cells per unit volume and, therefore, leukocytes are not usually considered to be rheologically important [Wolstenholme and Knight (1969) p. 102]. Possibly, in pathological conditions such as infections and leukemia, the role of white cells becomes much more important to the rheology of the blood.

Red cells are biconcave disks whose dimensions have been most recently and thoroughly studied by Canham and Burton (1968). In eight subjects a mean red cell diameter of 8.065μ with a standard deviation of 0.429 was found. Average values of area and volume computed from photographs were $138.1 \pm 17.4\mu^2$ and $107.5 \pm 16.8\mu^3$. The characteristic red cell shape is shown in Fig. 5 (Ponder, 1948).

The red blood cell consists of a very flexible membrane filled with a concentrated solution of hemoglobin. The viscosity of this solution has only recently been measured (Cokelet and Meiselman, 1968; Schmidt-Nielsen and Taylor, 1968; Dintenfass, 1968). The data of Cokelet and Meiselman show that hemoglobin solutions are Newtonian with a viscosity of about 6.0 cp at the concentration found in red cells, about 33% by weight. Their measurements indicate that blood viscosity is not lowered by the packaging of the

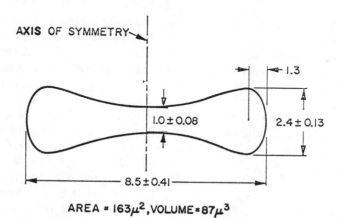

Fig. 5: Geometry of human red blood cell according to Ponder (1948). (Reproduced from Y. C. Fung, 1966.)

hemoglobin into red blood cells as compared to uniform suspensions of the same hemoglobin per unit volume as blood.

The notion that the red cell consists of a flexible membrane enclosing liquid without interior structure is supported by a number of observations and deductions. Red cells are capable of very large extensions and severe deformations (Bloch, 1962; Branemark and Lindstrom, 1963). In diapedesis it is clear that red cells can pass through openings in the endothelial wall of the order of 0.5μ in diameter (Skalak et al, 1970). Fung (1969) points out that the capability of the red cell to swell into a sphere could not be achieved by a solid unless some complex anisotropy is assumed. He also advances the argument that the sharp trailing edge observed on red cells flowing in capillaries can be readily achieved by a fluid-filled membrane, but it would require a highly nonuniform distribution of surface traction to produce the same deformation in a solid. The possibility of production of red cell "ghosts" consisting only of the membrane, but still showing the biconcave disk shape, reinforces the idea that the unstressed red cell is the equilibrium shape of a thin shell with no appreciable pressure inside it.

A great deal more is known about the red cell with regard to its chemistry, production, life span, metabolism and pathology (see for example, Harris (1965) with 2,074 references). But the elastic properties of the red cell wall are not well established.

The thickness of the red cell membrane is approximately 75 Å (Bishop and Surgenor, 1964). Using a typical radius of curvature of 4μ for the red cell, the thickness ratio is of the order of 1/400, so thin shell theory may be applicable. This means that membrane stresses may be expected to predominate over bending stresses in situations such as osmotic sphering and in sieving experiments (Gregersen et al, 1967). On the other hand, the unstressed shape of the red cell is such that it can deform isochorically (without change of volume) into an infinite variety of applicable shapes not requiring any stretching or tearing of the surface (Fung, 1966). For such deformations no membrane stresses are generated and the return to the unstressed shape after such a deformation must be effected by bending stresses, as in a rubber glove.

Hoeber and Hochmuth (1970) have utilized the motion of a red cell returning to its natural shape after a deformation to estimate the modulus of elasticity of the red cell wall by comparison to models made of rubber. The models were 4.0 cm in diameter and had a thickness of 68.6μ which would correspond to a thickness of 145 Å for a red cell. The model cells were filled with oil of $9.6p$ viscosity and tested in oil of $2.3p$ viscosity giving a ratio similar to that of red blood cells in saline. The tests consisted of drawing the model cell into a tube and then expelling it. The typical deformed shape in the tube is shown in Fig. 6. After expulsion, the time required to recover the normal shape was measured from motion pictures. Similar tests

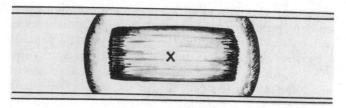

Fig. 6: Characteristic constrained configuration of a model cell in a circular tube (from Hoeber and Hochmuth, 1970).

were made on red cells expelled from pipettes 3.2–4.2μ in diameter. By a dimensional analysis, the data was used to estimate the modulus of elasticity of the red cell membrane (Young's modulus) as 7.2×10^5 dynes/cm² with a range of 9.6×10^4 to 1.1×10^7 dynes/cm². An even lower estimate of 5×10^3 to 1.5×10^4 dynes/cm² has been derived by Hochmuth and Das (1970) using the technique of stretching a single red cell. The value of 7.2×10^5 dynes/cm² is about one order of magnitude lower than the values suggested by other methods (Rand, 1964; Katchalsky et al, 1960). The difference may be due partly to the fact that these earlier estimates involved large strains near rupture. Since most tissues exhibit an increasing modulus with increasing strain, the same kind of nonlinear behavior may hold for the red cell membrane also. The estimate of Young's modulus by Katchalsky et al (1960) is based on a viscoelastic model and the internal pressure and volume in sphered, hemolyzing red cells. The value derived is 2.4×10^7 dynes/cm² and is, at best, applicable near rupture. Rand (1964) did not measure strains, but measured the time required to produce rupture of a cell partially drawn into a micropipette at constant pressure. The analysis assumes a viscoelastic model and that a uniform, isotropic tension exists in the membrane, as in a a liquid interface. Fung and Tong (1968) have pointed out that this is a dubious assumption.

No attempt at a more precise analysis of the deformation produced by the pipette technique of Rand (1964) has been made, but a consistent analysis of the finite deformations involved in sphering a red cell has been given by Fung and Tong (1968). By assuming the end result is a perfect sphere, it is found that the transformation from the original biconcave disk shape to the sphere is determined solely by the requirement that the principal strains at any point be equal, independent of elastic moduli, wall thickness or surface tension. The analysis leads to a requirement of unequal strain distribution, the strain being less at the rim of red cell than for the area originally on the axis of the cell. This requires an increased modulus and/or thickness of the membrane at the rim. The analysis does not provide any estimate of the elastic modulus; one of the unknowns is the magnitude of the surface tension which may result when red cells are placed in hypotonic solution to sphere them.

Another source of information concerning the flexibility of red cells comes from experiments on their passage through micropores of polycarbonate sieves (Gregersen et al, 1967; Gregersen and Bryant, 1968). The sieves used in these experiments were 12μ in thickness with pores produced by nuclear bombardment and etching. The standard deviation of the pore diameter in a given sieve is less than 10% of the mean value. Figure 7 shows the measured pressure flow relations at low pressures through sieves having pore diameters of 6.8μ and 4.8μ. It was found that sieves with pores of 3μ diameter or larger allowed passage of 100% of the red blood cells. For 2.4μ pore diameter, the red cell count was reduced to about 70% in the filtrate compared to the original suspension. A critical pore diameter of 2.84μ was computed as the one at which the minimum area required to enclose the red cell volume of $100\mu^3$ would be equal to the unstressed red cell area of $150\mu^2$. The shape of the deformed red cell was taken to be a cylinder capped at each end by a hemisphere of the same diameter. This model was used also by Canham and Burton (1968) to compute the minimum cylindrical diameter possible from their data on area and volume of red blood cells. They found an average of 3.33μ for the critical diameter.

Sieving tests have been extended to pores as small as 2.2μ in diameter and pressures as high as 70 cm Hg by Chien et al (1970). Varying degrees of hemolysis result as measured by the release of hemoglobin and potassium from the erythrocytes and by changes in the number and volume of the cells in passage through the sieves. The relative resistance to the flow of the suspension of red cells in Ringer solution to that of the Ringer solution itself is shown in Fig. 8. The flat portions of the curves at high pressure correspond to considerable hemolysis, except in the case of the 4.4μ sieves. Such data have not yet led to specific estimates of elastic moduli but are potentially a method of arriving at quantitative data for large samples. These results have also been used to estimate the resistance of a single cell in a pore relative to the suspending Ringer solution by assuming the fraction of pores occupied by cells is the same as in the filtrate. This yields a relative resistance that varies from about 2×10^2 to 1×10^4 times that of Ringer solution alone when the pore diameter decreased from 4.4μ to 2.2μ. (Chien et al, 1970). These high resistances are associated with the stretching of the membrane rather than any viscous flow of the liquids inside or outside the membrane.

The general picture of the red blood cell membrane which emerges from the various experiments described above suggests that any failure criterion should be expressed at least in terms of a function of the two principal strains at any point, rather than in terms of a strain in any one direction. For example, a simple criterion would be

$$(3.1) \qquad \frac{(\epsilon_1 + \epsilon_2)^2}{2a^2} + \frac{(\epsilon_1 - \epsilon_2)^2}{2b^2} = 1$$

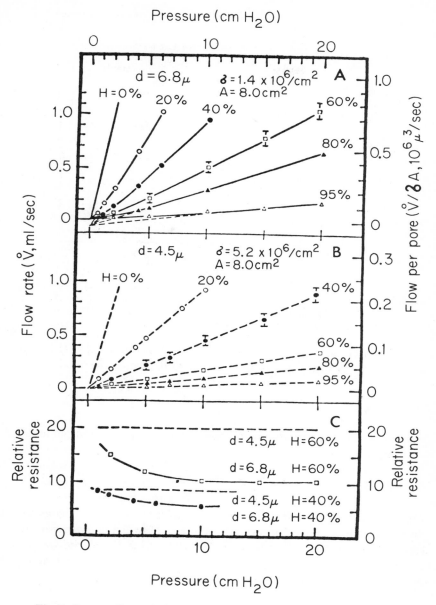

Fig. 7: Pressure-flow relationships of 0–95% human *RBC* suspensions through 6.8μ (A) and 4.5μ sieves (B). The ordinates at left give total flow through the sieve; those at right give the average flow through each pore. From the results in A and B on 40 and 60% *RBC* suspensions, relative resistance (flow of Ringer solution/flow of suspension) values are calculated (C).

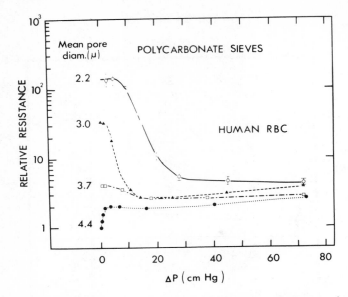

Fig. 8: Flow resistance (= pressure/flow rate) of red blood cell suspensions relative to that of the suspending medium as a function of filtration pressure. Hematocrit about 2%. (From Chien, et al, 1970).

where ϵ_1 and ϵ_2 are the principal logarithmic or natural strains. The logarithmic strain ϵ is defined in terms of the conventional strain e by (Malvern (1969) p. 151)

$$(3.2) \qquad \epsilon = \ln(1 + e)$$

where e is defined, for any line element stretching from an initial length L_0 to a length L, by

$$(3.3) \qquad e = \frac{L - L_0}{L_0}$$

Equation (3.1) is plotted in Fig. 9 with a particular choice of constants as discussed below. Points inside the ellipse in Fig. 9 represent permissible states of strain. Points on the boundary or outside the ellipse would lead to rupture. The fractional change in area of any small element under the principal strains ϵ_1 and ϵ_2 may be shown to be

$$(3.4) \qquad \frac{A - A_0}{A_0} = e^{(\epsilon_1 + \epsilon_2)} - 1 = (1 + e_1)(1 + e_2) - 1$$

where A is the final area which was initially A_0. The constant a in Eq. 3.1

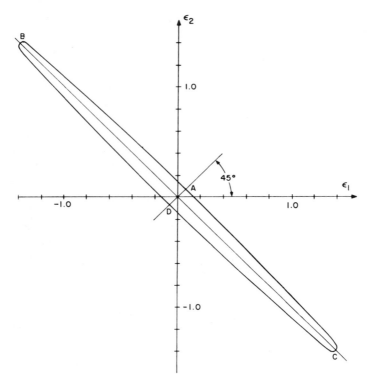

Fig. 9: Ellipse of failure proposed for red cell membrane.

is chosen so that $\epsilon_1 = \epsilon_2 = a = 0.0786$ at point A in Fig. 9 and $e_1 = e_2 = 0.082$ and $(A - A_0)/A_0$ is 0.172, which was measured by Rand and Burton (1963) in sphering tests. They found the area increased from $146\mu^2$ to $171\mu^2$, on the average, prior to hemolysis. The constant b is chosen in Eq. 3.1 so that at point B in Fig. 9, $-\epsilon_1 = \epsilon_2 = b = 1.386$, $(A - A_0)/A_0 = 0$; $\epsilon_1 = -3/4$; $e_2 = 3$. This value is suggested by tests made by Kochen (1968) in which filaments were drawn from red cells having a mean length of 24.6μ. Allowing for the remainder of the cell itself being about 8μ, the final length is about four times the original length.

Negative values of ϵ_1 and ϵ_2 correspond to compression. The point D in Fig. 9 might be interpreted as failure by buckling as in crenation, but this is only a qualitative suggestion at this time. It is possible that an accurate failure criterion should be expressed in terms of stresses and include stress or strain rates.

IV. PARTICLE FLOW

A. Rigid particles.

In capillary blood vessels which are the same order of magnitude in diameter as red blood cells, an accurate description of the flow requires consideration of the red blood cells as discrete particles. When the red blood cells move in a steady flow in a vessel of constant diameter, the shape of each cell remains more or less constant, and hence, each moves as a rigid particle of the same shape. For this reason, results of theories and tests on rigid particles are of interest. The shape a red cell will take in a given flow cannot be predicted by rigid particle analysis, but the velocity of the cell and pressure drop can be computed if the shape is known.

Most studies have assumed particles are simple analytical shapes such as spheres, spheroids, ellipsoids or disks. Since the Reynold's number of capillary flow is of the order of 10^{-2} or less, creeping flow equations have been adopted in most cases. The equations of slow viscous flow or Stokes flow are the Navier-Stokes equations with the acceleration terms neglected

$$(4.1) \qquad 0 = -\nabla p + \rho f + \mu \nabla^2 u$$

and the equation of continuity

$$(4.2) \qquad \nabla \cdot u = 0$$

where u is the velocity vector, p is the pressure, f is the body force per unit mass and ρ and μ are the density and viscosity of the fluid, which is plasma in the case of capillary blood flow.

Since the specific gravity of red blood cells is about 1.10 and that of plasma is about 1.03, effect of gravity on capillary blood flow is small and is usually neglected by assuming the red cells are neutrally buoyant. For a neutrally buoyant particle in Stokes flow, the resultant force and moment on the particle due to pressures and viscous stresses exerted by the fluid must be zero. This is an important condition that must be satisfied by any solution. For rigid particles this zero drag condition can be arrived at by superposition of various solutions since the problem is completely linear. For example, for a particle in a tube flow, the two basic solutions may be taken to be the flow past a fixed particle and the flow due to a moving particle with zero discharge as in sedimentation in a closed-end tube. Only results for the neutrally buoyant case will be discussed here.

Solutions for a line of spherical particles have been given by Wang and Skalak (1969) and for a line of spheroidal particles by Chen and Skalak (1970). The particles are assumed to be axisymmetrically located as indicated in

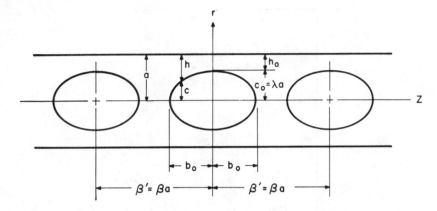

Fig. 10: Definition sketch of dimensions and spacing of spheres, spheroids and flexible, axisymmetric particles in a cylindrical tube.

Fig. 10. For spheres, the two semiaxes are equal, $c_0 = b_0$. Some of the typical results of interest are shown in Figs. 11, 12 and 13. The additional pressure gradient coefficient G for a particle spacing $\beta' = a\beta$ (Fig. 10) is plotted in Fig. 11 against the diameter ratio $\lambda = c_0/a$. The total pressure drop, P, in a length L is expressed as

(4.3) $$P = \frac{8\mu VL(1 + G)}{a^2}$$

where μ is the kinematic viscosity of the suspending fluid and V is the mean velocity of the suspension including both liquid and solid phases. When G is zero, (Eq. 4.3) Poiseuille's law results; the factor $(1 + G)$ is the relative apparent viscosity of the suspension compared to the suspending fluid. The coefficient G depends strongly on the diameter ratio λ. For values of $\lambda < 0.4$, the apparent viscosity is increased by less than one percent. A relative apparent viscosity of two is not reached until about $\lambda = 0.9$.

For particle spacing greater than $\beta = 2$ ($\beta = \beta'/a$), the additional pressure drop coefficient G varies approximately as $1/\beta$ which indicates that the additional pressure drop per particle is relatively constant if the particle spacing is more than one tube diameter. This is illustrated in Fig. 12 which shows the additional pressure drop coefficient, K, plotted against the particle spacing β. For each curve in Fig. 12, the left hand end of the curve represents the case of particles touching, like a string of beads. The additional pressure drop per particle, ΔP, is given by

(4.4) $$\Delta P = \frac{2\mu VK}{a}$$

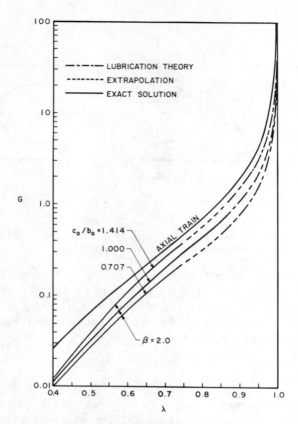

Fig. 11: Additional pressure gradient coefficient G for spheres ($c_0/b_0 = 1$), spheroids ($c_0/b_0 = 0.707$ or 1.414) and axial train (stacked coins) models as a function of λ. Dimensionless spacing $\beta = 2$ for spheres and spheroids. See Fig. 10 and Eq. 4.3. (Based on Wang and Skalak, 1969 and Chen and Skalak. 1970.)

where the coefficient $K = 4\beta G$. The fact that K varies little with β suggests that results computed for a single particle in an infinite tube may be used to estimate results for a line of particles by multiplying the pressure drop for one particle by the number of particles in a given length L if $\beta > 2$.

Figure 13 shows the particle velocity, U, divided by the mean velocity of the suspension, V, as a function of the diameter ratio λ. The results shown are for spheres. For spheroids of the same diameter ratio the results are within one or two percent of those shown for spheres (Chen and Skalak, 1970).

The curves in Figs. 11 and 13 are based on exact series solutions for $\lambda < 0.9$. The series involved converge slowly for $\lambda > 0.9$ and only approximate results based on lubrication theory are available in this range. The basic

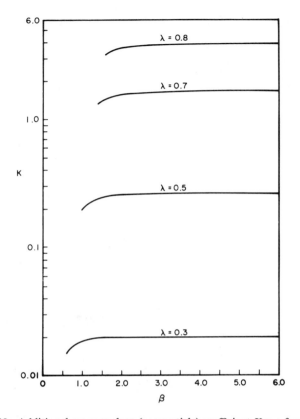

Fig. 12: Additional pressure drop (per particle) coefficient K as a function of the spacing β for a line of axisymmetric spheres. See Fig. 10 and Eq. 4.4. (Based on Wang and Skalak, 1969.)

assumption made in lubrication theory is that the gap width $(a-c_0)$ is small compared to its length so that the equations of motion and continuity (Eqs. 4.1, 4.2) may be reduced to

(4.5) $$0 = \frac{\partial p}{\partial z} + \mu\left(\frac{\partial^2 u_z}{\partial r^2} + \frac{1}{r}\frac{\partial u_z}{\partial r}\right)$$

and

(4.6) $$Q = 2\pi \int_0^a r u_z \, dr = \text{constant}$$

where (r, z) are radial and axial coordinates and u_z is the velocity in the axial direction. The approximation afforded by lubrication theory is probably quite good in the vicinity of the maximum diameter of a close-fitting particle,

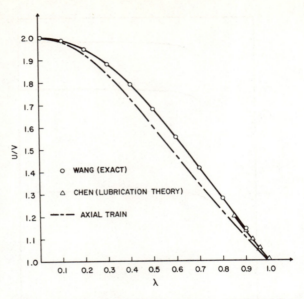

Fig. 13: U/V = ratio of particle velocity to mean velocity of flow of the suspension vs diameter ratio λ for spheres at spacing $\beta = 2$ and axial train (stacked coins) model.

but is likely to be poor near the ends of the particle. Between particles, the lubrication equations (4.5, 4.6) yield Poiseuille flow and no effect of interaction between particles is incorporated. The zero drag condition is satisfied in the lubrication theory by approximating the forces due to pressure differences and to shear stresses separately, and setting them equal and opposite to each other. The force due to the pressure difference across the particle is assumed to act on the entire projected area of the particle and the shear stress force is integrated over the surface of the particle. In the notation indicated in Fig. 10, the equation of zero drag is

$$(4.7) \qquad [p(-b_0) - p(b_0)]\pi c_0^2 = 2\pi \int_{-b_0}^{b_0} \mu \left[\frac{\partial u_z}{\partial r}\right]_{r=c} c \, dz$$

An asymptotic formula for close-fitting spheres ($h_0 \ll a$) was derived by Hochmuth (1967) on the basis of the lubrication theory (see also Hochmuth and Sutera, 1970). The result for the coefficient K in Eq. 4.4 is

$$(4.8) \qquad K = 2\pi \left(\frac{2}{1-\lambda}\right)^{1/2} - 15.75$$

The asymptotic formula and numerical results computed without further approximations from Eqs. 4.5, 4.6 (Chen and Skalak, 1970) agree within

10% only for $\lambda \geq 0.995$. In Hochmuth's treatment, the surface of the sphere is approximated by $c(z)$ of the form

$$c(z) = c_0 - \tfrac{1}{2}kz^2 \tag{4.9}$$

where $k = 1/c_0$ in this case. The approximation (Eq. 4.9) is quite accurate in the neighborhood of minimum gap width but not elsewhere.

The idea of using lubrication theory approximations for the flows in which a close-fitting particle is involved was used previously by McNown et al (1948) and Christopherson and Dawson (1959), but these authors treated the case of a particle settling in a tube rather than the zero drag case of a neutrally bouyant particle.

Another approximation has been developed by Bungay and Brenner (1969) using a singular perturbation method based on the full equation (4.1) and expansions in appropriate stretched coordinates, assuming again that the gap width is small compared to the tube radius. This method reproduces the leading term of lubrication theory results (Eq. 4.8) as a first approximation and can, in principle, be extended to higher orders of approximation. The technique has been used also to obtain approximate solutions to the problems of a concentric spheroid, an eccentric sphere and cylindrical train (stacked coins) eccentrically located (Bungay, 1970). Asymptotic formulas for the particle velocity and drag are given in each case. An eccentric sphere rotates in addition to translating in the axial direction. This case is of interest in capillaries whose diameters are somewhat larger than that of the red cells. It appears that in narrow capillaries the flexibility of the red cells tends to make them self-centering, and they do not appear to rotate.

The approximate equations (4.5, 4.6) become exact for the so-called stacked coins or axial train model (Whitmore, 1967) which consists of a cylinder of radius λa made up of coins or disks. The suspending fluid occupies the annulus $r = \lambda a$ to $r = a$, and u_z is assumed independent of z. In this case the additional pressure gradient coefficient in Eq. 4.3 is

$$G = \frac{\lambda^4}{1 - \lambda^4} \tag{4.10}$$

Equation 4.10 is plotted in Fig. 11, and forms an upper boundary for any train of particles of maximum diameter ratio λ. The velocity U of the disks making up the axial train is given in terms of the mean velocity V of the suspension as a whole by

$$\frac{U}{V} = \frac{2}{1 + \lambda^2} \tag{4.11}$$

Equation 4.11 is plotted in Fig. 13.

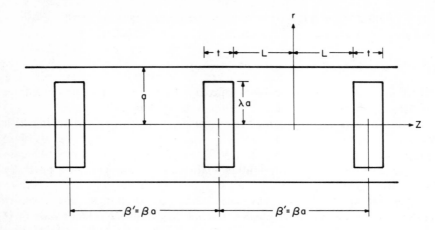

Fig. 14: Definition sketch for rigid disks as a model of capillary blood flow.

Another model of capillary blood flow which has been explored is that of a line of axisymmetric disks as shown in Fig. 14. Bloor (1968) computed streamlines and pressure gradients for a range of spacings $2L/a$ and particle thickness t/a, assuming the gap width $(1 - \lambda)a$ was small compared to the tube radius. Under this assumption and the requirements of zero drag, the flow in the gap is the same as in Whitmore's (1967) axial train solution. Using this gap solution as a boundary condition on $z = \pm L$, a numerical solution was obtained for the flow between disks. In coordinates moving with the disks, the flow shows the typical circulation of bolus flow first pointed out by Prothero and Burton (1961). This circulation persists, although weakly, for particles which do not fill the tube as shown in Fig. 15b which was computed from an exact solution for a line of spheres.

The bolus flow between two disks with $\lambda = 1$, with the disks extending completely across the tube, has been computed by Bugliarello and Hsiao (1967); Lew and Fung (1969a); Aroesty and Gross (1970). Lew and Fung show the velocity profile changes from the uniform velocity required on the disk face to within 1% of a Poiseuille flow in a distance of 1.3 tube radii from the disk. This is also the lower limit of the entrance flow length as the Reynolds number approaches zero (Lew and Fung, 1969b; 1970b). The additional pressure drop for one such face is $16\mu V/3a$. If the particles are spaced more closely, $(L < 1.3a)$ the total pressure drop over Poiseuille flow in the length $2L$ decreases (Lew and Fung, 1969a).

The flow in the space between two disks has been further studied by Lew and Fung (1970a) for the limiting case of disks of zero thickness ($t = 0$, Fig. 14). They conclude that for this model the pressure drop in the space between cells $(-L < z < L)$ is greater than that for Poiseuille flow by a

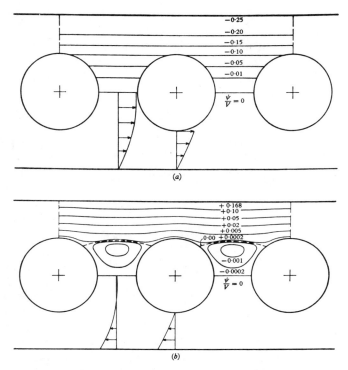

Fig. 15: Streamlines and velocity profiles for spacing $\beta = 1.5$ and diameter ratio $\lambda = 0.5$ for neutrally bouyant spheres. (a) Streamlines and velocity profiles in coordinates fixed to the tube. (b) Streamlines and velocity profiles in coordinates moving with the spheres. (From Wang and Skalak, 1969.)

significant amount, in the range $0.5 \leq \lambda \leq 1$ and $0 \leq \beta \leq 3.0$. However, since the disks are of zero thickness, a comparison of the computed effect of the increased pressure drop, which would be attributed to the space occupied by the cells ($L < z < L + t$), is not possible from these solutions.

The toroidal vortex motion which takes place in the plasma gap between two cells with $\lambda = 1$ has been computed by Aroesty and Gross (1970). These data have been utilized to determine the effect of this circulation on the transport of substances across the capillary wall and through the plasma to the red cells by a combination of diffusion and convection. Their computations show that the circulation augments mass transfer to the downstream cell, but reduces it for upstream cell so that the net effect is small. The net effect is a function of the Peclet number $Pe = UL/D$, where D is the diffusivity of the particular molecular species considered. For capillary blood flow, Pe is estimated to be ten at most for dissolved gases; very little augmentation of mass transfer due to convection is predicted by the model.

Experimental observations of the motion and pressure drop due to the flow of neutrally bouyant rigid particles in tubes are surprisingly scarce. For suspensions in which the particle size is much smaller than the tube diameter, there is an extensive literature (Goldsmith and Mason, 1967; Mason and Goldsmith, 1969). Tests on large rigid particles ($\lambda > 0.5$) having the shapes of disks, discoids, spherical caps and red blood cells have been reported only recently (Sutera and Hochmuth 1968; Hochmuth and Sutera, 1970). None of these tests corresponds to any case for which exact theoretical results are available, but data for hemispheres multiplied by two are in close agreement with theoretical values computed for spheres (Hochmuth and Sutera, 1970). The prediction of various theories that particles spaced more than one or two tube diameters apart do not interact is confirmed by these experiments. The additional pressure drop for closely spaced planoconvex disks was 30–40% less than for the same particles spaced widely apart. The data also closely confirm the predictions of the ratio U/V (particle velocity/mean velocity of the suspension) shown in Fig. 13. The results primarily depend on the diameter ratio, λ, and to a much smaller degree on the shape of the particle.

An interesting aspect of the experimental observations is the orientation which various models assume. For a disk or red blood cell shape, an orientation in which the plane of the disk includes the axis of the tube will be called "edge-on." When the plane of the disk is perpendicular to the axis of the tube, the orientation will be called "transverse." Spherical caps (planoconvex disks) with diameter ratios $\lambda = 0.732$–0.9982 were found to be self-centering and stable in the transverse orientation with the flat side upstream. Biconcave disks resembling red cells with a diameter ratio $\lambda = 0.8$ were observed only in the edge-on orientation. For larger values of λ, stable transverse orientations appeared also. Hemispheres were less stable and tended to rotate slowly.

Solid spheres with large diameter ratios ($\lambda > 0.9$) have been observed, in sedimentation experiments, to rotate in an eccentric position (McNown, 1948; Christopherson and Dowson, 1959). These experiments suggest that a symmetric shape such as a sphere may be unstable in translation when centered on the tube axis, but a shape which has a predominant convex curvature facing downstream is stable.

Burton (1969) has pointed out that there is literature on particle flow in tubes which is concerned with transport of solid substances in pipes (see Jensen and Ellis, 1967; Ellis et al, 1963). In most engineering applications, the pipe flows involved are almost always turbulent, but it is interesting that a solid such as coal in a spherical or cylindrical capsule form is found to have a most favorable diameter ratio 0.90–0.95. The driving of a slurry calls for greater consumption of pumping power because the stream velocity must be raised to the level of turbulence required to keep the particles suspended. A capsule generates its own lubricating film by a planing action in the case of a cylinder or by rolling if it is a sphere.

B. FLEXIBLE PARTICLES.

When blood flows through capillaries whose diameter is less than that of a red blood cell, it is obvious that the red cells must be deformed. Observations show that cells may also be deformed in somewhat larger vessels (see, for example, Skalak and Branemark, 1969). In either case, it is to be expected that deformations, dependent on the magnitude of the viscous stresses and hence on the velocity or pressure gradient, will result during flow. The shape of the particle will then be velocity dependent. Since the increased pressure drop is a function of shape, as discussed above for rigid particles, it may be expected that the apparent viscosity of a suspension of deformable particles is a function of mean velocity or pressure gradient in a given size tube, and it will vary in a nonlinear manner with tube diameter.

These qualitative ideas were first incorporated into a quantitative theory by Lighthill (1968) using the approximations of lubrication theory and a simplified linear relation of the cell deflection to the local pressure. The basic ingredients of the theory are similar to those used in the theory of compliant surface bearings (Dowson and Higginson, 1960; Castelli et al, 1967; Benjamin, 1969). In such bearings, the deformation of the elastic boundary due to the lubrication pressures appreciably influences the pressure distribution, giving rise to the coupled "elastohydrodynamic" problem. In the case of a flexible particle forced through a tube by a driving pressure, the requirement of zero drag (Eq. 4.7) is an additional condition which is, in fact, necessary to make the problem determinate.

The original formulation by Lighthill (1968) has been modified by Fitzgerald (1969a, b) to make it more accurate and realistic with respect to the blood flow problem; the qualitative results are the same. The principal differences introduced by Fitzgerald are first, that Eq. 4.5 is used to describe the fluid flow rather than the two-dimensional equivalent which neglects the last term of the equation. Secondly, in using the zero drag condition (Eq. 4.7), Fitzgerald uses finite limits of integration as shown in Eq. 4.7 rather than the approximation of these as $\pm \infty$. Thirdly, the shape of the particle in longitudinal cross-section was taken to be an ellipse rather than the parabola (Eq. 4.9). Fourthly, a more elaborate analysis of the red cell deformation was made, but the deflection of the cell at any point was still taken to be a linear function of the pressure at the point.

Fitzgerald assumes that at some reference pressure p_0 the particle just fills the tube, and its form is then given by

$$c(z) = a\left(\frac{1 - kz^2}{a}\right)^{1/2} \tag{4.12}$$

where $c(z)$ is the particle radius (Fig. 10), a is the radius of the tube, and k is a characteristic curvature of the particle. Assuming the deflection at any point is proportional to the pressure difference $p - p_0$, the gap width $h(z)$ is

$$h(z) = a\left[1 - \left(\frac{1 - kz^2}{a}\right)^{1/2}\right] + \alpha(p - p_0) \tag{4.13}$$

where α is a compliance coefficient which, in the general case, includes the compliance of both the wall and the particle. For the blood flow case, as pointed out by Fung (1966), the deflection of the capillary wall may be neglected. Then α is a compliance of the particle only.

It is convenient to use coordinates moving with the particle velocity U (Fig. 10) so that the particle appears stationary and the tube wall has a velocity $-U$. In these coordinates there will generally be a discharge, Q', of the fluid in the negative z direction. With respect to the particles, this is a "leakback" and is an important unknown of the problem. In terms of the particle velocity U and the mean velocity of the suspension V in coordinates fixed to the tube it follows that

$$Q' = \pi a^2 (U - V) \tag{4.14}$$

Since U is generally greater than V (Fig. 13), leakback is usually positive. The analysis is written in terms of the leakback and particle velocity by integrating Eq. 4.5 as usual in lubrication theory to yield a Reynolds equation

$$\begin{aligned}\frac{1}{16\mu}\frac{dp}{dz}(2Rh + h^2)&\left[2R^2 + 2Rh + h^2 - \frac{2Rh + h^2}{\ln(1 + h/R)}\right] \\ &= -U\left[\frac{(R + h)^2}{2} - \frac{2Rh + h^2}{4\ln(1 + h/R)}\right] + \frac{Q'}{2\pi}\end{aligned} \tag{4.15}$$

where R is the deformed particle radius

$$R = a - h \tag{4.16}$$

The zero drag condition (Eq. 4.7) is written in terms of the same variables with the further approximation $c_0 \cong a$

(4.17) $$= \int_{-b}^{b} \left[\frac{dp}{dz}\left(R^2 - \frac{2Rh + h^2}{2 \ln(1 + h/R)}\right) + \frac{2U}{\ln(1 + h/R)}\right] dz$$

Equations 4.13, 4.15–4.17 are cast into dimensionless form and solutions are obtained by numerical integration. The results are expressed in terms of dimensionless parameters $A, C, E,$ and F defined by

(4.18) $$A = \frac{\mu U \alpha}{a^2 (ka)^{1/2}}$$

(4.19) $$C = \frac{Q'}{\pi a^2 U} = \frac{U - V}{U}$$

(4.20) $$E = \frac{\alpha}{a}[p(-b) - p(b)]$$

(4.21) $$F = \frac{\alpha}{a}\left[\frac{1}{2}(p(b) + p(-b)) - p_0\right]$$

A is a dimensionless particle velocity; C is a dimensionless leakback and also is a measure of the ratio of the typical gap thickness to the tube radius; E is a dimensionless pressure drop; F is a measure of the particle deformation under the mean pressure. Given sufficient data to determine any two of the parameters A, C, E, F, any other may be determined by the results shown in Fig. 16.

A resistance parameter D is further defined by

(4.22) $$D = \frac{E}{A} = [p(-b) - p(b)]\frac{a(ka)^{1/2}}{\mu V}$$

From a physical standpoint, the values of D are the most easily interpreted. For the range of parameters shown in Fig. 16, D varies from about 100 to 1000. A Poiseuille flow of the suspending fluid alone at velocity U corresponds to a value of D equal to sixteen. It follows that D is related to the additional pressure drop coefficient K (Eq. 4.4) and the additional pressure gradient coefficient G in Eq. 4.3 by

(4.23) $$(D - 16)(ka)^{1/2} = \frac{2KV}{U} = \frac{8\beta GV}{U}$$

Assuming a particle spacing $\beta = 2$ (Fig. 4) and that the initial shape of the particles is spherical ($ka = 1$), the range $100 < D < 1000$ corresponds to

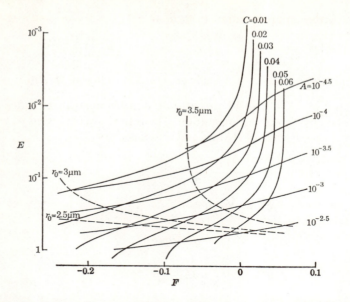

Fig. 16: Curves of constant velocity parameter A and constant film thickness parameter C on an F-E diagram. The broken lines represent compatibility curves for a 7.5μ red cell in capillaries of various radii. (From Fitzgerald, 1969.)

$5.6 < G < 61.5$. If the particles were rigid, this would require a diameter ratio $\lambda > 0.98$ (Fig. 11). These values suggest that the range covered in Fig. 16 is for comparatively tight-fitting particles and probably does not allow for sufficient flexibility of the particles to be realistic in all cases of blood flow. This may be due to an overestimation of the rigidity of the red cell. Fitzgerald assumes that the red blood cells initially will be transversely located and bent into a parachute shape with the rim of the red cell touching the tube wall when at rest. Recent observations in models and of red cells in glass capillaries (Hochmuth et al, 1970) as well as *in vivo* (Skalak and Branemark, 1969) have suggested that the starting position at rest is a squashed edge-on position in which the stiffness is probably less than first thought. Moreover, Fitzgerald (1969) uses a stiffness formula for the red cell bending which, as Fung and Tong (1968) have pointed out, is probably not correct.

Quite apart from the particular numerical values suggested for red blood cells, the qualitative results of theory in either the original version (Lighthill, 1968) or as modified (Fitzgerald 1969a) are of interest and pertinent to blood flow to some extent. First, there is the demonstration that the model assumed yields a velocity dependent resistance. The coefficient D varies as $U^{-1/2}$ for small velocities and the pressure drop per particle varies as $U^{1/2}$ rather than with U as usual in the Stokes flow. It is suggested (Fitzgerald, 1969b) that this is a possible mechanism of the nonlinear resistance observed in the perfusion experiments at low flow rates.

Another effect which Fitzgerald (1969b) suggests is that at low velocity the clearance between the red blood cells and the capillary wall becomes so small that the cell would effectively come in contact with the wall. It is suggested this might further increase the resistance by a large factor and constitute a mechanism by which the control of blood flow to a capillary bed by upstream arterioles would be further enhanced. This is a possible mechanism which would require measurements *in vivo* for verification. It appears from *in vivo* observations that red cells normally slide quite freely against the endothelial wall, so the apparent limiting coefficient of friction may be quite low.

The shapes of flexible particles predicted by lubrication theory have a characteristic form in that the gap width narrows gradually from the leading end of the cell and tapers to a minimum near the tail end of the particle. One case computed is shown in Fig. 17. The necking of the gap at the rear end of the particle would produce, in a more accentuated case, a "fishtail" appearance of the back end of the particle. Such tails are sometimes observed on red blood cells in narrow capillaries (Branemark and Lindstrom, 1963).

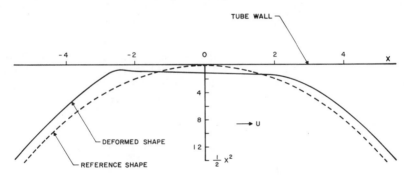

Fig. 17: Particle shape in the vicinity of the tube wall as computed for a flexible model of a red blood cell. The difference between the deformed and reference shapes is proportional to the pressure distribution. (Based on Lighthill, 1968.)

The pressure distribution may also be visualized in Fig. 17 since the deflection of the particle is proportional to $p - p_0$. The distance between the deformed and reference shape in Fig. 17 is a measure of $p - p_0$. The pressure goes through a local minimum near the rear of the cell at the point of the maximum necking. Such a feature of the pressure distribution may be seen in records taken with large models (see Fig. 17 in Lee and Fung, 1969). Photographs taken of the models also show gaps gradually decreasing to the rear as in Fig. 17.

The above discussion has implied that the capillary diameters under discussion are smaller than the red blood cell diameter; however, the theory is valid also for the case where the red cell is the smaller. In this case, the additional pressure drop for a single axisymmetric particle falls off rapidly even if the cell is not deformable, as shown in Fig. 11.

To date, one of the deficiencies of the lubrication theory as applied to flexible particles is in the simple elastic behavior assumed. The dependence of the deflection at any point on the local pressure may be a reasonable approximation in the narrow lubricating gap, but it cannot reproduce the large deflection of the rear of a red cell which gives the so-called "parachute" appearance. To reproduce such features as well as the lubrication phenomena, a more complete and accurate treatment of both the red cell and the fluid motion are required. The lubrication theory makes no attempt to conserve either volume or area of the red blood cell which are likely to be important constraints.

No theories more exact than the lubrication theory are available for flexible particles which resemble red cells. However, an exact solution for neutrally buoyant liquid drops with a surface tension has been evaluated for a number of cases (Hyman and Skalak, 1970). The presence of surface tension at the interface gives the particle a certain amount of elasticity and restores the particle to a spherical shape in the absence of any other stresses. The problem considered and the final shape of the drops for one case are shown in Fig. 18. At rest the drops form a line of equally spaced spheres along the axis of the cylindrical tube, as shown by dashed curves in Fig. 18. To begin the computation, it is assumed a pressure gradient is suddenly applied so that

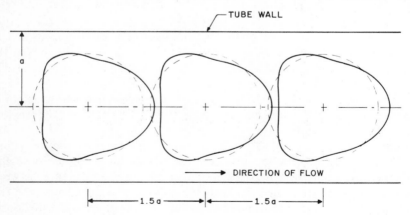

Fig. 18: Drop shapes computed for neutrally bouyant liquid drops suspended in a viscous fluid. Diameter ratio at rest is $\lambda = 0.70$, shown by dashed circles. The viscosity of the interior fluid is zero. The "stiffness" of the drops depends on the dimensionless parameter $\delta = T/\mu V$ where T is the surface tension and μ is the viscosity of the exterior fluid. Here $\delta = 1$.

the suspension is set into motion while the drop shapes are spherical. The velocity field is computed, including the velocity at the surfaces of the drops. If these velocities are such that the shape of the drop is not constant, the shape is adjusted by a displacement proportional to the computed velocities corresponding to a small time-step. The velocity field is next recomputed for the new shape, and the entire process is repeated until a steady-state shape is found. In the process of solution for the velocity fields, two infinite series of exact solutions to Eqs. 4.1 and 4.2 are used. One series applies to the fluid inside of the drop, and the other applies to the fluid outside of it. At the surface of the drop the two series are matched by a number of boundary conditions. The velocity vector and the tangential component of stress are required to be continuous across the boundary, but the normal component of the stress is required to be discontinuous so as to counterbalance the effect of surface tension. These boundary conditions generally lead to some nonzero tangential velocity at the boundary and to a circulation within the drop. This is not realistic as a model of the red cell. Another deficiency is the fact that the surface tension T is a constant for the entire surface. Nevertheless, in the absence of exact solutions to any other models, the results are of some interest.

As shown in Fig. 18, a neutrally bouyant liquid drop may be deformed by a flow into a shape which shows some necking at the rear as in lubrication theory, but the drop also shows the flattening and beginning of concavity of the back end. This concavity is a feature that appears prominently in models of red cells that will be discussed later.

The liquid drop model also demonstrates a nonlinearity of the additional pressure drop with increasing velocity. If the velocity is sufficiently low, the surface tension holds the drops spherical. As the velocity increases, the deformation increases and in all cases computed, the coefficients K and G in Eqs. 4.3 and 4.4 decrease. For a deformation such as shown in Fig. 18, the decrease is about 20–30% even though the maximum drop diameter is not appreciably reduced.

A number of studies have also considered large bubbles whose length is much greater than the tube diameter (Bretherton, 1961; Davies and Taylor, 1949; Cox, 1962; Goldsmith and Mason, 1962). These papers demonstrate some of the same qualitative features as described above, such as formation of a concave cavity at the rear end and a gap width increasing with velocity. However, these long bubbles also are not good models of red blood cells because the interface boundary conditions do not correspond.

Direct measurements of the apparent viscosity of whole blood or suspensions of red cells in glass tubes less than 8μ in diameter have been reported by Prothero and Burton (1961) and Braasch and Jennett (1969). The tests by Prothero and Burton (1961) used micropipettes with a mean diameter of 8.4μ. They found the apparent viscosity of blood relative to that of plasma to vary from 1–2 linearly as the hematocrit varied from 0–50%. This would correspond to a diameter ratio for rigid particles of about 0.85 or 0.90 (Fig. 11, $G \cong 1$). This is a reasonable figure since these tests were carried out at velocities about ten times greater than expected *in vivo*; therefore, the cell deformation was probably quite large. The measurements by Braasch and Jennett (1968) employ an oscillating flow apparatus. They find that in glass capillaries 6μ in diameter all blood samples exhibited an apparent relative viscosity between 2.2–2.7 with hematocrits ranging from 43–65%. They did not report the velocity of flow, but presumably there was again a considerable deformation of the red cells.

The dimensions of red blood cells in flow through glass capillaries from 4 to 10μ diameter have recently been measured at cell velocities from 0–2 mm/sec (Hochmuth et al, 1970). These observations confirm that the edge-on orientation of red cells is the usual position in tubes below 9μ in diameter. As the velocity of flow increases, the rear of the cell develops the typical concavity; it is suggested that the cells become more nearly axisymmetric. The cell diameter and length were measured as a function of velocity and the plasma gap was computed. It had a range of 0.6–1.9μ. A significant aspect of the data was that above a velocity of 1 mm/sec the shape of the cell and the plasma gap appeared to be independent of the velocity. For lower velocities, the gap thickness decreased with velocity. These observations are in accord with expectations based on elastic particle theory, except that the lack of dependence of the shape on velocity above 1 mm/sec suggests that elastic forces have become completely negligible. The behavior of the membrane might then be represented as a certain limiting case as illustrated below.

Generally, in terms of principal stresses and strains defined in Sec. 3, an elastic behavior of the membrane, neglecting bending stresses, would be

(4.24) $$\sigma_1 = C_1(\epsilon_1 + \epsilon_2) + C_2(\epsilon_1 - \epsilon_2)$$
(4.25) $$\sigma_2 = C_1(\epsilon_1 + \epsilon_2) + C_2(\epsilon_2 - \epsilon_1)$$

where C_1 and C_2 are elastic moduli which might also depend on the invariants of strain for nonlinear elasticity; C_1 measuring stiffness under an increase of area without shear strain; C_2 the stiffness in a deformation without change of area. It would appear that for a red cell membrane C_1 is much larger than C_2. The limiting behavior mentioned above would be achieved with $C_1 \to \infty$ and $C_2 \to 0$. Then the area change and the difference between σ_1 and σ_2 would be negligible at any point. To apply this case directly, it would be assumed $\sigma_1 = \sigma_2 = T$ at each point, T being a function of position in general, and it would be required that $\epsilon_1 + \epsilon_2 = 0$ everywhere.

Thin rubber models of red blood cells filled with oil have been constructed and tested by Lee and Fung (1969) and Sutera et al (1970). The models of Lee and Fung were apparently somewhat thicker and stiffer, but certain general conclusions can be drawn from both sets of tests: The orientation of model cells in tubes whose diameter is less than that of the cell is predominantly edge-on. During flow the cell is progressively deformed, developing a concavity at the rear and becoming very nearly axisymmetric by bulging of the cell until only a thin plasma gap remains between the cell and the tube. The additional pressure drop due to the model cell is not a linear function of velocity as would be predicted by creeping flow theory for rigid particles. The slope $\partial(\Delta P)/\partial U$ decreases with increasing cell velocity. In the tests of Lee and Fung (1969) the data may be fitted fairly well by assuming the pressure proportional to the square root of velocity as predicted by Lighthill's lubrication theory. The model tests of Sutera et al (1970) show a similar dependence of additional pressure drop on velocity, but at sufficiently high velocities, the slope $\partial(\Delta P)/\partial U$ becomes constant, indicating again a region in which the elasticity of the cell has no further effect. Seshadri et al (1970), doing tests on properly scaled rubber models, predicted a relative apparent viscosity of the red cell suspension at 65% hematocrit in a 6μ tube to be 2.6 which compares to measured values of 2.2–2.7 reported by Braasch and Jennett (1969). The shapes of the rubber models and red blood cells in glass capillaries are quite similar, as shown, for example, in Fig. 19. The great variety of shapes under various circumstances *in vivo* in man are shown by Branemark (1971). Two views of blood flow in man are seen in Fig. 20. The apparant "parachute" shape may be seen in Fig. 20a. In Fig. 20b the cells

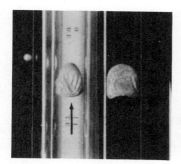 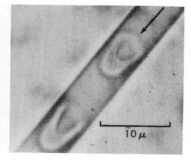

Fig. 19: Deformed shapes of model cell and erythrocytes. a. Model cell at mean flow velocity = 0.5 cm/sec. Diameter of cell (undeformed)/diameter of tube = 1.3. b. Erythrocyte in 6.3μ capillary, velocity of cell = 0.1 mm/sec. Diameter of cell/diameter tube = 1.35. (From Sheshadri et al, 1970.)

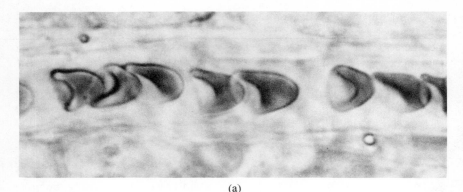

(a)

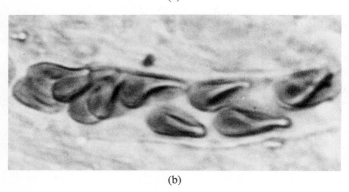

(b)

Fig. 20: *a.* Shapes of red blood cells *in vivo* in man in a capillary of about 7μ diameter. Most of the cells show the parachute-like shape; the cell on the left shows in addition the tail-flap appearance. *b.* Red bood cells *in vivo* in man in a vessel of about 12μ diameter. Note flattening of cells at the rear. (From Skalak and Branemark, 1969.)

appear to be in an edge-on orientation flattened at the rear and bulged at the front under the influence of the pressure gradient.

V. MECHANICS OF THE MICROCIRCULATION AS A WHOLE

The previous section has considered blood flow in vessels that are of the same order diameter as red blood cells, an important category in the microcirculation. Equally important or more so, depending on what phenomena are of interest, are the larger vessels ($10-500\mu$) of the microcirculation. For information of the rheology of blood in this vessel range and in shear viscometry, the reader is referred to the chapter by Cokelet in this volume. The remainder of the present chapter will be directed toward some items which are pertinent to the microcirculation and may form some bridge between various viewpoints.

The logical starting point in the theory is to investigate Stokes flows containing two or more particles which do not fill the tube and may be located eccentrically. Brenner (1970a) has derived a general formula for the additional pressure drop caused by the presence of a single eccentrically situated, neutrally bouyant particle in a duct when there is a small ratio of particle size to duct size. An explicit formula is given for the case of a spherical particle in a cylindrical duct. The coefficient K in Eq. 4.4 is then

$$（5.1） \qquad K = \frac{80E^2\lambda^3}{3} + 8\lambda^5 + 0(\lambda^6)$$

where λ is the diameter ratio and $E = e/a$. The eccentricity is e, the distance from the sphere center to the tube axis. It may be anticipated that, if the particles are more than one tube diameter apart, they would not interact any more than in the axisymmetric case. But in the usual situation of interest, particles are quite close together. Brenner (1970a) investigated the possibility of building expressions for the pressure drop due to the flow of a dilute suspension in a tube from Eq. 5.1. He showed that this cannot be properly done because particle to particle interactions are required. He concluded that the computation of pressure drop for a suspension in tube flow is an open question.

Brenner (1970b) has considered the rheology of two-phase systems in more detail, including the case of suspensions of solid particles in which there are sufficiently large numbers of particles so that it is reasonable to define macrocontinuum field variables in terms of averages over microcontinuum variables. For example, the microvelocity field is the detailed velocity in the vicinity of a single particle. The macrovelocity is an average over a volume that may contain many particles. Brenner treats a dilute system of spherical particles in detail and shows that if the spheres are neutrally bouyant, the mean speed of rotation Ω of the particles is the same as the local macroscopic fluid rate of rotation, $\omega = \nabla x u/2$ where u is the macroscopic fluid velocity. One of the unusual aspects of the results is the possibility that, if $\Omega \neq \omega$, then the macroscopic stress tensor may not be symmetric as usual in ordinary continuum theory. This may occur if an external field exerts couples on the particles. However, it may also occur if the particles are neutrally bouyant but are of sufficient concentration to transmit couple stresses. The notions of a nonsymmetric stress tensor and a microrotation different from the macrorotation are incorporated into continuum theories of so-called micropolar fluids.

Theories of polar fluids have been reviewed by Cowin (1968). Considering only the simplest case of incompressible creeping flow with no body forces or body couples exerted by external agencies, the equations of conservation of linear and angular momentum are

(5.2) $$0 = -\nabla p + \mu \nabla^2 u + 2\tau \nabla \times (\Omega - \omega)$$
(5.3) $$0 = \alpha \nabla \nabla \cdot \Omega + \delta \nabla^2 \Omega - 4\tau (\Omega - \omega)$$

where p and u are the macroscopic velocity and pressure and μ, τ, α and δ are coefficients of viscosity of the fluid. When relative rotational viscosity τ is zero, Eq. 5.2 reduces the Navier-Stokes equations (4.1) with $f = 0$. The stress tensor σ_{ij} is given in Cartesian tensors by

(5.4) $$\sigma_{ij} = -p\delta_{ij} + \mu\left(\frac{\partial u_j}{\partial x_i} + \frac{\partial u_i}{\partial x_j}\right) - 2\tau e_{ijk}(\Omega_k - \omega_k)$$

where e_{ijk} is the alternating tensor. In a tube flow, with z as the axial direction, Eq. 5.4 yields $\sigma_{rz} \neq \sigma_{zr}$ unless the relative angular velocity $(\Omega - \omega)$ is zero.

Polar fluids have a characteristic length, ℓ, which is defined in terms of the viscosity coefficients by

(5.5) $$\ell = \left(\frac{\delta}{4\mu}\right)^{1/2}$$

The solution of a macroscopic problem with typical length L_0 depends on the length ratio, L, defined by

(5.6) $$L = \frac{L_0}{\ell}$$

and a coupling number N

(5.7) $$N = \left(\frac{\tau}{\mu + \tau}\right)^{1/2}$$

which measures the relative influence of the coupling coefficient τ.

Solutions of Eqs. 5.2 and 5.3 are known (Cowin, 1968) for Couette and Poiseuille flow assuming that Ω vanishes at the boundaries. It is not clear that this is the correct boundary condition in all cases.

The use of micropolar theories for blood flow has been suggested in various forms by several authors (Allen et al, 1967; Bleustein and Green, 1967; Ariman, 1969). In each case it is shown that the mean velocity profile for a polar fluid in a tube flow is flatter than the parabolic profile for a Newtonian fluid. This is qualitatively in agreement with experimental data on blood flow in tubes from 30–130μ in diameter (Gaehtgens et al, 1970a). But polar fluid theory is not needed for this purpose. Gaehtgens et al (1970a) show a good correspondence with the experimental data obtained if a Casson equation is used and a cell-free plasma layer at the tube wall of less than 1μ is assumed. A more stringent test would be to see whether polar fluid theory can correctly predict the rate of rotation of red cells, assuming this can be identified with the microrotation, Ω.

The rate of rotation of red cells in the flow of citrated blood has been measured in glass tubes 40 and 70μ in diameter over a range of hematocrits and velocities by Sevilla-Larrea (1968). The results show a great deal of scatter, but it is nevertheless clear that the rate of rotation Ω of the red cells is generally less than that of the blood as a whole ($\omega = \nabla x u/2$). The red cell rotation is impeded by neighboring cells, and hence, a relative microrotation exists. There does not appear to be any comparison of this data to theoretical results.

There is a qualitative difference between predictions of polar fluid theory and experimental results on blood flow as regards pressure drop in tube flow that may be difficult to resolve. It is a general rule (Cowin, 1968) that the resistance of polar fluids to any motion will increase as the length ratio L in Eq. 5.6 decreases. This means that for a fluid of a given microscale the apparent viscosity should increase as the tube diameter decreases. But the well-known Fahraeus-Lindquist effect shows that the apparent viscosity of blood decreases with tube diameter.

The contradiction may well be resolved if the existence of a peripheral plasma gap, free of red cells, is postulated at the tube wall when applying the polar fluid theory. To predict such behavior from first principles would require a theory in which the microstructure of the polar fluid was self-adjusting, depending on the geometry and flow conditions. No such theory is available at present.

Another direction of extension of the theory of viscous flow containing particles is upwards in Reynolds number. The Reynolds numbers in capillary blood vessels are small (10^{-2} or less) computed with respect to either the cell diameter or the vessel diameter. For the larger vessels of the microcirculation, the Reynolds number, with respect to the individual red cell, may still be small enough to make creeping flow assumptions valid, but at some point the Reynolds number, with respect to the tube diameter, must become sufficiently large to make the effects of inertia appreciable. Even in steady flow, if the geometry is variable, as in a sudden taper or curve, effects of inertia on the velocity and pressure drop are not negligible if the Reynolds number is of the order of unity. In the large arteries Reynolds numbers of the order 10^3 are common, and the transition from wave propagation in the arteries to the creeping flow in capillaries takes place in a few centimeters at most. The details of this transition have been gradually clarified as a result of systematic measurements of velocities and pressures in microcirculations *in vivo*.

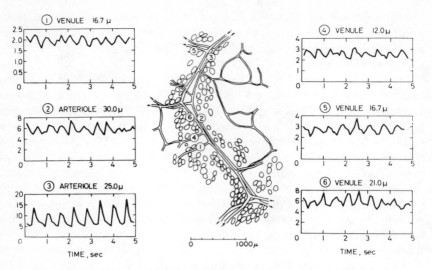

Fig. 21: Typical examples of velocity patterns obtained in arterioles and venules of one experiment by Gaehtgens et al (1970b). Flow velocities were obtained by visual analysis of recorded phototransistor signals. Time resolution was 100 msec. Heart rate in this experiment was approximately 100/min. The drawing of the vasculature was made after photomicrographs. (From Gaehtgens et al, 1970b.)

The most recent measurements of red blood cell velocities in the mesenteric microvessels of the cat by Gaehtgens et al (1970b) are particularly informative with respect to the pulsations observed. These measurements show definite pulsations in venules as well as in arterioles at the heart rate as illustrated in Fig. 21. Velocities were also measured in capillaries and found to be 1.7 mm/sec and less. At this velocity the resolution of the photometric dual slit technique used was not sufficient to observe an oscillatory component. This leaves open the possibility that the flow is also pulsatile in the capillaries. Pulsatile capillary flow has been demonstrated in the lungs (Morkin et al, 1965). The measurements of Gaehtgens et al (1970b) suggest this may be common to some extent in the systemic circulation as well.

Pressure measurements in venules and arterioles from 10 to 60μ in diameter have been made by Wiederhielm et al (1964) and by Intaglietta et al, (1970). These measurements show that the pressure is pulsatile in arterioles, but not, as a rule, in the venules. This does not, of course, necessarily contradict the velocity data cited above since the pressure in the venules is a function of the impedance downstream which is very low.

All these measurements suggest that a more detailed understanding of blood flow in the microcirculation might be aided by considering the microvasculature as an elastic network with pulsatile flow, just as in the arterial tree itself. Inertia and compliance in the microcirculation will have less importance than in the arterial system, but cannot be entirely negligible as will be shown.

The compliance of the vessels in the microcirculation is small (Fung, 1966) and is generally quite properly neglected when considering the details of the flow within such vessels. But with regard to the pulsatile part of blood flow in the microcirculation, the existence of some nonzero compliance is essential if the pattern of pulsatility is to vary at all from one generation of vessels to another. If all the vessels were perfectly rigid, the arteriolar flow into and the venular flow out of any portion of a capillary bed would be identical at every instant.

The role of inertia is more subtle. Its influence would appear to be negligible if judged by the Womersley parameter $\alpha = (a^2\omega/\nu)^{1/2}$. This parameter is about 0.05 at the fundamental heart rate for the 25μ arteriole in Fig. 21. For this range of α the flow is in phase with the pressure gradient and velocity profile is parabolic (McDonald, 1960). This implies that inertia can be neglected, and the network may be regarded as having capacitance and resistance only. In such a network the hyperbolic equations of wave propagation become parabolic, and disturbances propagate as in a diffusion or heat conduction.

On the other hand, attention has recently been given to the role of inertial effects in pressure-flow relations in the perfused isolated hind paw of the dog when the flow is perfectly steady (Benis et al, 1970). The effects are of the nature of an increased pressure drop in laminar flows at Reynolds numbers of the order of unity which occur in curved and tapered tubes and at bifurcations. Such losses for a Newtonian fluid are nonlinear in the velocity and are approximated by Benis et al (1970) by an expression of the form

(5.8) $$\Delta P = A\mu Q + BQ^2$$

where A and B are constants depending on the vascular geometry, μ is the viscosity, and Q is the total flow rate into the paw. The term BQ^2 is the additional dissipation due to inertial effects. By perfusing with two different Newtonian fluids, the constants A and B may, in effect, be evaluated. Applying these coefficients to perfusion with blood allows an estimate of the ratio of the quadratic term ("inertial loss") to the linear term ("viscous loss") in Eq. 5.8. The ratio of the inertial to the viscous pressure loss ranges from about 0.005–1.0 for an albumin–Ringer solution, while the applied pressure drop is increased from about 6 to 200 mm Hg. These experiments clearly demonstrate that inertia is not negligible in a complete bed such as the hind paw.

It might appear from the above discussion that pulsatile flow theory implies that inertia is negligible in the microcirulation but that experiments show it is not. The resolution lies in the fact that two different kinds of acceleration effects are being considered. The total acceleration effects are being considered. The total acceleration in a fluid motion is

$$(5.9) \qquad \frac{du}{dt} = \frac{\partial u}{\partial t} + u \cdot \nabla u$$

Linearized wave propagation theory includes the local acceleration term $\partial u/\partial t$ but not the convective acceleration $u \cdot \nabla u$. The kind of inertial effects measured by Benis et al (1970) are necessarily derived from the nonlinear term $u \cdot \nabla u$ since the flow is steady. In conclusion, it would appear that in the microcirculation $\partial u/\partial t$ is negligible but $u \cdot \nabla u$ is not.

This research was supported by the Office of Naval Research, Project NR 062–393 and by the U.S. Public Health Service, National Institutes of Health, Research Grant HE 13083. It is a pleasure to acknowledge also the helpful discussions with Dr. Shu Chien.

REFERENCES

Allen, S. J., C. N. DeSilva and K. A. Kline 1967. Blood flow: a study in the theory of deformable supensions in fluids. *In* B. Jacbson (ed.), Digest 7th Inter. Conf. Med. Biol. Eng. Stockholm, Sweden. pp. 375.

Ariman, T. 1969. On the analysis of blood flow. Presented at Soc. Engr. Science Nov. 1969. Inter. J. Eng. Science. In press.

Aroesty, J. and J. F. Gross 1970. Convection and diffusion in the microcirculation. Microvascular Res. In press.

Benis, A. M., S. Usami and S. Chien 1970. Role of viscosity and internal losses in pressure-flow relations studied on perfused canine hind paw. Fed. Proc. 29: 259 Abs. Microvascular Res. In press.

Benjamin, M. K. 1969. Compliant surface bearings: an analytic investigation. Doctoral thesis, Columbia University, New York.

Bishop, C. and D. M. Surgenor (eds.). 1964. The red blood cell. Academic Press, New York.

Bleustein, J. L. and A. E. Green 1967. Dipolar fluids. Inter. J. Engng. Sci. 5: 323–40.

Bloch, E. H. 1962. A quantitative study of the hemodynamics in the living microvasular system. Amer. J. Anat. 110: 125–53.

Bloor, M. I. G. 1968. The flow of blood in the capillaries. Phys. Med. Biol. 13: 443–50.

Braasch, D. and W. Jennet 1969. Erythrocyte flexibility, hemoconcentration and blood flow resistance in glass capillaries with diameter between 6 and 50 microns. *In* H. Harders (ed.), 5th European Conference on Microcirculation. Karger, Basel. pp. 109–12.

Branemark, P. I. and J. Lindström 1963. Shape of circulating blood corpuscles. Biorheol. 1: 139–42.

Branemark, P. I. 1971. Intravascular anatomy of blood cells in man. Karger, Basel.

Brenner, H. 1970a. Pressure drop due to the motion of neutrally buoyant particles in duct flows. J. Fluid Mech. In press.

Brenner, H. 1970b. Rheology of two-phase systems. *In* M. Van Dyke and W. G. Vincenti (eds.), Annual Review of Fluid Mechanics. II. Annual Reviews Inc., Palto Alto, California. pp. 137–76.

Bretherton, F. P. 1961. The motion of long bubbles in tubes. J. Fluid Mech. 10: 166–88.

Bugliarello, G. and C. C. Hsiao 1967. Numerical simulation of three-dimensional flow in the axial plasmatic gaps of capillaries. 7th Int. Cong. Med. Biol. Eng. Stockholm, Sweden.

Bugliarello, G. 1969. Some aspects of the biomechanics of the microcirculation, anno 1969. Proc. Midwestern Mechanics 11th Conf. Developments in Mechanics 5: 921–62.

Bungay, P. M. and H. Brenner 1969. Modeling of blood flow in the microcirculation—tube flow of rigid particle suspensions. Pressented at the 7th Annual SES Meeting Nov. 1969. Inter. J. Eng. Sci. In press.

Bungay, P. M. 1970. The motion of closely-fitting particles through fluid-filled tubes—modeling of capillary blood flow. Doctoral Thesis. Carnegie Institute of Technology, Carnegie-Mellon University, Pittsburgh, Pennsylvania.

Burton, A. C. 1969. The mechanics of the red cell in relation to its carrier function. *In* G. E. W. Wolstenholme and J. Knight (eds.), Circulatory and respiratory mass transport. Little, Brown and Company, Boston. pp. 67–84.

Canham, P. B. and A. C. Burton 1968. Distribution of size and shape in populations of normal human red cells. Circul. Res. 22: 405–22.

Castelli, V., G. K. Rightmire and D. D. Fuller 1967. On the analytical and experimental investigation of a hydrostatic, axisymmetric compliant-surface thrust bearing. Trans. A. S. M. E. J. of Lubrication Tech. 89: 510–20.

Chen, T. C. and R. Skalak 1970. Spheroidal particle flow in a cylindrical tube. Appl. Sci. Res. 22: 403–41.

Chien, S., C. A. Bryant and S. A. Luse 1970. Mechanical hemolysis of erythrocytes during passage through micropores. Presented 18th Annual Meeting Microcirculatory Soc. April 1970. Microvascular Res. In press.

Christopherson, D. G. and D. Dowson 1959. An example of minimum energy dissipation in viscous flow. Proc. Roy. Soc. of London, A, 251: 550–64.

Cokelet, G. R. and H. J. Meiselman, 1968. Rheological comparison of hemoglobin solutions and erythrocyte suspension. Science 162: 275–77.

Cokelet, G. R. 1970. Rheology of the blood. Chapter in this volume.

Cournand, A. 1964. Air and heart. *In* A. P. Fishman and D. W. Richards (eds.), Circulation of the Blood: Men and Ideas. Oxford University Press, New York. pp. 3–70.

Cowin, S. C. 1968. Polar fluids. The Physics of Fluids 11: 1919–927.

Cox, B. G. 1962. On driving a viscous fluid out of a tube. J. Fluid Mech. 14: 81–96.

Davis, R. M. and G. I. Taylor 1949. The mechanics of large bubbles rising through extended liquids and through tubes. Proc. Roy. Soc., A, 200: 375–90.

Dintenfass, L. 1968. Internal viscosity of the red cell and a blood viscosity equation. Nature 219: 956–58.

Dowson, D. and G. R. Higginson 1960. The effect of material properties on the lubrication of elastic roller. J. Mech. Eng. Sci. 2: 188–94.

Ellis, H. S., P. J. Redberger and L. H. Bolt 1963. Capsules and slugs. Indust. Eng. Chem. 55: 29–34.

Fishman, A. P. and D. W. Richards (eds.). 1964. Circulation of the blood: men and ideas. Oxford University Press, New York.

Fitzgerald, J. M. 1969a. Mechanics of red-cell motion through very narrow capillaries. Proc. Roy. Soc. Lond. B. 174: 193–227.

Fitzgerald, J. M. 1969b. Implications of a theory of erythrocyte motion in narrow capillaries. J. Appl. Physiol. 27: 912–18.

Fung, Y. C. 1966. Theoretical considerations of the elasticity of red cells and small blood vessels. Fed. Proc. 25: 1761–772.

Fung, Y. C., B. W. Zweifach and M. Intaglietta 1966. Elastic environment of the capillary bed. Circul. Res. 19: 441–61.

Fung, Y. C. and P. Tong 1968. Theory of the sphering of red blood cells. Biophys. J. 8: 175–98.

Fung, Y. C. 1969. Blood flow in the capillary bed. J. Biomech. 2: 353–72.

Gabe, I. T. and L. Zazt 1968. Studies of plasma viscosity under conditions of oscillatory flow. Biorheol. 5: 86.

Gaehtgens, P., H. J. Meiselman and H. Wayland 1970a. Velocity profiles of human blood at normal and reduced hematocrit in glass tubes up to 130μ diameter. Microvascular Res. 2: 13–23.

Gaehtgens, P., H. J. Meiselman and H. Wayland 1970b. Erythrocyte flow velocities in mesenteric microvessels of the cat. Microvascular Res. 2: 151–62.

Goldsmith, H. L. and S. G. Mason 1962. The movement of single large bubbles in closed vertical tubes. J. Fluid Mech. 14: 42–58.

Goldsmith, H. L. and S. G. Mason, 1967. The microrheology of dispersion. In F. R. Eirich (ed.), Rheology. V. Academic Press, New York. pp. 86–250.

Gregersen, M. I., C. A. Bryant, W. E. Hammerle, S. Usami and S. Chien 1967. Flow characteristics of human erythrocytes through polycarbonate sieves. Science 157, 825–27.

Gregersen, M. I. and C. A. Bryant 1968. Evaluation of deformability of red cells by sieving tests. In A. L. Copley (ed.), Hemorheology. Pergamon Press. pp. 539–49.

Hagen, G. 1839. Ueber die Bewegung des Wassers in engen cylindrischen Röhren. Poggendorfs Annalen der Physik und Chemie 46: 423–42.

Hales, S. 1733. Statical essays. Haemastaticks II. Innays and Manby, London. Reprinted by Hafner Publishing Co., New York, 1964.

Harvey, W. 1628. Exercitatis anatomica de motu cordis et sanguinis in animalibus. [English translation, annotations by C. D. Leake.] 4th ed. Charles C. Thomas, Springfield, Ill., 1958.

Hochmuth, R. M. 1967. Large spherical caps in low Reynolds number tube flow: a model of blood flow in capillaries. Doctoral thesis. Div. Eng. Brown University.

Hochmuth, R. M. and N. M. Das 1970. Uniaxial stretching of the red cell membrane. ASME Winter Annual Meeting 1970, In press.

Hochmuth, R. M. and S. P. Sutera 1970. Spherical caps in low Reynolds number tube flow. Chem. Eng. Sci. 25: 593–604.

Hochmuth, R. M., R. N. Marple and S. P. Sutera 1970. Capillary blood flow: I. Erythrocyte deformation in glass capillaries. Microvascular Res. In press.

Hoeber, T. W. and R. M. Hochmuth 1970. Measurement of red cell modulus of elasticity by in vitro and model cell experiments. ASME Paper No. 70-BHF-4. Presented at Biomechanics and Human Factors Conf., June 1, 1970, ASME.

Hyman, W. A. and R. Skalak 1970. Viscous flow of a suspension of deformable liquid drops in a cylindrical tube. Tech. Rep. No. 5, Project NR 062-393, Depart. of Civil Eng. & Eng. Mech., Columbia University.

Intaglietta, M., R. F. Pawula and W. R. Tompkins 1970. Pressure measurements in the mammalian microvasculature. Miscrovascular Res. 2: 212–20.

Jensen, E. J. and H. S. Ellis 1967. Pipelines. Scientific American 216: 62–72.

Katchalsky, A., O. Keden, C. Klibensky and A. de Vries 1960. Rheological considerations of the hemolysing red blood cell. In A. L. Copley and G. Stainsby (eds.), Flow properties of blood and other biological system. Pergamon Press, New York. pp. 155–64.

Kochen, J. A. 1968. Viscoelastic properties of the red cell membrane. In A. L. Copley (ed.), Hemorheology. Pergamon Press. New York. pp. 455–63.

Kroph, A. 1930. The Anatomy and Physiology of Capillaries. Hafner Publishing Co., New York.

Landis, E. M. 1964. The capillary circulation. In A. P. Fishman and D. W. Richards (eds.), Circulation of the Blood: Men and Ideas. Oxford University Press, New York. pp. 355–406.

Lee, J. S. and Y. C. Fung 1969. Modeling experiments of a single red blood cell moving in a capillary blood vessel. Microvascular Res. 1: 221–43.

Leeuwenhoek, A. V. 1688. On the circulation of the blood. Latin texts, 65th letter to the Royal Society. Facsimile with introduction by A. Schierbeek, N. B. de Graaf, 1962.

Lew, H. S. and Y. C. Fung 1969a. The motion of the plasma between the red cells in the bolus flow. Biorheol. 6: 109–19.

Lew, H. S. and Y. C. Fung 1969b. On the low Reynolds number entry flow into a circular cylindrical tube. J. Biomech. 2: 105–19.

Lew, H. S. and Y. C. Fung 1970a. Plug effect of erythrocytes in capillary blood vessels. Biophys. J. 10: 80–99.

Lew, H. S. and Y. C. Fung 1970b. Entry flow into blood vessels at arbitrary Reynolds number. J. Biomech. 3: 23–38.

Lighthill, M. J. 1968. Pressure forcing of tightly fitting pellets along fluid-filled elastic tubes. J. Fluid Mech. 34: 113–43.
McDonald, D. A. 1960. Blood flow in arteries. Williams and Wilkens, Baltimore.
McNown, J. S., H. M. Lee, M. B. McPherson and S. M. Engez 1948. Influence of boundary proximity on the drag of spheres. Proc. 7th Inter. Cong. App. Mech. London. pp. 17–29.
Malpighi, M. 1661. De pulmonibus. Observationes Anatomicae, Bologna.
Malvern, L. E. 1969. Introduction to the Mechanics of a Continuous Medium. Prentice-Hall, Inc., New Jersey.
Mason, S. G. and H. L. Goldsmith 1969. The flow behaviour of particulate suspensions. In G. E. W. Wolstenholme and J. Knight (eds.), Circulatory and Respiratory Mass Transport. Little, Brown and Company, Boston. pp. 105–29.
Morkin, E., J. A. Collins, H. S. Goldman and A. P. Fishman 1965. Pattern of blood flow in the pulmonary veins of the dog. J. Appl. Physiol. 20: 1118–128.
Poiseuille, J. L. M. 1840. Recherches experimentales sur le mouvement des liquides dans les tubes de tres petits diametres. Academie des Sciences, Comptes Rendes 11: 961–67; 11: 1041–048; (1841) 12: 112–15 (1842) 15: 1167–186.
Ponder, E. 1948. Hemolysis and Related Phenomena. Grune and Stratton, Inc., New York.
Prothero, J. and A. C. Burton 1961. The physics of blood flow in capillaries: I. The nature of the motion. Biophys. J. 1: 565–79. 1962. II. The capillary resistance to flow. Biophys. J. 2: 199–212. III. The pressure required to deform erythrocytes in acid-citrated dextrose. Biophys. J. 2: 213–22.
Rand, R. P. and A. C. Burton 1963. Area and volume changes in hemolysis of single erythrocytes. J. Cell and Comp. Physiol. 61: 245.
Rand, R. P. 1964. Mechanical properties of the red cell membrane: II Viscoelastic breakdown of the membrane. Biophys. J. 4: 303–16.
Rouse, H. and S. Ince 1957. History of hydraulics. Iowa Inst. Hydraulic Res. State Univ. of Iowa, Iowa.
Schiller, L. 1933. Drei Klassiker der Strömungslehre: Hagen, Poiseuille, Hagenbach. Akademische Verlag, Leipzig.
Schmidt-Nielsen, K. and C. R. Taylor 1968. Red blood cells: Why or why not? Science 162: 274–75.
Seshadri, V., R. M. Hochmuth, P. A. Croce and S. P. Sutera 1970. Capillary blood flow: III. Deformable model cells compared to erythrocytes in vitro. Microvascular Res. In press.
Sevilla-Larrea, J. F. 1968. Detailed characteristics of pulsatile blood flow in small glass capillaries. Doctoral thesis. Biomed Eng. Carnegie-Mellon University, Pitts., Penna.
Skalak, R. and P. I. Branemark 1969. Deformation of red blood cells in capillaries. Science 164: 717–12.
Skalak, R., P. I. Branemark and R. Ekholm 1970. Erythrocyte adherence and diapedesis. Angiol. 21: 224–39.
Sobin, S. S. 1966. The architecture and function of the microvasculature. In Y. C. Fung (ed.), Biomechanics ASME, New York. pp. 132–50.
Sutera, S. P. and R. M. Hochmuth 1968. Large scale modeling of blood flow in the capillaries. Biorheol. 5: 45–73.
Sutera, S. P., V. Seshadri, P. A. Croce and R. M. Hochmuth 1970. Capillary blood flow: II. Deformable model cells in tube flow. Microvascular Res. In press.
Wang, H. and R. Skalak 1969. Viscous flow in a cylindrical tube containing a line of spherical particles. J. Fluid Mech. 38: 75–96.
Whitmore, R. L. 1967. A theory of blood flow small vessels. J. Appl. Physiol. 22: 767–71.
Wiederhielm, C. A., J. W. Woodbury, S. Kirk and R. F. Rushmer 1964. Pulsatile pressures in the microcirculation of frog's mesentery. Am. J. Physiol. 207: 173–76.
Wolstenholme, G. E. and J. Knight (eds.). Circulatory and Respiratory Mass Transport. Little, Brown and Company, Boston.
Young, T. 1808. Hydraulic investigations, subservient to an intended Croonian lecture on the motion of the blood. Phil. Trans. of Royal Society of London 98: 164–86.
Zweifach, B. W. 1961. Functional Behavior of the Microcirculation. Charles C. Thomas, Publisher. Springfield, Ill.

19
Mechanical Hemolysis in Flowing Blood

PERRY L. BLACKSHEAR, JR.

Hemolysis, thought to be of mechanical origin, occurs in virtually every recipient of a prosthetic valve. Evidence for the presence of intravascular hemolysis in these patients includes elevated serum hemoglobin, virtual absence of haptaglobin, reduction in red blood cell life span, and the presence of fragmented cells in the circulating blood. Tests of prosthetic devices *in vitro* as well as *in vivo* suggest that hemolysis in these prosthetic devices is again of mechanical origin.

In tests designed to shed light on the mechanism of mechanical hemolysis in flowing blood, the following classifications appear convenient. When prosthetic materials are present, at virtually every level of shear rate there is a finite rate of hemolysis due to cells interacting with the wall. This phenomenon appears to be time dependent and very intimately dependent upon the characteristic of the wall. It is also influenced by plasma components, most particularly the concentration of chylomicra. There are species differences, sex differences, and influences of diet.

Evidence points to two distinct types of in-bulk hemolysis unrelated to interaction with the wall. The first type occurs when shear stress is approximately 1000 dynes/cm^2. Hemolysis commences when the shear stress is maintained for time durations in excess of several seconds. In most prosthetic devices this shear stress is encountered only briefly, that is, for milliseconds. For brief encounters in the neighborhood of 1–100 msecs., a second type of hemolysis occurs which requires a shear stress on the order of magnitude of 40,000 dynes/cm^2 to cause lysis. These levels may be approached in turbulent separated flows associated with prosthetic valve recipients who have high blood pressure.

An examination of the mechanism for the three classes of hemolysis has led to the following hypotheses. Cells that come in physical contact with surfaces to which they will adhere become anchored to the surfaces by long thin processes which either are mechanically torn or collapse upon themselves. This hemolysis is proportional to the surface area, dependent upon the stickiness of the surface coating and correlates with flow parameters according to $\mu(du/dy)^2$. It was shown that the likelihood of the cell touching the wall is proportional to du/dy, and the force causing lysis is proportional to the product $\mu(du/dy)$. The mechanism for lysis at intermediate shear stresses in the neighborhood of 1 to 2 thousand dynes/cm^2 is still unclear. A tentative explanation is that the cells exposed to the prolonged intermediate shear stress become distorted in such a way that the membranes collapse upon themselves yielding osmotically fragile hemoglobin-filled cell fragments. Indeed, such cell fragments are observed in the residue from tests in which the cells have been subjected to prolonged exposures at such shear stresses. Hemolysis caused by short duration exposures at shear stresses in excess of 40,000 dynes/cm^2 are thought to be caused by the yield stress of the membrane being exceeded. The only fragments observed in these tests are either ghosts or membrane fragments. No hemoglobin-filled cell fragments have been observed. The value of 40,000 dynes/cm^2 is consistent with the short duration micropipette tests reported by Rand in which the effective surface tension at lysis is given at 30 dynes/cm.

The results of these basic *in vitro* studies are applied to an analysis of hemolysis in prosthetic valves. The evidence suggests that hemolysis is caused by the first mechanism described, that is, cells interacting with surfaces. A mounting amount of evidence is on hand to suggest that valves tested *in vitro* produce greater rates of hemolysis when installed in the patient. This suggests the possibility that additional mechanisms of mechanical hemolysis are present in the incompletely anticoagulated human. One explanation is that there is an enhancement of cell-wall adhesivity due to the presence of clotting factors. This is supported by the role played by red cells in clotting. The second is that microclots generated at the prosthetic valve become lodged in the microvasculature and there serve as secondary sources for lysis quite similar to the mechanical hemolysis that is thought to accompany microangiopathic hemolytic anemia. Perhaps both mechanisms are operable.

To the extent that they exist, rules for the design of nonhemolytic blood handling equipment will be presented below.

* * *

I. INTRODUCTION

In this volume papers by Anliker, Bergel, Cokelet, Jones, Kenner, and Skalak describe the influence of pumps, blood vessels and blood component properties on pressure and velocity distributions of blood in the circulation. This paper discusses the way pressure and velocity distribution influence the longevity and state of health of a blood component.

In normal, healthy humans the motion and resulting forces acting on blood components do not destroy them but even facilitate their function. There are an orderly production and utilization or recycling of these blood components which are only mildly influenced by the stress produced by motion. In some disease states, however, or with the insertion of prosthetic devices, blood component destruction occurs and the orderly recycling process is altered. For example, the red blood cell, which is normally produced in the bone marrow and lives for 120 days after which it is scavenged by the reticuloendothelium system, can find its physical properties and morphology so altered that it expires prematurely and dumps its constituents in the circulating blood. A similar shortening of life span and altering of material pathways can occur in other blood components. In this paper we will focus on the mechanical factors that influence RBC destruction.

When the cells lose their contents in the vasculature (intravascular hemolysis), there are several stages of this hemolysis that are of direct interest.

1) Cell contents can trigger the extrinsic clotting reaction (Johnson, 1967) by the release of ATP which is converted to ADP causing platelet stickiness (Hjelm et al, 1966). Indeed, Johnson considers hemolysis an integral part of hemostatic plug formation. Brain (1969) asserts that clots entrap and fragment rapidly moving red cells so that a hemolytic-clot inducing cycle can be self-perpetuating. These events may occur on a local level and involve only a small fraction of the total red cell mass. They may have serious consequences without taxing the kidney's ability to clear debris or the RE system's ability to replace the lysed cells.
2) Cell contents, when released intravascularly, can saturate the kidney. Bernstein and co-workers (1965) have shown that the limit of tolerance is 0.1 mg of hemoglobin per 100 cc of blood. This is approximately one part in 150,000 for each circuit through the body, equivalent to a halving of the RBC life span.
3) The limit on the production of red cells in the bone marrow is less severe. However, if the combination of intra and extravascular hemolysis is such that the RBC life span is 1/5–1/10 the normal value of 120 days, anemia results.

Thus, hemolysis is of practical consequence over a wide range of resulting RBC life spans. In the following sections we will briefly review some observations on mechanical hemolysis *in vivo*, the mechanics of hemolysis, and some techniques for predicting the occurrence of mechanical hemolysis.

II. MECHANICAL HEMOLYSIS *IN VIVO*

A. GENERAL OBSERVATIONS

There are numerous indications of intravascular hemolysis (reduced haptoglobin levels, hemoglobin or iron in the urine, and other changes) but limited methods for quantitatively assessing the rates of hemolysis.

Methods of assessing rates of hemolysis. Perhaps the most frequently used method is to tag cells with a radioactive label, ^{51}Cr (Gehrman et al, 1966) or DF^{32}P (Adam et al, 1968). There are difficulties in interpretation, however, since there is elution of the ^{51}Cr tag. Considerable support for the tagging method is given by the good correlation with other indicators. For example, Walsh et al (1969) shows an excellent correlation between levels of serum lactic dehydrogenase, SLDH, an enzyme found in abundance in the RBC, and the reductions in ^{51}Cr cell half-life spans. In addition, there is an excellent correlation between excreted urinary iron and reduced cell life spans. These latter measurements are indicators of intravascular hemolysis. The correlation (in prosthetic valve recipients) strongly suggests that the reduced red cell life spans are caused at least in part by intravascular hemolysis and, if accelerated scavenging occurs in the RE system, it is proportional to the intravascular hemolysis.

Many investigators have shown that cell fragmentation accompanies several forms of mechanically induced intravascular hemolysis (Brain, 1969; Lynch and Alfrey, 1968). In section III of this paper it is suggested that fragmentation is suggestive of a special mechanism of hemolysis and may be used as a diagnostic tool. In the following paragraphs the ^{51}Cr or DF^{32}P life spans will be used as measures of the rate of lysis. Fragmentation, when found, suggests a possible mechanism of lysis.

B. HEMOLYSIS IN THE ABSENCE OF PROSTHETIC MATERIAL

1. Valvular heart disease. Gehrman et al (1966) in a study of thirty-five patients with valvular heart disease find a significant reduction in ^{51}Cr RBC half-life in seventeen subjects. Data for aortic stenosis (restricted flow area) and insufficiency (back flow) are shown in Fig. 1. Data from patients with defective mitral valves showed far less severe hemolysis. In a similar study Eyster et al (1968) observe cell fragmentation occurred in five out of twelve valvular heart disease patients, but it is not clear what fraction of valvular heart disease patients show this phenomenon. In a study of 600 patients Forshaw and Harwood (1967) find only 3% with significantly elevated cell fragments, whereas over half of the patients tested showed evidence of intravascular hemolysis by an absence of or reduction in the serum haptoglobin. It is clear that some but not all patients with valvular heart disease experience intravascular hemolysis and a lesser number also show elevated cell fragmentation. In each case the investigators call attention to the high

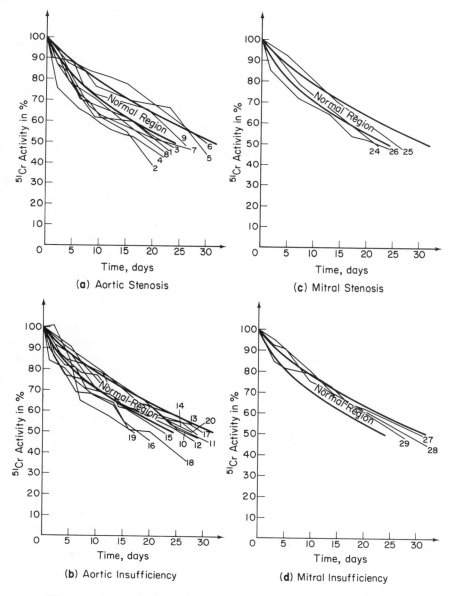

Fig. 1: ^{51}Cr decay curves for valvular heart disease compared to normal (from Gehrman et al, 1966). a) Aortic stenosis. b) Aortic insufficiency. c) Mitral stenosis. d) Mitral insufficiency.

velocities, high level of turbulence, and the existence of abnormal surfaces (though not prosthetic) and state belief that hemolysis is of mechanical origin.

2. *Microangiopathic hemolytic anemia* This term, introduced by Brain (1969) and recently reviewed, refers to disease states that result in fibrin filaments appearing in small blood vessels to which cells become attached and fragmented. There is some controversy over whether cells are trapped or simply stick to these strands (Bull and Kuhn, 1970). In all cases abnormal fibrin is thought to be present, and this is associated with a depletion of the supply of its precursor, fibrinogen. The fragments observed are similar in appearance to those described above (where fibrinogen levels are not depleted) in some cases of valvular heart disease. Brain (1969) and Bull and Kuhn (1970) consider this form of lysis mechanical and assert that both sticky surface and high velocity are required to form the cell fragments. Here hemolysis is clearly related to the interaction between flowing cells and a stationary surface. In valvular heart disease, hemolysis in bulk has been suggested (Gehrman et al, 1966).

C. HEMOLYSIS IN THE PRESENCE OF PROSTHETIC MATERIALS

By far the richest source of information for our present purposes is the prosthetic heart valve. Hemolysis caused by prosthetic valves was first reported by Ross et al (1954) and subsequently in numerous other publications (see references). Study after study report evidence of intravascular hemolysis. Most of the studies show a significant reduction in RBC life spans, and increased fragmentation (Hjelm et al, 1966; Bernstein et al, 1966).

As in the case of valvular heart disease, hemolysis is more severe in aortic than in mitral prostheses. This is shown in Fig. 2. Alfrey and Lynch (1968) and Brain (1969) have called attention to the similarity in fragments found in patients with prosthetic valves and patients with microangiopathic hemolytic anemia. Once again it is not certain whether hemolysis occurs in bulk or at the prosthetic surfaces, and many speculative explanations are put forward.

Of special interest in interpreting the observed reduction in RBC life spans are the observations (Stevenson and Baker, 1964; Marsh, 1964) that blood from prosthetic valve patients displays a normal life span when transferred into a normal, compatible recipient. This suggests that the hemolysis observed is probably intravascular and is not preceeded by irreversible sublethal injury to the cell. In the literature cited above, reference is made to hemolysis severe enough to cause anemia with prosthetic valves. Under these circumstances there is usually a leak such that backflow can occur. This observation,

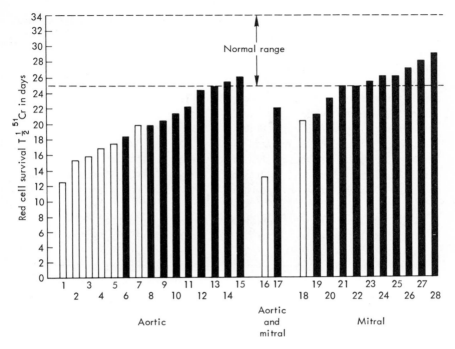

Fig. 2: Red cell survival in 28 patients following replacement of the aortic, aortic and mitral, or mitral valve by a Starr-Edwards prosthesis. The black columns indicate patients in whom the valve was competent. The crosshatched columns indicate patients with clinical evidence of regurgitation.

compared with the fact that in leaky natural valves hemolysis is slight, is relevant in interpreting the role of the prosthetic material. In reviewing the levels of hemolysis in prosthetic valves we conclude that:

1) all prosthetic valves produce sufficient hemolysis to exacerbate an already severe clotting problem.
2) most prosthetic valves presently used do not tax the kidney or the ability of the bone marrow to compensate.
3) in the presence of paravalvular leaks prosthetic valves can cause hemolysis so severe that fatal anemia may result.

III. MECHANICAL HEMOLYSIS *IN VITRO*

A. Mechanisms of hemolysis

1. Hemolysis occurs at a critical strain. When a red blood cell membrane is stretched, it behaves as though pores were open, to the extent that protein molecules may pass through (Katchalsky et al, 1960). These investigators have likened these pores to those that appear in a woven fabric such that even though the fabric may be stretched only 10–20%, the pore diameter may increase ten-fold. Rand (1964) has mechanically stretched RBC membranes by sucking on them with a micropipette. He shows that the pattern of hemolysis is consistent with a model of the membrane behaving as a viscoelastic solid in which hemolysis occurs when the strain exceeds some limiting value. Subsequently, Seeman et al (1969) have performed osmotic lysis experiments giving results consistent with the existence of a critical strain for lysis. That pores form upon osmotic lysis has also been demonstrated by Seeman (1967). It is of interest that the time required to reach maximum pore size upon adding hypotonic solution to packed cells is 15 seconds, corresponding closely to the time constant of the viscoelastic element in Rand's model.

That a filamentous mesh may in fact be present on the red cell membrane is supported by the extraction of actin and myosin-like proteins from the membrane by Onishi (1962).

Rand's data can be expressed as an effective membrane tension *vs* time required to achieve the critical strain. This is shown in Fig. 3. The shape of the curve reflects the viscoelastic properties of the membrane. Parenthetically, Rand (1968) points out that this viscoelastic behavior is more akin to the properties of a monomolecular protein film as studied by Biswas and Haydon (1963) than to those of a bilipid layer.

In a velocity gradient or where cells become anchored to a wall in a flowing stream, fluid forces stretch the membrane. (The problem of relating fluid forces to membrane tension will be considered later.) Lysis resulting from

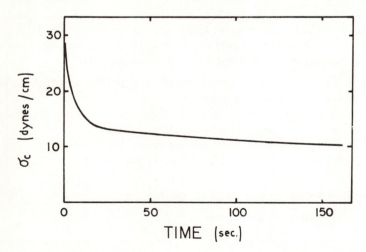

Fig. 3: Critical membrane tension, σ_c, leading to lysis *vs* duration of time (from Rand, 1964).

membrane stretching, if performed gently, should result in ghosts. If done hastily such that the stretching exceeds that required for lysis, the membrane tears, spilling the cell contents and leaving behind membrane fragments. It is not expected that hemoglobin-filled cell fragments as observed *in vivo* (see Sec. 11) would result from this form of lysis.

2. *Cell contents may be altered when the membrane is stretched.* Seeman et al (1969) have shown that potassium ions escape before hemoglobin in gradual lysis. Hjelm et al (1966) infer that cell contents considered vital to membrane stability, for example ATP and 2-3-DPG, may also leak out before hemoglobin does.

If we assume the pore size *vs* surface tension to cause lysis (in the limit of vanishingly small times) to be linear, we may infer the membrane tension (and ultimately the fluid shear stress required to cause leakage of various substances). Figure 4 gives such a plot. It can be seen that K^+, ATP, and 2-3-DPG could leak at tensions well below those for hemoglobin. Bernstein et al (1966) have introduced the term "sublethal" RBC damage in relationship to mechanical damage (Shea et al, 1967; Indeglia et al, 1966, 1968). It may be that the loss of cell constituents short of hemolysis impairs cell functions and increases the cell's vulnerability to subsequent stress. It is not clear what form the products of lysis from such a process would take. Weed and Reed (1966) have discussed cellular alterations that predispose a cell to spontaneous lysis.

3. *Fragmentation.* Rand (1964) found that, when a tongue of cell membrane of length πd was sucked into a micropipette of internal diameter d the membrane collapsed upon itself (the way an oil-water interface would)

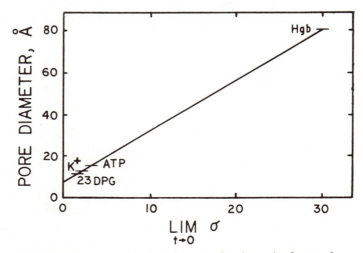

Fig. 4: An estimate of the pore diameter that forms in the membrane due to the elastic component only *vs* membrane tension.

to yield a small and a large hemoglobin-filled cellule. The larger cellule retained the biconcave shape. Similar observations have been reported by Weed and Reed (1966) who cite references to this phenomenon dating back to 1864. There are several confusing aspects to fragmentation. First, it does not always occur when a tongue of cell membrane of diameter d and length πd is pulled from the cell in the absence of a constraining tube. Examples are published by Kochen (1966) and Blackshear and Watters (1969).

It is possible to reconcile these conflicting observations by considering the model of membrane tension introduced by Fung and Tong (1968). They assume the resultant tension in a membrane to be made up of a viscoelastic component σ_v which depends upon strain and can therefore be anisotropic, and a surface tension σ_s which is presumably locally isotropic. The resultant membrane tension $\sigma = \sigma_v + \sigma_s$ will be expected to be locally isotropic for $\sigma_v \ll \sigma_s$.

In Rand's experiments there is no evidence that the membrane had been strained so that fluid-like behavior may be expected. In Kochen's experiments the cell is anchored at a small site on a glass slide and stretched by a retreating miniscus. It seems far more likely that anisotropic membrane strain, hence tension, should develop. The longitudinal strain (hence longitudinal σ_v) would be expected to exceed the lateral strain (hence lateral σ_v) and membrane collapse would require a far longer filament.

A second anomaly arises from the observation that Rand's fragments resume the biconcave shape, whereas fragments produced by Brain (1969) and Bull and Kuhn (1970) remain distorted. To this author's knowledge there is no report of distorted ghosts (see for example Kochen, 1966). It would appear that the distorted shape is due to changes in the cell content and presumably requires a longer stress time than in the Rand experiments. It may be significant that Rand's cellules formed rapidly whereas those observed to form by Brain (1969) took minutes to form. If formation of a hemoglobin-filled fragment should correlate with irreversible shape changes, the existence of cell fragments would lead to the suspicion of a mechanism of lysis involving relatively immobilized, distorted cells.

4. Hemolysis in flowing blood remote from a foreign wall. In the passage through a prosthetic valve, a cell will experience a sharp change in pressure, and if cavitation occurs, there will be velocity gradients, turbulence, turbulent shear stress and the possibility of gas bubbles.

The relative importance of these exposures has been studied by a number of authors (Bernstein et al, 1965; Blackshear and Watters, 1969; and others). In brief the factors that are rejected as not contributing to hemolysis are large scale pressure fluctuations, small scale pressure fluctuations accompanying turbulence, and cavitation. Harvey et al (1944) showed that lungs are perfect filters of cavitation nucleation sites. Those factors that are presently in dispute as possible causes of lysis are turbulent shear stress and laminar shear stress.

The tests whereby the tolerance to shear stress is established are described in Sect. III B below. For brief exposures ($t < 10^{-3}$–10^{-4} sec) the lethal turbulent shear stress for normal red cells from healthy male humans is of order 50,000 dynes/cm^2 (Bernstein et al, 1967; Forstrom, 1969).

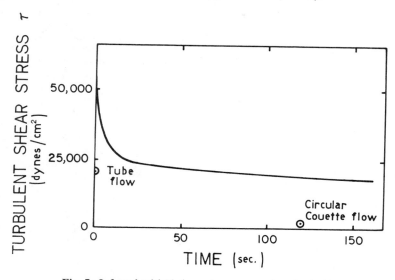

Fig. 5: Inferred critical shear stress versus time for lysis.

There is no satisfactory model for relating this shear stress to a membrane tension in a flaccid particle with virtually fixed surface area (Forstrom, 1969). If we use the treatment for the disruption of fluid particles (emulsion) available in the literature (Goldsmith and Mason, 1965), the resulting tension at rupture is approximately 30 dynes/cm^2, the value for short duration tension found by Rand (1964). If the fluid stress for lysis is linear with tension at lysis, then we may predict the shear stress-duration curve shown in Fig. 5. This curve will be discussed further in III B. The magnitude of shear stress encountered near the wall in a Starr-Edwards prosthetic valve has been estimated by Wieting (1968) to be approximately 7 dynes/cm^2. Turbulent jets with a vena contracta velocity of approximately 1200 cm/sec (Forstrom et al, 1970) are required to cause measureable hemolysis. This requires a pressure drop approximately three-fold higher than any available in the human circulation. It seems unlikely to this author that in-bulk hemolysis occurs in prosthetic valves. This point of view is disputed by Nevaril et al (1968). A discussion of the data underlying this dispute will follow in Sect. III B.

In the tests involving in-bulk lysis by turbulent shear flows, there is a trace amount of hemolysis at lower stress. Very small amounts of hemolysis may release enough of the cell contents to cause clotting processes to be locally excited.

5. Hemolysis at prosthetic surfaces

a. Hemolysis caused by occluding surfaces. In vitro experiments in roller pumps (Bernstein et al, 1967) show a marked reduction in hemolysis when the pumps were set nonocclusively. Direct microscope observations of occlusion between a silicone rubber finger and a glass slide showed virtually no cell entrapment (Bernstein et al, 1965). Experiments in occluding silicone rubber ventricles show that in the limit of vanishing occlusive pressures fifteen cells per cm^2/occlusion lysed (Bernstein et al, 1965). The conclusion advanced in Bernstein et al (1967) is that occlusive phenomena probably disturb the protein deposits on the surface causing sites to form to which cells may adhere when readmitted, and lysis follows as described in Blackshear et al (1966).

b. Hemolysis at prosthetic walls exposed to flowing blood. In vivo and in vitro tests were conducted at shear stress loads where in-bulk hemolysis is unlikely, to yield lysis that is thought to be due to interaction of cells with the prosthetic surface (Bernstein et al, 1967; Blackshear et al, 1966; Blackshear et al, 1965; Shapiro, 1968; Stewart and Sturridge, 1959; Steinbach, 1970). Dramatic evidence that such lysis is associated with wall contact is afforded in experiments where hemolysis was virtually eliminated by siliconizing (Blackshear et al, 1965) or albuminating (Indeglia, 1969) the vessel walls. This problem has been reviewed elsewhere (Blackshear, 1971).

The model proposed for cell lysis at walls is based on observations that cells flowing near walls sometimes become anchored to the wall, and draw out long processes. These processes break or collapse leaving a visibly altered cell flowing downstream (Blackshear and Watters, 1969) as shown in the following sketch.

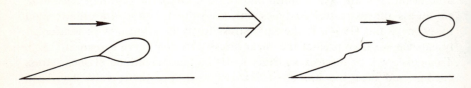

The cells so attached have been labeled tethered cells. Sometimes multiple tethers form as shown in Fig. 6. The tethers observed frequently have length to diameter ratios far in excess of π and do not spontaneously collapse immediately. This is in contrast to the micropipette experiments of Rand. Similar tethers were reported by Kochen. However, in his experiments cells were observed to lyse at the downstream extremity of the tethered cell.

Once again the discrepancies may be reconciled by taking recourse to a

Fig. 6: Washed cells in saline tethered to glass.

mechanical model. If the membrane is pictured as a surface containing pores, then the pore diameter may be expected to increase with isotropic strain and to be distorted with anisotropic strain. One may picture the tether itself to be strained more in the longitudinal than in the lateral direction while the downstream end of the tethered globule strained isotropically. Then, whether the tether breaks or the globule leaks hemoglobin will depend on the diameter of the tether and the force acting on the globule. Kochen estimates his tether diameters to be approximately 1.6 μ and also the total surface of the filament formed upon lysis to be approximately equal to the area of the undistorted cell. If it is assumed that both the area and volume of a tethered cell are the same as those of the undistorted cell and the globule is spherical, we may estimate the maximum tether length to be 20 μ and the spherical globule 4.4 μ in diameter.

The tethers reported in Blackshear and Watters (1969) were 50–160 μ in length, with diameters smaller than 1.6 μ. If we consider the minimum attainable tether diameter to be 500 Å, the maximum attainable tether length is 660 μ, and the diameter of the globule is 5.5 μ. A 160 μ long tether would have a diameter of approximately 0.15 μ. We may model the lysis mechanism in the following way (Fig. 7). If it is assumed as asserted by Kochen that the tension is distributed uniformly about the membrane and that critical strain is required according to Rand, the stress for lysis (Goldsmith and Mason, 1965) is given

$$(1) \qquad \mu \frac{\partial u}{\partial y}\bigg|_{y=0} = \tau_{\text{wall}} > \frac{1}{3} \frac{a}{b} \frac{\sigma_c}{R}$$

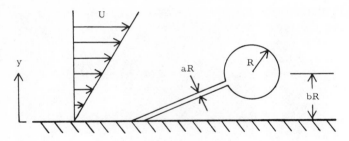

Fig. 7: Model for wall lysis.

where σ_c is critical stress for the duration of exposure (Fig. 3).

For example, if the tethers have dimension such that $aR = 1.6\ \mu$, $bR = 20\ \mu$, $R = 2.2\ \mu$ then $\tau_w = 7000$ dynes/cm². If $a = 0.15\ \mu$, $b = 160\ \mu$, and $R = 2.25\ \mu$, Eq. (1) yields $\tau_w = 40$ dynes/cm². In both the examples lysis should occur by critical strain at the downstream surface of the globule.

If, however, the tether became sufficiently long for collapse despite the anisotropy, tether collapse would be more likely the smaller the diameter (hence the longer the tether). It is clear that where tethering occurs, far lower shear stress is required to produce a critical strain or cause filament collapse than is required in bulk flow.

The threshold for lysis at critical strain (and perhaps for fragmentation) should be some fixed value $\tau_w = \mu\,(du/dy)_w$. The rate of lysis on the other hand should depend upon the number of sticky sites on the wall (and perhaps on the cell) and the rate of cell wall encounter, in addition to the number of cells having σ_c small enough for lysis to occur. Some of the factors that influence cell wall adhesion will be discussed in the next section. The mechanics of cells touching walls has been discussed in Blackshear (1971). Two models, a diffusion model as well as a rough wall model, both yield a rate of wall collision proportional to $(\partial u/\partial y)_w$. Thus, for a given surface area and a given population of sticky sites, the rate of hemolysis should be proportional to $\mu(\partial u/\partial y)_w^2$. Steinbach (1970) has critically reviewed the experiments in which hemolysis is thought to be by wall encounter and finds the rate of hemolysis correlates well with $(\partial u/\partial y)^2$. Viscosity was very little changed in these experiments.

6. *The influence of turbulence on wall hemolysis.* In laminar flow there are forces acting on the red cell to force it away from the wall. These are opposed by as yet unquantitated forces thought to be due to the cell-cell interaction in the bulk flow. In laminar flow there results a cell depleted layer in the neighborhood of the wall which tends to reduce cell-wall enounter (Bugliarello et al, 1965). The picture should be similar in turbulent flow so long as the laminar sublayer is large compared to the particle-depleted layer. Studies by Phibbs (1966) have shown that population in a 1 mm arteriole

begins to fall about 100 microns from the wall. Blackshear et al (1969) have shown that the perturbation velocities perpendicular to the wall show a maximum 15 μ from the wall and then fall abruptly. Although the cell population in laminar flow is somewhat lower at 15 μ than it is at 100, the influence of turbulence would be felt most pronouncedly in the way the turbulence influences the velocity components close to the wall and in particular the region within 15 μ of the wall. If the Reynolds number is of the order of magnitude of 10,000, the laminar shear layer, that is, where $y^* = 5$, is equal to 15 μ or less when the mean velocity is 600 cm/sec or more. Thus, turbulence should increase wall encounter at mean velocities in excess of 600 cm/sec. There will be an influence of turbulence at velocities less than that in the following way. The critical stress for a particular cell tethering is proportional to $\mu(du/dy)$, and the frequency of cell-wall encounter is proportional to du/dy. For given mean velocities, the velocity gradient will, generally speaking, be greater at the wall when turbulent flow exists than it will in laminar flow situations. Thus, even though the laminar sublayer in a turbulent flow may be thicker than the skimming layer, the existence of turbulence with its accompanying higher friction factor and thus higher du/dy_w should enhance cell-wall encounter and wall lysis.

B. Tests for determining the limits of tolerance of the RBC in flowing blood

1. The jet test. The jet test arose from the observations of a number of investigators who found that an isotonic fluid injected at high velocities intravascularly caused vasodilation. Andres et al (1954) suggested such a test would constitute an excellent mechanical fragility test. Blackshear et al (1969) found that lysis occurred at a critical velocity independent of tube size (for Re > 3000). Forstrom (1969) exhaustively explored the jet test, developing it as a diagnostic tool suitable for discerning membrane defects in a number of diseases. He found that in healthy male adults hemolysis was negligible until a jet velocity of 1700 cm/sec was reached, increased gradually until 2000 cm/sec was reached and then increased rapidly as shown in Fig. 8. In examining mammalian and a few nucleated red cells, he found a dependence on size such that the critical velocity varied with particle diameter to the third power. He also was able to show that by strengthening the membrane with trace quantities of glutaraldehyde he could double the apparent membrane strength (0.15% glutaraldehyde). It is significant that following a test no distorted cells or hemoglobin-filled fragments were observed. Where lysis occurred ghosts resulted, as did large membrane fragments and small membrane fragments in that order, as the velocity increased.

It is a source of some frustration that Forstrom has not been able to correlate the dependence on particle size with any current theory for liquid

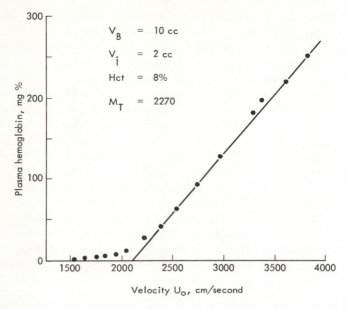

Fig. 8: Typical jet test data for human blood (from Forstrom et al, 1970).

droplet breakup. It is thought that hemolysis in this test results from a critical strain having been reached at times so short that only the elastic component in the membrane responds. According to Rand (1964) the critical strain in such short time is achieved when the effective surface tension reaches 30 dynes/cm. At the critical velocity = 2000 cm/sec, according to turbulent jet theory, the maximum shear stress occurring in the jet is 68,000 dynes/cm². The pressure inside the cell at osmotic lysis has been estimated at from 10,000–120,000 dynes/cm². According to Rand immediate hemolysis occurs in the micropipette experiment at an internal pressure of 12,000 dynes/cm². A shear stress of 68,000 dynes/cm² applied to a constant area liquid-filled red cell could conceivably develop such pressures.

The fact that the intact cells following an exposure to the jet test appear normal and the fragments consist of ghosts and cell membrane fragments rather than hemoglobin-filled cells and fragments strongly suggests that lysis occurs due to the mechanism of exceeding the critical strain or the membrane yield strength. The data would be far easier to interpret had the shear stress been laminar. In an attempt to clarify the magnitude of laminar shear stress that red cells can withstand without lysis, Forstrom et al (1970) have operated the jet test such that cells are forced through the capillary into saline instead of the other way around. This study gives an interesting insight

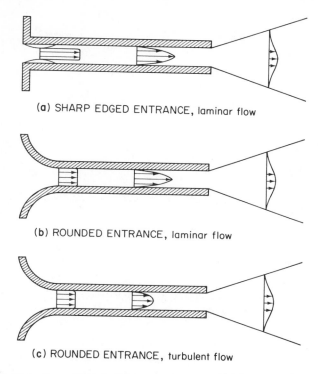

Fig. 9: Tube flow: a) Sharp-edged entrance, laminar tube flow. b) Rounded entrance, laminar tube flow. c) Rounded entrance, turbulent tube flow.

into the locus of lysis in tube flow. Figure 9 shows three situations encountered in these experiments. In the top figure a sharp-edged entrance is employed. It can be seen that flow separation at the entrance occurs. Of the several possible loci of lysis in this system, this turns out to be the most lytic since the tubes can be shortened to the point where only the orifice remains and the lysis remains the same. A typical result for flow into a sharp-edged tube is shown in Fig. 10. The lower two configurations shown in Fig. 9 are tubes having well-rounded entrances, one of which yields a laminar flow and one a turbulent flow. At low hematocrits where large skimming layers prevent cells from encountering the tube wall, the cells find themselves near the wall and stressed at virtually wall shear rate for the entire length of the tube. Similarly, this is found in turbulent flow conditions where the cells are stressed in the laminar sublayer but do not touch the wall. The duration of the stress can be estimated from the length of the tube and the velocity of the cell near the wall. In the test with the well-rounded inlet, hemolysis occurs at 1700 cm/sec, or, in other words, at conditions where it would be expected in the jet issuing from the tube. It is thought that this test has not yielded the upper limit of red

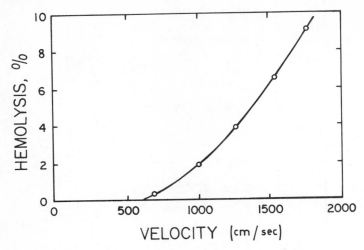

Fig. 10: Hemolysis in a tube (or orifice) with a sharp-edged entrance (from Forstrom et al, 1970).

cell tolerance to laminar shear stress, but it has illustrated that it is sufficiently high to be consistent with the values that can be predicted from Rand's tension for immediate lysis and the Taylor theory for droplet breakup (Taylor, 1934). Some of the data obtained in this way are represented in Fig. 3. These datum points do not represent the thresholds of lysis but laminar stresses that cells have been able to withstand without lysis.

2. *Tests for measuring hemolysis by wall contact.* It is the contention of the author, his collaborators and others (Shapiro, 1968; Yarborough et al, 1966), but not all others (Nevaril et al, 1969; Kusserow et al, 1963; Nevaril et al, 1968), that in systems where shear stress is small compared, say, to 20,000 dynes/cm², that lysis in a chemically inert prosthetic system is due to wall contact.

Steinbach (1970) has recently surveyed his own and other data representing a variety of configurations and flow conditions and was able to correlate the results with a parameter based on the model for RBC lysis at sticky walls outlined above. The results of this survey are shown in Fig. 11. The critical parameter is the rate of change of hemoglobin

$$\frac{dHb}{dt} = 2 \cdot 10^{-7} \cdot \frac{H^2}{1-H} \left(\frac{du}{dy}\right)^2 \text{ (surface area) (time factor)}$$

where H is the hematocrit, the time factor depends upon the way the test is conducted, dHb/dt is in units of mg/min, and du/dy is in sec^{-1}. There is a tendency for surface hemolysis in *in vitro* tests to show a high rate at first and then fall within approximately twenty minutes to a small fraction of the initial

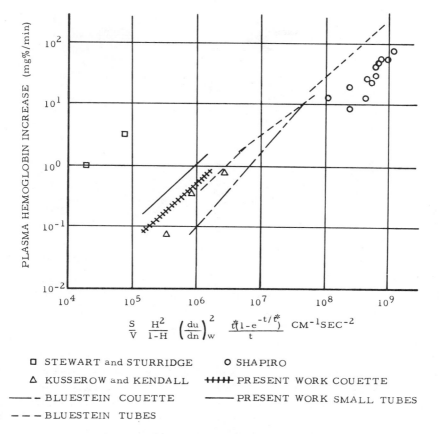

Fig. 11: Correlation of hemolysis (from Steinbach, 1970).

value. The reason for this is not clear at present. Our group believes that it is due to the formation of a stable nonadhesive protein layer, in keeping with the observations of Ponder (1964, 1966) that nearly all nonantigenic proteins reduce cell adhesivity. For short tests and for *in vitro* tests with no anticoagulants, the time factor is believed to be unity.

One test requiring relatively small quantities of blood from which abundant data is available, is performed with the Fleisch Hemoresistometer (Fleisch, 1960). The part of this device that contacts blood is shown in Fig. 12. Blood fills the space between the cup and impeller. The impeller is rotated at varying speeds and produces varying amounts of hemolysis. Garfin et al (1968) have shown the influence of the wall by demonstrating that identical tests with and without albumin-coated walls gave virtually no hemolysis and normal hemolysis respectively. Albumin added to the blood was not protective.

Fig. 12: Fleisch hemoresistometer impeller and cup.

In the next section the results of the Fleisch test will be compared with those of the jet test to show the variables that influence wall hemolysis (and thus cell-wall adhesivity) as opposed to in, bulk lysis (and thus RBC membrane elasticity or fragility). As a basis for comparison, we will take ACD cells in plasma from a normal human male who has fasted for sixteen hours. An increase in fragility indicated by the jet test is represented by a decrease in the critical velocity and in the Fleisch test by an increase in the plasma hemoglobin, for a standardized run. Table I gives a comparison between healthy human males and the results obtained with human female and dog blood, blood at reduced temperature, postalimentary lipemic blood, lipemic blood cleared *in vivo* with heparin, and blood in which the plasma has been removed and the cells resuspended in saline. A "+" indicates that the fragility has increased in the test; "−" fragility significantly decreased, and "0" no significant difference. The most dramatic changes in fragility were

TABLE I
Factors That Influence Bulk and Wall Hemolysis

Parameter	Jet Test	Fleisch Test
Females (human)	+	−
Dogs	0	−
Reduce temperature	−	−
Postalimentary lipemia	0	+
Postalimentary lipemic blood cleared with heparin *in vivo*	0	0
Saline instead of plasma	0	+

+ = Increase in hemolysis
− = Decrease in hemolysis
0 = No significant change

brought about by the presence of chylomicra in postalimentary blood in the Fleisch test. Jasper and Jain (1965) have shown that chylomicra and mechanical agitation are required to cause this enhanced hemolysis in the presence of foreign surfaces. Blackshear (1966) first recognized the importance of postalimentary lipemia in hemolysis due to artificial organ experiments. Indeglia (1969) showed that heparin clearing of chylomicra *in vivo* produced blood as rich in tryglyceride as the uncleared blood, but when tested *in vitro*, it was not lytic and the differences were striking. Sometimes increases as high as 300% in hemolysis were measured in the Fleisch test when chylomicra were present. No effect was observed in the jet test. This is interpreted to mean the chylomicra enhance wall adhesivity. Another striking observation is that plasma proteins appear to have a protective effect in the Fleisch test but do not alter the results of the jet test. Blackshear and Watters (1969) have shown that tethering is frequent when washed cells are exposed to glass or plastic slides, but when the slides are first coated with a bovine serum albumin no tethering is observed. Recall that serum albumin coating eliminates hemolysis in the Fleisch hemoresistometer test (Garfin et al, 1968).

Forstrom (1969) has published the results of the jet fragility tests in a number of disease states in which membrane disorders are suspected. These results are shown in Tables 2 and 3. The jet fragility parameter is τ and its control value is one. In a number of the disorders, osmotic fragility tests have not been abnormal. It would appear that purely mechanical fragility tests may have some possible application in the diagnosis of membrane defects.

There is another mechanical fragility test that is reported in the literature by Nevaril et al (1969) that is conducted in a laminar Couette-flow viscometer with a wall clearance of approximately 64 microns. Typical results of this test are shown in Fig. 13. Subsequent work by Hellums and his colleagues (1968) have demonstrated that in this device there is a marked increase in

TABLE II
Pathologic Blood Results—Jet Fragility Test

Defect and Sex		τ	Analysis
G-6-PD (M)		0.77	Fragile
PNH (M)		0.74	Fragile
Acute lymphatic leukemia (M)		0.98	Normal
Myelogenous leukemia (M)		0.73	Fragile
Acute leukemia (M)		0.68	Fragile
Unknown hemolytic anemia (M)		0.75	Normal
	(M)	0.86	Normal
	(M)	0.88	Normal
	(M)	0.91	Normal
Sickle cells	(M)	0.72	Fragile
	(M)q	0.65	Fragile

Normal Results: $\tau = 0.88$–1.14 males
$= 0.78$–1.04 females

Table III
Aortic Valve Defect and Valve Prothesis Results—Jet Fragility Test

Description and Sex	τ	Analysis
Preop. (M)	0.77	Fragile
Preop. (M)	0.53	Fragile
Preop. (M)	0.86	Normal
Postop.	0.98	Normal
Postop. hemolytic anemia (M)	0.89	Normal
Postop. (F)	0.86	Normal
Postop. hemolytic anemia and ball variance (M)	0.80	Fragile
Ball replaced	0.77	Fragile

Normal Results: $\tau = 0.88$–1.14 males
$\phantom{\text{Normal Results: }\tau} = 0.78$–$1.04$ females

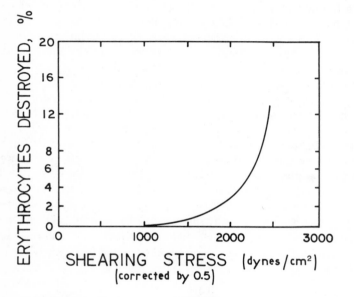

Fig. 13: Hemolysis versus shear stress for Couette flow (from Nevaril et al, 1969).

hemolysis and a wall shear rate of approximately 1000 dynes/cm². Changes in gap widths and suspending phase viscosity have demonstrated that this threshold value is indeed a shear stress dependent phenomenon. Changing the surface to volume ratio of the device does not alter the fraction of the red cell destroyed at a particular shear stress. Hellums and his colleagues believe this represents a test in which the cells are lysed in bulk by laminar stress. Blackshear and his colleagues believe that the duration of the test is sufficiently long for each cell to find its way to one of the walls and there to adhere and to be, in part, lysed by the mechanism described above. No matter what the interpretation may be, this data is of importance in the design of prosthetic devices. That is, far smaller stress is seen to lyse cells than would be anticipated from Rand's critical membrane as shown in Fig. 3. The test shows either a threshold for in-bulk lysis in prolonged shear ($t \approx 120$ sec), or a threshold for greatly accelerated wall lysis—the % value to yield lysis from Eq. (1) is 1/40, a number very close to the tether dimensions observed by Kochen (1966). Fragments resulting from this test at lower shear stress are quite similar to the fragments that are obtained by Brain (1969) and are observed in prosthetic valve recipients. Similar fragments are obtained in the Fleisch hemoresistometer test, while no hemoglobin-filled fragments are observed in the jet test. As has been seen, a shear stress of 1000 dynes/cm² is not found in prosthetic valves, and those stresses that are found do not persist for more than a fraction of a second in bulk. Thus, if this is a bulk flow hemolysis, it would have importance only in special prosthetic devices in which it is desired to impose very high shear stresses on blood for prolonged periods of time. If it is a wall phenomenon, it could represent the threshold for the lysis of a particularly common tether.

IV. HEMOLYSIS IN PROSTHETIC VALVES

When considering the mechanism of hemolysis in prosthetic valves, the following observations appear to be relevant: In valvular heart disease, hemolysis seems to be greatest at the aortic valve which is stenotic, less in the mitral valve which is stenotic, and not detectable in ^{51}Cr half-life measurements in either mitral or aortic valves which are purely insufficient (not stenotic) (Gehrman et al, 1966). Red cell fragments containing hemoglobin are found in the circulating blood in approximately 30% of the patients reported. Patients with prosthetic valves in the aortic position have the most hemolysis, less in valves in the mitral position in which insufficiency is not marked. In patients with marked valvular insufficiency in both the mitral and the aortic position, massive hemolysis occurs such that reoperation is frequently indicated. Fragments are seen in approximately half of the patients examined. An examination of *in vitro* test results suggests that in-bulk lysis could not occur in either valvular heart disease or in the prosthetic valve except for trace amounts. On the other hand, a continuous hemolysis is thought to occur at the diseased surface in the case of valvular heart disease and at the prosthetic surface in the prosthetic valve. This hemolysis should be worse the higher the wall velocity gradient. For given velocities these velocity gradients should be higher the more intense the turbulence. A little hemolysis can release clotting factors that enhance wall adhesion (Johnson, 1967).

We conclude that hemolysis in prosthetic valves is by wall hemolysis in which the excitation of the clotting factors by the local release of cell contents makes wall adhesion worse. In valvular heart disease the site for adhesion is thought to be the nonendothelialized surfaces; in the prosthetic valve, the prosthetic surfaces.

V. PREDICTING MECHANICAL LYSIS

A. Tube flow no entry or exit effects

The foregoing considerations have been summarized in Fig. 14 for the case of blood flow in a circular tube in which neither the entering nor exiting conditions cause lysis. The figure based on the discussion in the text is self-explanatory except for the region where shear rate is less than 10^3 sec^{-1}.

Red cells and other debris have been seen to accumulate in regions of low shear (Patschek and Madras, 1969; Keller and Yum, 1970). The exact circumstances where accumulation occurs is not clearly known. The value given was derived from observations of Keller and Yum (1970) on cholesterol-coated walls. It is too high for most protein-coated walls and is given as an extreme case to point out the existence of a problem.

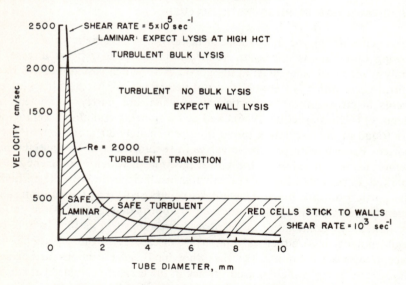

Fig. 14: Domains of fully developed flow of blood in a tube.

B. Hemolysis in separated flow

This problem has been discussed in connection with the jet test. On the basis of the work of Forstrom (1969) and Forstrom et al (1970) we estimate that the threshold of lysis in orifice flow or in the vena contracta of a sharp tube entrance is 1200 cm/sec while at the exiting jet it is 1700–2000 cm/sec.

REFERENCES

Adam, V., H. Heimpel, M. Wegener, H. Stein and E. Kratt 1968. Hyperhämolyse bei Patienten mit operierten und nicht operierten Herzklappenfehlern. Med. Klin. 63: 414.
Anderson, N. N., E. Gabrieli and J. A. Zizzi 1965. Chronic hemolysis in patients with ball valve prostheses. J. Thoracic and C-V Surg. 50: 501.
Andres, R., K. L. Aierler, H. M. Anderson, W. N. Stainsby, G. Cader, A. S. Ghrayyib and J. L. Lilienthal, Jr. 1954. Measurement of blood flow and volume in the forearm of man. J. Clin. Invest. 33: 483.
Bernstein, E. F., A. R. Castaneda, P. L. Blackshear Jr. and R. L. Varco 1965. Prolonged mechanical circulatory support: analysis of certain physical and physiologic considerations. Surgery. 57: 103.
Bernstein, E., R. Indeglia, M. Shea and R. Varco 1966. Sublethal damage to the red blood cell from pumping. C-V Surg. Suppl. I to Circ. 35 and 36: 1–226.
Bernstein, E. F., P. L. Blackshear, Jr. and K. H. Keller 1967. Factors influencing erythrocyte destruction in artificial organs. Am. J. Surg. 114: 126.
Biswas, B. and D. Haydon 1963. The rheology of some interfacial adsorbed films of macromolecules. I. Elastic creep phenomena. Proc. Roy. Soc. (London), Ser. A. 271: 296.
Blackshear, G. 1966. Personal communication.
Blackshear, P. L., Jr., F. D. Dorman and J. H. Steinbach 1965. Some mechanical effects that influence hemolysis. Trans. Amer. Soc. Artif. Int. Organs 11: 112.
Blackshear, P. L., Jr., F. D. Dorman, J. H. Steinbach, E. J. Maybach, A. Singh and R. E. Collingham 1966. Shear, wall interaction and hemolysis. Trans. Amer. Soc. Artif. Int. Organs 12: 113.
Blackshear, P. L., Jr. and C. Watters 1969. Observations of red blood cells hitting solid walls. Amer. Inst. of Chem. Eng. 65th Nat. Meeting, Cleveland, Ohio.
Blackshear, P. L., Jr. 1969. The behavior of blood cells near walls. Presented Artif. Heart Conf., Washington, D. C.
Blackshear, P. 1971. Hemolysis at prosthetic surfaces. In M. Hair (ed.), Marcel Dekker Publishing Co. In print.
Bluestein, M. 1968. Ph. D. Thesis. Northwestern Univ.
Brain, M. C. 1969. Microangiopathic hemolytic anemia. New Eng. J. Med. 281: 15.
Bonnabear, R. C. and C. W. Lillehei 1968. Mechanical "ball failure" in Starr-Edwards prosthetic valves. J. Thor. and C-V Surg. 56: 258.
Brodeur, M. T. H., D. W. Sutherland, R. D. Koler, J. A. Kimsey and H. E. Griswold 1960. Red cell survival in patients with aortic valvular disease and ball valve prosthesis, Circulation. 30: Suppl. III-55.
Brodeur, M. T. H., D. W. Sutherland, R. D. Koler and H. D. Griswold 1965. Hemolytic anemia and valvular disease. New Eng. J. Med. 272: 104.
Brodeur, M. T. H., D. W. Sutherland, R. D. Koler, A. Starr, J. A. Kimsey and H. F. Griswold 1965. Red blood cell survival in patients with aortic valvular disease and ball valve prostheses. Circulation. 32: 570.
Bugliarello, G., C. Kapur and G. Hsia 1965. The profile viscosity and other characteristics of blood flow in a nonuniform shear field. In A. Copley (ed.), Proc. 4th Int. Conf. Rheol. Part 4. Interscience. p. 351.
Bull, B. and I. Kuhn 1970. The production of schistocytes by fibrin strands (a scanning electron microscope study). Blood. 35: 104.
Cooley, M. H. 1968. Intravascular hemolytic syndrome following aortic valve replacement. Complete hematologic and cardiac recovery without surgical intervention. Arch. Int. Med. 118: 486.
Dameshek, W. 1964. Hemolytic anemia and valvular heart disease. C. P. C., New Eng. J. Med. 271: 898.

DeCesare, W., C. Roth and C. Hufnagel 1965. Hemolytic anemia of mechanical origin with aortic valve prosthesis. N. Eng. J. Med. 272: 1045.
Dennis, E., P. C. Johnson, S. Kinard, S. McCall and M. E. DeBakey 1964. The pattern of red blood cell survival after prosthetic ball valve replacement. C-V Res. Center. Bull. 3: 62.
Eyster, E., K. Mayer and S. McKenzie 1968. Traumatic hemolysis with iron deficiency anemia in patients with aortic valve lesions. Ann. Int. Med. 68: 995.
Fleisch, A. 1960. Der Hämoresistometer: Ein Gerat zur Bestimmung der Mechanischen Resistenz der Erythrocyten. Schweiz. med. Wschr. 90: 186.
Forshaw, J. and L. Harwood 1967. Red blood cell abnormalities in cardiac valvular disease. J. Clin. Path. 20: 848.
Forstrom, R. J. 1969. A new measure of erythrocyte membrane strength—The jet fragility test. Ph. D. Thesis. Univ. of Minn.
Forstrom, R., P. L. Blackshear, Jr., P. Keshaviah and F. Dorman 1970. Fluid dynamic lysis of red cells. Presented 3rd Joint Meeting AIChE, San Juan, Puerto Rico.
Fung, Y. and P. Tong 1968. Theory of the sphering of red blood cells. Biophys. J. 8: 175.
Furuhjelm, et al 1964. Hemolytic anemia after repair of ostium primum defect. Nordisk Medicum. 72: 1446.
Garfin, S., R. Indeglia, M. Shea and E. Bernstein 1968. Effect of albumin-coated surfaces on erythrocyte mechanical destruction. Surg. Forum 19: 135.
Gehrman, G., W. Bleifeld and D. Kaulen 1966. Hertzklappenfehler und Hämolyse. Klinische Wochenshrift, 44: 1229.
Goldsmith, H. L. and S. G. Mason 1965. The microrheology of dispersions. Pulp and Paper Res. Inst. Canada, Pointe Claire, P. Q., Canada.
Harvey, E. N., D. K. Barnes, W. D. McElroy, A. H. Whiteley, D. C. Pease and K. W. Cooper 1944. Bubble formation in animals. I. Physical factors. J. Cell Comp. Physiol. 24: 1.
Herr, R. H., A. Starr, W. R. Pierie, J. A. Wood and J. C. Bigelow 1968. Aortic valve replacement: A review of six years' experience with the ball valve prosthesis. Ann. Thor. Surg. 6: 199.
Hjelm, M., S. Ostling and A. Persson 1966. The loss of certain cellular components from human erythrocytes during hypotonic hemolysis in the presence of dextran. Acta Physiol. Scand. 67: 43.
Indeglia, R. A. 1969. Erythrocyte response to mechanical trauma. Ph. D. Thesis, Univ. of Minn.
Indeglia, R. A., F. D. Dorman, A. R. Castaneda, R. L. Varco and E. F. Bernstein 1966. Use of GBH-coated Tygon tubing for experimental prolonged perfusions without systemic heparinization. Trans. Amer. Soc. Artif. Int. Organs. 12: 116.
Indeglia, R. A., M. A. Shea, R. Forstrom and E. F. Berstein 1968. Influence of mechanical factors on erythrocyte sublethal damage. Trans. Amer. Soc. Artif. Int. Organs. 14: 264,
Jasper, D. and N. Jain, 1965. Effects of lipemia upon RBC fragility, sedimentation rate and plasma refractometer indexes in the dog. Amer. J. Vet. Res. 26: 332.
Johnson, S. A. 1967. Platelets in Hemostasis. In W. H. Seegers (ed.), Blood clotting enzymology. Academic Press, New York.
Kastor, J. A., M. Akbarian, M. J. Buckley, R. E. Dinsmore, C. A. Sanders, J. G. Scannel and W. G. Austen 1968. Paravalvular leaks and hemolytic anemia following insertion of Starr-Edwards aortic and mitral valves. J. Thor. and C-V Surg. 56: 279.
Katchalsky, A., O. Kedem, C. Klibansky and A. DeVries 1960. Rheological considerations of the hemolysing red blood cell. In A.L. Copley and G. Stainsby (eds.), Flow Properties in Blood and Other Biological Systems. Pergamon Press, New York.
Keller, K. and S. Yum 1970. Erythrocyte tube-wall interactions in laminar flow of blood suspensions. Trans. Am. Soc. Artif. Int. Organs 16:42.
Kochen, J. A. 1966. Viscoelastic properties of the red cell membrane. In Hemorheology. Proc. 1st Internat. Conf. Univ. of Iceland, Reykjavik.
Kusserow, B. K. and L. W. Kendall 1963. In vitro changes in the corpuscular elements of blood flowing in tubular conduits. Trans. Amer. Soc. Artif. Int. Organs. 9: 262.

Lynch, E. and C. Alfrey 1968. Erythrocyte fragmentation: A diagnostic clue in hemolytic anemias. Ann. Thor. Surg. 6: 199.
McHenry, M., E. Smeloff and P. Hatterslay 1968. Complications of heart valve replacement. Calif. Med. 109: 1.
Marsh, G. W. 1964. Intravascular hemolytic anemia after aortic valve replacement. Lancet. 2: 986.
Nevaril, C., E. Lynch, C. Alfrey and J. Hellums 1968. Erythrocyte damage and destruction induced by shearing stress., J. Lab. Clin. Med. 71: 784.
Nevaril, C. G., J. D. Hellums, C. P. Alfrey, Jr. and E. C. Lynch 1969. Physical effects in red blood cell trauma. AIChE J. 15: 707.
Ohnishe, T. 1962. Extraction of actin and myosin-like proteins from erythrocyte membrane. J. Biochem. (Tokyo). 52: 307.
Petschek, H. and P. Madras 1969. Thrombus formation on artificial surfaces. Art. Heart Prog. Cong., Washington, D.C.
Phibbs, R. 1966. Orientation and distribution of erythrocytes in blood flowing through medium-sized arteries. Hemorheol. Proc. 1st Int. Conf., Reykjavik, Iceland.
Ponder, E. 1964. L'Adhesivité des Globules Rouges. Nouvelle Revue Francaise d'Hematologie. 4: 609.
Ponder, E. 1966. Effect of basic proteins on the adhesiveness of red cells. Nature 209: 307.
Rand, R. 1964. Mechanical properties of the red blood cell membrane. II. Viscoelastic breakdown of the membrane. Biophys. J. 4: 308.
Rand, R. 1968. The structure of a model membrane in relation to the viscoelastic properties of the red cell membrane. J. Gen. Phys. 52: 1735.
Reed, W. A. and M. Dunn 1964. Fatal hemolysis following ball valve replacement of the aortic valve. J. Thor. and C. V. Surg. 48: 436.
Reynolds, R. D., C. A. Coltman, Jr. and B. M. Beller 1967. Iron treatment of sideropenic intravascular hemolysis due to insufficiency of Starr-Edwards valve prostheses. Ann. Int. Med. 66: 659.
Ross, J., C. Hufnagel, C. Freis, W. Harvey and E. Partenope 1954. The hemodynamic alterations produced by a plastic valvular prosthesis for severe aortic insufficiency in man. J. Clin. Invest. 33: 891.
Sarnoff, S. J. and R. B. Case 1955. Physiologic considerations relating to the Hufnagel operations with special reference to postoperative anemia. Int. Symp. on Cardiovascular Surgery. W. B. Saunders Co., Phila., Penna. p. 328.
Sears, D. A. and W. H. Crosby 1965. Intravascular hemolysis due to intracardiac prosthetic devices: Diurnal variations related to activity. Am. J. Med. 39: 341.
Seeman, P. 1967. Transient holes in the erythrocyte membrane during hypotonic hemolysis and stable holes in the membrane after lysis by saponin and lysolecithin. J. Cell Biol. 32: 55.
Seeman, P., T. Sauks, W. Argent and W. Kwant 1969. The effect of membrane-strain rate and of temperature on erythrocyte fragility and critical hemolytic volume. Biochem. Biophys. Acta. 183: 476.
Shapiro, S. I. 1968. M. S. Thesis. Univ. Calif. Berkeley.
Shea, M. A., R. A. Indeglia, F. D. Dorman, J. F. Haleen, P. L. Blackshear, Jr., R. L. Varco and E. F. Bernstein 1967. The biologic response to pumping blood. Trans. Amer. Soc. Artif. Int. Organs. 13: 116.
Steinbach, J. 1970. M. S. Thesis. Univ. of Minn.
Stevenson, T. D. and H. J. Baker 1964. Hemolytic anemia following insertion of Starr-Edwards valve prosthesis. Lancet. 2: 982.
Stewart, J. W. and M. F. Sturridge 1959. Hemolysis caused by tubing in extracorporeal circulation. Lancet. 1: 340.
Stohlman, F., Jr., S. J. Sarnoff, R. B. Case and A. T. Ness 1956. Hemolytic syndrome following insertion of lucite ball valve prosthesis into cardiovascular system. Circulation. 13: 586.
Taylor, G. 1934. The formation of emulsions in defineable fields of flow. Proc. Roy. Soc., A146: 501.

Veneziale, C. M., W. F. McGuckin, P. E. Hermans and H. T. Mankin 1966. Hyperhaptologlobinemia and valvular heart disease: Association with hemolysis after insertion of valvular prostheses and in cases in which operation had not been performed. Proc. Mayo Clin. 41: 657.

Viner, E. D. and J. W. Frost 1965. Hemolytic anemia due to a defective Teflon aortic valve prosthesis. Ann. Int. Med. 63: 295.

Walsh, J., A. Staff and L. Ritzmann 1969. Intravascular hemolysis in patients with prosthetic valves and valvular heart disease. Circulation. 39: 1–135.

Weed, R. and C. Reed 1966. Membrane alterations leading to red cell destruction. Am. J. Med. 41: 681.

Westring, D. W. 1966. Aortic valve disease and hemolytic anemia. Ann. Int. Med. 65: 203.

Wieting, D. 1968. A method for analyzing the dynamic flow characteristics of prosthetic heart valves. ASME Paper 68-WA/BHF-3.

Yacoub, M. and D. Keeling 1968. Chronic hemolysis following insertion of ball valve prosthesis. Brit. Heart J. 30: 676.

Yarborough, K. A., L. F. Mockros and F. J. Lewis 1966. J. of Thoracic and C-V Surg. 52: 550.

PART IV

Injury and Prosthesis

20

Biomechanical Compatibility of Prosthetic Devices

J. P. PAUL
J. HUGHES
R. M. KENEDI

I. DEFINITIONS AND GENERAL CONSIDERATIONS

By definition, a prosthesis replaces a missing body part structurally, functionally or cosmetically, but the dividing line between a prosthetic and an orthotic device may be very narrow. By this definition, wigs, dentures, heart valves, hydrocephalus tubes and many other devices may be considered as prostheses as well as the more obvious artificial limb. In some cases, the prostheses function without any obvious biomechanical relationship to the host. In other cases, the devices appear to function despite gross incompatibility. Generally, the biomechanical characteristics of a device comprise one of many factors controlling the acceptability and function of the device. It is worth listing some of the other factors before looking more closely at biomechanics.

> Biochemistry—the action of biological material on the implant with respect to corrosion or degradation. This may be important in altering the mechanical strength, stiffness or ductility.
> Physiology—the device may alter natural body fluid circulation leading to cell destruction, clotting, or edema, or it may locally inhibit evaporation from the skin surface leading to perspiration problems.
> Histology—the shape and surface of an internal device must be such that normal growth of tissue surrounding it is not impaired.
> Pharmacology—the material, its corrosion products, or any materials such as solvents or plasticizers which may be released during use must have no toxic or irritant action.

Sterility—the materials of the device and its shape should be such as to allow cleaning and sterilization.

Some general references in connection with the above are: Little (1968); Scales (1965); Bloch and Hastings (1967); Braley (1967).

II. STRUCTURAL COMPATIBILITY

By far the greatest number of cases where prosthetic devices are used require the mechanical strength of the device. These include artificial legs and joint replacements. In considering internal structural prostheses, it is apparent that the force actions which they transmit depend directly on the forces transmitted by the adjacent musculature. Apart from the classical studies of Pauwels (1935) and Inman (1947), little work has been done until recently on the determination of the magnitudes and directions of the forces which may be involved. The values of forces transmitted at some joints in some activities are now available (Rydell (1966); Paul (1967a, b, c; 1969a, b); Morrison (1967, 1968). The curves of variation with time of the values of hip knee joint forces during level walking have, respectively, two and three maximum turning values in the stance phase of each cycle. An indication of the variation in joint force with body weight W and the ratio of stride length L to height H in level walking is given in Figs. 1 and 2. Here the average of the maxima is plotted to a base of WL/H (Paul, 1970a). In view of the magnitudes of the normal loads at the joints, these may be as relevant a factor as

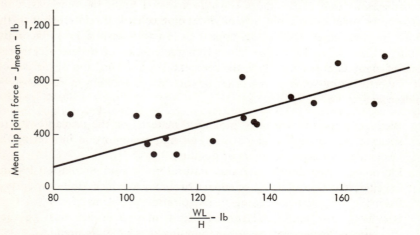

Fig. 1: Variation of mean hip joint force in level walking with body weight W, stride length L and height H (Paul, 1970a). Reproduced from Proc. Roy. Soc. Med. with permission of editors.

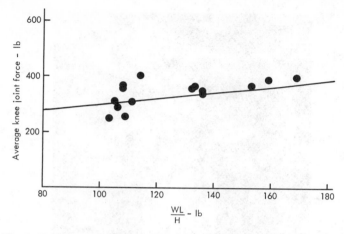

Fig. 2: Variation of mean knee joint force in level walking with body weight W, stride length L and height H (Paul, 1970a). Reproduced from Proc. Roy. Soc. Med. with permission of editors.

material degeneration and stress concentration in the failure of joint replacements such as the Judet as shown in Fig. 3 from Devas (1954).

The satisfactory functioning of complete or partial joint replacements depends on the establishment of some form of lubrication. Unfortunately, there is still considerable variation of opinion on the mechanism of lubrication of normal joints Dowson (1967); Longfield et al (1969); McCutchen (1966); Linn (1968). It appears that further study of the mechanical, physical, and chemical properties of the lubricating fluid and bearing surfaces will be required to throw light on this problem and assist in the design of prosthetic devices. The conclusion of the tests forecast by Carlsen (1970) to obtain the variation in hip joint fluid pressure are awaited with interest. He uses an Austin Moore type prosthesis carrying fourteen pressure transducers whose signals are to be transmitted by radio telemetry to a recorder. In this connection microscopic studies (Kenedi et al, 1970) have shown a differentiation in the structure of the cartilage on the femoral head and in the acetabulum resulting in different mechanical characteristics which appear relevant to their lubricating performance. This demonstrates an intrinsic biomechanical compatibility in the normal function which may be difficult to obtain in the prosthesis.

Most of the problems with structural internal prostheses occur, however, at the interface between the implanted device and the biological material— loosening of screws for nail plates or of intramedullary stems. Charnley (1965) shows an elementary analysis of the relative movement to be expected between

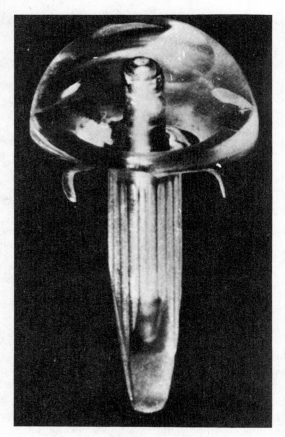

Fig. 3: Worn surface of Judet prosthesis. (Devas, 1954). Reproduced with permission from J. B. and Jt. Surg.

a femoral head prosthetic replacement and the shaft as shown in Fig. 4a. Under axial loading there is a calculated relative displacement of some 10 μ between the end of the stem of the prosthesis and the bore. It is equally interesting to look at bending deformations since a femoral head replacement is inevitably subjected to these. In the situation shown in Fig. 4a the bone is stressed but not the stem of the prosthesis. If, due to the obliquity of the joint load, a bending moment of about 400 lb in is transmitted to the bone end, then the displacement of the tip of the stem relative to the marrow cavity might be 100μ. Alternatively, if the free body diagram for the prosthesis is as shown in Fig. 4b, where the ratio of flexural rigidity EI for the stem of the prosthesis exceeds twice the EI value for the shaft of the femur, the contact faces at the shoulder of the prosthesis will tend to open at position P! The EI value for the bone might reasonably be 0.2×10^6 lb in.[2]

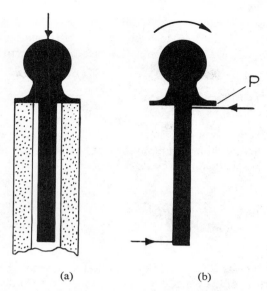

Fig. 4: (*a*) Idealized assembly of total hip prosthesis with shaft of femur (after Charnley, 1965). (*b*) Free body diagram of prosthesis transmitting moment through two localized contact areas.

while typically the value for the prosthesis is 1.8×10^6 lb in.² It is apparent that a matching of the stiffness of the prosthesis to that of the host would be advantageous (Murphy, 1954). Charnley recommends cementing the prosthesis to the bone, and in these circumstances, there is no relative displacement. But, unless the prosthesis stiffness is considerably reduced at its tip, there will be high stress gradients in the neighbouring bone.

Similarly in the design of devices to be secured to long bones by screws, the use of nail plates whose cross section varied along the length in a suitable manner could minimize the relative movements between them. Since the tensile modulus for steel is typically 29×10^6 lb/in.² and that for bone 2×10^6 lb/in.², it appears that the use of organic based or ceramic materials with their lower elastic moduli would be advantageous.

In some cases, a prosthetic device has no structural function. Examples might be hydrocephalus tubes or cardiac pacemakers from the presently available devices or, looking into the future, complete mechanical heart replacements or implanted renal dialysis units. The structural restraint of these devices must, however, be considered or device failure may result—by fracture of the electrode leads in early pacemaker designs or by kinking of the hydrocephalus tube with neck movement. It is interesting to note that at least one group (Mammorelli et al, 1970) is already looking into the problems of placing a 3 lb total heart replacement prosthesis inside the rib cage. The device must function in a satisfactory manner when placed in any of the attitudes the host may adopt and also during any period of acceleration which may be experienced. The rigid device is to be constrained by the flexible rib cage without prejudicing the function of the heart or the surrounding organs.

III. ARTIFICIAL LIMBS

a) GENERAL

A patient who has suffered amputation at any level has, to some degree, suffered a loss in four major respects

1) Structure
2) Power Source
3) Information Source
4) Cosmesis

The structural loss is particularly important in lower limb amputees since the major function of the leg and foot is to support the trunk. The loss is not solely that of the skeleton since generally the areas of the foot adapted to load-bearing are also removed. Equally, the arm amputee suffers from the loss of structural components since generally the function of the hand and arm is to place in a desired position at the required attitude some object such as a spoon or a pencil. Invariably, the prosthetic replacement has fewer kinematic degrees of freedom, and controlled placing of objects can generally be achieved only by the use of compensatory trunk movements.

For the lower limb amputee the loss of power source means that the remaining musculature has to take over the work previously performed by the amputated muscles in resisting the moments which act at the junction of adjacent segments during stance and giving the appropriate positive and negative accelerations to the limb replacement during swing phase. Similarly, the arm amputee must either have external power to operate his prosthesis or must use accessory shoulder girdle movements to supply the power which is transmitted by cable or other connection to the required site. Alternatively, a surgical technique, cineplasty, is available and was, for a time, popular (Klopsteg and Wilson, 1954). In this technique a muscle, whose insertion has been removed in the amputation, is looped into a skin tunnel so that an external connection passing through the skin tunnel can be pulled by the muscle. Obviously, the device attached must match the length-tension characteristics of the muscle. This procedure is infrequently used now, however, because skin breakdowns in the tunnel region were frequent, generally due to difficulties in cleaning and conditioning the skin to transmit the required forces.

The extremities, in common with other body parts, possess specialized nerves transmitting information such as pressure, temperature, pain, angle of joint flexion, (hence extremity position), and tension in ligament or tendon. All this information, which is used subconsciously in locomotion or manipulation of objects with the hands, is lost to the amputee. One of his major difficulties is replacing these missing signals with sufficient information from other sources to allow satisfactory performance.

The static appearance of a prosthesis may vary from the normal by amounts which may range from slight to considerable. The chief loss in the appearance of the amputee is dynamic; the limp of the leg amputee, the limited function of the arm amputee or the noise of a powered prosthesis.

Thus, in these four respects, the amputee is at a disadvantage, and it is of interest to look at the manner in which biomechanical matching of amputee and device can minimize them. All four factors generally interact, and it is not possible, indeed it is most inadvisable, to consider them separately.

b) INTERFACE PROBLEMS

Most difficulties in lower limb prosthetics arise at the stump/socket interface. The load actions to be transmitted correspond to the ground to foot contact forces and the gravity and inertia forces on the prosthesis. If complete dynamic cosmesis is specified, the user walks with a "normal" gait, that is with trunk movements which are in the range adopted by healthy humans without a physical handicap. If the displacements are specified, the accelerations and force actions are constrained to the range of normal values. To get an indication of the probable order of magnitude of these force actions in normal level walking, Table 1 has been compiled. It must be emphasized that these are to be considered only as orders of magnitude

TABLE I*

Typical dimensionless numbers representing force actions at leg segment junctions during level walking.
Units for force lb/lb body weight
Units for moment lb-in/lb (body weight)·in (of stature)

Segment Junction	Ankle	knee	Hip
Resultant Force Externally Applied Moment Tending to:	1.2	1.2	1.2
Extend (Plantar Flex)	0.10	0.04	0.08
Flex (Dorsi Flex)	0.02	0.06	0.10
Adduct (Invert)	0	0.04	0.07
Abduct (Evert)	0.01	0	0.01
Rotate Medially	0.01	0.01	0.01
Rotate Laterally	0.02	0.01	0.01

* From Bresler and Frankel (1950); Paul (1967b); Morrison (1967).

since the values of the quantities depend on anatomical dimensions, style and speed of locomotion as well as many other factors. The amputee/prosthesis interface must, however, transmit force actions corresponding to these, and it is essential in the design of the socket of a prosthesis to take cognizance of them.

The classical works on socket design (Radcliffe, 1957; Radcliffe and Foort, 1961; McLaurin, 1957) have been insistent on the necessity for consideration of those anatomical areas which are available to transmit the necessary force actions. Figures 5 and 6 from Radcliffe and Foort indicate the manner in which the moment actions are transmitted to a below knee patellar tendon bearing (PTB) prosthesis. It is of interest to consider the possible order of magnitude of these forces for a 5 in stump length. Using the factors of Table 1 applied to a 70 in. amputee weighing 150 lb, the flexing and extending moments to be transmitted are 630 and 420 lb in respectively, and the adducting moment is of a like order of magnitude. If the lines of action of the resultant horizontal forces transmitting these moments are 5 in apart, the forces must be of the order of 85–125 lb, almost as much as body weight. It is apparent, therefore, that socket design must be based on dynamic considerations as well as static. Attempts to reproduce uniform contact pressures are certain to fail since, as the moments vary in magnitude and direction during the walking cycle, so the transmitting forces will change in magnitude and area of application.

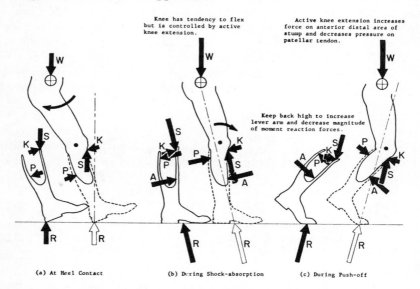

Fig. 5: Lateral view of amputee and PTB prosthesis showing "free body" force actions. (Radcliffe and Foort, 1961). Reproduced with authors' permission.

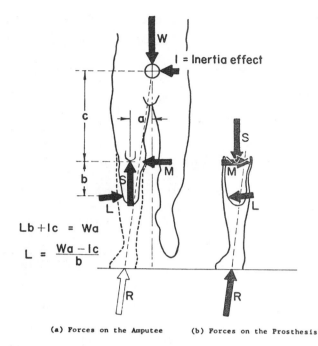

Fig. 6: View from the rear of amputee and PTB prosthesis showing "free body" force actions. (Radcliffe and Foort, 1961). Reproduced with authors' permission.

These resultant forces are transmitted to the stump by pressure transmitted through the skin to underlying structures. The skill of the prosthetist is involved in the selection of the appropriate areas through which to transmit pressure and the relieving of the other areas where pressure would lead to discomfort and skin break down. Schrock et al (1968) show the results of tests where pressure transducers were fitted to below knee stumps and Fig. 7 shows the cyclical pressure variations obtained. Figure 8 shows corresponding results obtained by Appoldt et al (1968) from tests on an above knee amputee. The high values of contact pressure measured are worthy of particular note.

The question of biomechanical compatibility arises for the surfaces of the socket and the surface and substructure of the stump. While there has been considerable research performed on the mechanical properties of biological tissue when loaded in its own plane, little is known of the mechanics of the transmission to the skeletal structure of loads applied normally or obliquely to the skin surface. It is a matter of common experience that pressure normal to the skin surface occludes the local blood supply and, over an extended period of time, will lead to skin breakdown and ulceration. Simi-

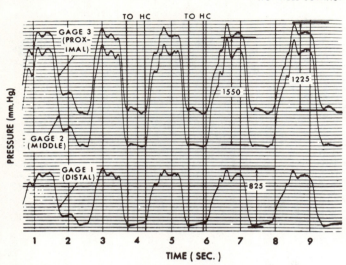

Fig. 7: Cyclical variation of pressure intensity between stump and prosthesis for a walking patient. *TO* and *HC* refer respectively to toe-off and heel contact (Shrock et al, 1968). Reproduced with permission from Bull. Prosth. Res.

larly, tangential movements of an area in contact with skin lead to discomfort and formation of a blister. There is a need for quantitative information on the allowable normal and shear stresses which various areas of the skin surface can tolerate. In general, the prosthesis will apply to the stump a time varying system of stress and the information obtained must therefore be dynamic, that is tolerable stress levels for intermittent or fluctuating loading with various wave forms. The general practice in prosthetics is to fit "hard" sockets on the principle that the mechanical matching can be achieved by the stump tissues. For particular problems, however, such as bony protrusions or other painful areas a resilient interface may be built into the prosthesis (Wilson et al, 1968), which forms an "air cushion socket" with the resilient material acting as a membrane between the stump and air in the distal part of the socket. Alternatively, removable pads of foamed silicone rubber can be fitted as described by Hampton (1966). In the supply of a leg prosthesis to the patient, the greatest cause for delay is the requirement for a custom-made socket. Any procedures which could produce a durable, accurately formed socket without the use of highly-skilled prosthetists and fabrication would signify a major improvement.

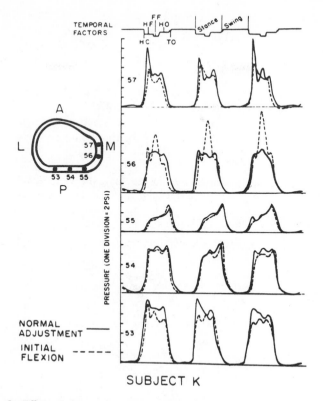

Fig. 8: Effect on pressures at the ischial brim of an above knee prosthesis of variation of alignment (Appoldt et al, 1968). Reproduced with permission from J. Biomechan. I, p. 256. Pergamon Press.

c) GROSS MECHANICAL PERFORMANCE

In addition to the local biomechanical compatibility corresponding to contact at the interface, a good leg prosthesis requires compatibility in overall mechanical performance. Firstly and obviously, the alignment must be such as to attain comfort, stability and cosmesis of gait, and these factors have usually been assessed by the prosthetist while the patient is walking in front of him in the amputee clinic. Adjustment requires the patient to interrupt his walking. Foort and Hobson (1970) describe a motorized device for adjustment of alignment during locomotion. Initial tests have been performed with the adjustment being controlled by the prosthetist. It appears that more consistent results are obtained, however, by allowing the amputee himself to control the device. This, of course, corresponds to best automatic control practice, by closing the control loop in a simpler manner and by eliminating time delays.

The design of an above knee prosthesis generally takes account of four situations in the walking cycle in which the amputee may experience difficulty

- A In maintaining stability early in the stance phase
- B In initiating knee flexion prior to toe-off
- C In preventing excessive heel rise at the start of the swing phase
- D In decelerating the forward swing of the shank at the end of stance phase

Problem A is best illustrated by Fig. 9 which shows the classical diagram from Radcliffe (1957) illustrating the relation between the moment which the amputee exerts at the hip to prevent knee flexion and collapse and the dimensions b and d defining the centre of the knee joint. Obviously, as dimension d is diminished less hip moment action is required for stability, but this raises

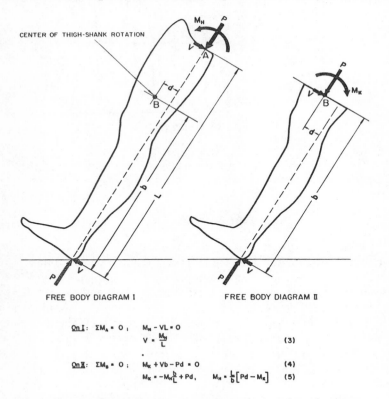

Fig. 9: Force actions acting on an amputee using an above knee prosthesis (Radcliffe, 1957). Reproduced with author's permission.

problem *B*. A prosthesis with a knee axis well back will require excessive hip flexor activity to initiate knee bend. If a single axis joint of this type is used, a compromise must be reached, based, ideally, on the ability of the amputee to exert these moments. Radcliffe's paper illustrates the design of one of the so-called polycentric mechanisms which varies the position of the instantaneous centre of rotation and solves both of these problems. Cosmetic appearance may not be optimal due to projection of the upper part of the mechanism forward of the shank when the amputee sits down.

Lowe (1969*a*, *b*) and Hughes et al (1969) describe tests to assess the effect of different types of mechanisms aimed at assisting the amputee at this phase of his locomotion. The shank of the prosthesis is used as a dynamometer to measure the six quantities describing the force actions transmitted. The galvanometer recorder also carries one channel transmitting a signal corresponding to the angle of knee flexion. The information is translated to punch tape and analysed by digital computer to give the moments about axes through the amputee's hip. These moments are transmitted to the amputee by contact pressure with the prosthesis and are resisted by tension in suitable groups of muscles acting at the hip. The value of the moment is, therefore, a measure of the intensity of contact pressure. The amount of energy required to perform this process is not directly obtainable but the mechanisms may be compared experimentally by comparing the values of $\int_{HS}^{TO} |M| \, dt$ where *M* is hip moment, *t* is time, and *HS* and *TO* denote heel strike and toe-off respectively. Preliminary results for one test subject are shown in Fig. 10. Further tests are obviously necessary to investigate the effect of alignment on these values. Work is currently in progress to make this a useful device clinically. The computation of hip moment is being performed by a compact analogue computer, and the output is immediately available in the same form as Fig. 10. This device could be used as a guide in adjusting the prosthesis to optimum alignment. Prior to this, however, a selection must be made on some basis of the device which is best suited to the needs and capability of the amputee.

Obviously, the long-term work capacity of the amputee cannot reasonably be measured at a limb fitting center. It is hoped, however, to take quantitative measurements of the isometric moments which the amputee can exert on the principle that this information will guide the selection of the prosthesis and its alignment when fitted in order to attain the maximum biomechanical compatibility. Allied to this problem of work capacity is that of optimal performance of the disabled. Bard and Ralston (1959) have shown, from expired air analysis tests, that in walking the amputee energy consumption is greater than the normal and that it increases rapidly with speed changes from a preferred value. It is contended, but not proven, that the closer the amputee gait is to the normal the less the energy expenditure. It would be

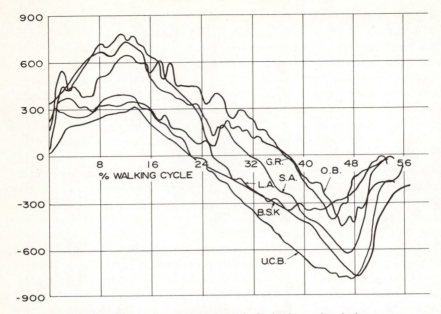

Fig. 10: Amputee leg to trunk moments in flexion/extension during stance when using different knee mechanisms (Lowe, 1969a). The curve markings represent: BSK Blatchford stabilized knee; GK Greissinger knee; LA Lammers knee; OB Otto Bock stabilized knee; SA Conventional single axis constant friction knee; UCB "Berkeley" polycentric knee.

interesting to see controlled comparisons between the dynamics and energetics of the gait adopted in the walking training clinic and that used when clinical and social pressures do not emphasize the cosmetic aspect of performance.

Problems C and D can be illustrated with reference to Fig. 11 from Radcliffe and Lamoreux (1968) which shows the variation of flexion extension moment with knee angle during level walking for a normal individual. The provision of a mechanism which can reproduce the moment pattern ABC to prevent excessive heel rise at the start of swing phase and the corresponding variation CDE to decelerate the extremity in preparation for heel strike will allow the amputee to acquire greater dynamic cosmesis in his gait pattern. The mechanism should also reproduce the variation of the moments with walking speed. A number of devices are available which correspond reasonably closely to these requirements (Mauch, 1958; Murphy, 1964; Contini et al, 1961; Hughes et al, 1970).

The basic effect of knee mechanisms is to absorb energy and replace the action of the knee muscles. Unfortunately, many of the muscles producing action at the human knee are two-joint muscles, and they may be used to

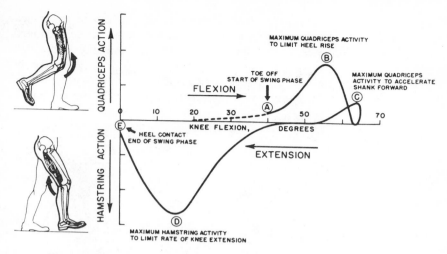

Fig. 11: Variation of flexion/extension moment at the knee with knee angle during the swing phase of level surface walking by a normal subject (Radcliffe and Lamoreux, 1968). Reproduced with permission from Bull. Prosth. Res.

match requirements for energy absorption at one joint and work output at the adjacent joint (Elftman, 1939; Paul, 1970). This effect is not possible between knee and hip although attempts have been made to link the knee and ankle of a prosthesis to achieve it (Staros and Murphy, 1964). Generally, however, in locomotion the limb segments follow a repetitive series of motions which vary slightly with velocity and the nature and inclination of the walking surface. The moments to be exerted are large, and the power requirements are such that present technology cannot supply an acceptable source within the limits placed on weight, size, noise, temperature and reliability.

d) UPPER EXTREMITY PROBLEMS

The situation is better in respect of upper extremity prostheses but only marginally so. The presently available power sources do not provide sufficient energy to undertake a full range of shoulder, elbow, wrist and hand movements, and, of course, no limbs are available which use power in so many locations. The reasons for this include weight, mechanical complexity and, particularly, control sites. If mechanical control is utilized, it is difficult to obtain more than three control movements. At its present state of development myoelectric control from single muscles is used routinely in below elbow prostheses to control grasp. Generally, a simple velocity control is used without feedback. Theoretically, other muscles can be used to give control signals, but more complicated prostheses on this basis are not routinely available. Scott (1967) shows that two functions can be controlled from one EMG channel, but this system is not yet available on a clinical basis. The control systems for automatic and conventionally controlled externally powered hands are described by Peizer (1967) and illustrated in Fig. 12. Although feedback is incorporated into some control systems, this is either in respect of gripping force or velocity of opening/closing (or both).

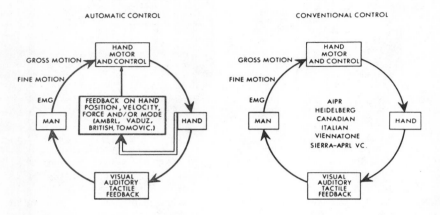

Fig. 12: Control systems for automatic and conventionally controlled externally powered hands (Peizer, 1967). Reproduced with permission from Bull. Prosth. Res.

For locating the arm in space the amputee is required to use his sight. For above elbow prostheses he develops, to some degree, an awareness of overall limb position from the magnitude and direction of the movements exerted on the trunk by the limb attachments. For complete control, feedback should provide the same information as the normal intact human hand. These include position, attitude, gripping force, slip, weight and temperature. It is relatively easy to devise transducers to measure these quantities, but the major problem is to transmit the information to the brain. Skin contacts have been used to transmit a sense of gripping force, and studies are under way at M.I.T. to simulate position awareness by multiple vibrating contacts. Detailed references to these topics can be obtained in Institution of Mechanical Engineers (1969). Generally, the control of upper extremity prostheses involves an intimate biomechanical adaptation. Unfortunately, most of the adaptation is in the cerebro-neuro-muscular system of the amputee. Research to improve control systems and transducers in the prostheses would appear to be a fruitful field of study.

REFERENCES

Appoldt, F., L. Bennett, R. Contini 1965. Stump-socket pressure in lower extremity prostheses. J. Biomech. 1: 247.

Bard, E. and H. J. Ralston 1959. Measurement of energy expenditure during ambulation with special reference to evaluation of assistive devices. Arch. Phys. Med. & Rehab. 40 (10): 415.

Bloch B. and G. W. Hastings 1967. Plastics in surgery. Charles C. Thomas, Publisher. Springfield, Ill.

Braley, S. 1967. The silicones in clinically oriented bioengineering. In B. L. Segal and D. G. Kilpatrick (eds.), Engineering in the Practice of Medicine. Williams and Wilkins, Baltimore.

Carlsen, C. E. 1970. Experimental measurement of pressure distribution on the human hip joint. SESA Spring Meeting, Huntsville, Ala.

Charnley, J. 1965. Biomechanics in orthopaedic surgery. In R. M. Kenedi (ed.), Biomechanics and Related Bioengineering Topics. Pergamon, London.

Devas, M. B. 1954. Arthroplasty of the hip. J. Bone Jt. Surg. 36B (4): 561.

Dowson, D. 1967. Modes of lubrication in living human joints. Proc. Inst. Mech. Eng. 181 (3J): 70.

Elftman, H. 1939. The function of muscles in locomotion. Amer. J. Physiol. 125: 357.

Foort, J. and D. Hobson 1970. Manitoba Rehabilitation Institute, Winnipeg. Personal communication.

Hampton, F. 1966. NU Suspension casting technique. Bull. Prosth. Res. 10-6: 52.

Hughes, J., P. J. Lowe and J. P. Paul 1969. Dynamic assessment of above knee prostheses. Proc. 3rd Int. Symp. External Control of Human Extremities, Dubrovnik, Yugoslavia.

Hughes, J., J. P. Paul and R. M. Kenedi 1970. Control and movement of the lower limbs. In D. C. Simpson (ed.), Modern trends in biomechanics. Butterworth, London.

Inman, V. T. 1947. Functional aspects of the abductor muscles of the hip. J. Bone Jt. Surg. 39 (3): 607.

Institution of Mechanical Engineers (1969) Basic problems of prehension, movement and control of artificial limbs. Proc. Inst. Mech. Eng. 183: 3J.

Kenedi, R. M., J. Graham and M. Mital 1970. Univ. Strathclyde. Glasgow, Scotland. Personal communication.

Klopsteg, P. E. and P. D. Wilson 1954. Human limbs and their substitutes. McGraw-Hill Book Company, New York.

Linn, F. C. 1968. Lubrication of animal joints. J. Biomech. 1: 193.

Little, K. 1968. The interactions of plastics and tissues. Biomed. Eng. 3 (9): 404.

Longfield, M. D., D. Dowson, P. S. Walker and V. Wright 1969. Boosted lubrication of human joints by fluid enrichment and entrapment. Biomed. Eng. 4: 517.

Lowe, P. J. 1969. Knee mechanism performance in amputee activity. Ph. D. Thesis, Univ. Strathclyde, Glasgow, Scotland.

McCutchen, C. W. 1966. Physiological lubrication. Proc. Inst. Mech. Eng. 181 (3J): 55.

Mammorelli, V. T., J. R. Venuto, V. P. Satinsky, O. F. Muller and G. E. McNeal 1970. Stresses involved in implanting a mechanical artificial heart. SESA Spring Meeting, Huntsville, Ala.

Morrison, J. B 1967. The forces transmitted by the human knee joint during activity. Ph. D. Thesis, Univ. Strathclyde, Glasgow, Scotland.

Murphy, E. F. 1954. Engineering principles in fracture fixation. Amer. Acad. Orthopedic Surgeons, Instructional Course Lectures. Vol. IX. Edwards, Michigan.

Murphy, E. F. 1964. The swing phase of walking with above knee prostheses. Bull. prosth. Res. 10-1.

Paul, J. P. 1967a. Forces transmitted by joints in the human body. Proc. Inst. Mech. Eng. 181 (3J): 8.

Paul, J. P. 1967b. Forces at the human hip joint. Ph. D. Thesis, Univ. Glasgow, Scotland.

Paul, J. P. 1967c. The patterns of hip joint force during walking. Paper 38–18, Proc. 7th Int. Conf. Med. Biol. Eng. Stockholm.
Paul, J. P. 1969 Loading on the head of the human femur. J. Anat. 105 (1): 187.
Paul, J. P. 1970a. The effect of walking speed on the force actions transmitted at the hip and knee joints. Proc. Roy. Soc. Med. 63 (2): 200.
Paul, J. P. 1970b. Load actions on the human femur in walking and some resultant stresses. SESA Spring Meeting, Huntsville, Ala.
Paul, J. P. 1970c. Action of some two-joint muscles of the hip in walking. J. Anat. 105 (1): 208.
Pauwels, F. 1935. Der schenkelhalsbruch, ein mechanisches problem. Ferdinand Enke, Stuttgart, Germany.
Peizer, E. 1967. VAPC Research Report. Bull. Prosth. Res. 10-8: 208.
Radcliffe, C. W. 1957. Biomechanical design of an improved leg prosthesis. Biomechan. Lab. Res. Re. Ser. II, Issue 33. Univ. Calif., Berkeley.
Radcliffe, C. W. and J. Foort 1961. Patellar-tendon-bearing below-knee prosthesis. Biomechan. Lab. Univ. Calif., Berkeley.
Radcliffe, C. W. and H. J. Ralston 1963. Performance characteristics of fluid controlled prosthetic knee mechanisms. Biomechan. Lab. Res. Rep. 49. Univ. Calif., Berkeley.
Radcliffe, C. W. and L. Lamoreux 1968. "UC-BL" swing control unit. Bull. Prosth. Res. 10-10: 75.
Rydell, N. W. 1966. Forces acting on the femoral head prosthesis. Acta. Orthop. Scand. Suppl. 88; 37: 1.
Scales, J. T. 1965. Some engineering and medical problems associated with massive bone replacement. *In* R. M. Kenedi (ed.), Proc. Symp. Biomechanics and related bioengineering topics. Butterworth, London.
Schrock, R. D., J. H. Zettl, E. M. Burgess, and R. L. Romano 1968. A Preliminary Report of Basic Studies from Prosthetics Research Study. Bull. Prosth. Res. 10-10: 90.
Scott, R. N. 1967. Myoelectric control of prostheses and orthoses. Bull. Prosth. Res. 10-7: 93.
Staros, A. and E. F. Murphy 1964. Properties of fluid flow applied to above knee prostheses. Bull. Prosth. Res. 10: 40.

21

Fluid Dynamics of Heart Assist Devices

ROBERT T. JONES

In this article certain hemodynamic phenomena that arise in connection with the use of artificial blood pumping devices will be reviewed. Among these are: 1) flows produced by collapsing bulbs, 2) the impedance presented by the aorta, 3) limiting velocities and instability of flow in elastic vessels, 4) effectiveness of valveless arterio-arterial pumps, and 5) wave reflection phenomena and instabilities associated with the intra-aortic balloon pump.

* * *

Maintenance of life by an artificial blood pump is an everyday occurrence in most large hospitals. The pump-oxygenator which takes over the function of both the heart and the lungs during surgery seems, however, to be limited in use to a maximum of several hours. Beyond this time one looks for certain symptoms which are taken as evidence of changes in the physiological constituents of the blood, that is, loss of red blood cells (hemolysis) and platelets as well as denaturation of protein components. In spite of this deficiency the pump-oxygenator is considered adequate for most surgical procedures, though not suited in its present form to chronic support of a weakened or failing heart.

Pending the development of an ideal total heart replacement, there have been numerous attempts to provide partial assistance to the heart (see references) for periods longer than the few hours available with the pump oxygenator. These have generally taken the form of pumps to assist the left ventricle, since left ventricular function is most critical and evidently most often deficient.

Several types of left ventricular assist pumps are illustrated in Fig. 1. In Fig. 1a the pump is equipped with inflow and outflow valves and is placed in parallel with the ventricle (atrio-arterial pump) where it may divert all or part of the flow around the left heart. By contrast, pumps placed in series with the ventricle may not significantly alter the flow but may be expected to relieve all or part of the systemic pressure from the heart. It is interesting that the series pump may accomplish this function even without valves if it is properly phased with the pulsatile outflow from the ventricle. The centrifugal pump described by Dorman et al (1969) operates by magnetic coupling through the intact skin and, of course, requires no valves.

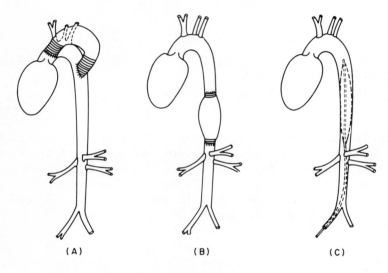

(A) THORACIC AUXILIARY VENTRICLE
(B) AUXILIARY VENTRICLE IN DESCENDING AORTA
(C) INTRA-AORTIC BALLOON PUMP

Fig.1: Heart assist devices

In order to understand the operation of such devices, it is first of all necessary to gain at least a crude understanding of the function of the left ventricle and the aorta in mechanical terms.

Our first curiosity here is in the flow produced by a collapsing or squeezing bulb since both the natural and the artificial ventricles operate in this fashion. As has been shown (Jones, 1969) potential flows of the type

(1) $$\Phi(x, y, z, t) = \alpha x^2 + \beta y^2 + \gamma z^2$$

provide perhaps the simplest starting point for mathematical modeling of the bulb flows. Here Φ is the velocity potential, x, y, z are cartesian coordinates and α, β, γ are constants or functions of the time such that

(2) $$\alpha + \beta + \gamma = 0$$

Such flows are trivial solutions of the Navier-Stokes equations and, of course, of Laplace's equation. If x, y and z are taken as the initial coordinates of a particle of fluid (or of the boundary) we have

(3) $$\zeta = xe^{\int \alpha \, dt}$$

(4) $$\eta = ye^{\int \beta \, dt}$$

(5) $$\xi = ze^{\int \gamma \, dt}$$

for the Lagrangian coordinates. The Lagrangian coordinates are thus related to the initial coordinates by simple affine transformations. Plane surfaces remain plane and a bulb of initially arbitrary shape undergoes distortions that can be represented by stretching its coordinates in different directions while preserving a constant volume.

Figure 2 shows streamlines emanating from a collapsing bulb together with pressures calculated by the foregoing method on the assumption of frictionless flow. The motion of the liquid in a bulb does not completely determine the pressures, but there remains an arbitrary function of the time to be determined by the impedance into which the bulb works, that is, the relation between pressure and flow in the aorta.

In a typical case the diastolic pressure in the aorta will be 70–80 mm Hg and the ejection of blood from the ventricle will raise this pressure by 60–70 mm. Pressures arising from dynamic motions within the bulb, however, are of the order of 3–4 mm only. Thus it seems that very little of the effort of the heart is directed against the inertia of the blood but appears initially in elastic distension of the arteries; later to be dissipated in frictional resistance through the capillaries.

Examination of the heart cavity shows a surface far from streamlined form, with deep furrows (trabeculae) and tendons to support the mitral valve. Similarly, the architecture of the aortic system shows little disposition toward flow dynamics as ordinarily understood. Branching arteries often show no preference for the flow direction. Here again we find but a small fraction of the pressure in kinetic form; $\rho u^2/2$ is of the order of a few millimeters.

In prosthetic heart pumps streamlining does not play an important role in the purely mechanical function, though the flow does seem most important in relation to clotting. The clotting phenomenon is at present a dominant factor in the success or failure of artificial heart devices including prosthetic valves and assist pumps. Though beyond the scope of the present discussion,

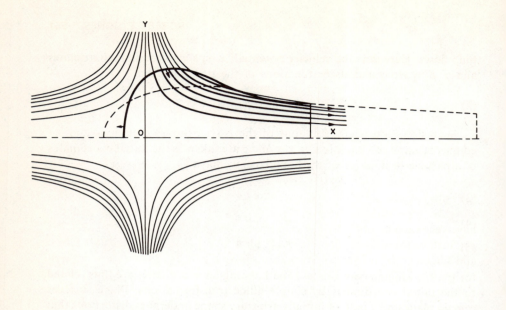

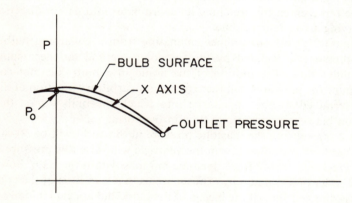

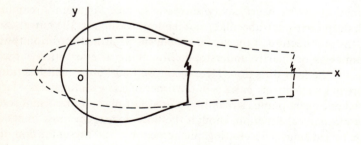

Fig. 2: Streamlines and pressures in collapsing bulb.

the relation between clotting and flow has become a major item of research at our laboratory and elsewhere.

Nearly all current heart assist pumps operate by pneumatically driven plastic bulbs. In our experience we have found it essential that the bulb not change its volume by stretching but rather by folding. The bulb must undergo a change of extrinsic geometry without altering the intrinsic geometry of the surface. The solution appears by reflection of the convex surface in a plane so that progressively larger areas of the surface change from convex to concave, along a singular curve determined by the intersection of the plane. In practice the bulb has finite thickness and the surface bends to a finite radius along this curve. In spite of the concentration of bending strain, bulbs of silicone rubber have very satisfactory fatigue life in this mode of deformation.

Such bulbs are occasionally made by fitting initially flat sheets of fabric and silicone rubber over a convex form. Here one encounters a problem well known to tailors and known to mathematicians from the properties of the Tschebyschev net. It seems that such a net (as a handkerchief) can be fitted over a convex surface whose integrated solid angle does not exceed 2π.

The function of the aorta and the major vessels as an elastic compliance is well characterized by the simplest theory of aortic impedance known as the "windkessel" theory (Taylor, 1965). Here the aorta acts simply as a lumped capacitance, and if the peripheral resistance is linear, one obtains an exponential decay of pressure following ejection of a volume of blood from the heart.

Figure 3 illustrates our attempt to make a direct physical model of the aortic system. In medical circles such devices are known by the term "mock circulation." We have found such a model useful in studying and isolating the purely mechanical effects of cardiac assist devices. The model aorta is made of silicone elastomer with compliance determined so as to lie within the physiological range. The substitute ventricle is made of the same material and is placed in an air tight box so that it can be actuated pneumatically by a linkage and piston drive as shown. Figure 4 shows the type of pressure pulse produced by our model compared with natural pulses taken from a textbook on physiology.

With such a model one has control of various parameters such as source impedance of the heart, cardiac output, wave speed and diameter of the aorta, and peripheral resistance. In addition, since the aorta is transparent, we have been able to observe visually the action of the intra-aortic balloon pump.

Fortunately, because of the high Reynolds number of the flow in the larger arteries, one is not required to duplicate the detailed rheological properties of the blood in such experiments. A rough calculation will show that the unsteady laminar boundary layer on the inner wall of the aorta has time to grow to a thickness of only about one millimeter during the heart beat. Separation vortices of the order of a centimeter in diameter may be expected

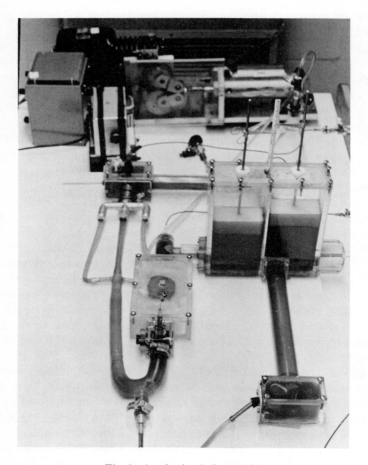

Fig. 3: Aortic circulation model.

to persist for two or more heart beats. It should be borne in mind, however, that the dynamic pressure associated with the flow is small compared with the actual pressure. Hence, flow irregularities do not normally have a great influence on the mechanical performance of heart assist pumps. They may, however, influence the behavior of artificial valves and, of course, the formation of thrombi.

In our early effort at Avco Everett Research Laboratory, we sought to develop and perfect the Kantrowitz auxiliary ventricle, illustrated in Fig. 1b. This device comprises a simple compliant pumping chamber which is fitted inside a more rigid plastic case. The space between the bulb and the case communicates by a small gas tube through a percutaneous connector to the control system outside the body. The bulb can then be alternately compressed and

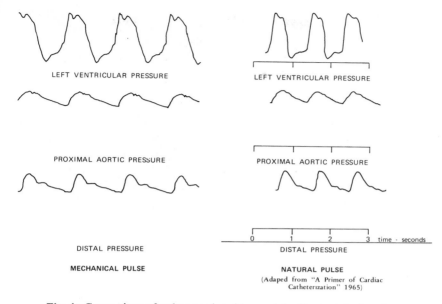

Fig. 4: Comparison of pulses produced by model with physiologic pulse.

expanded by pneumatic pressure from an external power source. Since the blood pumping chamber has no valves, it is necessary that its stroke be synchronized and properly phased with the action of the heart. This synchronization was accomplished with the aid of electrocardiac signals obtained from electrodes attached to the heart muscle. By withdrawing blood from the aorta prior to systole the pressure required for ejection by the heart can be reduced, thus lessening the effort of the heart. After the ejection from the natural heart is complete and the aortic valve is closed, the blood withdrawn into the auxiliary ventricle can be pumped back into the aorta, raising the pressure and supplying energy for the systemic circulation. This technique has the advantage of creating higher pressures in the aorta during diastole and hence the possibility of increased coronary perfusion (in practice the coronary flow has in some cases been observed to diminish, perhaps because the assisted heart requires less). If we suppose, for example, that the stroke volume of the artificial ventricle Q_b is as great as the output stroke of the natural ventricle Q_h then it is clear that the pressure rise during systole may be suppressed and the pressures in the natural ventricle need hardly exceed diastolic pressure. This special situation requires that the withdrawal stroke of the AV imitate the ejection stroke of the natural ventricle in reverse. In practice somewhat better results are obtained by reducing the pressure in the aorta prior to systole, permitting the aortic valve to open at a lower pressure and reducing the pressure during isometric contraction.

The intra-aortic balloon pump (Fig. 1c), functions by a similar principle (Moulopoulos et al, 1962) since inflation and deflation of the balloon affects aortic pressures in much the same way as injection and withdrawal of a corresponding quantity of blood. The balloon pump has the advantage that it can be inserted through a femoral artery under local anesthesia. In this case electric signals for timing are obtained from skin electrodes.

The reduction of outflow pressure during systole may, of course, permit the heart to eject a greater flow quantity. Hence, the auxiliary ventricle (or the balloon pump) may increase the cardiac output as well as effect a reduction in left ventricular pressure. The precise division of effort between flow and pressure will depend on what may be termed the "source-impedance" of the heart. Animal experiments with such pumps indicate that a healthy heart has a very high source impedance (is flow-limited). The primary effect of the auxiliary pump then appears as a reduction of pressure in the left ventricle. However, experience with human patients in cardiac distress indicates that the weakened heart may be pressure limited. In such cases the arterio-arterial pump will act primarily to increase the cardiac output.

Because of its dependence on pathology, it seems unlikely that any universal relation for cardiac impedance can be found. However, the reduction of ventricular pressure and the increase of flow can be combined linearly into a quantity which may be thought of as total effectiveness and which can be simply related to the stroke volume of the pump and to the unassisted aortic pulse (Jones et al, 1968). This cycle analysis of the arterio-arterial pump is based on the windkessel theory of aortic impedance. Figure 5 illustrates the essential idea of the cycle analysis. Here the pump injects its stroke

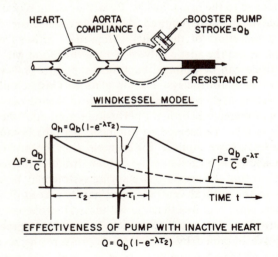

Fig. 5: Analysis of arterio-arterial pump.

volume into the aorta which distends as a windkessel. The pressure increment so produced then decays exponentially as the volume flows out through the peripheral resistances. All of the added volume would ultimately find its way into the periphery if enough time were available. However, the pump must withdraw volume and lower the pressure in the aorta prior to the next heart beat. It must therefore withdraw a portion of the blood previously injected, namely the fraction

(6)
$$e^{-\lambda t_d} \times Q_b$$

where $\lambda = 1/RC$. R is the peripheral resistance and C the aortic compliance or capacitance. Q_b is the stroke volume of the pump and t_d is the diastolic interval. The complete analysis, which entails the assumption that operation of the pump does not change systolic or diastolic intervals, leads to the formula

(7)
$$\frac{\Delta Q}{Q} - \frac{\Delta P}{P} = \frac{Q_b}{Q_h}(1 - e^{-\lambda t_d})$$

The quantity $(1 - e^{-\lambda t_d})$ may be conveniently replaced by the quantity $(P_1 - P_2)/P_1$ where P_1 and P_2 are systolic and diastolic pressures with the unaided heart. In this form it is evident that the operation of such valveless pumps depends on the pulsatile character of the aortic pressure. In order to obtain an upper bound for the effectiveness, it has been assumed that the pump displaces its volume Q_b instantaneously. Other variations of the stroke Q_b as a function of the time will require a convolution of Q_b with the decay factor $e^{-\lambda t}$.

The use of the windkessel theory for aortic impedance corresponds to an assumption that the arterio-arterial pump will be equally effective at any position along the aorta. Experiments with animals and with our artificial circulation model tend to support this assumption provided the timing of the pump cycle is adjusted to account for the delay in wave propagation.

Figure 6 shows the result of an experiment made in the circulation model to check this assumption. Here the pumping effect of an intra-aortic balloon placed in the descending thoracic aorta is compared with the effect of an auxiliary ventricle placed near the heart. In each case the same stroke volume was used, but the timing of the balloon was advanced slightly. The curves on the left of the chart recordings show the rather large aortic pressure produced by the pump during diastole. Near the center of each diagram the chart speed was reduced, compressing the pulses. The rise in left ventricular pressure when the pumps were turned off is clearly evident here and is approximately the same for the balloon and the AV. During the compressed portions of the records an integrating circuit was switched on to record mean aortic pressures. The mean pressures are an indication of the flow, and it will be noted that each

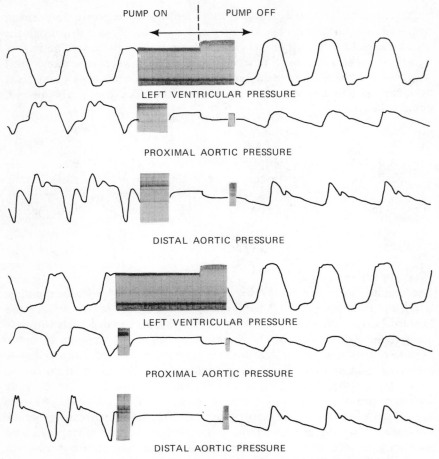

Fig. 6: Comparative effects of balloon pump and auxiliary ventricle.

pump increased the cardiac output by the same amount. In spite of the obstruction presented by the balloon in the aorta, the distal mean pressure is very effectively increased.

The action of the arterio-arterial pump will evidently depend not only on the source impedance of the heart, but also on the complex impedance presented by the aortic system. The impedance of the aorta is, of course, complicated by the phenomenon of pulse wave propagation. (McDonald, 1960; Young, 1808, 1809; Womersley; 1956 Anliker and Maxwell, 1966). The elastic tube, except for reflections, presents a characteristic resistive impedance. With an injection velocity u the pressure developed is

$$\Delta p = \rho u c \tag{8}$$

where c is the wave propagation velocity of the order of 500–1000 cm/sec. The flow quantity will be uA where A is the cross-sectional area of the artery. The characteristic impedance is then

$$Z_c = \frac{\rho c}{A} \tag{9}$$

For a wave speed of 600 cm/sec and an area A of 4 sq cm, we obtain

$$Z_c = 150 \text{ (cgs)} \tag{10}$$

Estimates of the capillary resistance associated with the various outlets may be made from data on mean pressures and flow to various organs given in physiology textbooks. Values of peripheral resistance vary over a wide range, especially in pathologic conditions such as shock. However, it is significant that the values are an order of magnitude greater than either the pulse impedance or the frictional resistance of the aorta. Hence, we may suppose that the pulse wave will be strongly reflected by large resistances which are rather closely coupled through the main branching arteries. Furthermore, since the time for a pulse wave to traverse the length of the aorta is short compared with the duration of the heart beat one can expect the pressure and flow to be dominated by multiple reflections from the periphery during a single heart beat.

The foregoing characteristics of the aortic system lead us to what might be termed a "slender windkessel" theory of aortic impedance. Clearly, if an elastic tube of finite length is inflated sufficiently slowly it will behave simply as a compliance. With more rapid inflation waves and their reflections from the ends will become prominent (Fig. 7). The important parameter here is the characteristic time for reflection along the length of the tube, ℓ/c, compared with the time scale of the inflation process. By expanding the wave reflection formulas in ascending powers of ℓ/c, one obtains, first of all, the result of the windkessel theory followed by a succession of terms which correct the windkessel theory for the finite length of the aorta (Jones, 1969). Curiously, the first correction does not involve the length of the windkessel at all but merely takes account of the fact that the windkessel is an elastic tube. By considering one of the two families of characteristics, there is obtained

$$\Delta p = \frac{1}{2} P_{\text{windkessels}} + \frac{1}{2} \rho u c + \cdots \tag{11}$$

Thus the first correcting term is simply 1/2 the characteristic pressure, $\rho u c$. Figure 7 illustrates the process of inflation of an elastic tube for the case of a stepwise inflow velocity $u(t)$ and shows the nature of the correction. Applying this calculation to the aorta, one should take account of outflow at the reflection sites. Such outflow leads to progressive diminution of the reflected

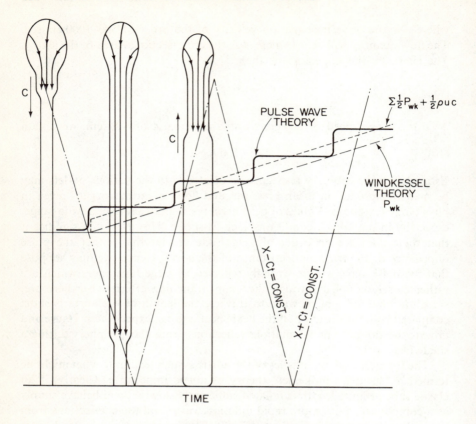

Fig. 7: Inflation of an elongated windkessel.

pulses by a fixed fraction (reflection coefficient K) and thence to an exponential decay of aortic pressure. The windkessel theory (P_{wk}) shows this behavior by a convolution of the inflow pulse with the factor $e^{-\lambda t}$. However, since it depends on the integral curve of the inflow pulse, P_{wk} shows too slow an initial rise of pressure. This behavior is illustrated in Fig. 8.

A method of arterio-arterial pumping not illustrated in Fig. 1 consists of withdrawing blood directly through cannulae inserted in the femoral arteries and reinjecting the same blood during diastole. While the effect of distance from the heart is rather small in the case of the balloon or the abdominal ventricle, such effects will certainly be larger here. In particular, the taper of the artery will be significant since the femoral vessels may be only 6 mm in diameter while the aorta near the heart is approximately 20–25 mm.

The pulse wave phenomenon in a tapered elastic tube leads, in the case of constant wave speed, to an equation similar in form to that employed in the analysis of acoustic horns. A treatment of the exponentially tapered tube has

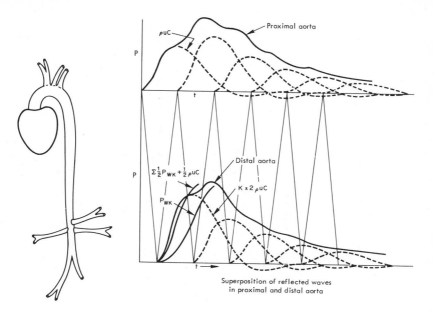

Fig. 8: Superposition of reflected waves in proximal and distal aorta.

been given (Evans, 1960). A transformation which yields a one-parameter family of solutions in which the wave speed and the tube diameter vary together is described by McMahon et al (1969).

Perhaps the simplest treatment is that employing the spherical wave equation as suggested by Thomas Young (1808, 1809) for a straight conical taper. Here the velocity potential will have the well-known form

(12) $$\phi = \frac{1}{R} f\left(t \pm \frac{R}{c}\right)$$

where R is the distance from the virtual apex of the aorta. Figure 9 shows a calculation on this basis which illustrates the transformation between pressure and flow effected by a conically tapered elastic tube. Here we have assumed that a quantity $\Delta Q = 10$ cc is withdrawn from the narrow (6 mm) end of the conical tube, following the pressure and velocity variations shown. At the upper, wider end of the tube the pressure pulse is much reduced, although the flow quantity temporarily withdrawn has increased to 40 cc.

Pulsatile pumping by direct withdrawal and injection through a small artery must evidently be restricted to certain levels of pressure and velocity in order to avoid blood damage. The flow velocity is in any case limited to values less than the pulse wave velocity. In the balloon pump the fluid withdrawn can be a light gas such as helium, enabling greater volumes to be

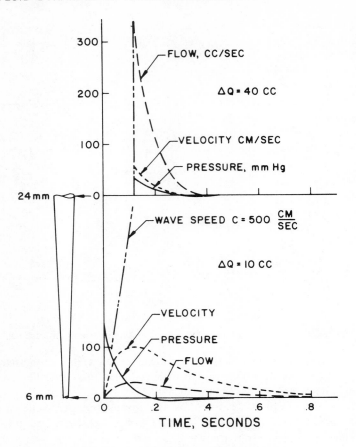

Fig. 9: Transmission of pulse in tapered artery.

pumped. The balloon, however, has its own peculiar instability which we have termed "bubble trapping." The origin of this phenomenon becomes evident when we attempt to calculate the flow and pressure produced by an expanding cylindrical balloon in a circular tube. For this purpose we use the potential

(13) $$\phi = A\left(x^2 - \frac{1}{2}r^2\right) + B \log r$$

which satisfies the boundary condition for a cylinder expanding within a fixed cylinder of radius $r_o = \sqrt{B/A}$. Calculation shows a parabolic pressure distribution along the length of the expanding cylinder with pressures at the center as much as 100 mm higher than the pressures at the ends. An ordinary limp balloon cannot, of course, support such pressure gradients, and the inflating gas simply rushes toward the ends of the balloon inflating them first and sealing off the aorta. As Professor Shapiro of M. I. T. has pointed out, any manner of expelling blood from the space around the balloon must involve such a falling pressure gradient and will lead to bubble trapping unless corrective measures are taken.

Figure 10 shows the bubble trapping phenomenon as it appears in our Avco circulation model. The blood trapped between the ends of the balloon is, of course, not effective in the pumping cycle, and the aorta in this region may be subjected to high pressures of the order of the drive pressure.

One method we have found successful in limiting this phenomenon is to divide the balloon along its length into a number of compartments, inflating the compartments selectively through orifices of different size. Figure 11

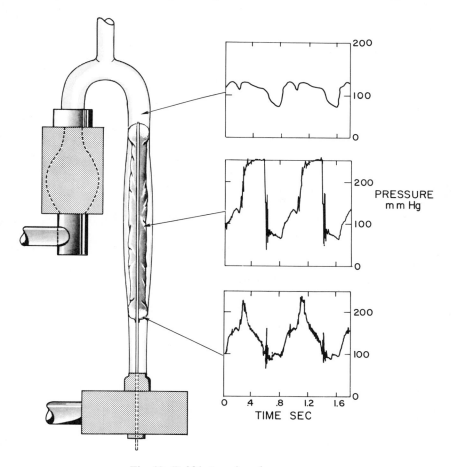

Fig. 10: Bubble trapping phenomenon.

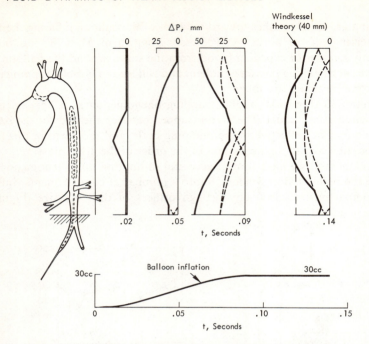

Fig. 11: Calculated pressures with controlled balloon inflation.

shows pressures calculated for such a controlled balloon in an elastic artery. Here it has been assumed that the pulse wave velocity is not modified by the presence of the balloon, an assumption permissable only if uncontrolled movements of the gas along the length of the balloon are effectively impeded.

Acknowledgment. *The work described herein was supported in part by the U.S. Air Force Office of Scientific Research. This paper was presented at a Specialists Meeting on "Fluid Dynamics of Blood Circulation and Respiratory Flow," May 4–6, 1970 in Naples, Italy, under the sponsorship of the Fluid Dynamics Panel of the AGARD, NATO, the Advisory Group for Aerospace Research and Development.*

REFERENCES

Anliker, M. and J. A. Maxwell 1966. The dispersion of waves in blood vessels. Biomechan. Sym. ASME, New York.

Birtwell, W. C., M. S. Soroff, M. Wall, A. Bisgerg, M. J. Levine and R. A. Deterling 1962. Assisted circulation-improved method for counterpulsation. Trans. Amer. Soc., Artif. Int. Organs. 8: 35.

Clauss, R. H., W. C. Birtwell, S. Albertal Lunzer, W. T. Taylor, A. F. Fosberg and D. E. Harken 1961. Assisted circulation: I. arterial counterpulsator. J. Thoracic and Cardiovas. Surg. 41: 447.

Dennis, C., J. R. Moreno, D. P. Hall, C. Grosz, S. M. Ross, S. A. Wesolowski and A. Senning 1963. Studies on external counterpulsation as a potential measure for acute left heart failure. Trans. Amer. Soc. Artif. Int. Organs 9: 186.

Dorman, F., E. F. Bernstein, P. L. Blackshear, R. Sovilj and D. R. Scott 1969. Progress in the design of a centrifugal cardiac assist pump with trans-cutaneous energy transmission by magnetic coupling. Trans. Amer. Soc. Artif. Int. Organs 15:

Evans, R. L. 1960 Nature, 186–290.

Gradel, F., T. Akutsu, P. A. Chaptal and A. Kantrowitz 1965. Successful hemodynamic results with a new U-shaped auxiliary ventricle. Trans. Amer. Soc. Artif. Int. Organs.

Jones, R. T. 1969. Blood Flow. Ann. Rev. Fluid Mechanics Vol. 1. Annual Rev. Inc., Palo Alto, Calif.

Jones, R. T., H. E. Petschek and A. R. Kantrowitz 1968. Elementary theory of synchronous arterio-arterial blood pumps. Med. & Biol. Eng. 6: 303–308.

Kantrowitz, A. and A. Kantrowitz, 1953. Experimental augmentation of coronary flow by retardation of the arterial pressure pulse. Surgery 34: No. 4: 678–87.

Liotta, D., C. W. Hall, D. A. Cooley and M. E. DeBakey 1964 Prolonged ventricular bypass with intrathoracic pumps. Trans. Amer. Soc. Artif. Int. Organs, 10: 154.

McDonald, D. 1960. Blood Flow in Arteries. Williams and Wilkins Co., Baltimore.

McMahon, T. A., V. S. Murthy, C. Clark, M. Y. Jaffrin and A. H. Shapiro 1969. The dynamics and fluid mechanics of the intra-aortic balloon heart assist device. M. I. T. Fluid Mech. Lab. Pub. No. 69-11.

Moulopoulos, S. D., S. Topaz and W. J. Kolff 1962. Diastolic balloon pumping (with carbon dioxide) in the aorta: A mechanical assistance to the failing circulation. Amer. Heart J., 63: 669.

Taylor, M. G. 1965. Wave travel in arteries and the design of the cardiovascular system. *In* E. Attinger (ed.), Pulsatile Blood Flow. McGraw-Hill Book Company, New York.

Womersley, J. R. 1956. Mathematical analysis of the arterial circulation. Wright Air Develop. Center Tech. Rept. WADC TR 56-614. Dayton, Ohio.

Young, T. 1808. Hydraulic investigations. Subservient to an intended Croonian lecture on the motion of the blood. Phil. Trans. Roy. Soc. (London) 98: 164–86.

Young, T. 1809. On the functions of the heart and arteries. The Croonian Lecture. Philos. Trans. Roy. Soc., London. M1355, 99: 1–31.

22

Biomechanical Problems Related To Vehicle Impact

NICHOLAS PERRONE

I. INTRODUCTION

The advent of the nonrail surface vehicle (mostly motor vehicles) in this century has had a dramatic effect on Western society. One might look upon this class of vehicles as a new and significantly different element in our environment that produces largely, but not entirely, beneficial effects. In excess of 100 million vehicles travel U.S. roads and highways at this time, and their number is increasing constantly.

Unlike the situation with pesticides and pollutants, the negative effects of vehicles on our ecology are clearly visible without waiting for time to pass to assess the damage. We need only tally up the fatalities and the seriously injured occurring daily in this country and others in the Western world. The dollar and pain damage is truly staggering. Yet, the effect of the vehicle on our environment has somehow not been noticed nearly as much as other factors.

It is indeed a regrettable situation in view of the fact that much of the damage that vehicles can inflict could be minimized by good engineering design. The wide margin between current vehicle and highway design and what is attainable, in the judgment of this writer, will not be bridged until much clearer recognition of the problem is acknowledged at a national level and greatly increased numbers of dollars and people are invested to correct the difficulty.

To be sure, there is no magic solution, no panacea that will immediately solve the problem of survival of vehicle impact. Research has to be conducted along many fronts concurrently, ranging from behavioral science questions that bear on the problem of the drinking driver, to design systems that will mitigate the terrible effects of impact via various energy absorbing devices.

Foremost among the problems that manifest themselves are those related to the biomechanical aspects of the impact situation. The essence of the design problem is to introduce or devise systems such that when an impact occurs, as inevitably is the case, the trauma attendant with a given a level of impact will be minimized or perhaps even completely negated.

In the next section a statistical profile of the entire vehicle impact problem is given. In the subsequent section biomechanical problems are examined in detail. Energy absorbing systems, or what is commonly referred to as crashworthiness design, are considered next. The air bag or inflatable restraint system as well as the problem of aircraft crashworthiness are also treated in this section. Discussions are presented in the final section.

II. STATISTICAL PROFILE OF THE AUTOMOBILE IMPACT PROBLEM

The incidence of accidental deaths jolts even the most complacent individual: in the United States accidents are the leading cause of death for the age group up to 44 years (Anon., 1969).

Some perspective of the statistical significance of accidents during much of this century is shown in Fig. 1 (McFarland, 1969). It should be noted that

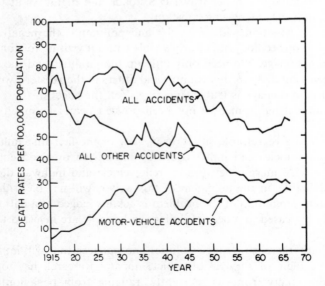

Fig. 1: Accidental death rates. From McFarland (1969) with permission of the author.

the ordinate indicates death rate per 100,000 population so that the effect of population growth is essentially nondimensionalized-out.

The trend for nonmotor vehicle accidents is very clearly downward during the five-decade spectrum shown. To be specific, the death rate has approximately halved; this is probably due to the increase in capabilities in the field of medicine and related technologies. For example, with better communication we can now more quickly suggest an antidote when someone suffers an accidental poisoning. With motor vehicles, however, it is quite clear that from approximately 1930 to date, the death rate level has held virtually constant. When we stop to consider that the population has been growing at a healthy rate during this period, it is obvious that the fatalities have also grown steadily.

In calendar year 1969 in the United States, approximately 60,000 fatalities were reported due to vehicle-related accidents. At the present point in time, approximately half of all accidental deaths are attributed to vehicle-related causes. The danger of accidents to the American public was implicitly underlined three years ago when the President appointed a special Commission on Product Safety at the behest of Congress; part of their reports are just being released (Heffron, 1970).

Two of the important resources that a nation has are its people and the time they can usefully and constructively devote to contributing in their collective way to the growing gross national product. In this connection, vehicle impact like no other "disease" takes a gargantuan bite out of our nation's people resources. For the 1965 calendar year, a careful examination was conducted of the age distribution of fatality victims of motor vehicle accidents (McFarland, 1969). Figure 2 is reproduced from that study. Clearly, the median age of the fatality victims is about 20 years. The ordinate of Fig. 2 represents the estimated number of years lost assuming normal lifetime expectancy. Figure 2 concentrates only on fatalities and still demonstrates a loss of 2 million man-years during 1965.

When one considers that injuries are of the order of 3 million a year and very serious debilitating injuries of the paralytic or irrevocably serious type ("economic deaths") probably number at least one quarter of a million, then the total number of man-years lost would surely reach the level of some 10 million or more. If we value a man-year as worth approximately $5000 in GNP, then the *net loss to the country* in current and future GNP *is approximately 50 billion dollars per year*. Of course, the pain and suffering of the victims and their families cannot be quantified but are just as important, if not much more so, than economic aspects.

It is noteworthy that a recently published study showed that families of fatality victims of the 1967 calendar year recovered approximately half of the actual losses suffered in accidents. Many other immediate economic consequences were felt such as moving to cheaper housing, borrowing money,

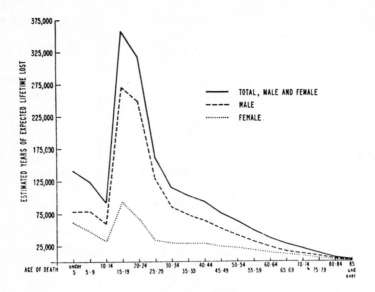

Fig. 2: Distribution by age and sex of estimated 1,806,000 years of expected life lost in motor vehicle accidents in the U.S. during 1965. (Based on data from Vital Statistics of U.S., 1965. Total estimated years of life lost during 1965: 1,806,095 by 49,163 fatalities.) From McFarland (1969) with permission of the author.

mothers going to work, and so forth (Anon., April, 1970). The cost suffered by the families of fatality victims for the 1967 calendar year was estimated at some 5 billion dollars.

To gain some appreciation of the impact speed for various accidents that occur throughout the country in any given year, Fig. 3 is reproduced from the results of Kihlberg (1965) which has been portrayed as shown in Ryan and Mackay (1969). From Fig. 3 it should be clear that approximately 80% of the accidents occur above 30 miles per hour at time of impact. In Ryan and Mackay, (1969) it was shown that the velocities at impact of accidents occurring in Australia and England are noticeably less than those occurring in the United States. This is not surprising nor unreasonable when we consider the much larger number of superhighways and high speed roadways in the United States.

It is worth noting that the kinetic energy of a 4000-lb vehicle increases appreciably (as the square of the velocity) with speed as shown in Fig. 4. Obviously, from Figs. 3 and 4, hundreds of thousands of foot pounds of kinetic energy must be absorbed in the vast majority of motor vehicle accidents (more than 90%).

Again reference is made to the data from Kihlberg (1965) to obtain an idea of the portion of the vehicle involved in accidents as well as the damage

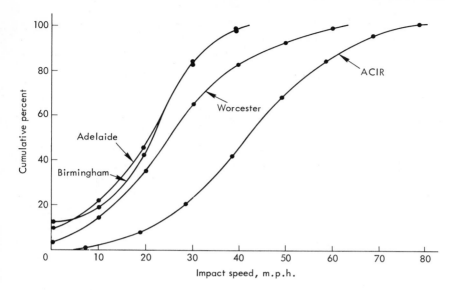

Fig. 3: Distribution of speed at impact by 10 mph intervals. From Ryan and Mackay (1969) with permission of the authors.

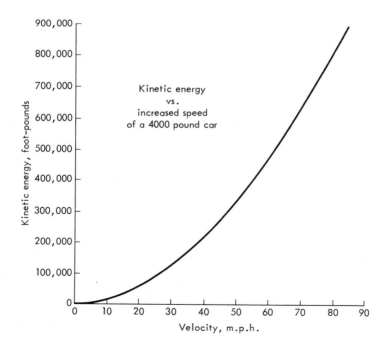

Fig. 4: Kinetic energy *vs* increased speed of a 4000 pound car.

level that occurs. This data is presented in Fig. 5. Covering half the total, frontal impact is clearly the dominant accident mechanism, with roll-over and side impact the next two closest categories. An examination of accident statistics reveals that in more than 95% of the cases vehicle damage level is moderate to very severe.

Let us focus next on the most critical and, essentially, the pay-off question, people damage. In Fig. 6 we see again a two-part picture showing in one half that portion of the vehicle causing injury and in the second half the part of the body that is injured. A quick perusal of the second half suggests the cumulative percentage of all injuries exceeds 100%, the reason being that in any given accident one frequently finds injuries to multiple portions of the body.

The portion of the vehicle that causes injury is very much a function of

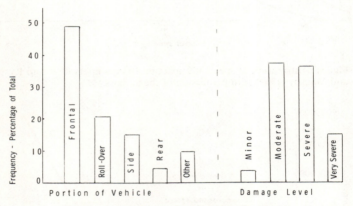

Fig. 5: Area and severity of vehicle damage.

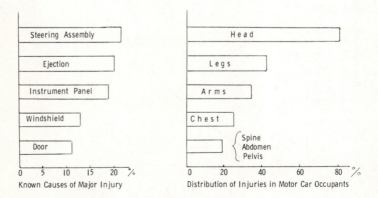

Fig. 6: Frequency of that portion of the vehicle causing injury and of the body that is injured.

time and place. For example, the data base for the distribution shown in Fig. 6 is for the mid-1960s in the United States (Kihlberg, 1965). For the same period of time the distribution may be somewhat different for other countries (Ryan and Mackay, 1969). Moreover, even within the same country, for example the United States, the situation today is certainly significantly different from what it was in the mid-60's when this data was collected. Ejection was shown to be a significant cause of injuries, but today we have much better latch systems and higher penetration velocity windshields. In addition, most cars are now equipped with collapsible steering columns which have been shown, from preliminary studies, to be very helpful in minimizing injury (Huelke and Chewning, 1969).

As far as fatalities are concerned, the critical portion of the body involved is very obvious. In approximately three-quarters of the cases where a fatality occurs, head injury is the cause. The next leading cause is chest injuries, and this is a very weak second covering about 5% of the cases.

In Fig. 6 the frequencies of injuries for various parts of the automobile are somewhat misleading. Daniel and Patrick (1965) have taken the Cornell data and manipulated it on a weighted basis in such a manner that a fatal injury would count twice as much as a dangerous injury, which in turn would be counted twice as much as a nondangerous injury which, in turn, would be twice as significant as a minor injury. With this weighting, the instrument panel and steering assembly appear to be the two most critical components followed by ejection, the windshield, and door structure in that order (Fig. 7).

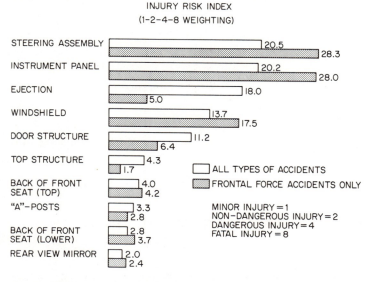

Fig. 7: ACIR rating of vehicle components in order of injury production based on the 1-2-4-8 weighting sequence.

The collapsible steering column mentioned above and the new high penetration velocity windshields (Fargo, 1968), which have come on the scene since the data of Fig. 7 was gathered, should both be instrumental in causing significantly less damage than had been the case in the past. *It appears likely, therefore, that the instrument panel is now the leading cause of serious injury, on a weighted basis.*

III. BIOMECHANICAL PROBLEMS

One of the primary problems in biomechanical research connected with vehicle impact is quantifying injury criteria in a meaningful manner. It was mentioned previously that head injury is the number one problem, certainly with respect to fatalities and injuries. Yet, to date we are incredibly far removed from a well-proven and precise design criterion that could be presented to the structural or safety engineer.

Much data on human volunteers and animals have been accumulated both by the Air Force group at Holloman Air Force Base and at Wayne State University. A so-called Wayne State University Tolerance Curve was suggested as is shown in Fig. 8 (Gurdjian et al, 1962). The ordinate is the deceleration level (or average deceleration level) and the abscissa displays the pulse time. It should be intuitively obvious as the figure indicates that the higher the deceleration level the shorter the pulse length and *vice versa*.

If we assumed the area under the force-time curve were constant for

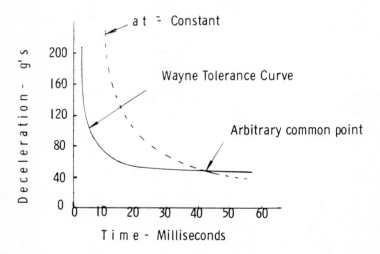

Fig. 8: Comparison of Wayne State Univ. Tolerance Curve with constant pulse area criterion. From Gurdjian et al (1962) with permission of the authors and the publisher.

injury initiation, this would suggest that the product of deceleration level times time is a constant as shown in Eq. (1) and would result in a tolerance curve which would be a perfect hyperbola.

(1) $$a\,t = \text{constant}$$

Superimposed as a dashed line on Fig. 8 is Eq. (1) wherein the constant is selected so that the curve passes through the suggested tolerance curve at the point noted.

The critical question in the selection of an appropriate criterion or tolerance curve is as follows: when the head impacts an object, be it the windshield, instrument panel, door post, or whatever the object may be, what impulse level could be tolerated without permanent brain damage or skull fracture? The "second collision" has been characterized as the one in which the person strikes the interior of the car, but in fact the critical collision might be termed the "third collision" in which the brain tissue sitting in the skull cavity sloshes to a halt.

What is perhaps the next best mathematical model (for engineering purposes) to describe the possibility of incipient head injury has been descibed by Gadd (1966). He reasoned as follows: if the deceleration-time pulse were the only determining factor and the impulse associated with the injury level were approximately constant, then Eq. (1) would be valid and could be plotted on a log-log curve as a 45° line, as shown in Fig. 9. Using all the data that was available, Gadd found that the line should not have a 45° slope but rather a 1:2.5 ratio as shown by the dash line in Fig. 9. For a square pulse the locus of points satisfying this criterion are given by

(2) $$a^n\,t = \text{constant}$$

where $n = 2.5$

Should the pulse have a varying amplitude, a minor modification of Eq. (2) results in the following:

(3) $$\int a^n\,dt = \text{Severity Index}$$

The constant is given the name Severity Index. If the deceleration is given in g's, then this impulse criterion is devised in such a way that a value of approximately *1000* units *is the borderline between fatal and nonfatal impacts.*

The Gadd criterion is probably slightly more preferable than the Wayne State Tolerance Curve in Fig. 8 in that it takes into account the size and shape of the pulse more readily and yet still retains a fair degree of flexibility. In a recent study Cook et al (1969) carried out a series of tests on the autos of accident victims who had been impaled against one side of the windshield of

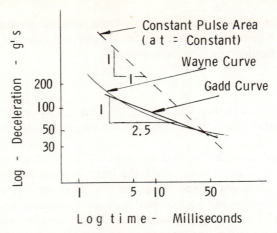

Fig. 9: Comparison of the Wayne State Univ., Gadd and Constant-Pulse-Area Curves. From Gadd (1968) with permission of the author.

their respective cars (Cook et al, 1969). They subsequently used headforms to impact the unbroken side of the windshield to simulate the actual accident situation and calculated the associated Severity Index from the test. The indices ranged between 175–250 which are well below the 1000 fatal threshold level. All of the two dozen or so victims involved received no permanent brain damage from their injuries.

While the Gadd approach does provide a reasonable semi-empirical criterion for head injuries, it still does not encompass the essential question of what is the fundamental mechanism that precipitates brain damage. Some interesting studies have been made and are continuing in this direction (Unterharnscheidt and Sellier, 1966). Elsewhere in this volume Werner Goldsmith discusses certain aspects of this question in further depth.

Various theories have been proposed including the so-called cavitation hypothesis, shock-wave hypothesis, and finally the shear-stress damage at the brain stem (where all the nerve fibers mushroom into the brain tissue). The last-named theory is the most appealing to the author. Further animal experiments as well as careful study of data associated with actual human accidents are indispensable in trying to determine the actual brain damage mechanisms. This understanding would certainly not be of academic interest as it would simultaneously provide a basis on which clinical practice could develop better therapeutic procedures for treatment, as well as providing design engineers with a reasonable performance set of specifications for structural configurations of possible impact targets (for example, an instrument panel or windshield).

A very fine survey paper which examines the concussion mechanism in dogs, cats, and monkeys should be noted (Hodgson et al, 1969). Apparently, in the best judgment of the Wayne State University group, shear strain at the brain stem is one of the primary causes of brain damage.

Realistic mathematical models governing head response, including the brain, are yet to be developed. With the advances made to date in numerical analysis of complex structural and structure–fluid systems, it should now be possible to at least attempt more sophisticated mathematical models of the skull-brain system. An accurate representation of the skull structure itself should not be a conceptually overwhelming job, although it could be somewhat tedious. When subjected to a blow, the skull structure deforms rapidly, in other terms, at high strain rates. During such a situation the physical properties of the skull material can vary significantly (McElhaney, 1966). Conceivably, without getting into the details of the highly nonlinear variation of the physical properties with strain rate, one could make engineering estimates that would account for these effects (Perrone, 1965).

Inclusion of the brain tissue inside the skull cavity could be accomplished by treating it as a fluid as was done in the past (Engin 1969) or by considering its physical properties as a viscoelastic material (Galford and McElhaney, 1969). In this regard, the effect of the brain stem, which has different properties than the bulk of the brain tissue, must be considered carefully and separately in order to obtain a good estimate of the shear strains that build up in this region during impact. In addition, the possibility of brain tissue being released through the foramen magnum must also be included within the mathematical model.

Other problems related to the brain damage question but which are, in fact, more applied in nature and not yet treated include the dynamic interaction of a skull structure with various components within the interior of vehicles, including, for example, the following: steering wheel, instrument panel, windshield or side window, main frame elements, and energy absorbing systems. These problem areas would require a fair amount of effort to evolve meaningful or applicable type solutions. In all these problems the plastic properties of the materials involved would be fundamental in the analysis to be performed. For example, for the interaction of a skull with a wheel assembly, a limit analysis of the collapse of a circular beam with out-of-plane loading would be a reasonable first approximation. The problem of impact with a curved windshield could be quite complex when one considers the very large deformations possible. A membrane-shell type of analysis would appear to be applicable.

An area of inquiry that would have application to head injury as well as impact with other parts of the body is an analytical and an experimental investigation of lacerative damage to surface tissue due to different impact levels on various shaped objects. Although a lacerative damage injury scale has been suggested (Cook et al, 1969), the designer could not normally gauge the lacerative potential of various objects which may be impacted by different parts of the human system without extensive cadaver or simulated headform type tests.

The thorax or chest region requires much more research starting with the static strength capabilities of the thorax. Many very useful experiments have been carried out containing data that suggest meaningful criterion on maximum dynamic loads that could be applied to the chest before causing fracture (Patrick and Mertz, 1970; Nahum et al, 1970). Recently, some improved mathematical models have been suggested and are being refined (Roberts and Chen, 1970).

The interaction of human bodies with the various restraint systems requires a great deal of further research effort. These restraint systems might consist of seat belts of the lap or shoulder harness type or of the so-called inflatable or air bag systems. With seat belts the essential problem is one of the physical properties and dimensions of the belt material needed to produce an optimal response of the occupant.

Problems with air bag systems are even more difficult in that the number of motions possible by a constrained person are virtually infinite. The interaction of a human subject with an inflating bag is one that requires an enormous amount of attention because there is no doubt that in the future, that is the next 3 or 4 years, all cars will be equipped with these systems (Anon., June 8, 1970). It is critically important that designers and developers of such air bag systems have the engineering tools at their disposal to make sensible judgments in the design process.

Kinematic and kinetic mathematical models of the response of human systems to various air bag systems and sizes will contain complexities such as direction of impact, size of occupant, time of inflation, as well as many other incidental factors. When developing computer based mathematical models of possible occupant motion, statistical aspects should be incorporated so that one could deduce what is the best type of system that would provide the greatest good for the bulk of the population (Segal and McHenry, 1968).

Another biomechanical problem directly caused by the air bag is the possibility of inner ear damage due to the high noise level at deployment (approximately 175 db). Although tests with normally healthy adult volunteers showed toleration of the "explosion" with no prolonged effects, it has been speculated by industrial representatives that the elderly and the young may not survive as readily (Anon., June 24–25, 1970).

Most of the attention to head injury in motor vehicle accidents has focused on frontal impact. Certainly all the animal testing work and tolerance curves apply to a frontal orientation. With air bag systems now on the verge of massive deployment, we could expect the relative incidence of frontal impact injuries to decrease but not side impact. Air bags will probably not offer very much protection when the collision force has its dominant component in the transverse direction. For these cases the parietal portions of the skull could strike against side windows, roof pillars, as well as the heads of other occupants.

Improved and more sophisticated mathematical models of the skull in concert with careful animal testing and human accident analysis should be most helpful in establishing tolerance levels (or Severity Indices) for non-frontal impact.

IV. ENERGY ABSORBING SYSTEMS

As pointed out by Stapp (1965) the automobile designers have been exerting most of their energies toward insulating occupants from vertical motions, that is making the ride increasingly smooth. Virtually no attention has been paid to the problem of decoupling the occupants from horizontal motion such as occurs during inadvertent impact. Undoubtedly, the vast majority of injury producing accidents that do occur may be traced to human rather than vehicle failure. Good design, however, must recognize this reality and properly prepare for a statistically significant number of occurrences of accidental impact that may take place during the life of a vehicle. Hence, the obvious need for energy absorbing systems.

As implied in the foregoing paragraph, the state of the art in design for energy absorbing structures is at least an order of magnitude behind the corresponding state regarding the rest of the vehicle including, for example, dynamic linear response due to vertical shock (road bumps), propulsive efficiency (excluding pollution), and other factors.

In vehicle-related industries (automobile and aircraft) the term *crashworthiness* is used to denote good energy absorbing capabilities during an impact situation. We may readily differentiate between *external* and *internal crashworthiness*. For example, with respect to the automobile during an impact situation, the front or side of the structure deforms in response to the collision and absorbs a certain amount of energy. This is categorized as external energy. During the second collision, the occupant strikes a portion of the interior of the vehicle which, if well-designed, absorbs a significant amount of the occupant kinetic energy. The environment immediately enclosing the occupant, with which he may make contact during the second collision, is referred to here as internally energy absorbing or alternatively, internally crashworthy.

The air bag is a relatively new element which obviously falls into the category of an internally energy absorbing device. During a collision a sensor detects the deceleration and immediately causes a plastic air bag to be deployed, making use of a resevoir of high pressure stored gas. The bag is fully inflated before the occupant begins moving forward, and his kinetic energy is dissipated by the venting of air from the bag as he deflates it.

In the judgment of the author, air bags will be fully operational on all new vehicles in the United States as standard equipment within about 4–5 years. Most but not all of the internal crashworthiness problems will be taken care of by these inflatable systems. For example, during side impact there will be a strong tendency, even with air bags, to have the side of the occupants head hit against door posts or side windows. In addition, rear-seated occupants would strike against air bags mounted in the back of the front seat and the force that would be exerted by them on the air bags would be transferred to the seat structure so that the front seat should be greatly strengthened to account for high force levels. [In fact, current seat designs are very much inadequate without air bags (Severy, et al, 1969).]

External crashworthiness aspects would require a concerted and continuous examination over a period of many years, independent of the question of whether there will be massive deployment of air bags on all vehicles. At present, automobiles have very inadequate external crashworthiness capabilities which should be significantly upgraded by of the order of 200,000 ft lbs more of energy absorbing capability. (A 3000-lb vehicle with a velocity of 30 miles per hour has an energy level of 90,000 ft lbs.) About 200 lbs of added metal is necessary to acquire this external crashworthiness capability.

Even without air bags a noticeable increase in energy absorbing material is necessary within the automobile. Instrument panels, hard internal mainframe elements such as windshield headers, door posts and roof frame elements and, as mentioned earlier, the front seat all require much improved crashworthiness properties. One has very little distance to work with for internal crashworthiness problems so that it is necessary to be very efficient and judicious when selecting materials and/or configurations. Although the use of plastic foams and padding has been widely observed, one cannot expect that these materials do any more than distribute the loading over a wider portion of the face or whatever part of the body is involved. A significant portion of structural material, preferably a metal, should be placed in series with the padding to absorb the amount of energy necessary in the short space alloted (Swearingen, 1966).

Another overriding factor in much of automobile design is cost, as millions of units are produced necessitating a close control of expense. A structural arrangement which permits a wide latitude of action and yet is still relatively inexpensive is the thin-walled tube (Fig. 10a) which is loaded transversly (Perrone, 1970). When collapse occurs, as shown in Fig. 10b, four well-defined yield hinges form at 90° from the loaded point (including the load point itself).

The problem of crashworthiness is not confined entirely to the automobile. In the area of general aviation, that is relatively light planes, the incidence of fatalities is approaching dangerous levels numbering close to 2000 per year at present (Bruce and Draper, 1969). A detailed discussion of the

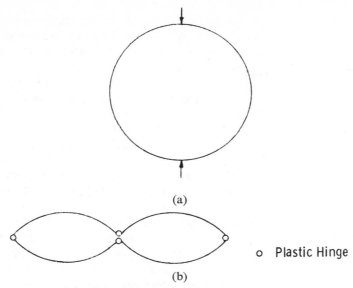

Fig. 10: (a) Tube with diametrically opposite loading. (b) Completely collapsed tube with plastic hinge lines.

entire problem of air crashworthiness can be found in an excellent Army publication (Turnbow et al, 1967). For a large fraction of the fatalities that occur in aviation accidents, the air-frame retains its structural integrity with perhaps minor deformation, and yet severe or fatal injury results. It has been observed that in the vast majority of general aviation accidents the velocity at impact (or skid landing) ranges from 20–75 miles per hour with the glide angles of less than 45° (Turnbow et al, 1967).

The occupant area for general aviation aircraft is probably more poorly designed than the automobile. When one observes the instrument panel being simply a solid flat plate with holes punched therein for instrumentation, one realizes that this is not the ideal structural element against which to strike a head at 30 or 40 miles an hour. In addition, many hard structural elements are openly exposed as, in fact, is the case with the automobile.

V. DISCUSSION

In this paper the many facets of the problem of vehicle impact are reviewed. Estimates of the cost of vehicle impact from a national viewpoint are made. Also, a detailed review of body injury and vehicle damage is presented. Special attention is focused on biomechanical aspects and numerous problems enumerated where further research efforts are required. The main mechanism of inhibiting or diminishing body injury from vehicle impact, namely energy absorbing systems, are discussed at some length. In this field as in perhaps few others, results obtained in research could very quickly find their way into immediate practical applications.

One of the many difficult aspects of the field are that the many recent safety innovations are causing helpful but confusing changes in the nature of the vehicle-related body damage that arises during impact. To assess the success of a given innovation frequently requires a number of years of painstaking and laborious data analysis from selected field accident cases.

Despite this difficulty of data gathering to assess system effectiveness, we could still make reasonable guesses as to the potential effectiveness of incipient systems and the way in which they would change the knowledge void spectrum. For example, air bags or inflatable restraint systems will surely become operational on all new cars in the mid-1970's. These will certainly change the relative importance of a number of biomechanical problems. Hitherto, frontal impact has been the dominant question, and associated skull impact has been a primary focus of attention. With working air bag systems it is likely that other directions of impact will occur with increasing frequency.

The question of developing a mathematical model to determine the fundamental injury mechanism within the brain during a blow still presents itself, but an oblique rather than a frontal direction should be the one of greater interest. What *is* the Severity Index for side impact (parietal)? Three-dimensional kinematic models are needed to provide inputs to air bag designers so they may anticipate possible motions of the driver and other occupants, considering fully size variations within the population.

Problems of immediate interest even before full deployment of air bag systems include impact of various body components against internal elements within the vehicle such as steering wheel, windshield, instrument panel, door post, roof and other main-frame elements. Solutions to these problems should be directly helpful in selecting and designing energy absorbing systems.

In summary, the vehicle impact problem presents a huge number of challenging and socially meaningful problems.

REFERENCES

Anon. 1969. A survey of current head injury research. A rep. by Sub-Committee on head injury of the Natl. Advisory Neurol. Diseases and Stroke Council for Natl. Inst. Health.
Anon. 1970. Economic consequences of automobile accident injuries. Rep. for Depart. Transp. by Westat Res. Corp.
Anon. 1970. Public Meeting on "Occupant Crash Protection," sponsored by Natl. Highway Safety Bur. Depart. Commerce, Washington, D.C.
Anon. 1970. Washington tells Detroit: Cure auto accidents now. Product Eng. pp. 11–14.
Bruce, J. and J. Draper 1970. Crash safety in general aviation aircraft. Rep. by 1969 Nadar Summer Group.

Cook, L. M., R. G. Rieser, A. W. Segal and A. M. Nahum 1969. Correlation between windshield head injuries and laboratory tests—Part I: Feasibility of relating headform impacts to clinical head injuries. Proc. 13th STAPP Car Crash Conf. pp. 167–90.

Daniel, R. P. and L. M. Patrick 1965. Instrument panel impact study. Proc. 9th STAPP Car Crash Conf.

Engin, A. E. 1969. Axisymmetric response of a fluid-filled spherical shell to a local radial impulse —A model for head injury. Paper No. 69-BHF-1. Presented Biomechan. Human Factors Conf. Ann Arbor, Michigan.

Fargo, R. D. 1968. Windshield glazing as an injury factor in automobile accidents. Cornell Aeronautical Lab. Rep. No. VJ-1823-R25.

Gadd, C. W. 1966. Use of a weighted impulse criterion for estimating injury hazard. Proc. 10th STAPP Car Crash Conf. pp. 95–100.

Galford, J. E. and J. H. McElhaney 1969. Some viscoelastic properties of scalp, brain, and dura. Paper No. 69-BHF-7. Presented Biomechan. Human Factors Conf. Ann Arbor, Michigan.

Gurdjian, E. S., H. R. Lissner and L. M. Patrick 1962. Protection of the head and neck in sports. J. Amer. Med. Assoc.

Heffron, H. A. 1970. Federal consumer safety legislation. Spec. Rep. for Natl. Comm. Product Safety.

Hodgson, V. R., L. M. Thomas, E. S. Gurdijian, O. U. Fernando, S. W. Greenberg and J. L. Chason 1969. Advances in understanding of experimental concussion mechanisms. Proc. 13th STAPP Car Crash Conf. pp. 18–37.

Huelke, D. F. and W. A. Chewning 1969. Accident investigations of the performance characteristics of energy absorbing steering columns. S. A. E. Paper No. 690184.

Kihlberg, J. K. 1965. Driver and his right front passenger in automobile accidents. Cornell Aeronautical Lab. Rep. No. VJ-1823-R16.

McElhaney, J. H. 1966. Dynamic response of bone and muscle tissue. J. Appl. Physiol. 21 (4): 1231–236.

McFarland, R. A. 1969. Significant trends in human factor research on motor vehicle accidents. Proc. 13th STAPP Car Crash Conf. pp. 1–17.

Nahum, A. M., C. W. Gadd, D. C. Schneider and C. K. Kroell 1970. Deflection of the human thorax under sternal impact. S. A. E. paper No. 700400. Proc. 1970 Int. Automobile Safety Conf. pp. 797–807.

Patrick, L. M. and H. J. Mertz 1970. Human tolerance to impact. S.tA. E. Manual No. 700195. Human Anatomy in Impact Injuries and Human Tolerances. pp. 90–103.

Perrone, N. 1965. On a simplified method for solving the impulsively loaded structures of rate-sensitive materials. J. Appl. Mechan.

Perrone, N. 1970. A new approach to impact attenuation. Tech. Rep. No. 15. Catholic Univ. Amer. Washington, D.C.

Roberts, S. B. and P. H. Chen 1970. Elastostatic analysis of human thoracic skeleton. Paper No. 70-BHF-2. Presented ASME Biomechan. Human Factors Conf. Washington, D.C.

Ryan, G. A. and G. N. Mackay 1969. Comparisons of car crashes in three countries. Proc. 13th STAPP Car Crash Conf. pp. 336–52.

Segal, D. J. and R. R. McHenry 1968. Computer simulation of the automobile crash victim—Revision No. 1. Cornell Aeronautical Lab. Rep. No. BJ-2492-V-1.

Severy, D. M., H. M. Brink, R. D. Baird and D. M. Blaisdell 1969. Safer seat designs. Proc. 13th STAPP Car Crash Conf. pp. 314–35.

Stapp, J. P. 1965. Impact attenuation for occupants of motor vehicles. Proc. 9th STAPP Car Crash Conf.

Swearingen, J. J. 1966. Evaluation of various padding materials for crash protection. Office Aviation Med. Fed. Aviation Agy. Rep. AM 66-40.

Turnbow, J. W., B. F. Carrolo, J. L. Haley, Jr., W. H. Reed, S. H. Robertson and L. W. T. Winberg 1967. Crash survival design guide. U. S. Army Mat. Lab. Rep. 67-22.

Unterharnscheidt, F. and K. Sellier 1966. In W. F. Caveness and A. E. Walker (eds.), Closed Brain Injuries: Mechanics and Pathomorphology. Proc. Conf. Head Injury. J. B. Lippincott Co., Phila. pp. 321–41.

23

Biomechanics of Head Injury

WERNER GOLDSMITH

The enormous incidence of head injuries has focused increased scientific attention on this problem with particular emphasis on the domain of biomechanics since an understanding of the physical processes and material response involved is of vital importance in both diagnosis and treatment of such trauma and in the design and construction of protective environments. Head injuries are usually classified both according to the severity of the trauma and according to the tissues affected, that is, scalp, skull, dura, brain, blood and blood vessels, and the head-neck junction. For obvious reasons the most serious problem, brain damage, has received the widest attention from the phenomenological, pathogenic, corrective and preventive viewpoints. Injuries of this type are produced either by a collision of the head with another object—physically representing an impact load—with or without skull perforation, or alternatively by an abrupt change in motion of the head not produced by contact, constituting an impulsive loading situation as occurs in the case of hyperextension or "whiplash." The characteristics of collisions and the behavior of engineering materials resembling those of head tissues have been summarized from the viewpoint of application to cases of cranial trauma.

During the last three decades a variety of mechanisms have been proposed as the direct cause of brain dysfunction, each allegedly supported by a host of experimental, clinical and pathological evidence. Considerable controversy prevails concerning the applicability of one or more of these mechanisms to a particular head injury. In an effort to substantiate these hypotheses, numerous tests have been conducted on animals, notably the Rhesus monkey, cadaver heads and inanimate models, which have been suitably instrumented to provide physical data and, in the first case, physiological information, under collision and abrupt deceleration. Consideration has also been given to the problem of correlation of data from both live and dead animals and from the human cadaver and its extrapolation to the level of the living human. A brief account of this topic will also be presented.

Acceptable mathematical models of the head injury process have been developed only recently and consist either of spherically-symmetric deformable systems featuring two components and subjected to a radial load, or multidegree-of-freedom point mass systems whose response under impulsive loading conditions was analyzed on the computer. Concomitantly, static and dynamic mechanical properties have been determined for the various cranial tissues under a variety of loading conditions. In addition to its basic value, this information can be used for the construction of improved analytical and physical models that would permit the observation of a more realistic response under simulated head injury loading situations. The mechanical characteristics of the cerebral blood vessels and of the craniospinal junction are also under scrutiny. Some of the results obtained are included in the present survey.

Human tolerance has been estimated in a variety of ways, including the use of volunteers, animal and cadaver experiments, with the problem of scaling from the animal to the human brain playing a vital role in the delineation of threshold levels. The information has been applied to the design of protective structures. A description of this subject as well as some suggestions for future work in the field of the biomechanics of head injury concludes the presentation.

* * *

I. INTRODUCTION

Head injuries are clinically regarded to consist of only externally produced damage to the cranial vault, its coverings and contents, with special reference to the central nervous system. This definition will also be adopted here; thus, extracranial (or maxella-facial) trauma such as damage to the eyes, facial bones, teeth, nose and jaw will not be further considered. The causes leading to head injuries can be divided into three categories on the basis of the manner of load application: (1) An impact or blow involving a collision of the head with another solid object at an appreciable velocity, such as resulting from a bullet, falling brick, or interior vehicular structure under crash conditions, (2) An impulsive load inducing a sudden head motion without direct contact, such as produced by the effects transmitted through the head-neck junction upon sudden changes in the motion of the torso due to rapidly applied acceleration or deceleration, and (3) Static or quasi-static loading produced by the relatively long-time crushing of a heavy object. The three situations encompass different physical phenomena; consequently, their mathematical representations emphasize different loading and response factors. There is also considerable evidence that clinical effects observed in patients can be similarly distinguished according to the type of load that caused the damage.

The magnitude of the head injury problem can be gauged from some statistical data gathered during the mid-60's in the United States. Of the approximately 58 million persons injured per year, 3.5–3.8 million occur in motor vehicle accidents, and, in about 70% of these, the head is involved (Kihlberg, 1966, 1970). Many of these are minor or nondangerous, involving contusions, lacerations and even brief unconsciousness, yet of the 45,000–50,000 annual fatalities resulting from motor vehicle accidents, about 62%, or 30,000, can be directly attributed to head injury. The total number of fatalities in the country due to cranial trauma thus can be expected at a level of about 100,000 per year (Gurdjian et al, 1968). Obviously, strenuous efforts should be exerted to ameliorate the medical problems created by the head injury situation.

TABLE I
Interrelation of the Mechanical, Physiological and Clinical Aspects of Head Injury

Discipline Concerned	Agent, Object or Condition	
Epidemiology	Environment	→ Protection
Biomechanics	- - - Mechanical cause ↓ Head and its components ↓ Brain	and → Prevention
Physiology and Biophysics	→ Mechanical input ↓ Cells ↓ Biologic output; effect on cells and tissues	
Clinical Medicine	Organ malperformance from tissue distortion and mechanical damage ↓ Head injury patient ↓ Diagnosis and treatment; Clinical management — — —	

The role of biomechanics is to provide a better understanding of the cause and effects, the mechanisms and processes involved. Specifically, the principles of continuum mechanics and dynamics of discrete systems can be applied to mathematical models of the response of the head and its attachments to a given external stimulus. This could lead to a prediction of the history of the field parameters, such as stress, strain, displacement, acceleration, pressure and the motion of specified elements. The tools of experimental mechanics and material science can also be employed to measure physical changes produced by impact or impulsive loading in replicas of the human head, involving either animal specimens, cadaver skulls, or inanimate models. Tests of this type on living animals are most frequently accompanied by concurrent physiological measurements and subsequent pathologic examination for the purpose of correlation. Data from such experiments can serve to substantiate or disprove theoretical hypotheses concerning damage mechanisms and patterns, to establish tolerance levels and failure limits for various types of loading, to assess the efficacy of protective devices and environments, to provide statistical data for injury patterns, and to provide information concerning the basic mechanical properties of the components involved. The intimate relationship of the field of biomechanics to other disciplines concerned with the head injury problem is indicated in Table 1.

II. THE COMPONENTS AND GEOMETRY OF THE CRANIAL SYSTEM

A median (or mid-saggital) view of the cranial vault exposing the solid tissues potentially involved in head injury considerations is shown in Fig. 1; Fig. 2 presents a more detailed cross section of the cerebral coverings. The components of the system include the skull, the dura, the pia-arachnoid complex, the brain, the blood vessels, the cerebrospinal fluid (CSF) and the blood. The craniospinal junction must also be included inasmuch as its physical characteristics significantly affect the reponse of the head either to a direct blow or to accelerations induced in some other parts of the body. Other portions of the head and face can be omitted from further consideration by the definition of the term "head injury." The only other tissue of possible pertinence, hair, appears to have a negligible influence with respect to the production of the cranial trauma and will be disregarded.

The scalp, comprising five tissues overlaying the cranial bone with a thickness of about 1/4 to 1/2 in., has a leathery consistency and is under constant slight tension in the living human. In descending order, it consists of (1) skin with its hairy covering, (2) the layer tela subcutanea, loose fibrous connective tissue that binds the skin to the deeper structures, (3) the aponeurotic layer, a fibrous membrane representing very much flattened tendon,

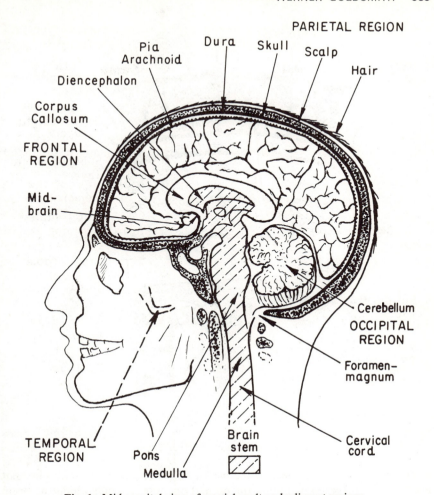

Fig. 1: Mid-saggital view of cranial vault and adjacent regions.

that connects the frontal and occipital muscles, (4) a very loose subaponeurotic layer of connective tissue, and (5) the pericranium, a tough vascular membrane with a somewhat looser zone, occasionally separately designated as the subpericranial layer, just covering the skull (Schaeffer, 1942). In the temporal region, the temporal muscle and its connective tissue are wedged in between the last two layers. The first three components are so interlaced by coarse fibers that they move as a unit; numerous blood vessels extend into the aponeurotic layer from below. Scalp is highly anisotropic and somewhat nonhomogeneous and may be described as fibers embedded in a matrix of fat.

The skull, virtually completely enclosing the brain except for the spinal cord opening (foramen magnum) is spheroidal in shape and consists of eight

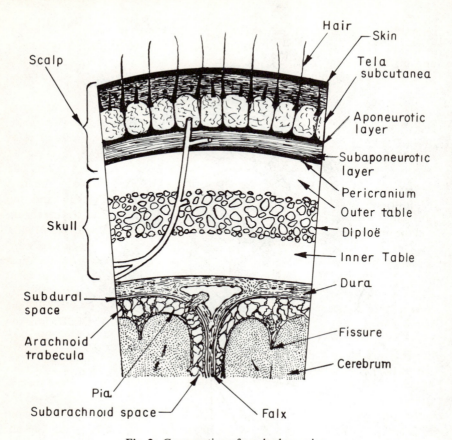

Fig. 2: Cross section of cerebral covering.

separate bones: the occipital at the back and lower part of the cranium that contains the foramen magnum, the cranial opening through which the brain stem connects with the spine; the frontal; two temporals, one at each side and at the skull base; two parietal bones, forming the sides and roof of the cavity; the sphenoid at the base of the skull; and the ethmoid at the anterior part of the cranial base. These bones are connected by sutures, but are immobile relative to one another; furthermore, the sutures calcify rapidly with age. The base of the skull, formed by the occipital, temporal, frontal, sphenoid and ethmoid bones closes the cranial cavity from below; it is a highly irregularly-shaped surface with three compartments or "fossa" housing various parts of the central nervous system.

In cross section, the skull varies in thickness from about 3/8 to 1/2 in, increasing in this dimension toward the base of the skull. It consists of an inner and outer table of compact bone with a highly vesicular middle layer resembling a honeycomb, called the diploë, and appears to be transversely isotropic. The composition of bone involves three types of cells (or osteons) and an intercellular matrix that contains collagenous fibers and calcium salts. All parts of the skull are traversed by a myriad of blood vessels connecting the dura to the scalp.

The dura is a tough, fibrous, dense, relatively inelastic membrane consisting of two layers; the outer, forming the lining of the skull, is a fibroconnective tissue rich in blood vessels and nerves, while the inner layer is of similar texture, but lined with a single layer of cells and less traversed by vessels. The two parts separate to form the venous sinuses of the brain. The inner layer also extends inward to form the partitions between the two cerebral and cerebellar hemispheres (falx cerebri and cerebelli) and the horizontal compartmentalization between the occipital lobes of the cerebrum and the upper surface of the cerebellum (tentorium cerebelli) (Schaeffer, 1942).

The next tissue inward, the arachnoid, a gossamer cobweb-like non-vascular membrane of interlacing reticular fibers, is separated from the dura by a narrow noncommunicating capillary called the subdural space, filled with lymphlike fluid. The arachnoid is attached to the pia, another highly delicate membrane closely investing the brain and consisting of white fibrous tissue. It contains blood vessels that send out branches into the nervous tissue by a set of connective tissue or trabecula that render a distinction between the two tissues impossible. The trabeculae form a series of subarachnoid spaces occupied by the cerebrospinal fluid that also invests the four ventricles of the brain, the central canal of the spinal cord and the perivascular spaces. It is a clear, colorless, nearly Newtonian fluid with a specific gravity of 1.004–1.007, containing small quantities of protein, glucose and inorganic salts (principally NaCl and KCl) and very few cells (Leeson and Leeson, 1966).

The brain, that portion of the central nervous system (CNS) enclosed by the cranial vault, is topographically divided into two convoluted hemispheres of the cerebrum, the two foliated hemispheres and a central region (vermis) of the cerebellum, the midbrain, and the medulla that connects to the cervical cord. The fossa of the skull base and the membranes of the falx and tentorum, consisting of dura and pia-arachnoid, assist in this physical division. In addition, each cerebral hemisphere is divided into the frontal, parietal, temporal, occipital, limbic and insula lobe by means of fissures and sulci, or superficial grooves, as shown in Figs. 1 and 3. The brain stem itself, shown in Fig. 3, is comprised of the regions identified as the diencephalon (involving the five thalamic parts), the midbrain, pons and medulla (or medulla oblangata), the latter passing through the foramen magnum into the cervical cavity. A structure called the reticular formation encompasses most of the last two tissues. The higher level functions are concentrated in the cerebrum while the lower functions, including respiration and vasomotor

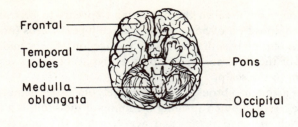

SECTION OF BRAIN

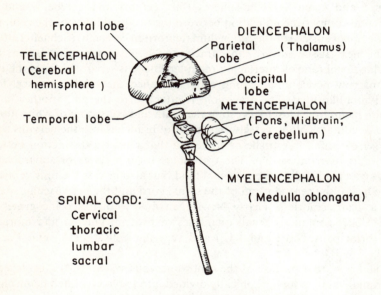

SPINAL CORD AND BRAIN

Fig. 3: View of cerebrum and brain stem.

control (heart rate and rhythm and blood pressure), emanate from the pons and the medulla. The reticular formation relates to consciousness.

Macroscopically, the egg-shaped brain is somewhat like a gel, although not as stiff, and consists of about 78% water, 10–12% phospholipids (a long-chained ester containing phosphate, fatty acids and nitrogenous compounds) 8% protein, and small amounts of carbohydrates, inorganic salts, and soluble organic substances (Ommaya, 1968). It is divided texturally into white and grey matter that denote conglomerations of the neural elements (axons) with and without fatty sheaths (myelin), respectively. This includes the cortex, an external layer of grey matter covering the fissures of the cerebral hemispheres, about 1.5–3.5 mm thick. Its density is slightly greater than water and cerebrospinal fluid and, when unsupported, it will creep under its own weight.

The brain is heavily infested with large and small veins and arteries connecting to the neck structures; arterioles, venules, capillaries, and vasa vasorum. These blood vessels connect to the overlying membranes. They are more or less elastic thin-walled tubes consisting either of a single layer of endothelial cells (for the smallest) or such a layer with two additional coats of muscle and connective tissue for the larger vessels. Their sizes range from the order of 1–2 mm to only a few micron in diameter. A series of neuroglial cells (bonding or connective tissue) forms a barrier between the blood and brain that permits an interchange of only a few selected substances between these tissues. The average weight of the human brain for a Caucasian male is 1500 grams (with a range of 1100–1700 grams); its length and transverse diameter are about 165 mm and 140 mm, respectively (Schaeffer, 1942). The characteristics of blood, a non-Newtonian fluid, are well known and are fully described in Cokelet's article (this book, Chap. 4).

The structure of the neck that represents the boundary conditions for head movement consists of the seven vertebrae of the cervical spine enclosing the spinal cord, a series of ligaments, notably the nuchae, blood vessels, the postural muscles, and the outer skin covering. Head motion with respect to the torso can occur in flexion-extension (up and down motion), lateral bending, torsion, or combinations of these displacements.

III. TYPES OF HEAD INJURIES

Damage to the components of the head, ranging from minor to fatal, can be categorized on a macroscopic basis by means of histopathological findings in case of fatality or by physiological malfunction or behavioral patterns for survivors. The first classification differentiates between blunt or penetrating blows on one hand and tissue types on the other as follows (Douglass et al, 1968; Goldsmith, 1966; Gurdjian et al, 1950, 1955; Thomas, 1970):

1. SCALP DAMAGE

   (a) Bruise, or subgaleal hematoma, involving leakage of blood from broken vessels into surrounding tissue below the skin.
   (b) Abrasion, or denuding of superficial skin layers by a sliding object.
   (c) Laceration, opening, or cutting of the skin by a sharp object; this may lead to an avulsion or peeling of tissue.

2. SKULL FRACTURE

(a) Depressed, due to
   i. High-speed projectile, accompanied by bone perforation and comminution.
   ii. Lower-speed missile (pistol bullet), producing perforation and fragmentation.
   iii. Slow-speed partial piercing by a sharp object with small attendant skull deformation.
(b) Linear or stellar, produced by a slowly moving blunt object; the first is frequently generated at points remote from the impact site due to tensile stresses involved in the snapback of a grossly deformed skull.
(c) Indented, a permanent deformation without fracture, occurring almost exclusively in young children.
(c) Crushed, due to static vise-like loading.

3. EXTRACEREBRAL BLEEDING OR HEMATOMA

(a) A clot in the epidural space.
(b) A subdural clot in the arachnoid space.

4. BRAIN DAMAGE

(a) Concussion, defined as an immediate and transient impairment of neural function, such as alteration of consciousness (defined in turn as a state of wakefulness and responsiveness to environment), disturbance of vision, and so forth, due to mechanical force (Gurdjian et al, 1955).
(b) Contusion, or bruising of the brain without a break in the surface or deeper tissues, but including rupture of small vessels and seepage of blood into adjacent tissues; this may occur at the point of impact (coup) or at distal stations, including the opposite pole (contrecoup).
(c) Intercerebral hematoma, due to rupture of larger vessels.
(d) Laceration, the most serious form of injury, involving the tearing of neural tissue.

The operational definition of concussion given here is physiological rather than mechanical, since the precise physical cause producing degeneration of the neural tissue of the brain stem has not been determined, and there is even some disagreement concerning the precise manifestation of concussive effects. For a more rigorous characterization, the following criteria for experimental cerebral concussion in subhuman primates may be cited in their order of importance (Ommaya, 1970):

1. Loss of coordinated responses to external stimuli.
2. A respiratory pause (apnea) of more than 3 seconds, followed by irregular slow breathing.
3. Decrease of heart beat by 20 to 30 per minute (bradycardia).
4. Loss of corneal and palpebral reflexes.
5. Loss of voluntary movements.

The severity of the concussion is to be determined on the basis both of the duration of characteristics (1) and (3) and of the abnormalities present in the electrocardiogram. This grading is analogous to the qualitative assessment of the degree of head trauma in humans. However, that is also measured by the period of recovery from a comatose or, alternatively, from a confused to a normal state for severe or light blows respectively, and further, by the time of recovery in cases of retrograde amnesia.

Brain damage can be produced either by a blow or by acceleration of the head without a collision, whereas the other tissue deficits are invariably generated by some type of impact. Furthermore, brain damage involving items $4b$, c and d may or may not involve concomitant concussion, depending on the location and severity of the trauma. Numerous histopathological investigations have been conducted on the primary effects of a fatal head injury involving contusions, lacerations, vascular lesions and nerve fiber degeneration, as well as on the secondary manifestations of edema (brain-swelling) or brain distortion due to the creation of an intercranial hematoma. These findings, as well as physiological and behavioral studies in surviving humans or experimental animals, have been summarized by Lindgren (1966); Ommaya (1970); Ommaya and Corrao (1969); Strich(1969); and Unterharnscheidt and Sellier (1966).

IV. THE CHARACTERISTICS OF COLLISIONS AND MATERIAL BEHAVIOR

The mathematical description of the complete mechanical effects produced on any object as the result of an impact requires a knowledge of the collision configuration, the relative impact velocity, and the geometry, composition and material properties of striker and target, both of these objects being considered as continuous structures. The forces produced between these bodies are generally quite large and of short duration; their magnitudes can only be calculated from a complete analysis of the actual process. These loads initiate stress waves that eventually encompass the entire volume of each body, as well as additional disturbances around the impact point called "contact phenomena." The latter include either temporary or permanent relative indentation of the two objects in the contact region, termed penetration (which is invariably present in an impact process); complete perforation in the case of thin targets pierced completely by a sharp object; cracking or fracture without perforation when brittle materials are involved; or fragmentation, shattering, comminution and pulverization under ultra-high velocity impact conditions.

In impulsive loading, where the rigidity of the striker can be neglected, only the nature and magnitude of a given time-dependent load, displacement, velocity, or acceleration acting over a given area or at a fixed point of the target needs to be specified; in such a process contact phenomena are not generated. Stress wave action can be ignored if the duration of the load is much longer than the time required for the traverse of the pulse across the object. If stress or strain distributions and hence wave propagation are of no concern, the target can also be modeled as a series of discrete masses connected by springs and dashpots that are intended to represent the deformation and dissipative characteristics of the material, respectively. The target can even be treated as a single rigid mass point for certain purposes. Either of these representations still permits the determination of the kinematic variables of the individual masses in a collision or an impulsive loading situation, and the force acting on each mass can then be calculated from Netwon's law.

The mechanical behavior of materials is described on a macroscopic basis by means of a constitutive equation which provides a relation between the stress σ_{ij} defined as the load per unit original area of the specimen, strain ϵ_{ij}, a change in linear dimension per unit magnitude, the time rates of such parameters, and possibly temperature Θ, entropy S or internal energy Z when thermodynamic considerations must be invoked. Since most collision processes occur very rapidly, the change in thermodynamic state variables is usually neglected and the temperature dependence of the parameters describing the specific material can be ignored. Material properties are considered to be homogeneous if identical at all state points of the object; they are isotropic if identical in all directions at the same point; completely anisotropic in the absence of any symmetry planes; or partially isotropic (that is orthotropic, transversely isotropic, and so on) for intermediate symmetry situations arising from definite directions of crystal or fiber orientation. In particular, the skull, scalp and dura exhibit significant anisotropic properties.

Classical and modern continuum mechanics, rheology, and theories of elasticity, viscoelasticity, wave propagation, dynamics, vibrations, impact phenomena, fracture mechanics and material science are all relevant to the problem of head injury. However, these subjects are too extensive to be summarized here; they have a long history of development, and a huge literature, including numerous surveys, abstracts of current periodicals, and texts at all levels of sophistication.

The utility of the principles of mechanics may be illustrated by a simple example. Pointed objects traveling at sufficient speed will perforate a thin target, such as the skull, frequently accompanied by comminution or fracture of the material from or surrounding the hole. In a closed fluid-filled shell, the transit of the projectile creates enormous increases in pressure due to the need for accommodating the extra volume of the projectile. For instance, if the skull is regarded as rigid and if its contents are considered to resemble water with a bulk modulus of 300,000 lb/in^2, the addition of a 1/8″ diameter sphere to the cranial cavity, assumed to occupy a volume of 61 in^3, leads to an increase in pressure of 40.4 lb/in^2, even if considered only statically. In actuality, larger gradients occur due to the motion of the cavity surrounding the projectile and stress wave propagation.

V. PROPOSED MECHANISMS OF BRAIN DAMAGE

The damage generated to the various covers of the brain in head injuries is usually quite evident and hence readily described on a mechanical basis, as indicated in Sect. III. Brain damage, on the other hand, is frequently not macroscopically detectable, and its presence must then be deduced from physiological or neurological malfunction, or from histopathological observation in fatal cases. In consequence, there has been a great deal of speculation as to the mechanics of the production of this trauma, and a number of theories proposed along this line have been scrutinized (Denny-Brown, 1945; Goldsmith, 1966; Gurdjian et al, 1955; Hayashi, 1969; Hodgson, 1970; Lindgren, 1966; Ommaya and Corrao, 1969; Sellier and Unterharnscheidt 1963, 1965; Thomas, 1970; Unterharnscheidt and Sellier, 1966; Ward et al, 1948). However, considerable controversy exists with regard to their acceptance, even at a qualitative level, since these hypotheses do not always fit clinical or pathological observations and are sometimes not in accord with the predictions of mathematical models. For example, the vast majority of these representations deny, without any real experimental evidence, the potential damaging effect of the initial transients, buttressed by the correct assertion that the duration of contact of most blows, 2 msec or longer, is about one order of magnitude greater than the transit time through either brain or skull to the antipole—about 0.12 msec. Yet many of the proposed mechanisms based on a steady state of motion disregard the constraining effects of the cerebrospinal junction—a direct contradiction. Furthermore, since the verification of these theories cannot be carried out in living humans, their efficacy must be deduced by a combination of mathematical and physical analogs, the latter consisting either of inorganic, animal or cadaver models. Experimental results on these devices must then be extrapolated to the living human—not a simple task!

Three major, relatively independent mechanisms incorporating some ancillary effects have been suggested by various investigators as producing traumatic neurological deficit:

(1) The establishment of large pressure gradients, possibly due in part to the propagation of steep-fronted waves in the brain and leading to damage as the result of linear acceleration involving both the absolute motion of the brain and its relative displacement with respect to the skull. Trauma could also be produced from concurrent large intercranial flows through the foramen magnum that initiate shear stresses in the brain stem, collapse of cavitation bubbles initially formed by the negative pressure at the antipole from the impact point (contrecoup), separation of the brain from the cranial wall in this region, and neurovascular friction (Edberg et al, 1963: Gurdjian and Lissner, 1961; Hayashi, 1969; Lindgren, 1966; Ommaya and Corrao, 1969; Rinder, 1969; Roberts et al, 1966a; Sjövall, 1943; Thomas et al, 1967; Unterharnscheidt and Sellier, 1966; Ward et al, 1948);

(2) Flexion-extension and/or bending of the upper cervical cord during motion of the head-neck junction (Friede, 1960, 1961; Hollister et al, 1958; von Gierke, 1966);

(3) The two simultaneous causes of (a) skull deformation due to local indentation or vibration of the entire shell, either with or without fracture, and (b) rotational acceleration of the head leading to relative angular motion of brain and skull, and producing sizable shear stresses in the process (Hirsch et al, 1970; Holbourn, 1943, 1945; Ommaya and Corrao, 1969).

The damaging aspects most frequently espoused by the pressure gradient theory are those of translational acceleration, shear stresses at the brain stem and pressures produced by the collapse of cavities near the contrecoup point. The last phenomenon was championed by Gross (1958a,b) and others, even though the analysis proposed for this mechanism is based solely on steady-state considerations. Hence questionable, experimental evidence was obtained from intermediate-speed-framing cameras of the formation of bubbles in a water-filled elliptical glass shell subjected to an impact. On the other hand, there is virtually no evidence of the existence of cavitation in actual biological tissue, and that which exists suggests that it is difficult to produce for the extremely brief period before the negative pressure vanishes (Hodgson, 1970; Ommaya, in press).

Damage to the cerebrospinal junction described by mechanism (2) may occur by means of either an impact or an impulsive loading of the torso. Damage by mechanism (3) was originally proposed by Holbourn (1943), one of the first researchers to provide a physically plausible explanation for the causes of brain damage. He reasoned that, since the bulk modulus of brain is much greater than its shear modulus, the local stresses are approximately proportional to the shear strain and can be regarded as a quantitative measure of tissue damage. He also suggested that for a period of load application of approximately 2–200 msec the impulse controlled the amount of damage, whereas for longer periods the level of force was the only significant damage criterion, and duration was not a factor. This behavior conforms to the characteristics of a linear spring-mass system for which this crossover in the pattern under transient acceleration occurs at about 1/3 to 1/4 of the period. The damaging angular velocity is equal to the damaging angular acceleration divided by the natural frequency of the brain (Hirsch et al, 1970). The generation of lesions in the central brain region have been attributed to actual increases in cranial volume that have been shown by Unterharnscheidt (Thomas, 1970) to occur under certain conditions as the result of impact-produced skull deformation. This spatial increment is believed to lead to cavitation, since it occurs much more rapidly than any possible inflow of fluid, particularly in the ventricular region.

All of these hypotheses have been subjected to both analytical and experimental examination. However, since some of the test results obtained and conclusions drawn are questionable, these investigations did not alter the opinions of the proponents of other theories. For example, the demonstration of the existence of shear near the brain stem (Gurdjian and Lissner, 1961), where the mechanisms controlling consciousness (and hence concussion) are located, shed little light on pathomorphological observations regarding lesions near the contrecoup point. Conversely, the hypothesis of cavitation did not explain the invariable presence of lesions in the fronto-occipital region regardless of the location of the blow; the latter, as well as coup injuries, might be due to the effects of skull snapback after inbending. Fibration of the skull has been ascribed to energy dissipation by spherical waves emanating from the impact point (Mazur, 1961). Under conditions of perforation, trauma can also be produced by (1) tearing or shearing of brain tissue in contact with the penetrating projectile or bone fragments; (2) large intercranial pressure changes due to the need to accomodate the additional volume of, for example, a bullet, and the generation of an oscillating cavity moving with the projectile.

VI. ANALYTICAL REPRESENTATIONS OF HEAD INJURY CONCEPTS

The first mechanical description of the response of a continuum type of head model subjected to a blow consisted of the analysis of a rigid, spherical shell containing an inviscid, compressible fluid of density ρ_F and bulk modulus K under axisymmetric conditions of irrotational flow. In one case, the shell traveling initially with constant velocity v_o is suddenly arrested (Anzelius, 1943), whereas in the second, the shell acquires a prescribed time-dependent velocity $v(t)$ either from rest or from a state of uniform motion (Güttinger, 1950).

Numerical evaluation of these models indicates that an initial compression wave emanates from the impact point and a tensile wave is simultaneously emitted from the antipole. This is a direct consequence of the assumption of the rigidity of the container that requires an infinite propagation velocity in the shell and also transforms each point of the shell surface into a continual source of energy of varying strength radiating into the field. The two waves propagate toward the geometric center of the fluid producing, upon superposition, large pressure gradients along a diameter in the vicinity of the center, although the equatorial plane including the sphere center is always stress-free. The negative pressures at the antipole and the large pressure gradients predicted by this analysis have been cited as substantive proof of the causes of brain trauma attributable to cavitation and translation acceleration in spite of the obvious shortcomings of the theory (Unterharnscheidt and Sellier, 1966).

The natural extension of these representations is the modelling of the container as a thin-walled shell of uniform thickness and of homogeneous, isotropic elastic material of density ρ_s, Young's modulus E, Poisson ratio μ, midsurface radius R, and thickness b. Solutions obtained by Benedict et al (1970), using extensional shell theory and an empirical axisymmetric pressure pulse over a given cap angle and elastic and fluid constants representative of skull and brain, respectively, indicate (a) a state of maximum compressive stress at the impact point followed by a maximum tensile stress at the antipole, (b) that the tensile stress at the impact point is more localized and greater in magnitude than that at the antipole, which is attributed to wave reflection in the shell producing outward buckling, and (c) that pressure fluctuations are largest near the interface and become progressively damped toward the center of the sphere where there is little variation in this parameter. The results obtained are indicative of the shell pulling away from the fluid and thus producing negative pressure as required by the cavitation hypothesis of the brain damage. Furthermore, the more diffuse and more penetrating presence of the negative pressure near the antipole as opposed to the highly local focus at the impact point is supposedly in accord with pathological findings. However, this does not explain the far more severe lesions observed near the former position nor those in other skull regions where the present model predicts minimal pressure fluctuations. It is clear that the very simple geometry, material behavior, and absence of any other damage mechanisms in the present analysis can provide only a tentative step in the direction of understanding of the total phenomenon.

A more sophisticated set of relations for the same physical model and geometrical symmetry including both extensional and bending effects of the shell has been developed by Engin (1969). In spherical coordinates r, φ and θ, representing the radius, colatitude and longitude corresponding to displacements w, u and v, respectively, the equation of motion for the fluid are given by

$$\left[\frac{1}{r^2}\frac{\partial}{\partial r}\left(r^2\frac{\partial}{\partial r}\right) + \frac{1}{r^2\sin^2\varphi}\frac{\partial}{\partial\varphi}\left(\sin\varphi\frac{\partial}{\partial\varphi}\right)\right]\Psi = \frac{1}{c_b^2}\frac{\partial^2\Psi}{\partial t^2} \text{ with } c_b^2 \equiv \frac{K}{\rho_F} \quad (1)$$

where Ψ is the velocity potential, and ρ_F is the fluid density.

The equations for the motion of the shell used by Engin are:

$$\left[1 + \frac{b^2}{12R^2}\right]\left[-\frac{\partial^2 u}{\partial\phi^2} - \cot\phi\frac{\partial u}{\partial\phi} + (\mu + \cot^2\phi)u\right] + \frac{b^2}{12R^2}\left[\frac{\partial^3 w}{\partial\phi^3} + \cot\phi\frac{\partial^2 w}{\partial\phi^2}\right]$$
$$(2) \quad -\left[\frac{b^2}{12R^2}(\cot^2\phi + \mu) + (1+\mu)\right]\frac{\partial w}{\partial\phi} + \frac{1-\mu^2}{E}\rho_s R^2\frac{\partial^2 u}{\partial t^2} = 0$$

$$\frac{b^2}{12R^2}\left[\frac{\partial^3 u}{\partial\phi^3} + 2\cot\phi\frac{\partial^2 u}{\partial\phi^2}\right] - \left[(1+\mu)\left(1+\frac{b^2}{12R^2}\right) + \frac{b^2}{12R^2}\cot^2\phi\right]\frac{\partial u}{\partial\phi}$$
$$+ \left[\frac{b^2}{12R^2}(\cot^3\phi + 3\cot\phi) - (1+\mu)\left(1+\frac{b^2}{12R^2}\right)\cot\phi\right]u$$
$$- \frac{b^2}{12R^2}\left[\frac{\partial^4 w}{\partial\phi^4} + 2\cot\phi\frac{\partial^3 w}{\partial\phi^3} - (1+\mu+\cot^2\phi)\frac{\partial^2 w}{\partial\phi^2}\right.$$
$$\left. + (2\cot\phi + \cot^3\phi - \mu\cot\phi)\frac{\partial w}{\partial\phi}\right] - 2(1+\mu)w - \frac{1-\mu^2}{E}\rho_s R^2\frac{\partial^2 w}{\partial t^2}$$
$$(3) \quad -\frac{(1-\mu^2)R^2}{Eh}\left[\rho_F\frac{\partial\Psi}{\partial t} - p_e\right] = 0$$

The symbols b, R, etc. are the same as those defined in the preceding paragraph. The term p_e represents the external loading, and is considered to be a function of ϕ and t only.

The axisymmetric loading conditions require independence of the displacements u, w, and the potential Ψ of the coordinate θ, and the vanishing of displacement v. The boundary condition at the interface is expressed by the continuity of the radial velocity.

The response of the system to a local radial impulse expressed as $\delta(t)$ has also been obtained by means of the Laplace transform technique—a natural tool for the solution of dynamic problems—requiring here a numerical inversion; physical constants were chosen to simulate the properties of the skull and brain. Numerical results (Engin, 1969) have been obtained both for the *in vacuo* and fluid-filled cases when a spatially uniform axisymmetric pressure is impulsively applied to the quiescent system over a polar cap angle of 15°, so that

$$p_e(\phi, t) = \begin{cases} 546.5\,\delta(t) & \text{for} \quad \phi \leq 15° \\ 0 & \text{otherwise} \end{cases} \quad (4)$$

The following values are assumed for the parameters:

Shell: $\rho_s = 2 \times 10^{-4}$ lb-sec²/in⁴; $E = 2 \times 10^6$ lb/in²; $\mu = 0.25$; $R = 3$ in; $b = 0.15$ in.

Fluid: $\rho_F = 0.938 \times 10^{-4}$ lb-sec²/in⁴; $c_b = 57{,}100$ in/sec

Figures 4 and 5 present the variation with ϕ at the midsurface of the shell,

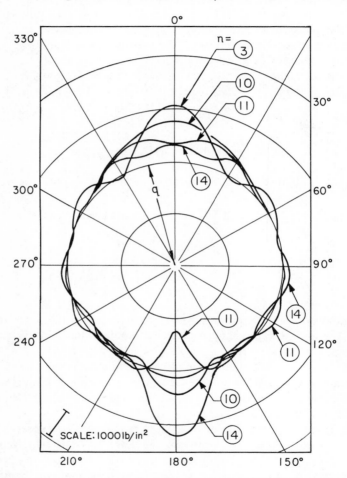

Fig. 4: Normal stress $\sigma_{\phi\phi}$ in the ϕ-direction as a function of polar angle ϕ at the midsurface of the elastic shell due to the application of a δ-Function of spatially uniform pressure of amplitude 546.5 lb/in² distributed over a polar cap of 15°, at various times $t = 9.125 \times 10^{-6} n$ seconds after loading (n is the number inside the circles). (After Engin, 1969; courtesy J. Biomech.)

$r = R$ of the normal stresses $\sigma_{\phi\phi}$ and $\sigma_{\theta\theta}$, respectively. Figure 6 presents the nondimensional excess pressure $\tilde{p} \equiv (p/E)(1 - \mu^2)(\rho_s/\rho_F)$ as a function of radial distance $\tilde{r} = r/R$ along the loading axis for various times after impact. The latter are indicated by numbers within circles denoting multiples of the basic time of $\Delta\tau = 9.125 \times 10^{-6}$ sec that represents one-tenth of the travel time in the shell to the antipole. The circle of radius q in the first two diagrams denotes the stress-free state; outward deviation indicates compression and inward deviation tension. The following results can be deduced from the diagrams: (a) Tensile stresses at the midsurface of the shell occur at the antipole in both directions, but only compressive stresses exist at the impact point; (b) significant tensile stresses also develop at $\phi = 35°$ in the θ direction; (c) the pulse amplitude in the fluid first increases and then decreases in its propagation towards the center; (d) the maximum negative pressure, occurring at $t = 15\Delta\tau$, is larger than the peak positive pressure and

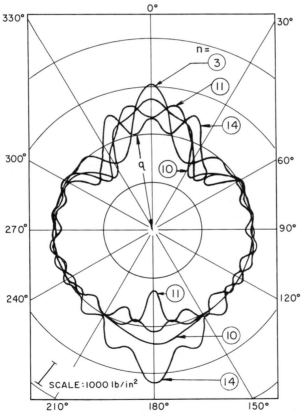

Fig. 5: Normal stress $\sigma_{\theta\theta}$ in the θ direction for the case shown in Fig. 4. (After Engin, 1969; courtesy J. Biomech.)

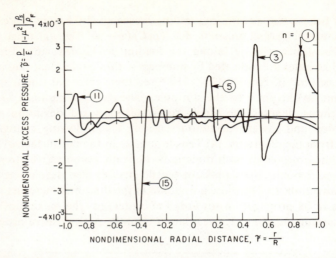

Fig. 6: Nondimensional pressure distributions at various times in the fluid along the diameter from the impact pole for the case shown in Fig. 4. (After Engin, 1969; courtesy J. Biomech.)

occurs approximately halfway from the center to the antipole. Thus, possible locations of skull damage include the pole $\phi = 0$ where the load is applied, with fracture being initiated at the inward-bent inner surface; the circular region defined by $\phi = 35°$ where tensile stresses result from outbending; and the antipole where tensile stresses are produced after the successive reflection of the pulses in the solid and fluid. High regions of negative pressure in the fluid, which are conceivable sources of brain damage, occur near the quarter positions along the axis of symmetry at two different times, as shown in Fig. 6. This differs significantly from the predictions for a rigid shell where this event is expected to occur at the center of the system.

A number of efforts have been directed towards placing the cavitation hypothesis of brain damage on a firmer analytical foundation, utilizing the basic assumption of a linear pressure gradient throughout the head. This implies that significant damaging processes do not occur during the passage of initial transient pressure waves, but only during the subsequent steady-state motion, a situation that has not been conclusively demonstrated. Gross (1958b) originally proposed an expression for the acceleration a_c required to initiate such cavitation, involving a hydrodynamic pressure drop p_c substantially equal to ambient pressure, based on a uniform acceleration. This formula is

(5) $$p_c = a_c \rho_B g L_n$$

where $\rho_B g$ is the weight density of the skull content, about 0.0385 lb/in³, and L_n is the distance from the nodal pressure point to the contrecoup position, believed to be affected by the location of the foramen magnum, the skull elasticity and the internal head volume. The existence of such linear pressure gradients in the saggital plane has been experimentally verified in cadaver heads after application of frontal blows resulting in accelerations of up to 80 g (Roberts et al, 1966a). It was also found in these experiments that the use of a rigid seal as opposed to an elastic one over the foramen magnum (simulating the elasticity at the cerebrospinal junction in the living human) and fluid flow through the opening, produced shifts in the nodal pressure point. The presence or absence of the falx or tentorial membranes involved about the same pressure distributions.

The use of a central nodal pressure point, corresponding to a rigid cylinder, was adopted by Sellier and Unterharnscheidt (1965) in the interpretation of both model experiments and the explanation of pathological lesions; damage by the cavitation mechanism was considered preeminent. The effect of vessel stiffness and fluid properties on the position of this nodal point has been tested using a vertical cylindrical shell containing a compressible fluid moved upward with constant acceleration (Kopecky and Ripperger, 1969). The results, asserted to be at least qualitatively similar to those for a spherical container under corresponding loading conditions, involve the solution of the equilibrium equation for an internally-pressured shell and conservation of mass for the fluid. They indicate the existence, at the top end of the cylinder, of a negative pressure that varies linearly with the acceleration, as required by the model, and nonlinearly both with shell stiffness and fluid bulk modulus; the nodal point of the gradient can occur anywhere along the cylinder axis. Experimental data for a plastic cylinder accelerated nonuniformly with a peak value of 102 g produced a negative pressure of only 2.1 lb/in^2, a value that is undoubtedly insufficient to initiate cavitation.

In a similar vein, Hayashi (1969) proposed a one-dimensional model of head injury consisting of a rigid, fluid-filled cylinder prefaced by a spring, intended to represent the elasticity of the biological components plus any protective cover, and striking a rigid wall with velocity v_o. For a very stiff spring relative to the elasticity of the fluid, the pulse propagates in the liquid just as in an elastic rod. This presumably produces cavitation at a negative pressure of one atmosphere relative to ambient conditions corresponding to an initial velocity $v_o \sim 2$ ft/sec or greater, a very doubtful situation. In the case of a soft spring, the pressure was found to be a harmonic function of time and linearly distributed across the skull with a central nodal point. These results could also be derived on the basis of a constant applied acceleration that, for values of 130–140 g, led to a cavitation phenomenon at the contrecoup position. The mobility of the spinal cord was taken into account by adding a branching cylinder to the rigid vessel representing the cranial cavity. Results computed from this model were reported to be in good agreement with the locations of lesions observed in autopsies, in spite of the previously cited shortcomings of this type of analysis, including its uniaxial nature.

Although apparently not nearly as applicable as the models described above, some investigators have suggested that useful information for head injury application could be deduced from studies of the dynamic loading of homogeneous spherical bodies. Axisymmetric wave propagation solutions in a viscoelastic sphere resulting from a single or else two equal, but oppositely directed, diametral loads have been derived (Valanis and Sun, 1967). The motion of an elastic and a viscoelastic sphere subject to a torsional step acceleration about a diametral axis, motivated by the rotational acceleration theory of brain damage, has also been analyzed (Lee and Advani, 1970).

Lumped-parameter models consisting of mass points, springs, and dashpots have been used to examine the relative motion of various parts of the human body to a prescribed impulsive loading (McHenry, 1970; Roberts et al, 1966b; von Gierke, 1964). Some of these have been specifically concerned with the response of the head and the deformation of the cerebrospinal junction (Hollister et al, 1958; Martinez et al, 1965; McHenry and Naab, 1966; Roberts et al, 1969). The number of degrees of freedom involved have ranged from two to eleven; the physical constants employed in the calculations were either obtained from human cadaver, anthropometric dummy data (Naab, 1966; Roberts et al, 1969; von Gierke, 1964), or else assumed to conform to measured experimental responses (Martinez et al, 1965; Severy et al, 1955). For example, the hyperextension induced in an animal seated on a cart upon sudden arrest of the vehicle has been modelled by a three degree of freedom system consisting of three mass points representing the head, the sum of the torso and vehicle seat, and the vehicle itself connected by two linear springs and a torsional spring for the neck subjected to a half-sine pulse of acceleration. The motion of the system is described by three coupled ordinary linear differential equations involving the head rotation and the relative displacements of the head to cart and torso (shoulder) to cart, respectively. The numerical results for one set of assumed spring constants are shown in Fig. 7a, together with the experimental data from a controlled rear-end collision of two cars containing human or dummy occupants (Severy et al, 1955) for which an attempt to match these constants was made (Fig. 7b). A three degree of freedom nonlinear model of the "whiplash" effect including damping has also recently been analyzed for a half-sine wave input of acceleration (Martinez and Garcia, 1968). The effect of the addition of a helmet and of the use of various seatbacks as represented by different spring and damping constants was numerically investigated.

Figure 8 presents results from a similar study involving nine degrees of freedom for the human body when a triangular force with an amplitude of 333 lb is applied through the mass center of the head in the posterior-anterior

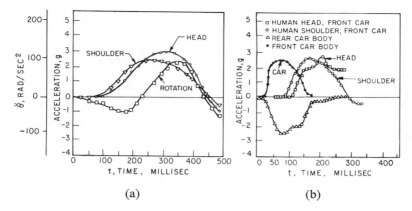

Fig. 7: Three degree of freedom system representing hyperextension effects in an animal seated in a suddenly arrested cart. (After Martinez and Garcia, 1968; courtesy J. Biomech.): (*a*) Calculated response for a half-sine wave of acceleration; (*b*) Corresponding experimental data.

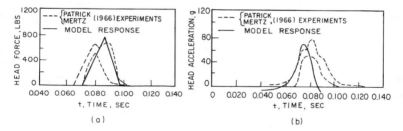

Fig. 8: Calculated force and accleration histories of the head for a 9 degree of freedom model of a seated human subjected to a triangular force in a sagittal direction through the mass center of the head and comparison with corresponding experiments. (After Roberts et al, 1969; courtesy, J. Biomech.): (a) Force history; (b) Acceleration history.

direction during the interval $60 \leq t \leq 90$ msec (Roberts et al, 1969). The calculated results are compared with pertinent experimental data obtained from transducers attached to occupants of rapidly decelerated vehicles (Patrick and Mertz, 1966) and show good correspondence. A sweep of the parameters indicated significant dependence of the response of the neck to the location and rise time of the applied load as well as to the rotational and translational stiffness of the neck.

The concept of the mechanical driving point impedance, a complex quantity defined as the ratio of a harmonic force to the velocity generated at its point of application in a linear system, has been utilized to assess the damage produced in experimental animals by such loads acting on the surface of the unprotected brain (von Gierke, 1966). The resultant contusions and lesions were correlated with the energy transferred by the blow, a quantity proportional to the Fourier transform of this impedance. It had been previously shown that a rigid sphere moving harmonically in an infinite viscoelastic medium possessing human tissue properties adequately modelled this process, and further, that the impedance is not significantly altered by the presence of boundaries removed somewhat from the loading position (Oestreicher, 1951; von Gierke et al, 1952). In addition, this parameter has been utilized to define appropriate properties, obtained from measurements in *in vivo* and *in vitro* animals, for linear lumped-parameter models of the head (Hodgson et al, 1967; Stalnaker et al, 1970).

VII. INANIMATE MODEL AND CADAVER EXPERIMENTS

In addition to the model experiments already cited in conjunction with specific theories of head injury, a number of other tests have been conducted both on fabricated structures and on cadavers to provide information concerning processes (as opposed to material properties, human tolerance, or design characteristics) involved in impact or impulsive loading of the head. Holbourn (1943), using a two-dimensional photoelastic 5% gelatin model surrounded by paraffin wax to simulate the brain encased by the skull, observed significant shear strains in the regions corresponding to the anterior portion of the temporal lobe when the system was given a sudden forward rotation. This observation tended to corroborate his rotational acceleration theory of brain damage. Gurdjian and Lissner (1961) employed the same experimental procedure to demonstrate the existence of shear strain attributable to intercranial flow at the simulated cerebrospinal junction when hammer blows were applied to a two-dimensional plastic skull model supplied with a tube to resemble the spinal canal. Three-dimensional birefringent plastic models of the skull have been examined photographically at intermediate framing rates by Edberg et al (1963) under conditions of free fall followed by an impact. During the descent of the unit, slight positive and negative pressures were observed in its upper and lower portions, respectively, whereas these trends were briefly reversed during the impact. Shear stresses in the region of the brain stem were also noted, but the negative contrecoup pressures did not manifest themselves when the model was embedded in a plaster-filled container. Other photoelastic measurements recorded by a high-speed framing camera (240,000 per second) were obtained by Flynn (1966) to demonstrate the process of stress wave propagation in a two-dimensional head model consisting of an outer ring of hard plastic (Homalite) snugly enclosing a soft urethane rubber disk. The unit was loaded by the explosion at the outer ring periphery of a one-grain electric primer with a peak pressure of 900 lb/in^2, a rise time of 60 μ sec and a total duration of 600 μ sec. It was found that stress wave propagation in the ring was not affected by the presence of the interior material and consisted principally of flexural waves that travelled more rapidly than any observable transient in the simulated brain. The results obtained in these various investigations must be regarded as qualitative in view of the lack of effort to duplicate the geometry and material properties of the actual biological components as well as the two-dimensional nature of most of the models.

Fluid-filled plastic spherical shells with thicknesses of about 1/4 of the radius were subjected to diametral impacts of unspecified magnitude in an attempt to experimentally obtain the pressure distribution by means of a fixed transducer at the contact point and a second sensor capable of being moved within the fluid along the impact axis (Sellier and Unterharnscheidt, 1963, 1965). Large positive and negative pressures were initially recorded at the coup and contrecoup points—at the latter position with an unrealistic amplitude of more than 2 atmospheres—followed by oscillations of decreasing amplitude, with the equatorial plane essentially stress free. However, the frequency response of the transducers was obviously inadequate to record the true transients (in view of a time scale of more than 2 msec for the first pressure oscillation) and the effect of the presence of the measuring device on the phenomenon was apparently not considered. Similar experiments were conducted in an extensive investigation involving 8-inch diameter empty and water-filled plastic and aluminum spherical shells, a plastic hemisphere, and a plastic hollow cylinder with attached tubes to represent spinal canals. These were supported in a variety of ways and loaded either by free-fall impact, collision of a falling weight, or a pendulum blow (Lindgren, 1966; Rinder, 1969). The overall pattern of the pressure distribution obtained from carefully calibrated devices was found to be in general accord with that noted by earlier investigators as cited above, although strong positive pressure peaks were observed at the contrecoup position. The shells were also supplied with accelerometers that recorded significantly different motions at the coup, contrecoup and equatorial positions although these were not correlated to the corresponding pressure data. Additional tests are currently being conducted with fluid-filled metallic and plastic spherical shells for the specific purpose of validating the predictions of the previously presented theories.

Numerous tests have also been conducted on cadaver skulls; this has the advantage of using a biological specimen with the actual geometry, but not necessarily the appropriate bone properties in view of changes occurring as the result of drying or embalming (Gurdjian et al, 1970a; McElhaney, 1966; McElhaney et al, 1964). Furthermore, since specimen acquisition is mostly a random process, reproducibility of the system is a much more difficult (if not impossible) condition to achieve than in the manufacture of an inanimate model. In one of the first tests of its type, both dry skulls and intact cadaver heads, with a strain-sensitive lacquer applied both to the exterior and interior surfaces of the calvarium to indicate the regions of maximum stress, were dropped onto a metallic anvil from various heights (Gurdjian et al, 1950). Linear fractures were produced with impact energies of 40 and 400–900 in-lb for the two cases. The inbending of the skull at the impact point causes outbending in adjacent regions where linear fractures are initiated in low velocity collisions, vulnerable areas being identifiable from the crack pattern of the brittle lacquer. Other tests involving the impact of skulls utilized accelerometers and strain gauges to obtain a relationship between kinematic variables and the initiation of fractures, including a study of skull deformation (Evans et al, 1958; Lissner et al, 1960; Sellier and Unterharnscheidt, 1965; Unterharnscheidt and Sellier, 1966). Measurements of the pressure resulting from either free fall or a controlled blow at various points in fluid-filled skulls sealed at the openings have also been variously executed (Gurdjian et al, 1961; Hodgson, 1968, 1970; Lindgren, 1966; Lissner et al, 1960; Roberts et al, 1966a; Sellier and Unterharnscheidt, 1965). The effect of an added layer of skin was found to result in significant damping. The measured intracranial pressure changes and accelerations produced in a cadaver skull due to a blow were combined with corresponding data from

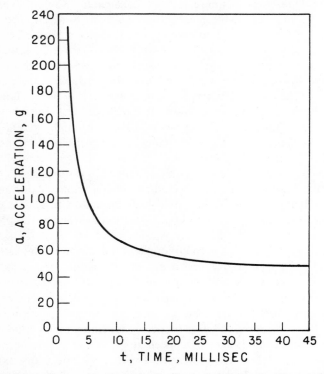

Fig. 9: Wayne State tolerance curve for the brain. (After Gurdjian et al, 1962.)

anesthetized animals struck in similar fashion for the purpose of developing a tolerance curve for brain injury, as shown in Fig. 9 (Gurdjian et al, 1961, 1962; Patrick, 1966b). The curve, based also on cadaver skull fracture and the observed onset of concussion in the animals, is purported to delineate the tolerance to linear acceleration before the occurrence of cerebral concussion in man. However, no clear description is given of how the data is scaled to the living human (Ommaya, in press), although correlations between results of similar tests on cadaver skulls and test animals have been attempted more recently (Hodgson, 1968, 1970). Finally, half-skulls filled with gelatin to resemble the brain and marked with a grid have been subjected both to angular and to combined linear and angular acceleration (Martinez, 1969). Intermediate-speed photographs of the event indicate the presence of large shearing strains produced solely by angular acceleration. It was suggested that a filler with properties more nearly resembling those of the cranial contents would provide better quantitative information on this subject.

VIII. EXPERIMENTS ON ANIMALS

A wide variety of experiments have been conducted with animals subjected to blows and/or accelerations under controlled conditions in an effort to relate the degree of concussion and/or skull fracture as determined from physiological, behavioral or histopathological measurement with mechanical impact parameters.* As in the case of inanimate models or cadavers, there are also significant disadvantages in the use of animal subjects including the variability of specimens, the need for scaling to human size, uncertainties concerning the criteria of concussion and their severity in both types of species, the greater difficulties of transducer attachment in the smaller animals, the need for anesthesia, and possible differences in reflex and behavioral patterns. Nevertheless, tests of this type can combine physical and biological data in a manner unattainable by any other means.

The principal animals selected for these tests were cats, dogs, rabbits and various types of monkeys. Mechanical variables were determined by means of strain gauges, accelerometers, piezoelectric pressure gauges, displacement and velocity transducers. Intermediate-speed framing cameras with rates up to 4000 per second recorded the motion of the subject either in one or simultaneously in two planes; these cameras were also employed to photo-

* See Denny-Brown and Russell (1941); Pudenz and Shelden (1946); Gurdjian et al (1953, 1954. 1955); Hollister (1958); Friede (1960, 1961); Sellier and Unterharnscheidt (1963, 1965); Unterharnscheidt (1962, 1970); Unterharnscheidt and Sellier (1966); Higgins et al (1967); Unterharnscheidt and Higgins (1969); Mundie et al (1967); Martinez (1963, 1970); Martinez et al (1965); Wickstrom et al (1965, 1967); Douglass et al (1968); Rinder (1969); Hodgson (1968, 1970); Ommaya (1966); Ommaya and Corrao (1969); Ommaya et al (1964, 1966a, b, 1967, 1968, in press); Hirsch et al (1970).

graph X-ray screens as alternatives to flash radiography. Not unexpectedly, the conglomerate test results indicate more of a diversity of opinion than any agreement concerning both the actual mechanisms principally responsible for cranial trauma and the magnitude of the mechanical variables associated with concussion thresholds. As an example, Fig. 10 presents a damage threshold curve for rabbits subjected to an approximately triangular pressure pulse produced by a fluid column mounted directly on the dura under the action of a piston (Rinder, 1969); the plot is clearly dependent on the duration of the load. Similarly the tolerance curve for monkeys subjected to "whiplash" based on rotational acceleration indicates a strong dependence on duration, as shown in Fig. 11. Although the manner of load application and the animal tested were different, the data presented by Hodgson (1970) appear to indicate, by contrast, that damage is substantially independent of pulse duration.

Some scaling law must also be invoked in order to transfer threshold data obtained from animals to human beings. The simplest relation of this type is predicated on the concept that intracranial pressure gradients control the damage. Since a given acceleration under steady-state dynamic conditions produces a fixed gradient in a given fluid, the magnitude of the pressure difference would be proportional to the length of the fluid column, that is the brain diameter. Thus, a given overpressure at the same acceleration is inversely proportional to the interior skull diameter, proclaimed as the appropriate scaling law by the advocates of the pressure difference theory (Unterharnscheidt and Sellier, 1966). Proponents of the shear strain theory of damage, resulting from angular rotation (Ommaya et al, 1967; Hirsch et

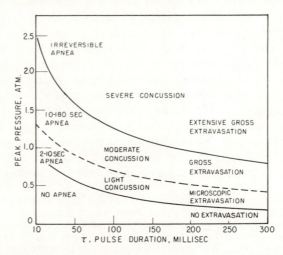

Fig. 10: Damage threshold curves for rabbits subjected to a triangular fluid pressure pulse applied directly to the dura (Rinder, 1969).

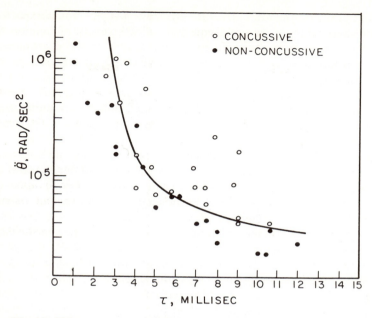

Fig. 11: Tolerance curve for Rhesus monkeys subjected to rotational acceleration (Ommaya et al, 1966.)

al, 1970) have adopted Holbourn's unpublished derivation based on a series of assumptions, most of them reasonable, leading to a relationship between angular acceleration $\ddot{\theta}$ and brain mass m for the model and prototype, designated by subscripts m and p, respectively, given by

$$\ddot{\theta}_p = \ddot{\theta}_m \left[\frac{m_m}{m_p} \right]^{2/3} \tag{6}$$

The same result is obtained when considering the more restricted hypothesis that angular acceleration (rather than a critical value of the shear stress) is the cause of injury.

IX. MECHANICAL PROPERTIES OF HEAD TISSUES

One of the vital links in an understanding of the biomechanics of head injury is a knowledge of the mechanical properties of the component tissues of the head. Until very recently, information in this area was very scarce (Ommaya, 1968), consisting solely of a few investigations involving the skull and the brain. In the first category, Franke (1956) carried out bending tests on whole cross sections of the dry, human cadaver skull from which moduli were derived ranging from 1.2–5.3 × 10^5 lb/in² for various regions.

Resonance tests on the same structure yielded a Young's modulus of 2.1×10^5 lb/in^2 and minimum natural frequencies of 500 and 600 Hz, respectively, for a gelatin-filled or empty skull. Evans and Lissner (1957) determined the tensile and compressive failure stress of embalmed human parietal bone tested as a structural unit without, however, concurrently measuring strain, and thus were not able to ascertain the modulus. Dempster (1967), in a study of the cortical grain structure of the human skull, observed random orientations in both the inner and outer tables, suggesting isotropy of the material in the tangential direction.

Franke (1956) carried out experiments on fresh pig brain and homogenate at body temperature using Oestreicher's theory (1951) to obtain the coefficient of shear viscosity from the measured driving point impedance of a glass sphere at frequencies up to 500 Hz. Surprisingly, a much smaller value was calculated for this parameter than for other tissues. The value was of the order of glycerin at room temperature, or about 15 poises, raising some questions concerning the interpretation of the data. Dodgson (1962), in an attempt to evaluate the Mises-Henke criterion of plastic flow for fresh and hydrated mouse brain, found a nearly linear logarithmic relation between the time of load application and the natural compressive strain, but without a definite yield value. He concluded from histological observations that the brain did not possess any continuous matrix exhibiting appreciable viscosity and that, except for an occasional droplet, it did not encompass any free fluid. Koeneman (1966) performed both creep and cyclic compression tests between 80–350 Hz, albeit at minimal amplitudes of 10 microstrain or less, in rabbit, rat and pig brains, measuring the deformation and calculating an elastic compressive modulus ranging from 1.2–2.2 lb/in^2 and absolute viscosities ranging from 35–44 poises, both parameters being relatively independent of frequency. On the basis of these experiments, he proposed a rheological model of brain tissue consisting of a system of closely packed elastic cells held together primarily by colloidal forces, in conformity with the observation that the viscosity increased with sample compression and elastic response was absent when only shear forces were applied. Ommaya (1968) reports values of the viscosity of fresh Rhesus monkey and cat brain homogenate of 407 poises at 10–15°C based on the fall of a steel sphere through the substance and data reduction using Stokes' equation. He ascribes the large differences in the values of this parameter reported by various investigators to the impotence of apparent viscosity measurements in quantifying the resistive properties of the brain. Furthermore, in assessing the validity of applying these data to the living human, it should be borne in mind that all tissues were examined in the *in vitro* state and that the brain material was derived from different animals without segregation into grey and white matter.

Since 1968 a massive effort has resulted in the generation of considerable additional data pertaining both to the individual tissues and to the composite human head by appropriate cross-correlations of experimental results from human *in vitro* samples and both *in vivo* and *in vitro* information from animals, notably from the Rhesus monkey. Both procedures considered standard for the testing of ordinary engineering materials and novel techniques, particularly suitable for the special behavioral characteristics of the biological specimens, were employed. The greatest amount of information has been obtained for the skull (Haynes et al, 1969; Hubbard, 1970; McElhaney et al, 1970; Melvin et al, 1970*a*, *b*; Robbins and Wood, 1969; Roberts and Melvin, 1969; Wood, 1969[1]) because of its ready accessibility, easier storage and handling, and most importantly, its greater similarity than other biological tissues to more conventional substances, in particular brittle ceramics, permitting its examination by standard apparatus and routines. It should be emphasized, however, that the mechanical properties of the skull combine the geometric, structural and material characteristics of the system and that a particular type of measurement may very well encompass a combination of two or perhaps even all three of these properties. The microscopic inhomogeneity of the trabecular system in the diploë layer, a structural characteristic, thus undoubtedly contributes to the variability from specimen to specimen of the results obtained in a particular macroscopic property evaluated on the hypothesis of homogeneity, more than variations in the magnitude of the parameter.

Human skull bones obtained from embalmed cadavers, craniotomies and autopsies have been tested both quasi-statically and dynamically in tension, compression and simple shear. Quasi-static torsion, Vickers hardness tests and geometric and density measurements have also been performed. The specimens consisted either of a whole cross section of the skull or samples only from either the inner or outer table or the diploë layer. Skull sections forming a beam, either with or without sutures in the test length, were subjected to three- and four-point bending both reversibly and to failure with results analyzed by layered beam theory (Hubbard, 1970). It was found here that simple strength-of-material considerations for the sandwich beam yielded computed properties for the compact bone in good agreement with test results solely from the compact area. Further, results indicated that bending failure occurred in the tensile section at a value near the ultimate tensile strain. Table 2 lists average quasi-static properties and their standard deviations obtained from up to 300 human samples (McElhaney et al, 1970). Figure 12 presents typical compressive stress-strain curves for the whole bone in the radial and two orthogonal tangential directions. These and other data confirm the transverse isotropy of the structures. It was noted that data from embalmed samples did not significantly differ from that of fresh

---

[1] Other publications by these authors and co-workers are in press or in preparation.

TABLE II
Properties of Human Cranial Bone*
(McElhaney et al, 1970)

Property	Mean	Standard Deviation
Skull thickness, in.	0.272	0.047
Diploë thickness, in.	0.108	0.042
Dry weight density, lb/in^3	0.051	0.019
Radial compression modulus, E_{C_r}, 10^5 lb/in^2	3.5	2.1
Tangential compression modulus, E_{C_t}, 10^5 lb/in^2	8.1	4.4
Poisson ratio in radial compression, μ_{C_r}	0.19	0.08
Poisson ratio in tangential compression, μ_{C_t}	0.22	0.11
Ultimate strength in radial compression, σ_{UC_r}, 10^3 lb/in^2	10.7	5.1
Ultimate strength in tangential compression, σ_{UC_t}, 10^3 lb/in^2	14	5.2
Ultimate strain in radial compression, ϵ_{UC_r}, 10^{-3}	97	80
Ultimate strain in tangential compression, ϵ_{UC_t}, 10^{-3}	51	32
Microhardness Vickers DPH, inner table	31.6	9.3
Microhardness Vickers DPH, outer table	34.2	8.0
Ultimate strength of diploë in direct shear, σ_{US}, 10^3 lb/in^2	3.1	0.5
Ultimate strength of diploë in torsion, σ_{US}, 10^3 lb/in^2	3.2	0.8
Modulus of torsion for diploë, G, 10^5 lb/in^2	2.0	1.4
Ultimate strength in tangential tension, σ_{UT_t}, 10^3 lb/in^2	6.3	2.7
Modulus in tangential tension, E_{T_t}, 10^5 lb/in^2	7.8	4.2
Ultimate strength in tangential tension for tables, σ_{UT_t}, 10^3 lb/in^2	11.5	3.8
Modulus in tangential tension for tables, E_{T_t}, 10^5 lb/in^2	1.78	0.3

* All properties refer to composite if not otherwise specified.

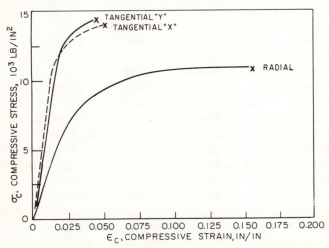

Fig. 12: Typical quasi-static compressive stress-strain curves for the whole human skull bone in the radial and two orthogonal tangential directions (McElhaney et al, 1970).

autopsy specimens. It was also observed that significant statistical fluctuations at the 5% level, not nearly as prominent in other tests, were found in all data involving the diploë layer in samples both from the same or from different skulls, apparently due to structural variations. Empirical correlation of the compressive Young's modulus, the ultimate compressive strength and the dry weight density were also obtained for the entire skull bone in the tangential and radial directions and for the diploë layer radially. When the stress-strain curves exhibited classical concave-downward behavior with monotonous increase in stress, indicative of a uniform load distribution throughout the specimen, the compressive stress and modulus exhibited an approximately linear relation, so that the failure strain could be taken as a constant value independent of porosity. When nonuniform load distribution occurred as manifested by progressive failure or buckling of the trabeculae, the failure strain was substantially higher. A model consisting of small cubical aggregates forming a large cube has been suggested for the skull bone. Quasi-static tests performed on seventy Rhesus monkeys are summarized in Table 3 for the purpose of illustrating the differences in properties relative to human cadavers.

TABLE III
Physical Properties of Rhesus Monkey (*Macaca mulatta*) Cranial Bone
(McElhaney et al, 1970)

Property	Mean	Std. Dev.
Thickness, in	0.101	0.010
Dry weight density, lb/in^3	0.065	0.010
Ultimate strength in tangential compression, $\sigma_{U_{C_t}}$, 10^3 lb/in^2	13.4	7
Modulus of compression, tangential E_{C_t}, 10^6 lb/in^2	9.4	4.6
Microhardness Vickers DPH, inner table	32.4	12.2
Microhardness Vickers DPH, outer table	34	10.2

Strain rate effects were observed in Young's modulus and in the failure stress and strain, but not in the absorbed energy to failure, during tensile tests on compact bone from the parietal, temporal and frontal regions. No variations were manifested with regard to either the location or the age and size of the donor. Figure 13 shows the average stress-strain curves at various strain-rates, the outer envelopes and the 90% failure boundary for these tests (Wood, 1969). Inclusion of the coronal suture in the sample yielded a much lower ultimate tensile strength in the tangential direction, $\sigma_{U_{T_t}} = 2140$ lb/in^2, smaller even than the corresponding value for the diploë layer, 4940 lb/in^2. Neither the shear nor the compressive properties were found to be significantly sensitive to strain rate in the range from $\dot{\epsilon} = 0.002$–500 per second.

Recent investigations of brain tissue have generally involved only human autopsy and *in vitro* monkey specimens. These have been subjected to (a) bulk modulus determination by superposition of an oscillating stress at frequencies up to 100 Hz on various levels of hydrostatic pressure and recording the displacements, (b) uniaxial and free-standing compression with

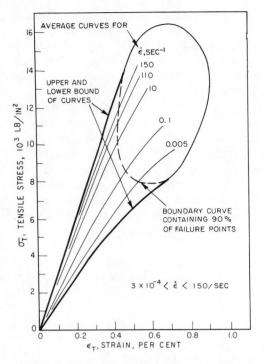

Fig. 13: Tensile stress-strain curves at various strain rates for compact human skull bone (Wood, 1969).

strain rates up to 65 per second, (c) harmonic shear and torsional tests at 10 Hz and up to 400 Hz, respectively, and (d) to creep, relaxation and free vibration tests. It was found (McElhaney et al, 1969; Estes and McElhaney, 1970) that human and monkey brain are substantially incompressible with a strain-rate independent bulk modulus of 300,000 lb/in², very close to that of water. A linear pressure-volume relation up to 100 Hz was also determined as well as a rate-dependent compressive modulus of the order of 10 lb/in², as indicated by the concave-upward true compressive stress-strain curves shown in Fig. 14. The latter could be empirically represented by the expressions

(7) $$\ln(\sigma/\dot{\epsilon}) = C_1 + C_2 \ln t \quad \text{or} \quad \sigma = e^{C_1} t^{C_2} \dot{\epsilon}$$

where constants C_1 and C_2 were evaluated as 0.50 and 0.782 for the human and 0.69 and 0.785 for the Rhesus monkey tissue. Values of the complex shear modulus for human autopsy specimens are at variance, yielding a range of G_1^* from 0.08–0.16 lb/in² and G_2^* from 0.04–0.095 lb/in² when determined by direct shear (Fallenstein et al, 1969) at 10 Hz as opposed to

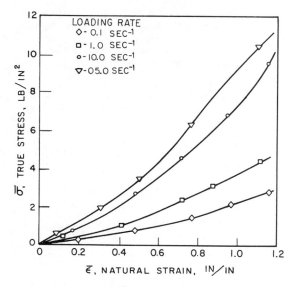

Fig. 14: True compressive stress-strain curves for human brain at various strain rates (McElhaney et al, 1969).

corresponding values of $G_1^* = 0.12$–20 lb/in² and $G_2^* = 0.05$–12 lb/in² obtained from torsion at 2–400 Hz (Shuck et al, 1970).

The data for the creep compliance of human and monkey brain and dura as well as monkey scalp has been fitted to the semilogarithmic expression

(8) $$J(t) = C_3 + C_4 \ln t$$

with constants C_3 and C_4 compiled in Table 4 (Galford and McElhaney, 1970). The curves for brain show large variations due to structural differences that overshadow effects of composition. They appear to be highly sensitive to stress level, and thus probably represent a nonlinear material. Both dura and scalp yield results independent of this quantity over a wide range and thus, most likely exhibit linear behavior. Table 4 also compiles the complex modulus values for these tissues under free vibration at a fixed frequency indicating the greater stiffness and viscosity of monkey brain relative to that of the human brain; the human dura is considerably stiffer than its anthropoid counterpart. A true compressive stress-strain curve for monkey scalp is presented in Fig. 15 (McElhaney et al, 1969). A linear viscoelastic model consisting of any desired combinations of springs and dashpots can be fitted to the data derived from a single type of test for any of these tissues, but the model would not be unique. The only proper representation would be that correctly predicting the material behavior for all types of loading,

TABLE IV

Parameters for the Empirical Semilogarithmic Creep Compliance Curves
$J(t) = C_3 + C_4 \ln t$ and complex Young's Moduli $E^* = E_1^*(\omega) + iE_2^*(\omega)$
from free vibration tests on various tissues
(Galford and McElhaney, 1970)

TISSUE	BRAIN		DURA		SCALP
Sample	Human	Monkey	Human	Monkey	Monkey
C_3	0.355	0.43	4.75×10^{-4}	1.0×10^{-5}	4.2×10^{-3}
C_4	0.026	0.026	0.97×10^{-5}	12.7×10^{-5}	6.5×10^{-3}
Valid after rise time of t, sec.	0.1	0.1	1.0	1.0	1.0
Time span, sec.	500	500	500	200	100
Stress level, lb/in²	0.5–1.0	0.5–1.0	75–225	85–140	10–60
E_1^*, lb/in²	9.68	13.2	4570	2400	210
E_2^*, lb/in²	3.80	7.8	500	475	74
ω, Hz	34.0	31.0	22	19	20
$|E^*|$, lb/in²	10.4	15.3	4600	2450	223

requiring the assimilation of results from a diversity of tests for its construction.

A series of *in vivo* experiments on brain tissue for the determination of transmitted force utilized a sinusoidally moving probe of fixed amplitude acting on the pia-arachnoid of a Rhesus monkey through a trephine hole in the skull and effectively yielded the driving point impedance (Fallenstein et al, 1969). A mathematical model of this operation has been constructed (Engin and Wang, 1970) consisting of the behavior of an elastic substance inside a rigid container with the interface free of shear stresses subjected to local radial harmonic excitation. The solution for the identical case involving a linear viscoelastic filler was obtained by the correspondence principle. This analysis permits the evaluation of the complex dynamic shear modulus from the phase relationship between displacement and force and the magnitude of the latter, obtained from the probe test, assuming only incompressibility of the animal brain. These results have thus far been interpreted only for low amplitude displacements where loss factors (tan δ^*) in the range from 0.6–0.9 were obtained; nonlinearities in material behavior at higher amplitudes will require a special analysis.

A thin elastic cylinder capable of expansion under pressure has also been inserted in live, immediate post mortem and fixed brain tissue from

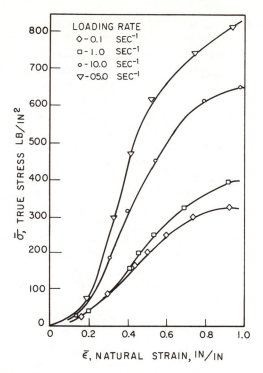

Fig. 15: True compressive stress-strain curves for monkey scalp at various strain rates (McElhaney et al, 1969).

a Rhesus monkey to ascertain their relative opposition to intercranial expansion (Metz et al, 1970). The modulus increased with strain level, but there was little difference between live and dead tissue or with regard to different regions or positions of the probe within the brain; the greatest variation of the modulus occurred in the case of fixed tissue. Additional tests were designed to compare the *in vivo* and *in vitro* properties of dogs and monkeys (Gurdjian et al, 1970*a*). These tests indicated that deformation patterns under identical loading conditions for the same species are quite similar in the live animal under anesthesia, in the dead animal with intact skull contents, and in the dry skull, although the patterns are different from the results of corresponding experiments on human cadaver skulls due to structural variations. While static deflection characteristics in the living dog skull and a freshly sacrificed specimen are nearly the same, both drying and immersion in formalin cause an increase in stiffness of the bone. Finally, the natural rotational frequency of the brain of the Rhesus monkey has been obtained both from cinefluoroscopic records of the motion of small radiopaque spheres of density equal to that of brain embedded in cerebral tissue obtained at 60–80 frames per second and from intermediate-speed photographs of the brain motion observed through a Lexan calvarium (Hirsch et al, 1970). The results indicate a fundamental rotational frequency in the range from 5–10 Hz that can be employed to assess the previously cited rotational damage criterion that is based on a linear spring-mass system.

Mechanical impedance measurements on dry cadaver skulls were reported by Franke to yield resonance at 820 Hz, with a corresponding value of 600 Hz given for the living human skull. Subsequent tests (Hodgson et al, 1967; Hodgson and Patrick, 1968; Gurdjian et al, 1970b) indicated a resonance at 900 Hz for a skull filled with silicon gel, an antiresonance at 313 Hz, and a damping constant of about 2 lb-sec/in. Stalnaker et al (1970) executed extensive tests of this type both on anesthetized Rhesus monkeys and on one whole human cadaver; the impedance of the head was first obtained with the body attached, then with the body removed, then with the skin and mandible removed, and finally with the brain excised. Measurements were made in the frequency range from 30–5000 Hz with a maximum acceleration of 20 g; the observed nonlinearities in the impedance were a maximum of 40%. In consequence, the data was fitted to a two degree of freedom system consisting of two masses and a connecting spring and dashpot that, for the cadaver head, exhibited weights of $m_1 g = 0.4$ lb for the parietal sector, $m_2 g = 9.0$ lb for the brain and other bones of the head, a spring constant $k = 2.6 \times 10^4$ lb/in for the skull stiffness, and a viscous damping coefficient $c = 2.4$ lb-sec/in for the tissues present, with a resonance and an antiresonance at 820 and 166 Hz, respectively.

Investigations are also in progress by Daly and Chalupnik with regard to the properties of the blood vessels of the head with strong indications of linear elastic behavior.[2] Burstein and Frankel found that, in flexion-extension of the head-neck junction, the instant center of motion of the head traverses upward along the cervical vertebrae. Based on a linear spring-mass-damper model with a torsional spring constant equal to 5 ft-lb/radian, they found a damping value of 17% critical for adult male volunteers with the effect of the postural muscles neglected.

X. HUMAN TOLERANCE AND PROTECTIVE DEVICES

The interrelated topics of human tolerance levels and the design and evaluation of protective systems and devices involve important biomechanical considerations pertaining to the head injury problem. A grading of the degree of tolerance has been suggested as voluntary, injury threshold, moderate damage permitting complete recovery, severe injury leading to permanent

---

[2] Personnal communication.

deficit, and fatality (Patrick and Sato, 1970). Attempts have been made to establish dosage levels for these categories using volunteers, cadavers, anthropomorphic dummies, experimental animals, physiological, pathological and clinical observations, and/or mathematical models. While the mechanical parameters producing dysfunction at the cellular level are undoubtedly the local stress, strain, pressure or volume expansion and perhaps their rates of application, these must bear some relation, currently unknown, to the macroscopic and measurable quantities that are generally regarded as being an index of head tolerance. For noncontact situations these are the excursion; linear and angular head velocity and acceleration, either average or peak; the applied torque; and the bending stress and stretch in the neck. In the case of a blow these are supplemented by the relative velocity, direction and duration of the impact; the impulse or energy; the location on the skull; the mass, geometry and contact area of the striker; and the hardness and toughness of its surface (Hodgson, 1970; Patrick and Grime, 1970). The use of a protective device designed to mitigate the injury or increase the tolerance threshold favorably affects nearly all of the factors cited.

Threshold values obtained by means of volunteers are generally geared to whole body tolerance that is, in many ways though not always, significantly below the level required to produce head injuries. Furthermore, all laboratory tests on humans must necessarily be designed below the point of impending deleterious effects (Lombard, 1951; Snively and Chichester, 1961). Stapp (1967, 1970), in a complete review of this subject, provides both physical and physiologic data on volunteers who assumed various attitudes and wore a variety of restraining devices when subjected to the acceleration conditions prevailing in high-speed sleds arrested by water or friction brakes or in swings striking a stop. Other devices of similar nature include drop towers and multidegree of freedom centrifuges occupied by volunteers.[3]

Although other investigators examining accidental falls noted the absence of damage either under ideal conditions for a deceleration of 140 g upon striking snow or a velocity of 100 ft/sec upon water entry, Stapp proposes a peak value of 30 g with an onset below 1500 g/sec as the limit of minimal or no injury when a 3" nylon or dacron lap belt is used. This value is low both in comparison with his personal experiences reflecting the limit of 40 g quoted by Patrick and Grime (1970) and relative to the value of 60–80 g frequently cited as the minimum required for the production of cerebral concussion. Tests by Mertz and Patrick (1967) on volunteers under conditions likely to produce hyperextension indicated that the limiting factor in this mode of damage was torque rather than shear or tensile stresses in the neck. Indications were that excursion angles up to 70° were safe, while those in excess of 80° would probably lead to injury. Patrick and Sato (1970) have also

---

[3] Located at the Wright-Patterson Air Force Base, Ohio.

compiled a human tolerance table based on experiments involving both volunteers and cadavers.

Cadavers, anthropomorphic dummies and head forms have been employed to ascertain damage thresholds upon contact with various structural components including solid steel blocks, clamped metal sheet panels, production instrument panels of automobiles and tempered and laminated glass panes. They have also been used to assess the protective design characteristics of these elements (Haynes and Lissner, 1962; Patrick, 1966a; Patrick and Daniel, 1966). The testing procedures involved the dropping of either an instrumented skull or a whole cadaver down a shaft onto the target, or else to observe the impact of a cadaver or dummy with a windshield after the sudden arrest of the transporting vehicle. Mild concussion was assumed to be equivalent to the production of a linear skull fracture—an admittedly tenuous hypothesis—as determined by subsequent X-rays of the calvarium: this did occur in gelatin-filled skull impacts on a steel block at 8.6 mph yielding a triangular deceleration with a peak value of 96 g lasting for 4 msec and producing an intracranial pressure increment of 36 lb/in^2. On the other hand, sudden vehicle stops did not produce penetration of 30 mil laminated windshields for initial speeds below 24 mph. The simulated response by cadavers of live human results has been improved by the addition of braces to resemble muscle action, and the differences in the reactions to impact of anthropomorphic dummies, while still considerable, are slowly being reduced as more information relative to biological material properties becomes available and more sophisticated methods of articulation are employed (Severy et al, 1968).

Tolerance curves for cerebral concussion resulting from pressure applied directly to the dura of rabbits and from angular acceleration to the head of Rhesus monkeys, obtained by direct comparison of physical measurements and physiological reponse or histopathological examination, have already been presented in Figs. 10 and 11, respectively. The curve shown in Fig. 9 is the only threshold relation thus far proposed for human cerebral concussion and it has been widely accepted. The relationship is in need of verification due to the hypothesis involved in its construction which states that the initiation of linear fractures in cadaver skulls corresponds to a concussive state, due to the correlations drawn between cadaver and live animal test results, and due to the extrapolation of the data to the living human. Based on this curve, a head weighing 10 lb can tolerate indefinitely a maximum acceleration of 50 g, independent of the value of the impulse. An impulse of 4.8 lb-sec can be supported without damage for periods of less than 5 msec regardless of the magnitude of the acceleration (Hirsch, 1966). Finally, a combination of the Holbourn theory of rotational acceleration damage involving the 2/3 power of the ratio of the brain masses involved as a scaling factor, together with data from Rhesus monkeys loaded impulsively, has provided a prediction of the limiting rotational velocity of 50 rad/sec for loads with durations of less than 20 msec, and a limiting rotational acceleration of 1800 rad/sec^2 for durations longer than this time before the 50% threshold level of cerebral concussion is reached in man (Ommaya, to appear).

Protective devices for the head must simultaneously serve the functions of broadly distributing the load and reducing the force, acceleration or impulse transmitted to the head by absorbing the energy of the blow. This has been accomplished by the use of protective and yielding coverings both for the head and over areas of likely contact with the head, such as the instrument panels of automobiles, by the employment of whole body and head restraint systems, and by inflatable cushioning mechanisms such as air bags. Helmets, a widely advocated from of head protection, do distribute the load by virtue of the hard exterior shell and have provisions for energy absorption by the presence of an interior liner. However, their benefits with regard to ameliorating one type of hazard is sometimes counterbalanced by augmenting the danger zone with respect to another, such as the greater moment of inertia of the system, or the injury potential to others in the case of the face guard of a football helmet.

The design limitations pertaining to helmet construction imposed by the mechanics and human tolerance involved in head injuries are illustrated by an example cited by Hirsch (1966). Practical considerations limit the thickness of a helmet to the order of 1 in.; thus, the deceleration time over this distance for decelerations of more than 200 g is less than 5 msec, placing a limit of 4.8 lb-sec that can be safely transmitted to the skull according to the data of Fig. 9. Conversely, a crash panel with yielding padding and metallic supports can increase the contact duration to the point of sustaining a deceleration not exceeding 50 g. This can be accomplished by material with a crushing or deforming strength of 50 lb/in^2 for the assumed 10-lb head exhibiting a contact area of 10 in^2.

Hirsch further calculated the upper limit of any proposed cushioning material as equal to the crushing strength of the skull, estimated as equivalent to a deceleration of 400–600 g based on the energy required to fracture cadaver skulls, 400 to 900 lb-in, the time of 1.2 msec from the beginning of contact to the initiation of fracture, and a 10 in^2 contact area. The deceleration time curve for a 1-in thick substance uniformly being crushed to 30% of its original thickness under constant load is shown in Fig. 16. The upper limits of deceleration permitted for cushioning helmet materials of 400 and 500 g specified by the U.S. military and British standards, respectively, correspond to energy absorptions of 2800 and 3500 in-lb and transmitted impulses of 13 and 12 lb-sec for the two cases. However, these values are considered injurious on the basis of the human cerebral concussion tolerance curve; adherence to its safety level provides an energy absorption of only 420 lb-in at a value of deceleration of 60 g. While proper panel design (Patrick, 1966a, 1967) can

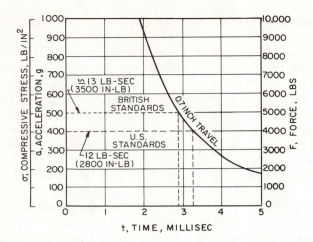

Fig. 16: Average deceleration and force as a function of time in crushing a 1-inch thick ideal material to 30% of its original dimension. Assumed weight of head, 10 lb.; area of striking surface, 10 in². (After Hirsch, 1966; courtesy J. B. Lippincott.)

increase the total energy absorbed in a vehicular impact without serious injury hazard, the transmission characteristics of the entire system must be analyzed to optimize conditions so that the level of impulse or acceleration transmitted to the skull will not lead to danger in the maximum number of cases.

Interesting information on the mechanics and the effectiveness of restraint systems has also been collected (Haley, 1970; Lange and van Kirk, 1970; Snyder, 1970); an account of these investigations is beyond the scope of the present review. Similarly, the enormous amount of data obtained for the proper design of vehicles and their components with regard to occupant safety, including biomechanical considerations of head injury, cannot be included here; the reader is referred to the various Proceedings of the Stapp Car Crash Conferences as an excellent source for this material.

XI. CONCLUDING REMARKS

In reviewing the state of the art of the biomechanics of head injuries, one might ask what has been the recent progress in this area and what are the prospects for the future? There is no doubt that significant advancements have occurred in this field since its last appraisal (Goldsmith, 1966). A number of material and structural properties of the skull and the material properties of other head components have been obtained from tests on live and dead animals and human autopsy material, an area which previously was essentially a vacuum. Structural models have been suggested for some of these tissues, and extrapolations of the data to humans have been proposed. Rudimentary two-component systems composed of continuous deformable media subjected to radial loading have been analyzed, and the sweeping of the parameters of discrete point mass systems on the computer has been continued. Improved anthropomorphic dummies are continuously being constructed whose response under actual impact situations can provide a better index of the stress histories experienced by humans under corresponding conditions than is now available. Further experimentation with animals is providing additional data on tolerance doses under controlled conditions of loading. Clinical examination of automobile victims are being correlated with measured mechanical variables obtained from laboratory tests involving cadavers that produce identical structural deformation to the same vehicle as the damage produced in the original car. This will provide the relationship between head acceleration and injury level outside the presently available spectrum (Patrick and Sato, 1970). The basic mechanisms producing brain damage are being explored further, although there seems to be no more agreement in this area among investigators than existed half a decade ago.

In what direction should further research proceed? There are still noticeable gaps in the list of properties of the cranial tissues even for the Rhesus monkey, the most common primate tested, including such important characteristics as the structural effects of the attachments between the components. Additional extrapolations from these data to functioning persons are needed and hence desirable, but perhaps the ingenuity required for such a procedure might be better devoted to the design and execution of truly harmless tests in the living human from which such information could be derived directly. As an example, one might measure surface waves in the scalp or in the skull by pinpoint probes and deduce material properties from an examination of the system, analogous to the procedure employed in the analysis of the driving-point impedance probe acting on the brain (Engin and Wang, 1970). It is perhaps not inconceivable that, with appropriate medical protocol, a probe of this type might be employed during actual surgical operations. Further use of radiographic techniques to monitor the motion of suitably treated particles implanted in the blood might provide more information of the motion of the cranial contents under low-level dynamic loading of humans.

In a different vein, the construction and analysis of mathematical models can proceed to a more sophisticated level. Use of nonradial loads acting on a spherically symmetric two-component system requires the introduction of an additional spatial variable and hence a more complicated set of partial differential equations, but their solution, while perhaps involving an order of magnitude of greater numerical effort, does not present a major problem in principle. Alternatively, three-component spherically symmetric models can be attacked, or else the geometry of the model can be altered to conform more closely to the actual shape of the human head. The shell simulating the cranium can be supplied with a hole to resemble the foramen magnum through which fluid could be pumped from a reservoir as a function of the intracranial pressure under dynamic loading conditions and attached to a device portraying the head-neck junction. In considering more advanced systems, there is an obvious practical limit to the introduction of geometrical refinements in the model. On the basis of the velocity of wave propagation and the shortest expected loading duration, it appears reasonable that questions of inhomogeneity in the brain need not be scaled below the level of one inch. Concurrently, the development of inanimate head models featuring materials with a closer resemblance to the physical properties of living human tissues should be pursued to assist in the validation of the analytical predictions concerning the response of the sytem under impact. Some progress has already been made in the selection or construction of such manufactured substances. Pursuit of this area might conceivably have the ancillary benefit of providing replacement parts for the human head with improved biocompatibility.

Finally, it may soon become practical to examine the mechanical and physiological responses of the individual cells to the controlled application of stresses either computed from a model or otherwise ascertained as acting in various regions of the cranium as the result of a given blow or loading. This information, when obtained, will serve as the bridge between tolerance limits based on macroscopic mechanical parameters and the functional effects, a vitally needed piece of information. The time span required to acquire a satisfactory level of understanding of these processes appears enormous but may be limited to tolerable proportions provided enough energy, talent and financial resources are applied to this task.

REFERENCES

Anzelius, A. 1943. The effect of an impact on a spherical liquid mass. Acta Pathol. Microbiol. Scand., Suppl. 48: 153–59.
Benedict, J. V., E. H. Harris and D. U. von Roseberg 1970. An analytical investigation of the cavitation hypothesis of brain damage. Trans. ASME. J. Basic Eng., 920: 597–603.
Denny-Brown, D. 1945. Cerebral concussion. Physiol. Rev. 25: 296–325.
Denny-Brown, D. and W. R. Russell 1941. Experimental cerebral concussion. Brain 64: 93–164.
Dodgson, M. C. H. 1962. Colloidal structure of the brain. Biorheol. 1: 21–30.
Douglass, J. M., A. M. Nahum and S. B. Roberts 1968. Applications of experimental head injury research. *In* Proceedings of the 12th Stapp Car Crash Conf. Soc. of Auto. Eng., New York. pp. 317–37.
Edberg, S., J. Rieker and A. Angrist 1963. Study of impact pressure and acceleration in plastic skull models. Lab. Invest. 12: 1305–1311.

Engin, A. E. 1969. The axisymmetric response of a fluid-filled spherical shell to a local radial impulse—a model for head injury. J. Biomech. 2: 324–41.
Engin A. E. and H.-C. Wang 1970. A mathematical model to determine viscoelastic behavior of *in vivo* primate brain. J. Biomech. 3: 283–96.
Estes, M. S. and J. H. McElhaney 1970. Response of brain tissue to compressive loading. ASME Paper No. 70-BHF-13.
Evans, F. G. and H. R. Lissner 1957. Tensile and compressive strength of human parietal bone. J. Appl. Physiol. 10: 493–97.
Evans, F. G., H. R. Lissner and M. Lebow 1958. The relation of energy, velocity and acceleration to skull deformation and fracture. Surg. Gyn. Obst. 107: 593–601.
Fallenstein, G. T., V. D. Hulce and J. W. Melvin 1969. Dynamic mechanical properties of human brain tissue. J. Biomech. 2: 217–26.
Flynn, P. D. 1966. Dynamic photoelastic stress patterns from a simplified model of the head. *In* W. F. Caveness and A. E. Walker (eds.), Head Injury Conference Proceedings. J. B. Lippincott, Philadelphia. pp. 344–49.
Franke, E. K. 1956. Response of the human skull to mechanical vibrations. J. Acoust. Soc. Amer. 28: 1277–284.
Friede, R. L. 1960. Specific cord damage at the atlas level as a pathogenic mechanism in cerebral concussion. J. Neuropath. Exper. Neurol. 19: 266–70.
Friede, R. L. 1961. The pathology and mechanics of experimental cerebral concussion. WADD Tech. Rep. 6-256. ASTIA Doc. No. 266210.
Galford, J. E. and J. H. McElhaney 1970. A viscoelastic study of scalp, brain, and dura. J. Biomech. 3: 211–22.
Goldsmith, W. 1966. The physical processes producing head injuries. *In* W. F. Caveness and A. E. Walker (eds.), Head Injury Conference Proceedings. J. B. Lippincott, Philadelphia. pp. 350–82.
Gross, A. G. 1958*a*. A new theory on the dynamics of brain concussion and brain injury. J. Neurosur. 15: 548–61.
Gross, A. G. 1958*b*. Impact thresholds of brain concussion. Aviation Med. 29: 725–32.
Gurdjian, E. S. and H. R. Lissner 1961. Photoelastic confirmation of the presence of shear strain at the craniospinal junction in closed head injury. J. Neurosurg. 18: 58–60.
Gurdjian, E. S., J. E. Webster and H. R. Lissner 1950. The mechanism of skull fracture. J. Neurosurg. 7: 106–14.
Gurdjian, E. S., H. R. Lissner, F. R. Latimer, B. F. Haddad and J. E. Webster 1953. Quantitative determination of acceleration and intracranial pressure in experimental head injury. Neurol. 3: 417–23.
Gurdjian, E. S., J. E. Webster, F. R. Latimer and B. F. Haddad 1954. Studies on experimental concussion. Relation of physiological effect to time duration of intracranial pressure increase at impact. Neurol. 4: 674–81.
Gurdjian, E. S., J. E. Webster and H. R. Lissner 1955. Observations on the mechanism of brain concussion, contusion, and laceration. Surg. Gyn. Obstet. 101: 680–90.
Gurdjian, E. S., H. R. Lissner, F. G. Evans, L. M. Patrick and W. G. Hardy 1961. Intracranial pressure and acceleration accompanying head impacts in human cadavers. Surg. Gyn. Obstet. 112: 185–90.
Gurdjian, E. S., H. R. Lissner and L. M. Patrick 1962. Protection of the head and neck in sports. J. Amer. Med. Assoc. 182: 509–12.
Gurdjian, E. S., L. M. Thomas, V. R. Hodgson and L. M. Patrick 1968. Impact head injury. Gen. Pract. 37: 78–87.
Gurdjian, E. S., D. Gonzales, V. R. Hodgson, L. M. Thomas and S. W. Greenberg 1970*a*. Comparisons of research in inanimate and biologic material: Artifacts and pitfalls. *In* E. S. Gurdjian, W. A. Lange, L. M. Patrick and L. M. Thomas (eds.), Impact Injury and Crash Protection. C. C. Thomas, Springfield. Ill. pp. 234–53.
Gurdjian, E. S., V. R. Hodgson and L. M. Thomas 1970*b*. Studies on mechanical impedance of the human skull: Preliminary report. J. Biomech. 3: 239–48.
Güttinger, W. 1950. Der Stosseffekt auf eine Flüssigkeitskugel als Grundlage einer physikalischen Theorie der Entstehung von Gehirnverletzungen. Zeit. Naturforsch. 5a: 622–28.

Haley, J. L. 1970. Fundamentals of kinetics and kinematics as applied to injury reduction, *In* E. S. Gurdjian, W. A. Lange, L. M. Patrick and L. M. Thomas (eds.), Impact injury and crash protection. C. C. Thomas, Springfield, Ill. pp. 423–41; discussion by G. Grime, pp. 442–43.

Hayashi, T. 1969. Study of intracranial pressure caused by head impact. J. Fac. of Eng. Univ. Tokyo (B) 30: 59–72; 117–24.

Haynes, A. L. and H. R. Lissner 1962. Experimental head impact studies. *In* Proceedings of the Fifth Stapp Automotive Crash and Field Demonstration Conference. Univ. Minn., Minneapolis. pp. 158–70.

Haynes, R. H., J. H. McElhaney and J. L. Fogle 1969. Mechanical properties of the skull. *In* Proceedings, 6th Annual Rocky Mountain Bioengineering Symposium. Laramie: 62–66.

Higgins, L. S., R. A. Schmall, C. P. Cain, P. E. Kielpinski, F. P. Primiano, T. W. Barber and J. A. Brockway 1967. The investigation of the parameters of head injury related to acceleration and deceleration. Technology, Inc., Life Sci. Div. San Antonio. Defense Documentation Center, Arlington, Virginia 659795.

Hirsch, A. E. 1966. Current problems in head protection. *In* W. F. Caveness and A. E. Walker (eds.), Head Injury Conference Proceedings. J. B. Lippincott, Philadelphia. pp. 37–40

Hirsch, A. E., A. K. Ommaya and R. H. Mahone 1970. Tolerance of subhuman primate brain to cerebral concussion. *In* E. S. Gurdjian, W. A. Lange, L. M. Patrick, and L. M. Thomas (eds.), Impact Injury and Crash Protection. C. C. Thomas, Springfield, Ill. pp. 352–69; discussion by E. J. Gurdjian, pp. 370–71.

Hodgson, V. R. 1968. Head impact response of several mammals including the human cadaver. Dissertation (Ph. D.). Wayne State Univ. Detroit.

Hodgson, V. R. 1970. Physical factors related to experimental concussion. *In* E. S. Gurdjian, W. A. Lange, L. M. Patrick and L.M. Thomas (eds.), Impact Injury and Crash Protection. C. C. Thomas, Springfield. Ill. pp. 275–302; discussion by A. K. Ommaya, pp. 303–307.

Hodgson, V. R. and L. M. Patrick 1968. Dynamic response of the human cadaver head compared to a simple mathematical model. *In* Proceedings of the 12th Stapp Car Crash Conference. Soc. Auto. Eng., New York. pp. 280–301.

Hodgson, V. R., E. S. Gurdjian and L. M. Thomas 1967. The determination of response characteristics of the head with emphasis on mechanical impedance techniques. *In* Proceedings of the 11th Stapp Car Crash Conference. Soc. Auto. Eng., New York. pp. 79–85.

Holbourn, A. H. S. 1943. Mechanics of head injuries. Lancet 2: 438–41.

Holbourn, A. H. S. 1945. Mechanics of brain injuries. Brit. Med. Bull. 3: 147–49.

Hollister, N. R., W. P. Jolley and R. G. Horne 1958. Biophysics of concussion. WADC Tech. Rep. 58–193. ASTIA Doc. No. AD 203385.

Hubbard, R. P. 1970. Flexure of cranial bone. Dissertation (Ph. D.). University of Michigan, Ann Arbor.

Kihlberg, J. K. 1966. Head injury in automobile accidents. *In* W. F. Caveness and A. E. Walker (eds.), Head Injury Conference Proceedings. J. B. Lippincott, Philadelphia. pp. 27–36.

Kihlberg, J. K. 1970. Multiplicity of injury in automobile accidents. *In* E. S. Gurdjian, W. A. Lange, L. M. Patrick and L. M. Thomas (eds.), Impact Injury and Crash Protection. C.C. Thomas, Springfield, Ill. pp. 5–24.

Koeneman, J. B. 1966. Viscoelastic properties of brain tissue. Thesis (M.S.). Case Inst. Tech. Cleveland, Ohio.

Kopecky, J. A. and E. A. Ripperger 1969. Closed brain injuries: An engineering analysis. J. Biomech. 2: 29–34.

Lange, W. A. and D. J. van Kirk 1970. The effectiveness of current methods and systems used to reduce injury. *In* E. S. Gurdjian, W. A. Lange, L. M. Patrick and L. M. Thomas (eds.), Impact Injury and Crash Protection. C. C. Thomas, Springfield, Ill. pp. 475–93; discussion by R. G. Snyder, pp. 494–95.

Lee, Y. C. and S. H. Advani 1970. Transient response of a sphere to symmetric torsional loading—a head injury model. Math. Biosci. 6: 473–487.
Leeson, C. R. and T. S. Leeson 1966. Histology. W. B. Saunders, Philadelphia.
Lindgren, S. O. 1966. Experimental studies of mechanical effects in head injury. Acta Chir. Scand. Suppl. 360. Stockholm.
Lissner, H. R., M. Lebow and F. G. Evans 1960. Experimental studies on the relation between acceleration and intracranial pressure changes in man. Surg. Gyn. Obstet. 111: 329–38.
Lombard, C. F. 1951. Voluntary tolerance of the human to impact accelerations of the head. J. Aviation Med. 22: 109–16.
Martinez, J. L. 1963. Study of whiplash injuries in animals. ASME Paper 63-WA-281.
Martinez, J. L. 1969. Personal communication.
Martinez, J. L. 1970. High-speed flash X-ray and cinematography in injury research. In E. S. Gurdjian, W. A. Lange, L. M. Patrick and L. M. Thomas (eds.), Impact Injury and Crash Protection. C. C. Thomas, Springfield, Ill. pp. 196–207; discussion by V. R. Hodgson, pp. 208–209. Discussion by D. J. Sass. Simple radiographic techniques for recording brain motion during acceleration. pp. 210–13.
Martinez, J. L. and D. J. Garcia 1968. A model for whiplash. J. Biomech. 1: 23–32.
Martinez, J. L., J. K. Wickstrom and B. T. Barcelo 1965. The whiplash injury—a study of head-neck action and injuries in animals. ASME Paper 65-WA/HuF-6. (Also appears in ASME Biomechanics Monograph 1967).
Mazur, L. 1961. Mechanism of production of hematomas and of contusion of the brain. Neurol. Neurochir. Psychiat. Pol. 11: 445–55.
McElhaney, J. H. 1966. Dynamic response of bone and muscle tissue. J. Appl. Physiol. 20: 1231–236.
McElhaney, J. H., J. L. Fogle, E. Byars and G. Weaver 1964. Effect of embalming on the mechanical properties of beef bone. J. Appl. Physiol. 19: 1234–236.
McElhaney, J. H., R. L. Stalnaker, M. S. Estes and L. S. Rose 1969. Dynamic mechanical properties of scalp and brain. In Proceedings of the 6th Annual Rocky Mountain Bioengineering Symposium. Laramie: 67–73.
McElhaney, J. H., J. L. Fogle, J. W. Melvin, R. R. Haynes, V. L. Roberts and N. M. Alem 1970. Mechanical properties of cranial bone. J. Biomech. 3: 495–512.
McHenry, R. R. 1970. Mathematical models for injury prediction. In E. S. Gurdjian, W. A. Lange, L. M. Patrick, and L. M. Thomas (eds.), Impact Injury and Crash Protection. C. C. Thomas, Springfield, Ill. pp. 214–33.
McHenry, R. R. and K. N. Naab 1966. Computer simulation of the crash victim—a validation study. In Proceedings of the 12th Stapp Car Crash Conference. Soc. Auto.. Eng., New York. pp. 126–63.
Melvin, J. W., D. H. Robbins and V. L. Roberts 1970a. The mechanical behavior of the diploë layer in the human skull in compression. In Developments in Mechanics. Vol. 5. Proc. 11th Midwest. Mech. Conf. Iowa State University Press pp. 811–18.
Melvin, J. W., P. M. Fuller and I. T. Barodawala 1970b. The mechanical properties the diploë layer in the human skull. SESA Paper, Spring Meeting. Huntsville, Ala.
Mertz, H. J. and L. M. Patrick 1967. Investigation of kinematics and kinetics of whiplash. In Proceedings of the 11th Stapp Car Crash Conference. Soc. Auto. Eng., New York. pp. 175–206.
Metz, H., J. H. McElhaney and A. K. Ommaya 1970. A comparison of the elasticity of live, dead, and fixed brain tissue. J. Biomech. 3: 453–458.
Mundie, J. R., R. L. Friede, L. O. Hoeft and H. E. von Gierke 1967. Correlation of pathology with physical factors in cerebral contusion. Aerospace Med. Res. Lab. Tech. Rep. Wright-Patterson Air Force Base. (cf. von Gierke, 1966).
Naab, K. N. 1966. Measurement of detailed inertial properties and dimensions of a 50th percentile anthropometric dummy. In Proceedings of the 12th Stapp Car Crash Conference. Soc. Auto. Eng. New York. pp. 187–95.
Oestreicher, H. L. 1951. Field and impedance of an oscillating sphere in a viscoelastic medium with application to biophysics. J. Acoust. Soc. Amer. 23: 707–14.

Ommaya, A. K. 1966. Experimental head injury in the monkey. *In* W. F. Caveness and A. E. Walker (eds.), Proceedings of the Head Injury Conference. J. B. Lippincott, Philadelphia. pp. 260–75.
Ommaya, A. K. 1968. Mechanical properties of tissues of the nervous system. J. Biomech. 1: 127–38.
Ommaya, A. K. 1970. The physiopathology of head injuries. Personal communication.
Ommaya, A. K. In press. The tolerance of man's brain to blunt injury. *In* Brain.
Ommaya, A. K. and P. Corrao 1969. Pathologic biomechanics of central nervous system injury in head impact and whiplash trauma. *In* Proceedings of the International Conference on Accident Pathology. Government Printing Office, Washington D.C.
Ommaya, A. K., S. D. Rockoff and M. Baldwin 1964. Experimental concussion. J. Neurosurg. 21: 249–65.
Ommaya, A. K., A. E. Hirsch, E. S. Flamm and R. H. Mahone 1966a. Cerebral concussion in the monkey: An experimental model. Science 153: 211–12.
Ommaya, A. K., A. E. Hirsch and J. Martinez 1966b. The role of whiplash in cerebral concussion. *In* Proceedings of the Tenth Stapp Car Crash Conference. Soc. Auto. Eng. New York. pp. 197–203.
Ommaya, A. K., P. Yarnell, A. E. Hirsch and E. H. Harris 1967. Scaling of experimental data on cerebral concussion in subhuman primates to concussion threshold for man. *In* Proceedings of the Eleventh Stapp Car Crash Conference. Soc. Auto. Eng., New York. pp. 47–52.
Ommaya, A. K., F. Fass and P. Yarnell 1968. Whiplash injury and brain damage: An experimental study. J. Amer. Med. Assoc. 204: 285–89.
Ommaya, A. K., J. W. Boretos and E. E. Beile 1969. The Lexan calvarium: An improved method for direct observation of the brain. J. Neurosurg. 30: 25–29.
Patrick, L. M. 1966a. Cadaver windshield impact research. Plastic Reconstructive Surg. 37: 314–23.
Patrick, L. M. 1966b. Head impact protection. *In* W. F. Caveness and A. E. Walker (eds.), Head Injury Conference Proceedings. J. B. Lippincott, Philadelphia. pp. 41–48.
Patrick, L. M. 1967. Prevention of instrument panel and windshield head injuries. *In* The Prevention of Highway Injury. Highway Safety Res. Inst., Univ. Mich. pp. 169–81.
Patrick, L. M. and R. P. Daniel 1966. Comparison of standard and experimental windshields. *In* L. M. Patrick (ed.), Proceedings of the Eighth Stapp Car Crash and Field Demonstration Conference. Wayne State Univ. Press, Detroit. pp. 147–66.
Patrick, L. M. and H. J. Mertz 1966. Impact dynamics of unrestrained, lap belted and lap and diagonal chest belted vehicle occupants. *In* Proceedings of the 12th Stapp Car Crash Conference. Soc. Auto. Eng., New York. pp. 46–93.
Patrick, L. M. and G. Grime 1970. Applications of human tolerance data to protective systems: Requirements for soft tissue, bone, and organ protective device. *In* E. S. Gurdjian, W. A. Lange, L. M. Patrick and L. M. Thomas (eds.), Impact Injury and Crash Protection. C. C. Thomas, Springfield, Ill. pp. 444–73; discussion by C. W. Gadd, p. 474.
Patrick, L. M. and T. B. Sato 1970. Methods of establishing human tolerance levels: Cadaver and animal research and clinical observations. *In* E. S. Gurdjian, W. A. Lange, L. M. Patrick and L. M. Thomas (eds.), Impact Injury and Crash Protection. C. C. Thomas, Springfield, Ill. pp. 259–73.
Pudenz, R. H. and C. H. Shelden 1946. The lucite calvarium—a method for direct observation of the brain. II. Cranial trauma and brain movement. J. Neurosurg. 3: 487–505.
Rinder, L. 1969. Experimental brain concussion by sudden intracranial input of fluid. Univ. Göteborg, Göteborg, Sweden.
Robbins, D. H. and J. L. Wood 1969. Determination of mechanical properties of the bones of the skull. Exper. Mech. 9: 236–40.
Roberts, S. B., C. C. Ward and A. M. Nahum 1969. Head trauma—a parametric dynamic study, J. Biomech. 2: 397–416.
Roberts, V. L. and J. W. Melvin 1969. The measurement of the dynamic mechanical properties of human skull bone. Appl. Poly. Symp. 12: 235–47.

Roberts, V. L., V. R. Hodgson and L. M. Thomas 1966a. Fluid pressure gradients caused by impact to the human skull. ASME Paper No. 66-HUF-1.
Roberts, V. L., E. L. Stech and C. T. Terry 1966b. Review of mathematical models which describe human response to acceleration. ASME Paper No. 66/WA/BHF-13.
Schaeffer, J. P. (ed.) 1942. Morris' Human Anatomy. 10th ed. The Blakiston Company, Philadelphia.
Sellier, K. and F. Unterharnscheidt, 1963. Mechanik und Pathomorphologie der Hirnschäden nach stumpfer Gewalteinwirkung auf den Schädel. In Hefte Unfallheilkunde. Vol. 76. Springer, Berlin. 124 pp.
Sellier, K. and F. Unterharnscheidt 1965. Mechanik der Gewalteinwirkung auf den Schädel. Excerpta Medica, Int. Cong. Ser. 93: 55–61.
Severy, D. M., J. H. Mathewson and C. O. Bechtol 1955. Controlled automobile rear-end collisions: an investigation of related engineering and medical phenomena. Can. Ser. Med. J. 11: 727–59.
Severy, D. M., J. M. Brink and J. D. Baird 1968. Back rest and head restraint design for rear end collision protection. SAE Paper No. 680079.
Shuck, L. Z., R. R. Haynes and J. L. Fogle 1970. Determination of viscoelastic properties of human brain tissue. ASME Paper No. 70-BHF-12.
Sjövall, H. 1943. The genesis of skull and brain injuries. Acta Pathol. Microbiol. Scand. Suppl. 48: 1–152.
Snively, G. G. and C. O. Chichester 1961. Impact survival levels of head acceleration in man. Aerospace Med. 87: 316–20.
Snyder, R. G. 1970. Occupant restraint systems of automotive, aircraft, and manned space vehicles: An evaluation of the state of the art and future concepts. In E. S. Gurdjian, W. A. Lange, L. M. Patrick and L. M. Thomas (eds.), Impact Injury and Crash Protection. C. C. Thomas, Springfield, Ill. pp. 496–561; discussion by J. P. Stapp, pp. 562–63.
Stalnaker, R. L., J. L. Fogle and J. H. McElhaney 1970. Driving point impedance characteristics of the head. ASME Paper 70-BHF-14.
Stapp, J. P. 1967. The problem: Biomechanics of injury. In Prevention of Highway Injury. Highway Safety Res. Inst., Univ. Mich. pp. 159–64.
Stapp, J. P. 1970. Voluntary human tolerance levels. In E. S. Gurdjian, W. A. Lange, L. M. Patrick and L. M. Thomas (eds.), Impact Injury and Crash Protection. C. C. Thomas, Springfield, Ill. pp. 308–49; discussion by C. L. Ewing, pp. 350–51.
Strich, S. J. 1969. The pathology of brain damage due to blunt head injuries. In A. E. Walker (ed.), The Late Effects of Head Injury. C. C. Thomas, Springfield, Ill. pp. 501–26.
Thomas, L. M. 1970. Mechanisms of head injury. In E. S. Gurdjian, W. A. Lange, L. M. Patrick and L. M. Thomas (eds.), Impact injury and crash protection. C. C. Thomas, Springfield. Ill. pp. 27–42; discussion by F. Unterharnscheidt, pp. 43–62.
Thomas, L. M., V. L. Roberts and E. S. Gurdjian 1967. Impact-induced pressure gradients along three orthogonal axes in the human skull. J. Neurosurg. 26:316–21.
Unterharnscheidt, F. 1962. Experimentelle Untersuchungen über gedeckte Schäden des Gehirns nach einmaliger und wiederholter stumpfer Gewalteinwirkung auf den Schädel. Fortschritte Med. 80: 369–78.
Unterharnscheidt, F. and K. Sellier 1966. Mechanics and pathomorphology of closed brain injuries. In W. F. Caveness and A. E. Walker (eds.), Head Injury Conference Proceedings. J. B. Lippincott, Philadelphia. pp. 321–41.
Unterharnscheidt, F. and L. S. Higgins 1969. Traumatic lesions of brain and spinal cord due to nondeforming angular acceleration of the head. Texas Rep. Biol. Med. 27: 127–66.
Valanis, K. C. and C. T. Sun 1967. Axisymmetric wave propagation in a solid viscoelastic sphere. Int. J. Eng. Sci. 5: 939–56.
von Gierke, H. E. 1964. Biodynamic response of the human body. Appl. Mech. Rev. 17: 951–58.
von Gierke, H. E. 1966. On the dynamics of some head injury mechanisms. In W. F. Caveness and A. E. Walker (eds.), Head Injury Conference Proceedings. J. B. Lippincott, Philadelphia. pp. 383–96.

von Gierke, H.E., H. L. Oestreicher, E. K. Franke, H. O. Parrack and W. W. von Wittern 1952. Physics of vibrations in living tissue. J. Appl. Physiol. 4: 886–900.
Ward, J. W., L. H. Montgomery and S. L. Clark 1948. A mechanism of concussion: A theory. Science 107: 349–53.
Wickstrom, J., J. L. Martinez, D. Johnston and N. C. Tappen 1965. Acceleration-deceleration injuries of the cervical spine in animals. In D. M. Severy (ed.), Proceedings of the Seventh Stapp Car Crash Conference. C. C. Thomas, Springfield Ill. pp. 276–84.
Wickstrom, J., J. Martinez and R. Rodriguez 1967. Cervical sprain syndrome: Experimental acceleration injuries of the head and neck. In The Prevention of Highway Injury. Highway Safety Res. Inst., Univ. Mich. pp. 182–87.
Wood, J. L. 1969. Mechanical properties of human cranial bone in tension. Dissertation (Ph. D.). Univ. Mich., Ann Arbor, Mich.

Index

A

Afterload, 306–08
Air-fluid surface, 320–21
Anemia, microangiopathic hemolytic, 506–07
Animals, experiments in, 611–13
Anisotropic arterial elasticity, 126–27
Anisotropy, skin, 154–55
Arteries
 anisotropic elasticity, 126–27
 branching effects, 392–96
 D'Alembert's solution, 386–89
 damping, 404–08
 equations of motion, 382–86
 flow in, 381–434
 flow pulses in conduits, 369–77
 input impedance, 132–33, 408–15
 law of similarity, 418–23
 leakage effects, 392–96
 longitudinal impedance, 396–401
 pressure in, 381–434
 pressure in conduits, 369–77
 shape of the distensibility curves of, 127–29
 simplified wave equations, 386–89

Arteries (*Contd*)
 tapering effects, 390–96
 transverse impedance, 401–04
 wave transmission, 404–08, 415–18
Artificial limbs, 536–46
Atrophy, functional, 260–61
Automobile impact problem, 568–74
Autoregulation, 7

B

Biomaterials, tissue compatibility and, 53–56
Biomechanics, defined, 29
Bleeding, extracerebral, 594
Blood
 description of, 64–66
 hemolysis in, 501–28
 liquid gas surface effects, 70–71
 problems in determining flow properties, 66–73
 red cells
  migrational effects at low shear rates, 83–89
  properties of, 464–71
  sedimentation, 71–72

Blood (*Contd*)
  rheology of, 63–103
  smooth-wall (Vand) effect, 72–73
  typical behavior, 73–83
  validity of the continuum model of, 95–100
  yield stress measurements, 89–95
Blood vessels, 105–39
  arterial input impedance, 132–33
  changes in the mechanical behavior of, 361–65
  deformation of elastic materials, 106–13
  describing mechanical properties of, 109–10
  elasticity of, 116–29
    anisotropic arterial, 126–27
    dynamic, 124–26
    shape of distensibility curves, 127–29
    static, 116–24
    visco-, 108–09
  physiological significance, 129–34
  pressurized isotropic homogeneous, 110–13
  pulse–wave velocity, 129–32
  sense organs in walls of, 133–34
  strain, 106–07
  stress, 106–07
  structure of, 113–16
  waves in, 340–61
  the well as an elastomer, 107–08
Bones, 237–71
  compound bar hypothesis, 239–41
  density of, 241–52
  functional adaptation of, 260–69
  functional construction of, 253–60
  long, gross shape of, 253–58
  as a material of construction, 238–52
  Pauwels' hypothesis, 260–66
  prestress hypothesis, 241
  spongy, functional architecture of, 258–60
  strength of, 241–52
  as a system, 261–63
  two-phase hypothesis, 239

Brain damage, 594–95
  proposed mechanisms of, 597–99
Branching effects, 392–96

C

Cadaver experiments, 608–11
Capillary-tube viscometers, 68–69
Cardiopulmonary biomechanics, 20–26
Cardiovascular system, 5
Cells
  contents altered, 509
  red
    migrational effects at low shear rates, 83–89
    properties of, 464–71
    sedimentation, 71–72
Circulatory system, 337–79
  changes in mechanical behavior of blood vessels, 361–65
  distensibility of heart ventricles, 366–69
  flow pulses in arterial conduits, 369–77
  pressure in arterial conduits, 369–77
  waves in blood vessels, 340–61
Collagen
  lungs, 319
  skin as, 143
  tendon as, 143
Collisions, characteristics of, 595–97
Compound bar hypothesis, 239–41
Compressive tests, 158
Concentric-cylinder viscometers, 67, 69
Conduits, arterial
  flow pulses in, 369–77
  pressure in, 369–77
Cone-and-plate viscometers, 67–68, 69–70
Contractility of heart, 289–302, 309–15
Cranial system, 588–93
Crashworthiness, 579
Creep function, reduced, 194

D

D'Alembert's solution, 386–89
Damping, 404–08

Data processors, development of, 56–57
Deformation
  of elastic materials, 106–13
  states of, 171–72
Diaphragm tests, 158
Distensibility
  curves, arterial, 127–29
  of heart ventricles, 366–69
Distensile tests, 158–59
Dynamic elasticity, 124–26

E

Elastic materials, deformation of, 106–13
Elastic response, 189, 193, 195–96
Elasticity
  blood vessels, 116–29
    anisotropic arterial, 126–27
    dynamic, 124–26
    shape of distensibility curves, 127–29
    static, 116–24
    visco-, 108–09
  modulus of, 239
  tissue and the postulates of, 164–65
Elastin, lungs, 319–20
Elastomechanics, finite, 162–75
Elastomer, vessel wall as, 107–08
Elastomeric polymers, constitutive laws for, 165–68
Electromagnetic catheter systems, 440–43
Electromagnetic flowmetering, 436–43
Energy absorbing systems, 579–81
Energy function in strain, 168–71
Environmental ions, 222–26
Equations
  motion, 382–86
  wave, 386–89
Extensional strain, 238–52
Extracerebral bleeding, 594

F

Flexible particles, 481–90
Flow
  in arteries, 381–434
  measurement, 435–55

Flow (Contd)
  particle, 472–90
  problems in determining properties, 66–73
  pulsatile, 423–27
  pulses in arterial conduits, 369–77
Flowmetering, electromagnetic, 436–43
Fragmentation, 509–10
Functional atrophy, 260–61
Functional hypertrophy, 260–61

G

Gas, liquid, 70–71

H

Hart–Smith materials, 171
Head injuries, 585–634
  analytical representations, 599–608
  brain damage, 594–95
    proposed mechanisms of, 597–99
  cadaver experiments, 608–11
  collisions and material behavior, 595–97
  the cranial system, 588–93
  experiments on animals, 611–13
  human tolerance, 622–26
  inanimate model, 608–11
  mechanical properties of head tissues, 613–22
  protective devices, 622–26
  types, 593, 595
Head tissues, mechanical properties of, 613–22
Heart
  afterload, 306–08
  assist devices, 549–65
  contractility, 289–302, 309–15
  determinants of performance, 303–16
  dimensional changes during contraction, 291–93
  distensibility of ventricles, 366–69
  mechanics of contraction, 289–302
  preload, 304–06
  stress–strain relationships, 296–99
  structure–function relationships, 289–91

Heart (*Contd*)
  valvular disease, 504–06
  ventricular wall stress, 293–96
Hematoma, 594
Hemolysis, 501–28
Human tolerance, 622–26
Hydrostatic pressure, 281–85
Hypertrophy, functional, 260–61

I

Impedance
  input, 132–33, 408–15
  longitudinal, 396–401
  transverse, 401–04
Inanimate models, 608–11
Indentation tests, 158, 159
Injury, 17–20
Input impedance, 132–33, 408–15
Intensive care units, 36–38
  prototype of, 57–58
Interstitial space, 273–86
  hydrostatic pressure, 281–85
  osmotic pressure, 281–85
  physiologic significance, 275–78
  structure of, 273–75
  swelling pressure, 278–81
Ions, environmental, 222–26
Isotropic homogeneous tube, pressurized, 110–13

J

Jet test, 515–18

L

Langer's lines, 154–55
Leakage effects, 392–96
Limbs, artificial, 536–46
Liquid gas, surface effects, 70–71
Living systems, nature of, 30–32
Loading, stress response in, 186–92
Longitudinal impedance, 396–401
Lungs, 317–35
  air–fluid surface, 320–21
  anatomy, 322–25

Lungs (*Contd*)
  collagen, 319
  continuum formulation, 330–32
  discrete element approaches, 332
  elastin, 319–20
  force bearing materials, 318–35
  single-degree-of-freedom model, 325–28
  spring network model, 329–30
  structure of, 318–25
  surfactant, 320
  volume–pressure behavior, 325–28

M

Macrometrology of tendons, 156–57
Mechanical lysis, predicting, 524–25
Mechanical response, general features of, 182–85
Medicine, 32–35, 36–57
  biomaterials and tissue compatibility, 53–56
  data processors, 56–57
  development of sensors, 56–57
  intensive care units, 36–38
    prototype of, 57–58
  need for sensors, 52–53
  oxygen transport system, 41–51
    application of, 51–52
  performance analysis, 40–41
  performance indicators, 38–40
  specific sensors, 56–57
Microangiopathic hemolytic anemia, 506–07
Microcirculation, 457–99
  historical background, 460–64
  mechanics as a whole, 490–96
  particle flow, 472–90
  plasma properties, 464–71
  red cell properties, 464–71
Mooney–Rivlin materials, 171
Motion, equations of, 382–86
Mucopolysaccharides, 275–78
Muscle
  mechanics, 6, 209–15
  –tendon systems, 152–54
Myogenic tone, 7

N

Neo-Hookean materials, 171
Nontensile test configurations, 157–59

O

Ompatibility, structural, 532–36
Osmotic pressure, 281–85
Oxygen transport system, 41–51
   applications of, 51–52

P

Particles
   flexible, 481–90
   flow, 472–90
   rigid, 472–81
Patient care, 29–60
Pauwels' hypothesis, 260–66
Physiology, 3–13
Plasma, properties of, 464–71
Preconditioning, 182, 183
Preload, 304–06
Pressor tests, 158
Pressure
   in arterial conduits, 369–77
   in arteries, 381–434
   hydrostatic, 281–85
   osmotic, 281–85
   pulsatile, 423–27
   swelling, 278–81
   tissue, 281–85
   –volume behavior, 325–28
Prestress hypothesis, 241
Prosthetic devices, 531–48
   artificial limbs, 536–46
   structural compatibility, 532–36
Prosthetic valves, hemolysis in, 523–24
Protective devices, 622–26
Pulmonary system, 6
Pulsatile flow, 423–27
Pulsatile pressure, 423–27
Pulse–wave velocity, 129–32
Puncture tests, 158

Q

Quasi-static stress, 147–49

R

Red cells
   migrational effects at low shear rates, 83–89
   properties of, 464–71
   sedimentation, 71–72
Reduced creep function, 194
Reduced relaxation function, 193, 196–203
Rehabilitation, 29–60
Relaxation function, 193
   reduced, 193, 196–203
Resting tension, 213
Rheology of blood, 63–103
Rigid particles, 472–81

S

Scalp damage, 593
Sedimentation, red blood cells, 71–72
Sense organs in blood vessel walls, 133–34
Sensors
   development of, 56–57
   need for, 52–53
   specific, 56–57
Shear rates, migrational effects, 83–89
Similarity, law of, 418–23
Skin
   anisotropy, 154–55
   as collagen, 143
   composition, 147
   correlations under quasi-static stress, 147–49
   distensile tests, 158–59
   indentation test, 158, 159
   Langer's lines, 154–55
   macrobehavior, 145–55
   mechanical indicators, 149–52
   microbehavior, 145–55
   nontensile test configurations, 157–59
   pressor tests, 158
   properties of, 141–79
   states of deformation, 171–72
   as structure, 142
   structure of, 147

Skin (*Contd*)
    suction test, 159
    swelling pressure, 278–81
Skull fracture, 594
Smooth-wall effect, 72–73
Static elasticity, 116–24
Strain
    blood vessels, 106–07
    energy function and its derivatives, 168–71
    extensional, 238–52
    –stress relationship
        heart, 296–99
        history, 181–208
Stress
    blood vessels, 106–07
    generalized, 167
    quasi-static, 147–49
    response in loading and unloading, 186–92
    –strain relationship
        heart, 296–99
        history, 181–208
    ventricular wall, 293–96
    yield measurements, 89–95
Suction tests, 159
Surfactant, 320
Surgery, 15–27
Surgical wound, the, 15–17
Swelling pressure, 278–81

T

Tapering, effects of, 390–96
Temperature, viscoelastic properties and, 219–21
Tendons
    boundary conditions in testing, 159–62
    as collagen, 143
    composition, 146–47
    correlations under quasi-static stress, 147–49
    macrobehavior, 145–55
    macrometrology of, 156–57
    mechanical indicators, 149–52
    microbehavior, 145–55
    –muscle systems, 152–54

Tendons (*Contd*)
    properties of, 141–79
    states of deformation, 171–72
    as structure, 141–42
    structure of, 146–47
Tensility, a continuum mechanics and, 143–44
Tension
    resting, 213
    uniaxial, 189–92
Thermodilution techniques, 444–51
Tissues
    biomaterials and compatibility of, 53–56
    head, mechanical properties of, 613–22
    postulates of elasticity and, 164–65
    pressures, 281–85
    soft, 162–75
        stress–strain history relations, 181–208
        viscoelasticity of, 192–95
    supporting, 6
    thermal properties of, 217–35
Trajectorial architecture theory, 259–60
Transverse impedance, 401–04
Tube viscometers, 66–67
Two-phase hypothesis, 239

U

Uniaxial tension, 189–92
Unloading, stress response in, 186–89

V

Valvular disease, 504–06
Vand effect, 72–73
Vehicle impact, 567–83
Velocity, pulse-wave, 129–32
Ventricular wall stress, 293–96
Viscoelastic properties, 221–22
    temperature and, 219–21
Visco-elasticity
    blood vessels, 108–09
    of soft tissues, 192–95
Viscometer
    capillary-tube, 68–69

Viscometer (*Contd*)
   concentric-cylinder, 67, 69
   cone-and-plate, 67–68, 69–70
   tube, 66–67
Volume–pressure behavior, 325–28

W

Waves
   in blood vessels, 340–61

Waves (*Contd*)
   equations, 386–89
   –pulse velocity, 129–32
   transmission, 404–08, 415–18

Y

Yield stress measurements, 89–95